Mark Hill

Best wishes and
I hope you find
it useful.

D1610031

Handbook of Alien Species in Europe

INVADING NATURE
SPRINGER SERIES IN INVASION ECOLOGY
Volume 3

Series Editor: JAMES A. DRAKE
University of Tennessee,
Knoxville, TN, USA

For other titles published in this series, go to
www.springer.com/series/7228

DAISIE

Handbook of Alien Species in Europe

Springer

ISBN: 978-1-4020-8279-5   e-ISBN: 978-1-4020-8280-1

Library of Congress Control Number: 2008933943

© 2009 Springer Science + Business Media B.V.
No part of this work may be reproduced, stored in a retrieval system, or transmitted in any form or by any means, electronic, mechanical, photocopying, microfilming, recording or otherwise, without written permission from the Publisher, with the exception of any material supplied specifically for the purpose of being entered and executed on a computer system, for exclusive use by the purchaser of the work.

Printed on acid-free paper

springer.com

Who is DAISIE?

DAISIE is not only the project's acronym, it also represents the consortium of 83 partners and 99 collaborators and their joint effort. To acknowledge this concerted work, we all accepted DAISIE as author of this *Handbook*. Consequently, each partner and collaborator may refer to *Handbook of Alien Species in Europe* as part of her/his own scientific output.

Partners

Pavlos Andriopoulos, Margarita Arianoutsou, Sylvie Augustin, Nicola Baccetti, Sven Bacher, Jim Bacon, Corina Başnou, Ioannis Bazos, Pavel Bolshagin, François Bretagnolle, François Chiron, Philippe Clergeau, Pierre-Olivier Cochard, Christian Cocquempot, Armelle Cœur d'Acier, Jonathan Cooper, Darius Daunys, Pinelopi Delipetrou, Viktoras Didžiulis, Franck Dorkeld, Franz Essl, Bella Galil, Jacques Gasquez, Piero Genovesi, Kyriakos Georghiou, Stephan Gollasch, Zigmantas Gudžinskas, Ohad Hatzofe, Martin Hejda, Mark Hill, Philip E Hulme, Vojtěch Jarošík, Melanie Josefsson, Salit Kark, Stefan Klotz, Manuel Kobelt, Yannis Kokkoris, Mladen Kotarac, Ingolf Kühn, Philip W Lambdon, Eugenia Lange, Carlos Lopez-Vaamonde, Marie-Laure Loustau, Arnald Marcer, Michel Martinez, David Matej, Mathew McLoughlin, Alain Migeon, Dan Minchin, Maria Navajas, Pierre Navajas, Wolfgang Nentwig, Sergej Olenin, Irina Olenina, Richard Ostler, Irina Ovcharenko, Vadim E Panov, Eirini Papacharalambous, Michel Pascal, Jan Pergl, Irena Perglová, Andrey Phillipov, Joan Pino, Katja Poboljsaj, Petr Pyšek, Wolfgang Rabitsch, Jean-Yves Rasplus, Natalia Rodionova, Alain Roques, David B Roy, Helen Roy, Daniel Sauvard, Riccardo Scalera, Assaf Schwartz, Ondřej Sedláček, Susan Shirley, Valter Trocchi, Montserrat Vilà, Marten Winter, Annie Yart, Artemios Yiannitsaros, Pierre Zagatti, Andreas Zikos

Collaborators

Borys Aleksandrov, Gianni Amori, Pedro Anastacio, Paulina Anastasiu, Jean-Nicolas Beisel, Sandro Bertolino, Alain Bertrand, Dimitra C Bobori, Marco Bodon, Laura Bonesi, Manuela Branco Simoes, Giuseppe Brundu, Vitor Carvalho, Sandra Casellato, Laura Celesti-Grapow, Jean-Louis Chapuis, Simone Cianfanelli, Kristijan Civic, Ejup Çota, Simon Devin, Yury Dgebaudze, Ovidijus Dumčius, Agustín Estrada-Peña, Massimo Faccoli, Helena Freitas, Jean-François Germain, Francesca Gherardi, Milka Glavendekić, Stanislas Gomboć, Morris Gosling, Michael Grabowski, Duncan Halley, Stephan Hennekens, Djuro Huber, Blagoj Ivanov, Jasna Jeremic, Nejc Jogan, Maria Kalapanida, Kaarina Kauhala, Marc Kenis, Ferenc Lakatos, Colin Lawton, Roland Libois, Åke Lindelöw, Elisabetta Lori, Andreja Lucic, Petros Lymberakis, Anne-Catherine Mailleux, Michael Majerus, Elizabete Marchante, Hélia Marchante, Marco Massetti, Joao Mayol-Serra, Robbie A McDonald, Dragos Micu, David Mifsud, Leen G Moraal, Didier Morin, Ilona B Muskó, Sterja Naceski, Gavril Negrean, Annamaria Nocita, Petri Nummi, Anna Occhipinti Ambrogi, Bjørn Økland, Nicolaï N Olenici, Irena Papazova-Anakieva, Maria Rosa Paiva, Milan Paunovic, Momir Paunovic, Giuseppina Pellizzari, Nicolas Pérez, Olivera Petrović-Obradović, Daniela Pilarska, Rory Putnam, Ana Isabel Queiroz, Dragan Roganovic, Philippe Reynaud, Nicoletta Riccardi, Louise Ross, Giampaolo Rossetti, Patrick Schembri, Emmanuel Sechet, Gabrijel Seljak, Vitaliy Semenchenko, Vadim E Sidorovich, Marcela Skuhravá, Václav Skuhravý, Wojciech Solarz, Andrea Stefan, Pavel Stoev, Jean-Claude Streito, Jan Stuyck, Catarina Tavares, Rumen Tomov, Katia Trencheva, Goedele Verbeylen, Claire Villemant, Baki Yokes

Prefaces

United Nations Environmental Programme: DAISIE is More Than a Scientific Reference

Among the 22 indicators that underpin the international target to 'reduce the rate of loss of biodiversity' is one covering trends in invasive alien species. This *Handbook*, the fruit of a three year European Union-funded project, will make a significant contribution to a wider understanding of these trends as they relate to the continent of Europe.

In doing so it will assist countries, including governments; business and industry; academia and civil society to deliver on commitments under the UNEP Convention on Biological Diversity as well as those emerging from such fora as the World Summit on Sustainable Development held in Johannesburg in 2002.

The *Handbook* should also serve to inspire others elsewhere in the world to carry out or support similar comprehensive and forward-looking assessments in order to get to grips with the issue of whether globally, as well as regionally, biodiversity is waxing or waning.

The *Handbook* takes a modern and digestible approach to the issues by flagging up 100 of the most invasive alien species in Europe and confirming the significant impact such invaders can have on native plant and animal communities. For example over 70% of these 100 invasive species have reduced native species diversity or have altered the invaded community and close to a fifth have affected the prospects for endangered species.

The *Handbook*, a product of the Delivering Alien Invasive Species Inventory for Europe or DAISIE Consortium, is however more than just a much needed scientific reference: it is also a fascinating and compelling read. Through words, graphs and images the book chronicles how wave after wave of curious and exotic species have been entering Europe's terrestrial, freshwater and marine environment stretching back well over a century and in some cases, such as the brown rat, six centuries or more.

The aliens, some of whom have hitched rides on ships, planes or trucks and others, lifted from one continent to another as a result of accidental or deliberate introductions, in part shed light on the Europe's changing patterns of trade with the rest of the world while also underlining the folly and in some cases the vanity of humanity.

The Canada goose for example, which is considered an economic liability by some and a nuisance by others, was introduced into Britain in the 17th century by King James II to add to his waterfowl collection in St James Park.

The *Handbook* also underlines the environmental and economic impacts that life-forms in the wrong place at the wrong time can make. The coypu, a native of Patagonia, has become established over large parts of Europe after being brought in for fur ranching. Damaged linked with the alien rodent includes massive destruction of reed swamps and predation of the eggs of aquatic and often endangered native birds. Control activities in Italy over a five year period managed to remove over 220,000 coypus at a cost of €2,614,408. However damage to the riverbanks from coypus over the same time exceeded €10 million and impact on agriculture reached €935,138.

The *Handbook* also underlines that aliens such as fungi and plants can be more widely destructive and perhaps more difficult to control if they take hold. The two forms of Dutch elm disease, both of which are thought to originate in Asia, have over recent decades decimated elms in many parts of Europe and a great deal of time and money has been spent trying to breed disease-resistant trees.

Meanwhile the common ragweed, a native of North America and introduced via agricultural products and later via horses during World War I, is now well established in many European countries including France, Switzerland, Hungary and the Balkans. It is linked to hay fever, asthma and other illnesses that can be a costly medical burden in some place such as the Rhone Valley. This is a proof, if proof were needed that it is far better and cheaper to prevent alien species from entering an ecosystem or another continent's environment than trying to eradicate them after the event.

The *Handbook* raises some profound concerns over the future not least from the way climate change – if unchecked – may aggravate the impact of some aliens in Europe. In doing so it underscores that in addressing one environmental challenge, in this case global warming, there are multiple benefits including the conservation of biodiversity upon which a great deal of human well-being including livelihoods, health and economic activity depend.

I would like to congratulate the many scientists who have contributed to the *Handbook* and would commend policy-makers but also the general public to put it in their ministerial in-trays and on their living room coffee tables as this is engaging and essential reading.

Achim Steiner
UN Under Secretary General and
Executive Director UN Environment Programme (UNEP)

The Council of Europe: DAISIE Is a Much-Needed Initiative

The impact of invasive alien species (IAS) on European ecosystems and native species is one of the most challenging issues in the field of conservation and wise use of biological diversity. While invasive alien species can affect all habitats, both terrestrial and aquatic, islands are particularly vulnerable to that threat, mainly because of their biological richness. They hold a high number of unique (endemic) species and their geographic isolation has created vulnerable habitats that can be easily invaded by new arrivals. Nearly half of the flora of the Canary Islands, for instance, is made up of non-native species. Invasive alien species are quoted as being one of the causes of extinction of species worldwide, mainly through their influence on island biodiversity through predation, competition, hybridisation or as vehicles for pathogens to which native species are not resistant.

Those are some reasons that moved the Council of Europe to promote, within its nature conservation programmes, action to avoid the intentional introduction and spread of alien species, to prevent accidental introductions and to build an information system on IAS. In 1984 the Committee of Ministers of the Council of Europe adopted a recommendation in that sense. Also the Bern Convention (*Convention on the Conservation of European Wildlife and Natural Habitats*), the main Council of Europe treaty in the field of biodiversity conservation, requires its 45 Contracting Parties "*to strictly control the introduction of non native species*". In 2003, the Bern Convention adopted the "European Strategy on Invasive Alien Species", aimed to provide precise guidance to European governments on IAS issues. The Strategy identifies European priorities and key actions, promotes awareness and information on IAS, strengthening of national and regional capacities to deal with IAS issues, taking of prevention measures and supports remedial responses such as reducing adverse impacts of IAS, recovering species and natural habitats affected. National strategies have been drafted and implemented following the priorities set in the European Strategy.

One of the points dealt at the European Strategy on Invasive Alien Species was the need to develop national and European information systems. It was recognised that information sharing between states and scientific institutions was a critical factor for prevention, both of new arrivals and spread of introduced aliens. The Strategy pointed out that there was already quite an important amount of expertise and information available and that any European inventory of IAS should count on existing databases. It proposed the European inventory to fill existing gaps and register systematically information on species taxonomy and biology, date and place of introduction, pathways of introduction, range and spread dynamics, risk of expansion to neighbouring countries, invaded ecosystems and impact, as well as the efficiency of measures taken for prevention, mitigation and restoration of affected ecosystems.

It was thus with great pleasure that the Council of Europe welcomed the DAISIE (Delivering Alien Invasive Species Inventories in Europe) project, which was fundamentally aimed to satisfy those recommendations made to governments. The DAISIE project has gathered in relatively short time a formidable amount of very relevant information on alien species in Europe, on existing expertise in our continent and, what is equally important, on trends. The systematisation of data permits to create a clear picture of the phenomenon of biological invasions, to develop prevention tools (such as Early Warning Systems on new invasions) and promote awareness on how alien species are distorting and affecting Europe's biological diversity. The project has been a scientific success and has also created a much needed instrument for governments and European institutions to control the problem.

As for the future, at the Council of Europe we feel it is vital to maintain a strong information system on IAS on the basis of DAISIE, extend it through similar national initiatives and encourage governments and scientist to continue collecting and analysing information. It is also vital to build stronger links between research and governmental actions and strengthen efforts to convince other partners (e.g., horticultural industry, pet and aquaria trade, hunters and anglers, forest community, etc.) to adopt voluntary codes that may limit the introduction of new alien species and the spread of already known invasive species.

Eladio Fernández-Galiano
Head of the Biological Diversity Unit of the Council of Europe

Prefaces

The European Commission: DAISIE is a Pioneering Work

The central importance of protecting biodiversity as a basis for safeguarding ecosystem goods and services, as well as ecosystem functions for supporting the life on the Earth, is broadly accepted throughout the world. There is no doubt that human activities have been behind the dramatic acceleration of biodiversity loss in last decades. Identifying and addressing the drivers of this process is an essential step to improve the situation.

Experience from nature conservation practice has shown that simple species based conservation approaches will not achieve the goal, due to the very complex relationships between species within habitats and ecosystems. After habitat loss and fragmentation, known as the most important causes of biodiversity loss, alien species which have developed invasive behaviour may cause very significant damage to natural ecosystems by outcompeting native species. These, so called invasive alien species (IAS), were indeed recognised as one of major drivers of biodiversity decline. It is therefore clear, that the goal of halting biodiversity loss within European Union by 2010 cannot be achieved without addressing the issue of invasive aliens.

In addition, in many cases it is the very strong economic impact of these species, which requires to be addressed. A few well known invasive species – such as the zebra mussel, the American crayfish, the American mink, the water hyacinth, the giant hogweed or the grey squirrel – cause yearly damages amounting to hundreds of millions of Euros over the European territory. If not prevented or controlled, these costs may multiply in the future as some of these species expand their area of distribution in response to climate change.

Invasive alien species do not recognise national boundaries and cannot be stopped without concerted efforts. Europeans today are more mobile than ever before. Increased mobility for people and goods has many benefits but it also increases opportunities for intentional introduction of highly invasive alien species imported originally as pets, ornamental plants or for another purpose, and for unintentional introductions of stowaways or contaminant organisms through trade or other pathways. Good practice in relation to policies and legislation relating to IAS is occurring in some European regions, but it remains scattered. There are at present no mechanisms to support harmonisation or basic consistency of approaches between neighbouring countries or countries in the same sub-region. The fragmented measures in place are unlikely to make a substantial contribution to lowering the risks posed by IAS to European ecosystems if pan-European policies are not implemented.

Therefore, it is critical to address this global threat at the European level as a shared problem of all Member states. Consequently, invasive alien species are included and addressed in the Action Plan adopted by European Commission together with the Biodiversity Communication in 2006. An EU framework on alien invasive species is under development and should be in place before 2010.

The DAISIE (Delivering Alien Invasive Species Inventories in Europe) project is an important element towards this ambitious endeavour. The project has put together an inventory which provides the first ever pan-European overview of over

11,000 alien animals, fish, birds, plants, insects and other species living in our environment – causing already or potentially damage to our natural heritage as well as to our economy.

I am sure that this pioneering work will serve as a good basis for addressing the problem in Europe, and will also contribute to global solutions. I also do hope that it will remain alive, permanently updated and as a flexible tool supporting efficient implementation of an agreed common approach to the management of IAS.

Ladislav Miko
Director, Protecting Natural Resources,
DG Environment, European Commission

The Editors: an End has a Start

A significant landmark in raising awareness of biological invasions among the public and policy makers in North America was the publication in 1993 of the monograph *Harmful Non-Indigenous Species in the United States* (Office of Technology Assessment 1993). This sweeping document, published by the US Congress, presented the first continental assessment of the degree to which introduced species had spread across the USA, the breadth of ecosystems subsequently impacted by these species and the policy options available to government and managers. Arguably, this synthesis played a pivotal role in the subsequent signing of Executive Order 13112 "Invasive Species" by President Bill Clinton in 1999. The Executive Order placed responsibilities upon Federal agencies "to prevent the introduction of invasive species and provide for their control and to minimise the economic, ecological, and human health impacts that invasive species cause" (USA 1999). Furthermore, Executive Order 13112 established the National Invasive Species Council to oversee implementation and required the initiation of National Invasive Species Management Plans. What is probably startling about the impact of *Harmful Non-Indigenous Species in the United States* is that at that time information on invasive species was remarkably poor with a minimum estimate of the number of alien species with origins outside the USA being little more than 4,500, the majority plants and invertebrates. Yet, while imprecise, these figures identified significant gaps in knowledge and highlighted that even imprecise estimates were sufficient to raise alarm bells.

Fast forward 15 years and to the other side of the Atlantic where another landmark publication appears: the *Handbook of Alien Species in Europe*. In the intervening period much has changed in the global perception of biological invasions, now widely recognised as one of the major pressures on ecosystems (Nentwig 2007). Yet it is only now that Europe has the first continent-wide snapshot of the scale and impact of biological invasions from the Mediterranean Sea to the Arctic tundra. On this occasion data are more robust, drawn from systematic searches and peer reviewed by experts, yet the astounding figure of at least 11,000 introduced species is acknowledged as only a first approximation and an undoubted underestimate (Olenin and Didžiulis 2009). Nevertheless, a picture emerges that is as remarkable as it is worrying.

Biological invasions are not a new phenomenon to Europe. For example, Corsica has been invaded more than twenty times in the last 2,500 years, first by the Phoenicians (565 BC), then Etruscans (540 BC), Carthaginians (270 BC), Romans (259 BC), Vandals (AD 455), Byzantines (AD 534), Goths (AD 549), Saracens (AD 704), Lombards (AD 725), Pisanos (AD 1015), Genoese (AD 1195), Aragonese (AD 1297), Genoese again (1358), Milanese (AD 1468) Franco-Ottomans (AD 1553), French (AD 1768), British (AD 1794) and the German-Italian Axis during the Second World War (Hulme 2004). These human invasions brought in their wake species from other parts of the world, either intentionally or by accident. Biotic homogenisation in Europe has probably been occurring over millennia, yet the long history of human invasion and trade blurs the distinction between native and alien species. Thus the origin of many species introduced in historical times is uncertain and especially so for marine ecosystems, where often the status and origin of species is unknown. Although this uncertainty frustrates analyses, it also indicates that many archaeophytes and archaeozoans have become integrated in native communities without clear evidence of detriment either to native species or ecosystem processes (e.g., Pyšek et al. 2005).

These historical trends should not encourage complacency regarding the ultimate impacts of invasive species. It is quite possible that recent introductions from outside Europe are likely to be more invasive than alien species that originate from another part of Europe (Lloret et al. 2004). The *Handbook of Alien Species in Europe* illustrates that for most taxa an increasing proportion of introduced species are from other continents, especially the Americas and Asia. Indeed, the trends in the cumulative records of alien species recorded in Europe reveal consistent increases with time (Hulme et al. 2009a). For example, on average 19 invertebrates (Roques et al. 2009), 16 plants (Pyšek et al. 2009) and one mammal (Genovesi et al. 2009) are newly introduced to one or more parts of Europe every year. These numbers may not sound especially threatening but for many taxa, recent rates are higher than those seen at the beginning of the 19th century indicating that the problem of invasions is not diminishing.

A clear signal is that global trade is a major driver of biological invasions in Europe. This is not surprising, since this signal is seen worldwide (Perrings et al. 2005). Accounting for the multitude of pathways by which an alien species is introduced is essential to disentangle the role of species and ecosystem traits in biological invasions as well as predict future trends and identify management options. The *Handbook of Alien Species in Europe* highlights that vertebrate introduction tend to be characterised as deliberate releases (often as game animals), invertebrates as contaminants of stored products or horticultural material, plants as escapes from gardens, while pathogenic fungi are generally introduced as contaminants of their hosts (Hulme et al. 2008b). Several major infrastructural projects linking together seas via freshwaters and canal networks in order to facilitate the movement of goods are a major source of introductions, for example into the Mediterranean from the Red Sea, and from the Caspian and Black Seas to the Baltic (Galil et al. 2009, Gherardi et al. 2009). Once introduced to Europe, species with tiny spores, such as fungi and bryophytes, may be able to spread across the continent without additional human assistance (Desprez-Loustau 2009; Essl and Lambdon 2009) and such unaided spread is likely to be the hardest to contain.

The anthropogenic signal on biological invasions persists after the initial introduction events in that even once established, many alien species remain associated with human modified ecosystems. Alien plants (Pyšek et al. 2009) and invertebrates (Roques et al. 2009) are proportionally more frequent in urban than semi-natural habitats, while birds and amphibians (Kark et al. 2009) as well as mammals

(Genovesi et al. 2009) are most frequently found in arable lands, gardens and parks. Clearly, in such habitats alien species are most likely to be perceived as having economic rather than ecological impacts. For example, in the UK alone, the cost to the timber industry of grey squirrel (*Sciurus carolinensis*) damage to beech, sycamore and oak is $15 million while two common grain contaminants, wild oat (*Avena fatua*) and field speedwell (*Veronica persica*), are significant agricultural weeds with annual costs of control running to $150 million (Williamson 2002). However, current appreciation of invasive species impacts on biodiversity in Europe is poor by comparison with North America (Levine et al. 2003 for plants, Roques et al. 2009 for invertebrates). This is evident in the percentage of species with known impacts recorded by the DAISIE (Delivering Alien Invasive Species Inventories in Europe) project that ranges from only 5% for plants, around 15% for invertebrates and marine taxa, to a high of 30% for vertebrates and freshwater species. These percentages most likely reflect the lack of information across large taxonomic groups but also the difficulty of quantifying subtle impacts on ecosystem processes.

So what are the response options to this worrying panorama? The *Handbook of Alien Species in Europe* provides generic information relating to management of particular taxonomic groups as well as more detailed information on control and eradication strategies for 100 of the worst species (Vilà et al. 2009). However such information is only of value if the mechanism for management and policy implementation exists. A key component of Executive Order 13112 in the USA was the establishment of a National Invasive Species Management Plan (USA 1999). The plan is focused upon five strategic goals: prevention; early detection and rapid response; control and management; restoration; and organisational collaboration. Each of the five strategic goals specifies ongoing objectives and the long-term vision for success in that area. Under each strategic goal, objectives describe what is to be accomplished over the next five years, and implementation tasks describe what agencies expect to do in order to accomplish that specific objective. To date, an estimated 67% of the first plan's 57 action items (encompassing over 100 separate elements) have been completed or are in progress.

Such a strategic document appears essential for Europe, effectively putting teeth onto the *European Strategy on Invasive Alien Species* (Council of Europe 2002). DAISIE has set a global precedent in the inventories of alien species and inspires others elsewhere in the world (Steiner 2009), fulfilled a pressing need within Europe (Fernández-Galiano 2009) and significantly raised awareness in European institutions (Miko 2009). While these are major milestones, we believe DAISIE and the *Handbook of Alien Species in Europe* is only the start: the start of the end to the fragmented legislative and regulatory requirements addressing invasive species (Miller et al. 2006). The start of the end to uncoordinated activities led by the different Directorates General (DG) of the European Union that do not appear to appreciate the cross-cutting nature of biological invasions (e.g., separate DGs for Agriculture, Environment, Health, Marine, Research, Transport etc.). The start of the end of piecemeal approaches to tackling invasive species across Europe that fail to coordinate pre- and post-border actions (Hulme et al. 2008b). The start of the end of underfunding taxonomy, management efforts and basic research on invasive species. And finally, hopefully in the not too distant future, the start of the end of the progressive homogenisation of Europe's flora and fauna.

Philip E Hulme, Wolfgang Nentwig,
Petr Pyšek and Montserrat Vilà

References

Council of Europe (2002) European strategy on invasive alien species. Council of Europe Publishing Strasbourgi ISBN 92-871-5488-0

Desprez-Loustau M-L (2009) The alien fungi of Europe. In: DAISIE Handbook of alien species in Europe. Springer, Dordrecht. 15–28

Essl F, Lambdon PW (2009) The alien bryophytes and lichens of Europe. In: DAISIE Handbook of alien species in Europe. Springer, Dordrecht. 29–41

Fernández-Galiano E (2009) The Council of Europe: DAISIE is a much-needed initiative. In: DAISIE Handbook of alien species in Europe. Springer, Dordrecht

Galil B, Gollasch S, Minchin D, Olenin O (2009) Alien marine biota of Europe. In: DAISIE Handbook of alien species in Europe. Springer, Dordrecht. 93–104

Genovesi P, Bacher S, Kobelt M, Pascal M, Scalera R (2009) Alien mammals of Europe. In: DAISIE Handbook of alien species in Europe. Springer, Dordrecht. 119–128

Gherardi F, Gollasch S, Minchin D, Olenin O, Panov V (2009) Alien invertebrates and fish in European inland waters. In: DAISIE Handbook of alien species in Europe. Springer, Dordrecht. 81–92

Hulme PE (2004) Invasions, islands and impacts: a Mediterranean perspective. In: Fernandez Palacios JM, Morici M (eds) Island ecology. Asociación Española de Ecología Terrestre, La Laguna, Spain

Hulme PE, Roy DB, Cunha T, Larsson T-B (2009a) A pan-European inventory of alien species: rationale, implementation and implications for managing biological invasions. In: DAISIE Handbook of alien species in Europe. Springer, Dordrecht. 1–14

Hulme PE, Bacher S, Kenis M, Klotz S, Kühn I, Minchin D, Nentwig W, Olenin S, Panov V, Pergl J, Pyšek P, Roques A, Sol D, Solarz W, Vilà M (2008b) Grasping at the routes of biological invasions: a framework to better integrate pathways into policy. J Appl Ecol 45, 403–414

Kark S, Solarz W, Chiron F, Clergeau P, Shirley S (2009) Alien birds, amphibians and reptiles of Europe. In: DAISIE Handbook of alien species in Europe. Springer, Dordrecht. 105–118

Levine JM, Vilà M, D'Antonio CM, Dukes JS, Grigulis K, Lavorel S (2003) Mechanisms underlying the impact of exotic plant invasions. Phil Trans R Soc Lond 270, 775–781

Lloret F, Médail F, Brundu G, Hulme PE (2004) Local and regional abundance of exotic plant species on Mediterranean islands: are species traits important? Global Ecol Biogeogr 13, 37–45

Miko L (2009) The European Commission: DAISIE is a pioneering work. In: DAISIE Handbook of alien species in Europe. Springer, Dordrecht

Miller C, Kettunen M, Shine C (2006) Scope options for EU action on invasive alien species (IAS). Final report for the European Commission. Institute for European Environmental Policy (IEEP), Brussels, Belgium

Nentwig W (ed) (2007) Biological invasions. Ecological Studies 193, Springer, Berlin

Office of Technology Assessment (1993) Harmful non-indigenous species in the United States. Report OTA-F-565. US Government Printing Office, Washington, DC

Olenin S, Didžiulis V (2009) Introduction to the species list. In: DAISIE Handbook of alien species in Europe. Springer, Dordrecht. 129–132

Perrings C, Dehnen-Schmutz K, Touza J, Williamson M (2005) How to manage invasive species under globalization. Trends Ecol Evol 20, 212–215

Pyšek P, Jarošík V, Chytrý M, Kropáč Z, Tichý L, Wild J (2005) Alien plants in temperate weed communities: prehistoric and recent invaders occupy different habitats. Ecology 86, 772–785

Pyšek P, Lambdon PW, Arianoutsou M, Kühn I, Pino J, Winter M (2009) Alien vascular plants of Europe. In: DAISIE Handbook of alien species in Europe. Springer, Dordrecht. 43–61

Roques A, Rabitsch W, Rasplus J-Y, Lopez-Vamonde C, Nentwig W, Kenis M (2009) Alien terrestrial invertebrates of Europe. In: DAISIE Handbook of alien species in Europe. Springer, Dordrecht. 63–79

Steiner A (2009) United Nations Environmental Programme: DAISIE is more than a scientific reference. In: DAISIE Handbook of alien species in Europe. Springer, Dordrecht

USA (1999) Executive Order 13112 of February 3, 1999: invasive species. Federal Register 64(25), 6183–6186

Vilà M, Başnou C, Gollasch S, Josefsson M, Pergl J, Scalera R (2009) One hundred of the most invasive alien species in Europe. In: DAISIE Handbook of alien species in Europe. Springer, Dordrecht. 265–268

Williamson MH (2002) Alien plants in the British Isles. In: Pimentel D (ed) Biological invasions: economic and environmental costs of alien plant, animal and microbe species. CRC Press, Boca Raton, FL

Acknowledgements

On behalf of all partners and collaborators we thank the European Commission's Sixth Framework Programme for supporting the DAISIE (Delivering Alien Invasive Species Inventories for Europe) project, contract SSPI-CT-2003-511202). We also thank Adrienne Käser and Rita Schneider for editorial assistance and Springer Publisher for their cooperation during the publishing process.

Contents

1. **A pan-European Inventory of Alien Species: Rationale, Implementation and Implications for Managing Biological Invasions**.. 1
   Philip E. Hulme, David B. Roy, Teresa Cunha, and Tor-Björn Larsson

2. **Alien Fungi of Europe** 15
   Marie-Laure Desprez-Loustau

3. **Alien Bryophytes and Lichens of Europe** 29
   Franz Essl and Philip W. Lambdon

4. **Alien Vascular Plants of Europe**... 43
   Petr Pyšek, Philip W. Lambdon, Margarita Arianoutsou,
   Ingolf Kühn, Joan Pino, and Marten Winter

5. **Alien Terrestrial Invertebrates of Europe** 63
   Alain Roques, Wolfgang Rabitsch, Jean-Yves Rasplus,
   Carlos Lopez-Vaamonde, Wolfgang Nentwig, and Marc Kenis

6. **Alien Invertebrates and Fish in European Inland Waters** 81
   Francesca Gherardi, Stephan Gollasch, Dan Minchin,
   Sergej Olenin, and Vadim E. Panov

7. **Alien Marine Biota of Europe**... 93
   Bella S. Galil, Stephan Gollasch, Dan Minchin, and Sergej Olenin

8. **Alien Birds, Amphibians and Reptiles of Europe** 105
   Salit Kark, Wojciech Solarz, François Chiron,
   Philippe Clergeau, and Susan Shirley

9. **Alien Mammals of Europe** 119
   Piero Genovesi, Sven Bacher, Manuel Kobelt, Michel Pascal,
   and Riccardo Scalera

10	**Introduction to the List of Alien Taxa** 129
	Sergej Olenin and Viktoras Didžiulis
11	**List of Species Alien in Europe and to Europe** 133
12	**One Hundred of the Most Invasive Alien Species in Europe** 265
	Montserrat Vilà, Corina Basnou, Stephan Gollasch, Melanie Josefsson, Jan Pergl, and Riccardo Scalera
13	**Species Accounts of 100 of the Most Invasive Alien Species in Europe** 269
	By individual authors
14	**Glossary of the Main Technical Terms Used in the Handbook** 375
	Petr Pyšek, Philip E. Hulme, and Wolfgang Nentwig

Index 381

Contributors

David Alderman
CEFAS Weymouth Laboratory, The Nothe, Weymouth, Dorset DT4 BUB, UK, d.j.alderman@cefas.co.uk

Paulina Anastasiu
University of Bucharest, Intrarea Portocalelor 1-3, Sector 6, 060101, Bucharest, Romania, anastasiup@yahoo.com

Margarita Arianoutsou
University of Athens, Faculty of Biology, Department of Ecology & Systematics, 15784 Athens, Greece, marianou@biol.uoa.gr

Sylvie Augustin
Institut National de la Recherche Agronomique, Zoologie Forestière, Centre de recherche d'Orléans, 2163 Avenue de la Pomme de Pin, CS 40001 Ardon, 45075 Orleans Cedex 2, France, sylvie.augustin@orleans.inra.fr

Sven Bacher
Department of Biology, Ecology & Evolution Unit, University of Fribourg, Chemin du Musée 10, CH-1700 Fribourg, Switzerland, sven.bacher@unifr.ch

Corina Başnou
Centre for Ecological Research and Forestry Applications, Campus de Bellaterra (Universitat Autònoma de Barcelona), 08193 Cerdanyola del Vallès, Barcelona, Spain, c.basnou@creaf.uab.es

Jean-Nicolas Beisel
Laboratory Biodiversity and Ecosysteme Functioning, University of Metz, Campus Bridoux, Avenue du Général Delestraint, 57070 Metz, France, beisel@univ-metz.fr

Sandro Bertolino
Laboratory of Entomology and Zoology, DIVAPRA, University of Turin, Via L da Vinci 44, 10095 Grugliasco (TO), Italy, sandro.bertolino@unito.it

Laura Bonesi
Dipartimento di Biologia, Università di Trieste, Via Weiss 2, 34127
Trieste, Italy, e-mail: lbonesi@units.it

François Bretagnolle
Institut de Biologie, Université de Neuchâtel, Rue Emile-Argand 11,
CH-2009 Neuchâtel, Switzerland, francois.bretagnolle@bota.unine.ch

Jean Louis Chapuis
Muséum National d'Histoire Naturelle, Département Ecologie et Gestion de la
Biodiversité, UMR 5173 MNHN-CNRS-P6, 61, rue Buffon, case postale 53, 75231
PARIS cedex 05, France, chapuis@mnhn.fr

Bruno Chauvel
UMR, INRA/ENESAD/UB, Biologie et Gestion des Adventices, 17 rue Sully
BP 86510, F-21065 Dijon cedex, France, bruno.chauvel@dijon.inra.fr

François Chiron
Department of Ecology, Systematics and Evolution, The Institute of Life Sciences,
The Hebrew University of Jerusalem, Jerusalem 91904, Israel,
chironfrancois@gmail.com

Philippe Clergeau
Institut National de la Recherche Agronomique, SCRIBE, IFR 140,
Campus de Beaulieu, 35 042 Rennes Cedex, France,
philippe.clergeau@rennes.inra.fr

Teresa Cunha
DG Research, European Commission, B-1049, Brussels, Belgium,
teresa.cunha@ec.europa.eu

Pinelopi Delipetrou
University of Athens, Faculty of Biology, Department of Botany,
15784 Athens, Greece, pindel@biol.uoa.gr

Marie-Laure Desprez-Loustau
Institut National de la Recherche Agronomique, Domaine de la Grande Ferrade,
UMR 1202 BIOGECO, Pathologie forestière, BP 81, 33883 Villenave d'Ornon
Cedex, France, loustau@bordeaux.inra.fr

Mathieu Détaint
Association Cistude Nature, Moulin du Moulinat, Chemin du Moulinat
– 33185 Le Haillan, France, mathieu.detaint@cistude.org

Simon Devin
Laboratory Biodiversity and Ecosysteme Functioning, University of Metz,
Campus Bridoux, Avenue du Général Delestraint, 57070 Metz,
France, s.devin@laposte.net

Contributors

Viktoras Didžiulis
Coastal Research and Planning Institute, Klaipėda University, Klaipėda,
LT 92294, H. Manto 84, Lithuania, viktoras@ekoinf.net

Franz Essl
Federal Environment Agency Ltd, Biodiversity and Nature Conservation,
Spittelauer Lände 5, A-1090 Wien, Austria, franz.essl@umweltbundesamt.at

Bella S. Galil
National Institute of Oceanography, P.O. Box 8030, Haifa, 31080, Israel,
bella@ocean.org.il

Piero Genovesi
Chair European Section IUCN SSC ISSG, INFS (National Wildlife Institute),
Via Ca Fornacetta 9, 40064 Ozzano Emilia BO, Italy,
piero.genovesi@infs.it

Francesca Gherardi
Dipartimento di Biologia Evoluzionistica, Università di Firenze,
Via Romana 17, 50125 Firenze, Italy, francesca.gherardi@unifi.it

Stephan Gollasch
GoConsult, Grosse Brunnenstr. 61, 22763 Hamburg, Germany,
sgollasch@aol.com

Martin Hejda
Department of Invasion Ecology, Institute of Botany, Academy of Sciences
of the Czech Republic, CZ-252 43 Průhonice, Czech Republic, hejda@ibot.cas.cz

Philip E. Hulme
National Centre for Advanced Bio-Protection Technologies, P.O. Box 84,
Lincoln University, Canterbury, New Zealand, hulmep@lincoln.ac.nz

Melanie Josefsson
Swedish Environmental Protection Agency, c/o Department of Environmental
Monitoring, P.O. Box 7050, SE 750 07 Uppsala, Sweden,
melanie.josefsson@snv.slu.se

Salit Kark
Department of Ecology, Systematics and Evolution, The Institute of
Life Sciences, The Hebrew University of Jerusalem, Jerusalem 91904, Israel,
e-mail: salit@cc.huji.ac.il

Kaarina Kauhala
Turku Game and Fisheries Research, Itäinen Pitkäkatu 3A, 20520 Turku,
Finland, kaarina.kauhala@rktl.fi

Marc Kenis
Forestry and Ornamental Pest Research, CABI Europe, Rue des Grillons 1, CH-
2800 Delemont, Switzerland, m.kenis@cabi.org

Stefan Klotz
Helmholtz Centre for Environmental Research – UFZ, Department of
Community Ecology, Theodor-Lieser-Str. 4, D-06120 Halle, Germany,
stefan.klotz@ufz.de

Manuel Kobelt
Community Ecology, Zoological Institute, University of Bern, Baltzerstrasse 6,
CH 3012 Bern, Switzerland, manuel.kobelt@zos.unibe.ch

Ingolf Kühn
Helmholtz Centre for Environmental Research – UFZ, Department of
Community Ecology, Theodor-Lieser-Str. 4, D-06120 Halle, Germany,
ingolf.kuehn@ufz.de

Philip W. Lambdon
RSPB, c/o Royal Botanic Gardens, Kew, Kew Herbarium, Richmond, Surrey,
TW9 3AB, UK, plambdon@googlemail.com

Tor-Björn Larsson
European Environment Agency, Kongenshytoro 6, DK 1050 Copenhagen K,
Denmark, tor-bjorn.larsson@eea.europa.eu

Carlos Lopez-Vaamonde
Institut National de la Recherche Agronomique, Zoologie Forestière,
Centre de recherche d'Orléans, 2163 Avenue de la Pomme de Pin, CS 40001
Ardon, 45075 Orleans Cedex 2, France, carlos.lopez-vaamonde@orleans.inra.fr

Olivier Lorvelec
Institut National de la Recherche Agronomique, SCRIBE – IFR 140,
Campus de Beaulieu – 35 000 Rennes, France, lorvelec@rennes.inra.fr

Hélia Marchante
Escola Superior Agrária de Coimbra Departamento de Ciências Exactas e
Ambiente Sector de Biologia e Ecologia, Bencanta. 3040-316 Coimbra Portugal,
hmarchante@mail.esac.pt

Dan Minchin
Marine Organism Investigations, 3, Marina Village, Ballina Killaloe,
Co Clare, Ireland, moiireland@yahoo.ie

Wolfgang Nentwig
Community Ecology, Zoological Institute, University of Bern, Baltzerstrasse 6,
CH 3012 Bern, Switzerland, wolfgang.nentwig@zos.unibe.ch

Anna Occhipinti-Ambrogi
Dipartimento di Ecologia del Territorio, Università di Pavia, Via S. Epifanio,
14 I-27100 Pavia, Italy, occhipin@unipv.it

Sergej Olenin
Coastal Research and Planning Institute, Klaipėda University, Klaipėda,
LT 92294, H. Manto 84, Lithuania, sergej@corpi.ku.lt

Contributors

Irina Olenina
Coastal Research and Planning Institute, Klaipeda University, H. Manto 84,
Klaipeda, 92294, Lithuania, e-mail: irina.olenina@balticum-tv.lt

Irina Ovcharenko
Coastal Research and Planning Institute, Klaipėda University, Klaipėda,
LT 92294, H. Manto 84, Lithuania, irinaovcarenko@yahoo.com

Vadim E Panov
Faculty of Geography and Geoecology, St. Petersburg State University,
10 linija VO 33/35, 199178 St. Petersburg, Russian Federation, rbic@mail.ru

Michel Pascal
Institut National de la Recherche Agronomique, UR SCRIBE, 16A allée Henri
Fabre, Campus de Beaulieu, CS 74205, 35042 Rennes Cedex, France,
michel.pascal@rennes.inra.fr

Jan Pergl
Department of Invasion Ecology, Institute of Botany, Academy of Sciences
of the Czech Republic, CZ-252 43 Průhonice,
Czech Republic, pergl@ibot.cas.cz

Irena Perglová
Department of Invasion Ecology, Institute of Botany,
Academy of Sciences of the Czech Republic, CZ-252 43 Průhonice,
Czech Republic, perglova@ibot.cas.cz

Joan Pino
Centre for Ecological Research and Forestry Applications, Campus de Bellaterra
(Uníversitat Autònoma de Barcelona), 08193 Cerdanyola del Vallès, Barcelona,
Spain, joan.pino@uab.es

Petr Pyšek
Department of Invasion Ecology, Institute of Botany, Academy of Sciences of the
Czech Republic, CZ-252 43 Průhonice and Department of Ecology, Faculty of
Science, Charles University Prague, Czech Republic,
pysek@ibot.cas.cz

Wolfgang Rabitsch
Federal Environment Agency Ltd, Biodiversity and Nature Conservation,
Spittelauer Lände 5, 1090 Wien, Austria, wolfgang.rabitsch@umweltbundesamt.at

Jean-Yves Rasplus
Institut National de la Recherche Agronomique, INRA, Centre de Biologie et
de Gestion des Populations, Campus International de Baillarguet, CS 30 016,
34988 Montferrier-sur-Lez, France, rasplus@ensam.inra.fr

Alain Roques
Institut National de la Recherche Agronomique, Zoologie Forestière,
Centre de recherche d'Orléans, 2163 Avenue de la Pomme de Pin, CS 40001
Ardon, 45075 Orleans Cedex 2, France, alain.roques@orleans.inra.fr

Helen Roy and David B. Roy
Biological Records Centre, CEH Monks Wood, Huntingdon CAMBS, PE28 2LS, UK, dbr@ceh.ac.uk

Daniel Sauvard
Institut National de la Recherche Agronomique, Zoologie Forestière, Centre de recherche d'Orléans, 2163 Avenue de la Pomme de Pin, CS 40001 Ardon, 45075 Orleans Cedex 2, France, Daniel.Sauvard@orleans.inra.fr

Riccardo Scalera
Via Torcegno 49 V1 A2, 00124 Rome, Italy, riccardo.scalera@alice.it

Tamara A. Shiganova
P.P. Shirshov Institute of Oceanology, Russian Academy of Sciences, Nakhimovsky Avenue, 36, 17997 Moscow, Russia, shiganov@ocean.ru

Susan Shirley
Department of Forest Science, Oregon State University, Corvallis, Oregon, USA, susanshir@gmail.com

Assaf Shwartz
Department of Ecology, Systematics and Evolution, The Institute of Life Sciences, The Hebrew University of Jerusalem, Jerusalem 91904, Israel, assaf.sh@mail.huji.ac.il

Wojciech Solarz
Institute of Nature Conservation, Polish Academy of Sciences, Mickiewicza 33, 31-120 Krakow, Poland, solarz@iop.krakow.pl

Montserrat Vilà
Estación Biológica de Doñana, Avd/María Luisa s/n, Pabellón del Perú, 41013 Sevilla, Spain, montse.vila@ebd.csic.es

Marten Winter
Helmholtz Centre for Environmental Research – UFZ, Department of Community Ecology, Theodor-Lieser-Str. 4, D-06120 Halle, Germany, marten.winter@ufz.de

Pierre Yésou
Office National de la Chasse et de la Faune Sauvage, 53, Rue Russeil, F 44 000 Nantes, Franc, p.yesou@oncfs.gouv.fr

Anastasija Zaiko
Coastal Research and Planning Institute, Klaipėda University, Klaipėda, LT 92294, H. Manto 84, Lithuania, nastiusha@takas.lt

Chapter 1
A pan-European Inventory of Alien Species: Rationale, Implementation and Implications for Managing Biological Invasions

Philip E. Hulme, David B. Roy, Teresa Cunha, and Tor-Björn Larsson

1.1 Introduction

Biological invasions by alien (c.f. non-native, non-indigenous, foreign, exotic) species are recognised as a significant component of global environmental change, often resulting in a significant loss in the economic value, biological diversity and function of invaded ecosystems (Wittenberg and Cock 2001). Numerous alien species, many introduced only in the last 200 years ago, have become successfully established over large areas of Europe (Hulme 2007). Future global biodiversity scenarios highlight potentially dramatic increases in biological invasions in European ecosystems (Sala et al. 2000). Interacting effects through rising atmospheric CO_2 concentrations, warmer temperatures, greater nitrogen deposition, altered disturbance regimes and increased habitat fragmentation may facilitate further invasions (Vilà et al. 2006). Early warning and prevention of the harmful impact of alien species on ecosystems is a fundamental requirement of the European Biodiversity Strategy and the EU Action Plan to 2010 and Beyond (European Commission 2006) yet, in the absence of reliable regional analyses, the European states have been unable to tackle this issue strategically (Miller et al. 2006; Hulme et al. 2007).

In the United States, the cost of biological invasions has been estimated to total $97 billion hitherto for 79 major bioinvasions (Pimentel et al. 2001). Although only limited monetary data are available at present for Europe, there is a similar indication that biological invasions have imposed losses on the economy. The strongest evidence is for alien pest and weeds that impact upon the agriculture, forestry, aquaculture and other sectors (Williamson 2002). Examples of direct economic impacts include the damage caused by Japanese knotweed *Fallopia japonica* to flood defences and the impact of bark stripping by grey squirrels *Sciurus carolinensis* on forestry production. The western corn rootworm *Diabrotica virgifera* was accidentally introduced in the 1990s into Serbia and is an important pest of maize and leads to yield losses. Preliminary studies on the potential of establishment of the western corn rootworm show that this pest is likely to survive and develop wherever maize is grown in Europe. Leaving aside introduced pests and diseases affecting agriculture, alien parasites such as

Gyrodactylus salaris (an ectoparasite of Atlantic salmon) and ***Anguillicola crassus*** (swimbladder nematode of eels) have led to dramatic decreases in fisheries sector incomes in several Nordic countries. The American oyster drill *Urosalpinx cinerea* is an important gastropod pest of the cultured oyster industry as it feeds preferably on oyster spat and is recorded as consuming more than half the oyster spat in certain European estuaries (Cole 1942). The muskrat ***Ondatra zibethicus*** and coypu ***Myocastor coypus***, both introduced by the European fur industry, damage river banks through digging and have increased the risk and severity of floods in many central and southern European countries. Notorious invasive alien weeds are of major economic significance, e.g., Mexican tea *Chenopodium ambrosioides*, knotgrass ***Paspalum paspaloides***, Canadian horseweed *Conyza canadensis*, Bermuda buttercup ***Oxalis pes-caprae***. While other alien plants act as hosts of plant pathogens e.g., rescuegrass *Bromus catharticus* as host for barley yellow dwarf virus and wheat stem rust. Invasive alien species can also affect human health e.g., phytophotodermatitis through contact with giant hogweed ***Heracleum mantegazzianum***, asthma and hay-fever arising from the pollen of annual ragweed ***Ambrosia artemisiifolia***, poisoning of humans through consumption of toxic fruit e.g., American pokeweed *Phytolacca americana*, silverleaf nightshade *Solanum elaeagnifolium,* or leptospirosis spread by the brown rat ***Rattus norvegicus***.

In addition, invasive alien species may also have profound environmental consequences, exacting a significant toll on ecosystems (European Commission 2004). These range from wholesale ecosystem changes e.g., colonisation of sand dunes by ***Acacia*** spp. and extinction of native species e.g., threats to endemic coastal plants following expansion of iceplant ***Carpobrotus edulis*** to more subtle ecological changes and increased biological homogeneity. For example, rhododendron ***Rhododendron ponticum*** reduces the biodiversity of Atlantic oakwoods and the American mink ***Mustela vison*** is held partially responsible for the decline in water vole *Arvicola terrestris* populations in the UK. The freshwater Asiatic clam ***Corbicula fluminea*** is a phytoplankton feeder, its dense populations may affect the structure of planktonic communities, competing with native clams, reducing fish stocks, and shifting primary production to benthic communities. It is a major macrofoulant of power-generating plants, and industrial and municipal water systems. A subtler, but potentially more serious impact of alien species is the possibility of hybridisation with native species. Hybridisation has occurred between alien sika ***Cervus nippon*** and native red *Cervus elaphus* deer, the alien ruddy ***Oxyura jamaicensis*** and native whiteheaded *Oxyura leucephala* ducks as well as between native and alien oaks *Quercus* spp. Hybridisation may introduce maladaptive genes to wild populations or result in a vigorous and invasive hybrid.

Several biological invasions now threatening Europe might have been prevented by a higher level of awareness of invasive species issues and a stronger commitment to address it e.g., the spread of the killer alga ***Caulerpa taxifolia***. Current inaction in many, though not all, countries is becoming increasingly disastrous for the region's biodiversity, health and economy (Hulme 2007).

European states should recognise the risk that activities within their jurisdiction or control may pose to other states as a potential source of invasions and take appropriate individual and cooperative actions to minimise that risk. This is particularly important within Europe as species introduced into the territory of one state can easily spread to neighbouring states, especially with its shared coastline, transboundary mountain ranges and international watercourses. It is also critical with regard to Europe's trading partners. Yet, historically the number and impact of harmful invasive alien species in Europe have been chronically underestimated, especially for species that do not damage agriculture or human health. Comparable estimates for Europe would play a pivotal role in informing policy and identifying resource priorities, yet until recently these data have not been available for any European region.

1.2 Rationale

Historically, invasive alien species issues have relatively low visibility in the European Community, outside specialist circles. However, in the late 1990s increasing awareness of the impact of biological invasions in Europe arose from clear evidence of impacts reported in regional environmental audits (Stanners and Bordeau 1995; European Environment Agency 1998, 2003). By 1998, the Community Biodiversity Strategy identified invasive alien species as an emerging issue of environmental importance (European Commission 1998) and in March 2002, the European Council (Environment) recognised that the introduction of invasive alien species was one of the main recorded causes of biodiversity loss and the cause of serious damage to economy and health (European Commission 2002). The European Council supported the use, as appropriate, of national, transboundary and international action. These include, as a matter of priority, measures to prevent such introduction occurring, and measures to control or eradicate those species following an invasion. Subsequently, under the auspices of the Bern Convention, the European Strategy on Invasive Alien Species was launched in 2002 (Council of Europe 2002).

With increasing awareness of the problem there followed recognition of policy and legislative commitments. A significant number of international policies and directives encompass alien species legislation in Europe (reviewed in detail by Miller et al. 2006). For example, Article 196(1) of the 1982 United Nations Convention on the Law of the Sea (UNCLOS) provides, that *"States shall take all measures necessary to prevent, reduce and control pollution of the marine environment resulting from the use of technologies under their jurisdiction or control, or the intentional or accidental introduction of species, alien or new, to a particular part of the marine environment, which may cause significant and harmful changes thereto"*. More generally, the European States have a commitment *"to strictly control the introduction of non-indigenous species"* (Bern Convention on the Conservation of European Wildlife and Natural Habitats) and *"eradicate those alien species which threaten ecosystems, habitats or species"* (UN Convention on Biological Diversity). The EU policy for the implementation of these conventions states that the European Community *"should take measures pursuing to prevent that alien species cause detrimental effects on ecosystems, priority species or the habitats they depend on and establish measures to control, manage and wherever possible remove the risks that they pose"*. This legislation also forms an integral element of the EU Habitats Directive which similarly contains provisions to ensure invasive alien species introductions do not prejudice the local flora and fauna. More recently, the EU Biodiversity Strategy (European Commission 1998) states that: *"The presence or introduction of alien species or sub-species can potentially cause imbalances and changes to ecosystems. It can have potentially irreversible impacts, by hybridisation or competition, on native components of biodiversity. Applying the precautionary principle, the Community should take measures to prevent that alien species cause detrimental effects on ecosystems, priority species or the habitats they depend on and establish measures to control, manage and wherever possible remove the risks that they pose"*.

Despite the Bern Convention efforts, Europe's practical programmes and coordination on invasive alien species lag behind many other regions of the world. Difficulties arise in the standardisation of the status of alien species. National studies often have access to far more detailed data, but classification of species may differ among countries. This is especially true in terms of the treatment of varieties, hybrids, reintroductions, translocations, feral species and naturally expanding populations. Guidelines for the classification of species status have only recently been suggested and have yet to be widely implemented (IUCN 2000) and the origin of ancient introductions prior to detailed floristic and faunal records is often uncertain. The heterogeneity in the degree to which different European nations are exposed to biological invasions may limit recognition of the risk that activities within their jurisdiction may pose to other nations. Species prioritised for management differ across Europe such that concerted actions should be planned at sub-regional scales. Finally, alien species in one European nation may be native in another. This poses considerable complexity on the development of regulations regarding trade within Europe. Whilst Europe's characteristics arguably make it harder to develop and implement common trade and movement policies, this should not be used as an excuse for failing to take decisive action (Council of Europe 2002).

Effective control of invasive alien species has been hampered in Europe by the lack of (1) monitoring for alien species at frequent enough intervals in regions of concern; (2) a means to report, verify the identifications, and warn of new sightings; and (3) risk assessments that predict the likelihood of a particular species becoming invasive. Information on the invasive alien species present in Europe is incomplete, and that which is available is scattered in a variety of published and unpublished accounts and databases. Anticipating invasions by alien species is difficult, because access to information on their previous invasive ability (one of the best predictors of whether a new species will become invasive) is mostly unavailable. A key recommendation of the European Strategy on Invasive Alien Species is the development of a regional inventory of alien species recorded in the wild (Council of Europe 2002).

The European Commission, under its Sixth Framework Programme of support to Community activities in research and technological development, launched a call in 2003 for an inventory of alien invasive species. The call was precise and exhibited considerable foresight and understanding of the needs of end-users and scientists alike:

> *Create an inventory of invasive species that threaten European terrestrial, fresh-water and marine environments and to provide the basis to prevent and control biological invasions through the understanding of the biological, social, economic and other factors involved. The inventory should be established using common definitions and criteria, and aims to cover all taxa known to be invasive, and all European countries, water bodies and seas. Where possible, the distribution of known invasions should be presented graphically. The work should also assess the ecological, economic and health risks and impacts of biological invasions in Europe as well as indicators for early warning.*

As a result of competitive bidding among several different research proposals, the contract for the alien invasive species inventory was awarded to a consortium of leading researchers of biological invasions in Europe, drawn from 18 institutions across 15 countries. The resulting project, DAISIE (Delivering Alien Invasive Species Inventories for Europe), was launched in February 2005 and ran for the three subsequent years with a European Commission contribution of €2.4 million. The general objectives of the project were:

1. To create an inventory of alien species that threaten European terrestrial, freshwater and marine environments
2. To structure the inventory to provide the basis for prevention and control of biological invasions through the understanding of the environmental, social, economic and other factors involved
3. To assess and summarise the ecological, economic and health risks and impacts of the most widespread and/or invasive species in Europe
4. To use distribution data and the experiences of the individual Member States as a framework for considering indicators for early warning.

By achieving these objectives, DAISIE aimed to deliver a European "one-stop-shop" for information on biological invasions in Europe.

1.3 Implementation

The European Strategy on Invasive Alien Species (Council of Europe 2002) encouraged the development of a pan-European inventory of invasive alien species to mobilise existing expertise for species inventory and review, link and integrate existing databases, include potentially invasive alien species that have a high likelihood of introduction or spontaneous spread from neighbouring countries, and identify priority species. Where available, information should include: species taxonomy and biology, date and place of introduction, means of arrival and spread, range and spread dynamics, risk of expansion to neighbouring countries, invaded ecosystems, population size and trends, impacts recorded and level of threat, other data relevant for risk analysis and, early warning systems, prevention, mitigation and restoration methods and their efficiency, references and contact details. In response to these requirements, DAISIE focused on three major areas of information gathering and dissemination:

1. The European Alien Species Expertise Registry: a directory of researchers and research
2. European Alien Species Database: including all known alien species in Europe
3. European Invasive Alien Species Information System: descriptions of key alien species known to be invasive in Europe that includes distribution maps of key invasive alien species in Europe known or suspected of having environmental or economic impacts

Each of these activities is briefly described below and they have been integrated together as a single internet portal for information on European alien species (www.europe-aliens.org).

1.3.1 European Alien Species Expertise Registry

Current expertise in biological invasions is distributed across research organisations throughout Europe and is funded mainly by national programmes. The European Expertise Registry represents a fundamental step towards linking these organisations and individuals in ways that provide added value at European level and provide the critical mass of expertise in invasive alien species research to meet European-scale requirements. The European Expertise Registry facilitates the clustering and information sharing among different national programmes targeting the same invasive alien species, helps establish teams of experts who can, once a new alien incursion has been reported, assess the situation and prepare an action plan for the invasive alien species at a particular site and enables the current breadth and scope of European knowledge on alien species to be assessed. The registry contains information on the field of expertise (distribution, conservation, ecology, economy, genetics, legislation, management, pathways, physiology, risk assessment and taxonomy) and on the taxonomic and geographic structure of the expertise. Within 12 months of its launch, the Registry contained information on 1,500 experts from nearly 90 countries for almost 3,000 higher taxa (family level or higher) and numbers have steadily increased since. These data already highlight a general paucity of expertise in the larger eastern European nations, as well as under-representation of expertise in alien fungi, moss and invertebrate species, especially insects.

1.3.2 European Alien Species Database

An up-to-date inventory of all alien species known to inhabit Europe is essential to building an early detection and warning system for the Europe's environmental managers. This critical step represented the major activity in DAISIE and involved compiling and peer-reviewing national lists of hundreds of species of fungi, plants, invertebrates, fish, amphibians, reptiles, birds and mammals. Data were collated for all 27 European Union member states, and where these states had significant island regions, data were collated separately for these as well. In addition, data were collated for European states that are not in the European Union such as Andorra, Iceland, Liechtenstein, Moldova, Monaco, Norway, the European part of Russia, Switzerland, Ukraine as well as former Yugoslavian states in the Balkans (Fig. 1.1). Finally, marine lists were referenced to the relevant maritime state and thus to have full coverage of the Mediterranean, marine data were included for North African and Near East countries. For each species, an attempt was made to gather information on native range, date of introduction, habitat, known impacts and population status. Considerable effort was required to ensure synonyms were accounted for accurately and all national lists were independently reviewed by experts.

Fig. 1.1 The geographic regions encompassed by the DAISIE database (shaded). Data were collated for the European Union as well as Andorra, Iceland, Israel Liechtenstein, Moldova, Monaco, Norway, the European part of Russia, Switzerland, Turkey, Ukraine as well as former Yugoslavian states in the Balkans. Marine lists were referenced to the relevant maritime state and thus include North African and Near East countries

Records of 11,000 alien species are included in the database (February 2008), the majority of records are for vascular plants with invertebrates also a significant component (Olenin and Didžiulis 2009).

1.3.3 European Invasive Alien Species Information System

The provision of selected species accounts covering high profile alien species not only delivers end users with relevant details for species identification and management but also helps raise public awareness of the issue of invasions. Accounts for a representative sample of 100 invasive alien species (given in **bold** throughout the book) have been produced and each includes information on biology, ecology, distribution, management information, references, links and images. The aim was to generate brief factsheets that might appeal to the general reader with links to more detailed information for specialists. The accounts cover three fungi, 18 terrestrial plants, 16 terrestrial invertebrates, 15 vertebrates, 16 inland and 32 coastal aquatic species invading natural and semi-natural habitats. Selection was based on ensuring a broad spectrum of life forms and functional types, a range of invaded ecosystems and clear examples of different impacts on European biodiversity, economy and health (Vilà et al. 2008). A key requirement for the effective management of invasive alien species is the ability to identify, map, and monitor invasions in order to assess their extent and dynamics (Hulme 2003). Unfortunately, there are no common global standards in terms of sample units (e.g., points, systematic grids, political boundaries), data collected (e.g., species occurrence, both species presence and absence, relative abundance), spatial extent (e.g., regional, national or continental) and resolution of the maps thus generated. This absence of common standards leads to a profusion of different maps that rarely facilitate comparison. Furthermore, biological invasions are dynamic, large scale phenomena and the spatial resolution and extent of a species map determine the degree to which the data are of use in addressing key issues in invasion ecology (Hulme 2003). This is especially of concern in the many attempts to characterise the spatial pattern of invasive alien species, identify invasion hotspots and predict rates of spread. DAISIE therefore had as an objective to establish a common European standard for the graphical presentation of the invasive alien species data as distribution maps. The Common European Chorological Grid Reference System with the size of the mapping grid ca. 50×50 km, depending on the latitude/longitude was used to produce distribution maps. This scheme employs a reasonably detailed resolution for Europe and is commonly used for species mapping. Data sources included European-wide and national atlases as well as regional checklists. For each species the known presence was plotted but areas where a species previously occurred but was eradicated were also considered. Where precise information on distribution was missing but the species was known to occur in a country/region/district, the distribution in these administrative units was recorded and mapped by using hatching. A different format was adopted for mapping invaders in aquatic habitats where linear distributions or maritime areas needed to be recorded. Distribution maps were generated for the 100 species for which accounts were produced and can be found in Vilà (2009).

1.4 Impact and Implications

It is hoped that the inventory, accounts, and distribution maps will provide a qualified reference system on invasive alien species in Europe, available online for environmental managers, legislators, researchers, students and all concerned. It should also encourage the exchange of data among different geographic regions and thereby to serve a node in the Global Information System for Invasive Species. Documenting current invasions, predicting new invasion sites, and preventing invasions are vital to the protection of biological diversity in Europe. Prediction of, and rapid response to, invasive alien species requires ready access to invasive alien species knowledge bases from many countries. It follows that internet-accessible knowledge bases are a precious tool which can provide crucial information for the early detection, eradication, and containment of invasive aliens which are most possible for species that have just arrived. With direct access to national knowledge bases throughout Europe, managers and policy-makers addressing the invasive alien species challenge should easily obtain data on which species are invasive or potentially invasive in particular habitats, and use this information in their planning efforts. Agencies responsible for pest control can quickly determine if a species of interest has been invasive elsewhere in Europe. Importers of new alien species (e.g., nurseries, botanical gardens, pet industry) can access data to make responsible business choices. Land managers can learn about control methods that have been useful in other areas, reducing the need to commit resources for experimentation and increasing the speed at which control efforts can begin.

The information available in the database also presents an outstanding resource to synthesise current knowledge and trends in biological invasions in Europe. The data will help identify the scale and spatial pattern of invasive alien species in Europe, understand the environmental, social, economic and other factors involved in invasions, and can be used as a framework for considering indicators for early warning. Describing in detail how these data can be applied to these questions is beyond the scope of this chapter. However, two examples of how these data can be mined to deliver policy relevant information and to disseminate information of invasion risk rapidly to stakeholders, policy makers and the public are presented below:

A key impact of the invasive species inventory is that it will provide an up-to-date view of the current status and distribution of alien taxa in Europe. Comparison between the estimates derived from previous datasets and those from the current inventory helps to identify major trends (Fig. 1.2). First, the distribution of alien species is heterogeneous across nations and this remains the case with the current data. The trends in numbers are strongly correlated between the historic and current

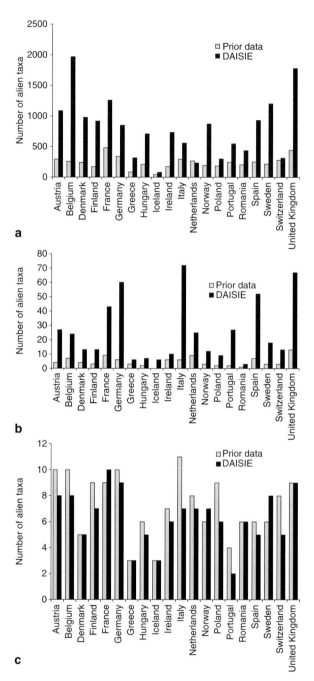

Fig. 1.2 National trends for 20 European states in the number of naturalised alien taxa in datasets pre-dating the DAISIE inventory (for sources see Hulme 2007) and datasets from the current (January 2008) DAISIE database (a) all higher plants, (b) neozoan birds, and (c) neozoan mammals. The two datasets are significantly correlated for plants ($r_s = 0.569$, df 18, $p < 0.01$), birds ($r_s = 0.717$, df 18, $p < 0.01$) and mammals ($r_s = 0.787$, df 18, $p < 0.01$)

datasets samples. There is some indication that numbers of alien taxa are correlated with national GDP but this only explains part of the international variation (Hulme 2007). However, the more recent data reveals a consistent increase in the average numbers of alien plants, birds and mammals found across Europe. This probably reflects a more thorough assessment in the recent data rather than a sudden increase in recently established aliens in Europe. These new data and their availability online will not only assist many European nations in the preparation of their National Strategy on Invasive Alien Species, but also herald the opportunity for discussion with neighbouring countries regarding regional and coordinated approaches for combating biological invasions.

Signatories to the Convention on Biological Diversity (CBD) have committed to achieve by 2010 a significant reduction of the current rate of biodiversity loss at the global, regional and national level (European Commission 2006). To facilitate the assessment of progress towards 2010 and communication of this assessment, a clear set of indicators have been proposed. The CBD has recognised an urgent need to address the impact of invasive alien species and has included 'Trends in invasive alien species' in the trial indicators to be developed and used for assessing global progress towards the 2010 target. The European Union (EU), in responding to the process for review of the EU Biodiversity Strategy and Biodiversity Action Plans, endorsed a first set of EU Headline Biodiversity Indicators in 2004 to monitor and evaluate progress towards the 2010 targets, including a general indicator 'Trends in invasive alien species in Europe'. A levelling off in the current increase in numbers of alien species and a reduction in the rate of establishment of alien species in new countries/regions, and/or a shrinking distribution of these within Europe would be a signal that this target is addressed successfully. The current inventory highlights that the rate of new naturalisations has consistently declined for some taxa most notably vertebrates such as inland fish, birds and mammals, while consistent increases are found for many invertebrates in both aquatic and terrestrial biomes (Fig. 1.3). However, even where rates of establishment are declining, the cumulative number of alien taxa is increasing and for plants, marine invertebrates and terrestrial insects, current rates of increase are over 10 new taxa per year. The pan-European inventory of alien species created through DAISIE provides a platform for European reporting on biodiversity indicators and highlights areas where Europe will need to direct resources to manage biological invasions.

1.5 Future Opportunities

Biological invasions are dynamic phenomena both in time and space and while DAISIE has assembled the most comprehensive dataset on alien taxa that Europe has ever seen, there is still a pressing need to update regularly the information on alien species, their biology, vectors of introduction, spread, impacts on environment and economy. The European Environment Agency (EEA) is responsible for environmental information exchange and dissemination and plays a key role in awareness-raising.

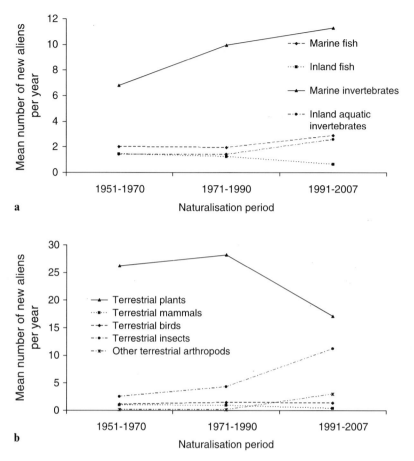

Fig. 1.3 Pan-European trends in the average number of new alien plants, invertebrate, fish, birds and mammals naturalising in Europe per year in three time periods 1951–1970; 1971–1990 and 1991–2007 in (a) aquatic and (b) terrestrial environments

It hosts the European Community Biodiversity Clearing-House Mechanism (EC-CHM), a regional CHM established in support of the CBD. This aims to make biodiversity-related information of Community institutions more easily accessible not only to these institutions but also to member states and the public. In its 2004–2008 strategy, the EEA identified as priorities both biodiversity information gathering under the 2010 process, and the need to build on strong partnerships with NGOs and the science community for such data and information gathering (European Environment Agency 2004). In delivering these priorities, the EEA collaborates with the European Topic Centre on Nature Protection and Biodiversity, which maintains and develops EUNIS (EEA Information System on Nature in Europe). The Topic Centre has intended to link EUNIS to the CBD/GISP system of interoperable databases and also to subregional or specialised databases and research networks in Europe on alien species.

The infrastructure established by DAISIE would fit well with the aims of both the EEA and the Topic Centre. If, as is hoped, DAISIE acts as a catalyst to generate greater awareness of alien species in Europe it is likely that data management will become increasingly complex. The future of the inventory may increasingly see a move away from a single database to the integration of national databases across the same infrastructure. The inventory will then become a tool that integrates different data sets as a seamless resource. This should allow users to access and use multiple separately owned sources of invasive species information and for data owners/custodians to control who they give access to within their own rules/terms and conditions. There will certainly be political and logistic challenges in updating and delivering such information across a region the size of Europe, DAISIE is just a first step in the right direction.

References

Cole HA (1942) The American whelk tingle, *Urosalpinx cinerea* (Say), on British oyster beds. J Mar Biol Assoc 25: 477–508

Council of Europe (2002) European strategy on invasive alien species. Council of Europe Publishing, Strasbourg. ISBN 92-871-5488-0

European Commission (1998) A European biodiversity strategy. COM, 42

European Commission (2002) Thematic report on alien invasive species. Second report of the European Community to the Conference of the Parties of the Convention on Biological Diversity. Office for Official Publications of the European Communities, Luxembourg

European Commission (2004) Alien species and nature conservation in the EU. The role of the LIFE program. Office for Official Publications of the European Communities, Luxembourg

European Commission (2006) Halting the loss of biodiversity by 2010 - and beyond - sustaining ecosystem services for human well-being. Communication from the Commission

European Environment Agency (1998) Europe's environment: the second assessment. Office for the Official Publications of the European Communities, Luxembourg/Elsevier Science, Amsterdam

European Environment Agency (2003) Europe's environment: the third assessment. Environmental Assessment Report No 10. Office for Official Publications of the European Communities, Luxembourg

European Environment Agency (2004) EEA strategy 2004–2008. EEA, Copenhagen

Hulme PE (2003) Biological invasions: winning the science battles but losing the conservation war? Oryx 37: 178–193

Hulme PE (2007) Biological invasions in Europe: drivers, pressures, states, impacts and responses. In: Hester R, Harrison RM (eds) Biodiversity under threat issues in environmental science and technology. Royal Society of Chemistry, Cambridge. 25:56–80

Hulme PE, Brundu G, Camarda I, Dalias P, Lambdon P, Lloret F, Medail F, Moragues E, Suehs C, Traveset A, Troumbis A (2008) Assessing the risks of alien plant invasions on Mediterranean islands. In: Tokarska-Guzik B, Brundu G, Brock JH, Child LE, Pyšek P, Daehler C (eds) Plant invasions: human perceptions, ecological impacts and management. Backhuys, Leiden. 39–56

IUCN (2000) Guidelines for the prevention of biodiversity loss caused by alien invasive species. IUCN, Gland

Miller C, Kettunen M, Shine C (2006) Scope options for EU action on invasive alien species (IAS) Final report for the European Commission. Institute for European Environmental Policy (IEEP), Brussels

Olenin S, Didžiulis V (2009) Introduction to the species list. In: DAISIE Handbook of alien species in Europe. Springer, Dordrecht. 129–132

Pimentel D, McNair S, Janecka J, Wightman J, Simmonds C, O'Connell C, Wong E, Russel L, Zern J, Aquino T, Tsomondo T (2001) Economic and environmental threats of alien plant, animal, and microbe invasions. Agric Ecosyst Environ 84: 1–20

Sala OE, Stuart Chapin F III, Armesto JJ, Berlow E, Bloomfield J, Davis F, Dirzo R, Froydis I, Huber-Sanwald E, Huenneke LF, Jackson R, Kinzig A, Leemans R, Lodge D, Malcolm J, Mooney HA, Oesterheld M, Poff L Sykes MT, Walker BH, Walker M, Wall D (2000) Global biodiversity scenarios for the year 2100. Science 287: 1770–1774

Stanners D, Bourdeau P (1995) Europe's environment: The Dobris assessment. Office for the Official Publications of the European Communities, Luxembourg

Vilà M, Corbin JD, Dukes JS, Pino J, Smith SD (2006) Linking plant invasions to environmental change. In: Canadell J, Pataki D, Pitelka L (eds) Terrestrial ecosystems in a changing world. Springer, Berlin. 115–124

Vilà M, Basnau C, Gollasch S, Josefsson M, Pergl J, Scalera R (2009) One hundred of the most invasive alien species in Europe. In: DAISIE Handbook of alien species in Europe. Springer, Dordrecht. 265–268

Williamson MH (2002) Alien plants in the British Isles. In: Pimentel D (ed) Biological invasions: economic and environmental costs of alien plant, animal and microbe species. CRC Press, Boca Raton, FL. 91–112

Wittenberg R, Cock MJW (2001) Invasive alien species: a toolkit for best prevention and management practices. CABI, Wallingford

Chapter 2
Alien Fungi of Europe

Marie-Laure Desprez-Loustau

2.1 Introduction

This chapter deals with fungi in the broad sense, i.e. organisms studied by mycologists, including species classified within three different eukaryotic kingdoms: Eumycota, Chromista (or Stramenipila) and Protozoa (Whittaker 1969; Cavalier-Smith 1986; Worral 1999). Lichen forming fungi are considered in Essl and Lambdon (2009).

Fungi are a major component of biodiversity on Earth, as the second largest group of Eukaryotes, after insects. The number of fungal species has been estimated to be at least 1.5 million, but less than 10% have been described (Hawksworth 2001). Fungi are inconspicuous organisms and have been far less studied than mammals or vascular plants. This also applies to invasion ecology: fungi are usually poorly, if at all, represented in alien species databases, with the exception of a few well-known examples of plant or animal pathogens. In particular, few comprehensive national checklists have been published for Europe (see below). The reasons for this low representation are most likely due to a poor knowledge of this group of organisms rather than a low invasion success (Desprez-Loustau et al. 2007). When undescribed species are found, how likely is it that they are native to the geographic location of their first record? The species concept itself is not easily handled in fungi. Fungal taxonomy has been evolving rapidly over the recent years, in particular with the use of molecular tools and phylogenetic analysis. This can be illustrated by the fact that more than 50% of the fungal species in the dataset compiled in DAISIE were described (or subjected to taxonomic revision) after 1950, and almost 20% only in the 2000s. Many fungal species previously defined on the grounds of morphology (or of symptoms on host plant for pathogens) have been shown to be a complex of several cryptic species differing in their ecology, and especially in their geographic range (Pringle et al. 2005). A recent example is *Mycosphaerella pini*, a foliar pathogen of pines, which was shown to be a complex of two phylogenetic species, with one found worldwide, while the other is restricted to the North-Central USA (Barnes et al. 2004). The poor knowledge in the biogeography of fungi can make it difficult to determine what is an alien species. Within the DAISIE dataset, more than 30% of fungal species are considered as cryptogenic, i.e., of unknown origin. New approaches, including phylogeography, might provide clues on the native range of species. For example, *Phytophthora ramorum*, recently discovered both in Europe and North America, is assumed to be alien in both regions, on the basis of a multilocus analysis demonstrating an extremely narrow genetic structure, most likely explained by the introduction of a few closely related genotypes (Ivors et al. 2006).

2.2 Inventory of Alien Fungi in Europe

Lists of alien fungi have recently been published for a few European countries/regions: Germany (Kreisel 2000, 90 species), Austria (Essl and Rabitsch 2002, 82 species), Lithuania (Kutorga 2004, 95 species), England (Hill et al. 2005, 197 species), northern Europe (NOBANIS 2007, 31 species), Poland (Solarz 2007, 89 species), Norway (Gederaas et al. 2007, 263 species) and France (Desprez-Loustau ML et al., 2008 unpublished, 227 species). Compiling these different lists and additional literature for other countries (principally from information in EPPO and CABI databases: www.eppo.org, www.cabi.org) gave a total of 688 species considered as alien in at least one European country. Synonymies were resolved by using the current names provided by the Index Fungorum (www.indexfungorum.org). It has to be noted that many species listed as alien in one country are considered as native in another European country (e.g., species of Alpine or Mediterranean origin in England and Norway).

However, the data in this compiled list are very heterogeneous among countries. Some national lists only include pathogenic fungi, by focusing on alien invasive fungi (e.g., in Lithuania) while most consider both pathogenic and non pathogenic fungi. The definition of alien itself might differ in different lists, e.g., including or not archaeomycetes (such as fungal agents of cereal rusts or other pathogens presumably introduced with their host plants before AD 1500) or alien species only found in glasshouses or artificial environments (such as human pathogens). Date and references of first occurrence and origin of the species are lacking in several national lists. Analyses among countries based on the compiled list would therefore be irrelevant at this stage, and anyhow limited to a few countries. In order to have a pan-European overview, we focused on a subset of species taken from the compiled list for which a systematic search for presence/absence in each of the geographic units considered in DAISIE was performed and the date and reference of first observation in Europe was recorded. This subset comprises 84 alien invasive fungi for Europe (1,062 invasion events) selected as following:

1. Species known to be alien to Europe
2. Species with a potential negative impact on biological diversity (invasive species, sensu IUCN), i.e. pathogenic on native (European) species and/or occurring in wild environments. Pathogens of crop and ornamental plants were not included, although they have high economic impact (Pimentel et al. 2001), except for a few ones also reported on wild plants. Non-pathogenic fungi were not included at this stage but their ecological impact has probably been under-appreciated (Desprez-Loustau et al. 2007).

The main sources for distributional data were CABI, EPPO, the USDA database for fungi (http://nt.ars-grin.gov/fungaldatabases/index.cfm), the New Disease Reports of the British Society for Plant Pathology (www.bspp.org.uk/ndr/), Viennot-Bourgin (1949) and Smith et al. (1988).

2.3 Taxonomy and Lifestyle of Invaders

In the compiled European list of 688 alien species, Ascomycota and Basidiomycota are nearly equally represented, with 46% and 40% of all species, which is close to figures for all described fungal species (approximately 60% Ascomycota and 40% Basidiomycota in Eumycota, Hawksworth et al. 1995). The under-representation of Basidiomycota in the subset of 84 species can be explained by the focus on pathogenic fungi (mostly belonging to Ascomycota). Major orders of Ascomycota, not including lichen forming species, are represented, except Laboulbeniales (mainly parasites of insects) and Meliolales (mostly tropical) (Hawksworth et al. 1995). The Oomycota are over represented both in the European compiled list and in the subset. This phylum comprises less than 1,000 described species (compared to more than 20,000 in Basidiomycota and Ascomycota) but includes many severe plant pathogens (Latijnhouwers et al. 2003). For example, *Phytophthora infestans*, the causal agent of potato late blight, was responsible for the Irish potato famine of 1845–1860 while the crown and root rot **Phytophthora cinnamomi** has caused severe damage in natural ecosystems worldwide (Zentmyer 1980). Among the 17 Oomycota included in the 84 species subset, it has to be noted that five have been described since 1999 in the *Phytophthora* genus, reflecting recent research focus on this group (Brasier 1999). As for *Phytophthora ramorum*, their native area is unknown and their alien status is usually assumed from the low genetic variation of European populations (Cooke et al. 2005). *Phytophthora alni* represents a particular case, resulting from hybridisation events, probably involving at least one alien species (Ioos et al. 2006).

Among the 84 species, 82 are plant pathogens, which is consistent with the much tighter association between fungi and plants than between fungi and animals (Berbee 2001). Parasites of insects might have been overlooked (cf. under-representation of Laboulbeniales). In the European compiled list, plant pathogens are also the most numerous, with 77% of all species. Other symbiotic fungi (animal pathogens or plant mutualists) represent 6% and saprobes 17%.

2.4 Temporal Trends in Invasions Since 1800

The number of newly recorded alien fungi, as well as the cumulative number, have been increasing exponentially since 1800 (Fig. 2.1). For France alone, the rate of introduction increased from less than 0.5 species/year before 1930 to approximately 2 species/year in the most recent period (since 1970).

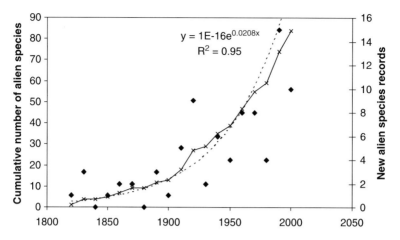

Fig. 2.1 Increase in the number of alien fungal species (84 species subset) recorded in Europe from 1820 to 2007, expressed for decades: left, cumulative records (crosses and solid line; significant exponential adjustment in dashed line); right, number of new records per period (diamonds)

An increased awareness of the problem of introductions (therefore increased reporting) and a better accessibility to records in recent years cannot be excluded as a partial explanation for this trend. For example, the New Disease Reports, an open-access online journal, was created in 2000 by the British Society for Plant Pathology. However, the CABI databases have a long history, dating back to the early 20th century (Pasiecznik et al. 2005).

2.5 Biogeographic Patterns

France, United Kingdom, Germany and Italy, i.e., large countries, possess the highest numbers of alien species (Fig. 2.2). These four countries are also those where first records for Europe were the most frequent, which may partially reflect detection and research efforts. Countries with the highest numbers of alien fungal species are also the ones with the highest level of imports (Fig. 2.3). This variable is a much better predictor for the number of alien fungal species than geographic variables such as country area, latitude or longitude (not shown). This feature is consistent with the main pathways of introduction of alien fungi (see below). The same was also observed for unintentionally introduced spider species alien to Europe (Kobelt and Nentwig 2008).

A majority of species originate from North America, followed by Asia. This general trend is observed both in the 84 species subset with only pathogenic fungi (51% and 28% of species, Fig. 2.4) and in the European compiled list also including non pathogenic fungi (43% and 21% of species). The proportion of species coming from Asia is higher in the last 30 years (from 24% to 35% in the 84 species

2 Alien Fungi of Europe

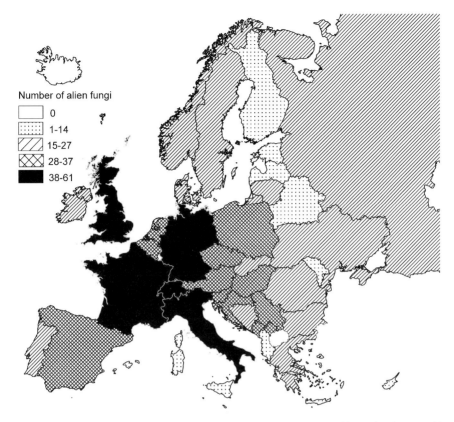

Fig. 2.2 Number of alien fungi reported for each European country (84 species dataset with documented distribution)

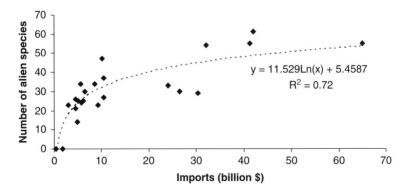

Fig. 2.3 Relationship between the number of recorded alien fungal species (84 species dataset with documented distribution) and the level of imports of goods in 2005 for European countries (OECD)

subset) and for plant parasites compared to saprobes (25% and 7% in the European compiled list). For non-pathogenic fungi, Australasia and tropical areas (without further specification) are relatively important regions of origin (21% and 23% in the European compiled list).

2.6 Main Pathways to Europe

A major feature of fungal invasions is that they are most, if not all, the result of accidental introductions. Desirable fungi deliberately introduced in new habitats mostly include fungi used for mycorrhization (especially in pine plantations of the southern hemisphere) and pathogenic fungi used for classical biological control (against invasive plants or insects). This concerns few species worldwide, and, among them, no documented case was found where the escape of these purposefully introduced pathogenic or mycorrhizal fungi caused substantial mortality to a non-target species or any other significant impact to the environment (Barton 2004; Schwartz et al. 2006; Hajek et al. 2007). A few species of edible fungi have also been introduced outside their native range into various parts of Europe. *Agaricus bisporus*, native to western and southern Europe, has escaped and established in other regions (without any apparent negative impact, Essl and Rabitsch 2002). Conversely, *Lentinula edodes*, the Asian shiitake mushroom cultivated in several European countries, has not hitherto been reported to occur in natural environments.

Most alien fungi, especially symbionts (including pathogenic and mycorrhizal fungi) therefore entered Europe as contaminants (sensu Hulme et al. 2008) in their host plant or animal (most often themselves deliberately introduced).

The history of plant diseases has long been linked to the transport of plants. It has been hypothesised for several diseases which were first reported in Portugal in previous centuries that they had been introduced with alien plants by early Portuguese explorers (with sometimes Azores islands on the route). This might be the case for *Phytophthora cambivora* and the crown and root rot **Phytophthora cinnamomi**, generalist pathogens with a probable centre of origin in Asia, for which the first

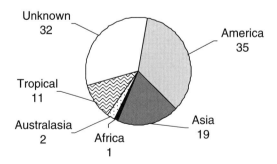

Fig. 2.4 Region of origin of alien fungi recorded in Europe (percentage of species for the 84 species subset)

records in Europe are from Portugal in the early 19th century, as the cause of ink disease of chestnut (Grente 1961).

Aphanomyces astaci, the agent of crayfish plague, is an example where the original host served as a vector for the pathogen to a related native species. North American crayfish, which are the natural host and healthy or subclinical carriers of *Aphanomyces astaci*, have been widely translocated to Europe, both intentionally and unintentionally. Conversely, European crayfish turned out to be highly susceptible and the epizootics spread all over Europe (Edgerton et al. 2004). Similarly, the American bullfrog was shown to asymptomatically carry the emerging pathogen *Batrachochytrium dendrobatidis*. Deliberate introductions of bullfrog especially in Europe could be a major way of introduction of the fungus, which has been implicated in global amphibian declines (Garner et al. 2006). There are many other examples of pathogen transfer from an introduced host (especially plant) to native species, e.g., *Cryphonectria parasitica*, the agent of chestnut blight, from Asian to American and European chestnut, or *Cronartium ribicola*, the agent of five-needle-pine rust, from Asian to American and European pines (Desprez-Loustau et al. 2007). The success of invasive pathogens in these "new encounters" or "novel interactions" following host jumps (or host switches) might be explained by a high aggressiveness against naïve host species that have not had an opportunity to evolve resistance (Parker and Gilbert 2004; Robinson 1996).

Although detailed pathways are rarely documented for alien fungi, a significant number of plant pathogens are obligate biotroph parasites (rusts, powdery mildews) and are therefore assumed to have been introduced with their living plant. More generally, plant trade, especially of ornamentals, is most likely a very important way of entry for pathogens. Indirect evidence of the role of trade routes through nurseries in the movement of *Phytophthora ramorum*, causal agent of sudden oak death in California and widely spread in European nurseries, was recently demonstrated by a multilocus genetic analysis (Ivors et al. 2006). Seeds are less frequently documented pathways, but may have been overlooked. Repeated introductions of pine seeds infected with *Sphaeropsis sapinea* probably explain the high genetic diversity of this fungus introduced to South Africa (Wingfield et al. 2001). Seed transmission might also be important for Europe where *Sphaeropsis sapinea* is mostly found in pine plantations and rarely on native pines. Timber imports are another important pathway. Dutch elm disease is a prominent example, with the introduction of *Ophiostoma ulmi*, the causal agent of the first pandemic, along with its bark beetle vector, in a shipment of logs from the Netherlands to the USA, and the re-introduction of the more virulent **Ophiostoma novo-ulmi**, responsible for the second pandemic, also on diseased elm logs from Canada into Britain (Brasier and Buck 2001). *Ceratocystis platani*, the agent of plane canker stain, is another famous example, presumably introduced from North America to Europe with military equipment during World War II landings in Provence and Italy, in 1944, although the fungus was identified only two decades later (Vigouroux 1986).

For many saprobes, compost, and more recently wood-chips, seem to be the main pathways. Several mushrooms, hitherto barely if at all present in Europe have been observed in prodigious numbers on wood-chips beds, such as *Stropharia aurantiaca*, presumed to have been introduced from Asia or Australasia (Marren 2006; Shaw et al. 2004). A relatively high number of tropical species have been first introduced in greenhouses, some of them later escaping in the wild, e.g., *Collybia luxurians* or *Leucocoprinus birnbaumii* (Essl and Rabitsch 2002; Pidlich-Aigner et al. 2002).

2.7 Most Invaded Ecosystems and Economic and Ecological Impacts

Introduction of pathogens are the major cause of emerging diseases, both in wild animals and plants (Dobson and Foufopoulos 2001; Anderson et al. 2004). Few fungi are pathogenic to animal species. Two alien fungi pathogenic to animals are reported in Europe in inland aquatic environments, causing severe decline in the populations of their host species: **Aphanomyces astaci** on crayfish and *Batrachochytrium dendrobatidis* on amphibians (Garner et al. 2005). Most fungi are associated with terrestrial plants and their impact is therefore highest in terrestrial environments.

Fungal pathogens of crop plants occasion important social and economic damage but estimates of yield and economic losses at national or European level are scarce. For the UK, economic loss due to plant pathogens was estimated as 6% of potential production, or about US$2 billion per year (in Pimentel et al. 2001). Economic damage might also be important for some forest tree pathogens, although precise assessments of economic losses are not available. For example, powdery mildew, the major disease of oaks over Europe, is caused by the alien species *Erysiphe alphitoides* (and possibly other related species; Mougou et al. 2008). This disease is responsible for losses in nurseries, regeneration failures, growth loss and declines in mature stands (Desprez-Loustau 2002).

The recent outbreak of *Ceratocystis platani* along the Canal du Midi in SW France is another example of economic impact through the mortality of high value trees. As the oldest canal still functioning in Europe, the Canal du Midi has been put on the World Patrimony list of the UNESCO and is now an important place for riverboat tourism. Plane trees planted along the canal banks are part of the landscape and the threat of a canker stain epidemics is a major concern for managers. A costly eradication program of the disease has been carried out. Canker stain of plane tree was also recently reported in a small area of SW Greece on the native oriental plane tree, *Platanus orientalis*, where it was probably secondary introduced from Italy or France (Ocasio-Morales et al. 2007). Eradication measures should be imposed before it spreads throughout the natural range of this ecologically and historically important host.

Cypress canker, caused by **Seiridium cardinale**, is another example of disease caused by an alien fungus affecting a native species with high social and cultural value. Italian cypress *Cupressus sempervirens* is a major feature of the Mediterranean landscape. Following the introduction of **Seiridium cardinale** from California, millions of cypress trees were killed in southern Europe, especially in some areas of Greece, Italy and southern France, with mortality reaching 25–75%. An after-effect of the disease is soil erosion in devastated hills. The disease has also had economic impacts through losses in cypress plantations, grown for highly valued timber and oils, in ornamental cypress trade, and indirectly in agriculture due to the destruction of windbreaks (Graniti 1998).

Other invasive fungal pathogens have had high environmental impacts through dramatic reductions in their host populations. The involvement of the emerging pathogen *Batrachochytrium dendrobatidis* in amphibian declines in several regions of the world, including Europe, has been hypothesised (Rachowicz et al. 2006). The recent epidemics of *Phytophthora alni* affecting alders are another example of potential high environmental impact, due to the important role of riparian forests.

Finally, it can be mentioned that a few human pathogenic fungi of alien (especially tropical) origin have been recorded in Europe (Kreisel 2000). The prevalence of imported mycoses is low, although higher in immunocompromised patients, but they can be highly virulent (Warnock et al. 1998).

2.8 Management Options and Their Feasibility

As for all invasive alien species, prevention, eradication and control are the three main types of measures aimed at limiting the negative impact of fungal invasions (Hulme 2006). At first sight, prevention appears as the most desirable measure and quarantine regulations against plant pests have been established for a long time. After the introduction of several destructive pathogens in the 19th century in Europe, especially *Phytophthora infestans*, the International Plant Protection Convention (IPPC) was developed in the 1920s (Schrader and Unger 2003). Phytosanitary regulations are then implemented at regional and national levels. For Europe, the Council Directive 2000/29/EC establishes the measures to be taken in order to prevent the introduction into, and spread within, the Community of serious pests and diseases of plants, with a list of species (including pathogenic fungi) the introduction of which is prohibited in all parts of Europe. *Ceratocystis fagacearum*, agent of oak wilt in North America, is one of these species. It is considered as a major threat for European oak forests, where it could cause a disease comparable to Dutch elm disease. Phytosanitary measures, including prohibition of the introduction of bark and the treatment of imported oak wood, have been effective up to now. However, recent disastrous introductions point to the limits of current quarantine measures based on species lists. *Phytophthora ramorum* is an example of pathogens which were completely unknown prior to their introduction (Rizzo et al. 2005). A revision of phytosanitary regulations towards more broadly defined targets is therefore necessary to prevent invasions. A pathway approach has already been adopted for wood packaging material (International Phytosanitary Standard ISPM15 of the IPPC). A similar approach could also be considered for nursery material (forestry and ornamental plants).

Very few examples of successful eradications have been reported for plant pathogens. *Synchytrium endobioticum*, agent of potato wart disease originating from the Andean region, was introduced into Europe in the late 19th century, where it quickly spread. It has been declared eradicated from a few countries (e.g., France). This success might be explained by intense concerted efforts (quarantine legislation and other statutory measures, use of resistant cultivars) triggered by the severity of the disease. Since early detection is the key to eradication success, eradication measures can only be taken if the organism is a target and therefore often already recognised as an invader in other regions. *Ceratocystis platani* in SW France and Greece and *Phytophthora ramorum* in Oregon are examples of eradication trials in secondary disease foci.

In many cases, fungal pathogens (or more generally invaders) are already well established when control measures are attempted. Eradication is no longer feasible and the aim is to mitigate the negative impact of invasive species and to help restore the ecosystem as quickly as possible. Research in Dutch elm disease control is exemplary by its integrative approach (Brasier 1996). Since the disease is vectored by beetles, much work was dedicated to investigate transmission but this option did not produce effective measures. The two major ways of restoring the balance between elms and the pathogen(s) have been: (1) breeding elms for resistance; (2) lowering the aggressiveness of the pathogen. Resistant elm cultivars have been produced by hybridisation of European elms with Asian species. Moreover, screening and propagating tolerant native elms is a major objective for which pan-European collaborations have been set up (Solla et al. 2005). A few clones show promise for resistance. It has even been proposed to produce transgenic elm trees to restore these "heritage trees" (Merkle et al. 2007). English elm plantlets transformed with antifungal genes are currently being tested (Gartland et al. 2005). The second option, i.e. decreasing aggressiveness of the fungal pathogen, has found support after the discovery of virus-like particles, called D-factors, negatively affecting the pathogen itself (Brasier 1996). The potential of these factors for artificial biological control has been investigated (Sutherland and Brasier 1997). Biological control using a hyperparasite virus has already been used for another alien fungal pathogen, *Cryphonectria parasitica*, the agent of chestnut blight. Natural hypovirulence, caused by the infection of *C. parasitica* by several strains of virus, was detected in Europe several years after the appearance of the disease (Grente 1965). Virus preparations (in compatible strains of *C. parasitica*) have been applied operationally in orchards in Italy and France (Robin et al. 2000). Research has been undertaken to select the most desirable viral strains for biological control in forest conditions, i.e. combining good transmission (which is low in currently used virus) and biocontrol potential (negative effect on the fungus).

2.9 Future Expected Trends

The rising trend of fungal invasions can hardly be expected to be reversed in a short term. No saturation effect can be seen from the temporal curves (Fig. 2.1) and global trade is still on an increasing trend, especially between Asia and Europe. Moreover, fungi have several traits which could make them successful invaders: many of them are especially suited for long-distance travel and have a high potential for evolutionary change (Parker and Gilbert 2004; Desprez-Loustau et al. 2007).

However, current control measures, mainly founded on quarantine exclusion on a species basis have proved to be poorly adapted and of limited efficacy. The species focus is especially little relevant for fungal introductions, by being at the same time too restrictive and too broad. Indeed, the analysis of past invasions shows that invaders were mostly unknown before they were observed in their introduced environment. Tropical areas or other regions, such as the Himalayas (potential centre of origin of *Ophiostoma ulmi*) may contain huge numbers of unidentified fungal pathogen threats (Brasier 1996). On the other hand, the species-level approach might overlook the population dimension of invaders. The introduction of additional genotypes of an already occurring species can have important negative consequences by increasing diversity and therefore potential for increased virulence, as well exemplified by the potato late blight fungus (Fry and Smart 1999). The pathway approach might be an interesting alternative or complement for quarantine measures. Development of databases such as in the DAISIE project could help identify the main pathways and species attributes to be targeted. Our list and database cannot be exhaustive today but will hopefully grow and promote the construction of a more comprehensive database.

Eradication and mitigation also need knowledge-based measures (Brasier 1996). More generally, the issue of fungal invasions points to the lack of baseline ecological data on fungal communities. Fungal diversity (species recognition, phylogeography, evolutionary potential) and the functioning of fungal communities deserve research efforts. This especially applies to non pathogenic fungi. The introduction and spread of saprophytic fungi have been documented, but their impacts are rarely known. *Clathrus archeri*, the octopus stinkhorn, is a typical example. After its accidental arrival in Europe in 1920, probably through wool import from Australia, it spread throughout Europe reaching high population levels (hundreds of fruiting bodies) in some locations (Parent et al. 2000). But the possible prejudice to native fungi has not yet been studied.

The more we begin to understand fungal ecology in general, the better we will be able to prevent introductions or at least to predict their ecological trajectories in recipient communities and therefore to mitigate their impact.

Acknowledgements The support of this study by the European Commission's Sixth Framework Programme project DAISIE (Delivering alien invasive species inventories for Europe, contract SSPI-CT-2003-511202) is gratefully acknowledged. I also would like to thank Phil Hulme and Wolfgang Nentwig for valuable comments on earlier versions of this manuscript. I am also very grateful to all fungi experts who provided data for the database, especially Adrien Bolay, Régis Courtecuisse, Ottmar Holdenreider, Pierre-Arthur Moreau, Irena Papasova, Cécile Robin, Ivan Sache, Joan Webber.

References

Anderson PK, Cunningham AA, Patel NG, Morales FJ, Epstein PR, Daszak P (2004) Emerging infectious diseases of plants: pathogen pollution, climate change and agrotechnology drivers. Trends Ecol Evol 19:535–544

Barnes I, Crous PW, Wingfield BD, Wingfield MJ (2004) Multigene phylogenies reveal that red band needle blight of Pinus is caused by two distinct species of *Dothistroma, D. septosporum* and *D. pini*. Stud Mycol 50:551–565

Barton (née Fröhlich) J (2004) How good are we at predicting the field host-range of fungal pathogens used for classical biological control of weeds. Biol Control 31:99–122

Berbee ML (2001) The phylogeny of plant and animal pathogens in the Ascomycota. Physiol Mol Plant Pathol 59:165–187

Brasier C (1999) *Phytophthora* pathogens of trees: their rising profile in Europe. Information note, Forestry Commission, London

Brasier CM (1996) New horizons in Dutch elm disease control. In: Forestry Commission, Report on Forest Research, London. 20–28

Brasier CM, Buck KW (2001) Rapid evolutionary changes in a globally invading fungal pathogen (Dutch elm disease). Biol Invasions 3:223–233

Cavalier-Smith T (1986) The kingdom Chromista: origin and systematics. In: Round FE, Chapman DJ (eds) Progress on phycological research. Biopress, Bristol. 4:309–347

Cooke DEL, Jung T, Williams NA, Schubert R, Obwald W, Duncan JM (2005) Genetic diversity of European populations of the oak fine-root pathogen *Phytophthora quercina*. Forest Pathol 35:57–70

Desprez-Loustau ML (2002) L'oïdium des chênes: Une maladie fréquente mais mal connue. Les Cahiers du DSF 1-2002. La Santé des Forêts [France] en 2000 et 2001: 95–99. Min Agri Alim Pêche Aff Rur (DERF), Paris

Desprez-Loustau ML, Robin C, Buée M, Courtecuisse R, Garbaye J, Suffert F, Sache I, Rizzo D (2007) The fungal dimension of biological invasions. Trends Ecol Evol 22:472–480

Dobson A, Foufopoulos J (2001) Emerging infectious pathogens of wildlife. Philos Trans R Soc Lond B 356:1001–1012

Edgerton BF, Henttonen P, Jussila J, Mannonen A, Paasonen P, Taugbíl T, Edsman L, Souty-Grosset C (2004) Understanding the cause of disease in European freshwater crayfish. Conserv Biol 18:1466–1474

Essl F, Lambdon PW (2009) The alien bryophytes and lichens of Europe. In: DAISIE Handbook of alien species in Europe. Springer, Dordrecht

Essl F, Rabitsch W (eds) (2002) Neobiota in Österreich. Umweltbundesamt, Wien

Fry WE, Smart CD (1999) The return of *Phytophthora infestans*, a potato pathogen that just won't quit. Potato Res 42:279–282

Garner TWJ, Walker S, Bosch J, Hyatt AD, Cunningham AA, Fisher MC (2005) Chytrid fungus in Europe. Emerg Infect Dis 11:1639–1641

Garner TWJ, Perkins MW, Govindarajulu P, Seglie D, Walker S, Cunningham AA, Fisher MC (2006) The emerging amphibian pathogen *Batrachochytrium dendrobatidis* globally infects introduced populations of the North American bullfrog, *Rana catesbeiana*. Biol Lett 2:455–459

Gartland KMA, McHugh AT, Crow RM, Garg A, Gartland J (2005) Biotechnological progress in dealing with Dutch elm disease. In Vitro Cell Dev Biol Plant 41:364–367

Gederaas L, Salvesen I, og Viken Å (eds) (2007) Norsk svarteliste 2007 – Økologiske risikovurderinger av fremmede arter. 2007 Norwegian black list – Ecological risk analysis of alien species. Artsdatabanken, Norway

Graniti A (1998) Cypress canker: a pandemic in progress. Annu Rev Phytopathol 36:91–114

Grente J (1961) La maladie de l'encre du châtaignier. Ann Epiphyt 12:5–59

Grente J (1965) Les formes hypovirulentes d'*Endothia parasitica* et les espoirs de lutte contre le chancre du châtaignier. C R Acad Agric France 51:1033–1037

Hajek AE, McManus ML, Delalibera I Jr (2007) A review of introductions of pathogens and nematodes for classical biological control of insects and mites. Biol Control 41:1–13

Hawksworth DL (2001) The magnitude of fungal diversity: the 1.5 million species estimate revisited. Mycol Res 105:1422–1432

Hawksworth DL, Pegler DN, Kirk PM, Sutton BC (1995) Ainsworth & Bisby's dictionary of the fungi. CABI, Kew

Hill M, Baker R, Broad G, Chandler PJ, Copp GH, Ellis J, Jones D, Hoyland C, Laing I, Longshaw M, Moore N, Parrott D, Pearman D, Preston C, Smith RM, Waters R (2005) Audit of non-native species in England. English Nature Research Reports, 662, Peterborough

Hulme PE (2006) Beyond control: wider implications for the management of biological invasions. J Appl Ecol 43:835–847

Hulme PE, Bacher S, Kenis M, Klotz S, Kühn I, Minchin D, Nentwig W, Olenin S, Panov V, Pergl J, Pyšek P, Roque A, Sol D, Solarz W, Vilà M (2008) Grasping at the routes of biological invasions: a framework for integrating pathways into policy. J Appl Ecol 45:403–414

Ioos R, Andrieux A, Marçais B, Frey P (2006) Genetic characterization of the natural hybrid species *Phytophthora alni* as inferred from nuclear and mitochondrial DNA analyses. Fungal Genet Biol 43:511–529

Ivors K, Garbelotto M, Vries DE, Ruyter-Spira C, Hekkert TE, Rosenzweig N, Bonants P (2006) Microsatellite markers identify three lineages of *Phytophthora ramorum* in US nurseries, yet single lineages in US forest and European nursery populations. Mol Ecol 15:1493–1505

Kobelt M, Nentwig W (2008) Alien spider introductions to Europe supported by global trade. Diversity Distrib 14:273–280

Kreisel H (2000) Ephemere und eingebürgerte Pilze in Deutschland. In: NABU, Ratgeber Neobiota. 73–77

Kutorga E (2004) Invasive fungi. Lithuanian invasive species database. www.ku.lt/lisd/species_lists/fungi_all.html. Cited Sept 2007

Latijnhouwers M, de Wit PJGM, Govers F (2003) Oomycetes and fungi: similar weaponry to attack plants. Trends Microbiol 11:462–469

Marren P (2006) The 'global fungal weeds': the toadstools of wood-chip beds. Br Wildlife 18:98–105

Merkle SA, Andrade GM, Nairn CJ, Powell WA, Maynard CA (2007) Restoration of threatened species: a noble cause for transgenic trees. Tree Genet Genomes 3:111–118

Mougou A, Dutech C, Desprez-Loustau ML (2008) New insights into the identity and origin of the causal agent of oak powdery mildew in Europe. Forest Pathol 38:275–287

NOBANIS (2007) North European and Baltic network on invasive alien species. www.nobanis.org. Cited Sept 2007

Ocasio-Morales RG, Tsopelas P, Harrington TC (2007) Origin of *Ceratocystis platani* on native *Platanus orientalis* in Greece and its impact on natural forests. Plant Dis 91:901–904

Parent GH, Thoen D, Calonge FD (2000) Nouvelles données sur la répartition de *Clathrus archeri* en particulier dans l'Ouest et le Sud-Ouest de l'Europe. Bull Soc Mycol France 116:241–266

Parker IM, Gilbert GS (2004) The evolutionary ecology of novel plant-pathogen interactions. Annu Rev Ecol Evol Syst 35:675–700

Pasiecznik NM, Smith IM, Watson GW, Brunt AA, Ritchie B, Charles LMF (2005) CABI/EPPO distribution maps of plant pests and plant diseases and their important role in plant quarantine. Bull OEPP 35:1–7

Pidlich-Aigner H, Hausknecht A, Scheuer Ch (2002) Macromycetes found in the greenhouse of the botanic garden of the Institute of Botany in Graz (Austria). Fritschiana 32:49–61

Pimentel D, McNair S, Janecka J, Wightman J, Simmonds C, O'Connell C, Wong E, Russel L, Zern J, Aquino T, Tsomondo T (2001) Economic and environmental threats of alien plant, animal and microbe invasions. Agr Ecosyst Environ 84:1–20

Pringle A, Baker DM, Platt JL, Wares JP, Latgé JP, Taylor JW (2005) Cryptic speciation in the cosmopolitan and clonal human pathogenic fungus *Aspergillus fumigatus*. Evolution 59:1886–1899

Rachowicz L, Knapp RA, Morgan JAT (2006) Emerging infectious disease as a proximate cause of amphibian mass mortality. Ecology 87:1671–1683

Rizzo DM, Garbelotto M, Hansen EM (2005) *Phytophthora ramorum*: integrative research and management of an emerging pathogen in California and Oregon forests. Annu Rev Phytopathol 43:309–335

Robin C, Anziani C, Cortesi P (2000) Relationship between biological control, incidence of hypovirulence and diversity of vegetative compatibility types of *Cryphonectria parasitica* in France. Phytopathology 90:730–737

Robinson, RA (1996) Return to resistance. agAccess, Davis, CA
Schrader G, Unger JG (2003) Plant quarantine as a measure against invasive alien species: the framework of the International Plant Protection Convention and the plant health regulations in the European Union. Biol Invasions 5:357–364
Schwartz MW, Hoeksema JD, Gehring CA, Johnson NC, Klironomos JN, Abbott LK, Pringle A (2006) The promise and the potential consequences of the global transport of mycorrhizal fungal inoculum. Ecol Lett 9:501–515
Shaw PJA, Butlin J, Kibby G (2004) Fungi of ornamental woodchips in Surrey. Mycologist 18:12–15
Smith IM, Dunez J, Lelliot RA, Phillips DH, Archer SA (1988) European handbook of plant diseases. Blackwell Scientific, Oxford
Solarz W (2007) Alien species in Poland. www.iop.krakow.pl/ias/default.asp. Cited Sept 2007
Solla A, Bohnens J, Collin E, Diamandis S, Franke A, Gil L, Buron M, Santini A, Mittempergher L, Pinon J, Vanden Broeck A (2005) Screening European elms for resistance to *Ophiostoma novo-ulmi*. Forest Sci 51:134–141
Sutherland ML, Brasier CM (1997) A comparison of thirteen d-factors as potential biological control agents of *Ophiostoma novo-ulmi*. Plant Pathol 46:680–693
Viennot-Bourgin G (1949) Les champignons parasites des plantes cultivées. Masson, Paris
Vigouroux A (1986) *Platanus* diseases, with special reference to canker stain; the present situation in France. Bull OEPP 16:527–532
Warnock DW, Dupont B, Kauffman CA, Sirisanthana T (1998) Imported mycoses in Europe. Med Mycol 36 (Suppl 1):87–94
Whittaker RH (1969) New concepts of kingdoms of organisms. Science 163:150–160
Wingfield MJ, Slippers B, Roux J, Wingfield BD (2001) Worldwide movement of exotic forest fungi, especially in the tropics and the southern hemisphere. BioScience 51:134–140
Worral JJ (1999) Brief introduction to fungi. In: Worral JJ (ed) Structure and dynamics of fungal populations. Population and Community Biology Series 25. Kluwer, New York. 1–18
Zentmyer GA (1980) *Phytophthora cinnamomi* and the diseases it causes. Monograph 10. American Phytopathological Society, St. Paul, MN

Chapter 3
Alien Bryophytes and Lichens of Europe

Franz Essl and Philip W. Lambdon

3.1 Introduction

Bryophytes (Bryophyta) include mosses (Bryopsida), liverworts (Hepaticopsida) and species-poor hornworts (Anthoceratopsida) (Söderström et al. 2002; Hill et al. 2006). Lichens are composite organisms, arising from a mutualistic association between a saprophytic fungus and a photosynthetic alga or bacterium (Ahmadjian 1993). The photosynthetic partner is usually also found as a common free-living species, and only the highly specific fungal partner is likely to be alien within Europe. Lichens are taxonomically disparate, united by common trophic strategy which has been adopted across a diverse range of fungal lineages. Lichens are distantly related to bryophytes, and biologically very different.

Why therefore do we consider the two groups together in this chapter? In the context of invasions they share a number of important features which present strong practical parallels in the issues they create: (1) they are poorly recorded, so we have little information to assess their invasion history; (2) they are dispersed efficiently by spores, and have much greater natural colonizing ability than other major taxa; (3) since they have few cultivated uses there is a near-absence of deliberate introductions; (4) being small organisms and rarely parasitic, their impacts tend to be measurable only on a micro-scale (5) the possibility of subtle but long-term effects of such invasions has yet to be considered by the scientific community.

3.2 The Dataset

Our checklist of alien bryophytes for the DAISIE database was built on the most recent checklists of European liverworts (Söderström et al. 2002) and mosses (Hill et al. 2006) and we have updated these with subsequent new records. Direct evidence for introduction is available for very few species, but we included a number of the most likely candidates for alien status (based on factors such as anomalous geographic distribution, association with some means of human transport, Söderström 1992) in the cryptogenic category. Similarly, we included alien and cryptogenic species in a particular European country, but native elsewhere in the continent. Due to the large number of gaps in status information, for some countries, native status had to be extrapolated from known parts of the range, but since very few species are widespread, this was rarely necessary. Species only recorded from glasshouses were excluded, as data for these taxa is very limited and spatially very heterogenous. For all alien bryophytes, we compiled the following additional data from the literature: habitat invaded, year of first occurrence, floristic status, region of origin, mode of introduction and impact. The list has been verified by experts.

For lichens, the availability of data is even worse; there is an annotated checklist only for Austria (Breuss 2002), and short overview of taxa for Switzerland (Wittenberg 2006) and the UK (Gilbert 2002). However, this appears genuinely to reflect the rarity of alien taxa.

Taxonomy and nomenclature of liverworts are based on Söderström et al. (2002), of mosses on Hill et al. (2006), and of lichens on Coppins (2002). However, some of the most recent records are not included in these checklists and we retain names as published in their first reports.

3.3 Species Numbers

We identified 45 bryophytes species which we consider to be alien at least in some parts of Europe. These comprise 21 alien mosses and 11 liverworts, but no hornworts. We also include 13 cryptogenic species (11 mosses and two liverworts), which are strong candidates to be alien although there is insufficient evidence to be certain. A considerable number of (sub)tropical species (e.g., *Marchantia planiloba*, *Vesicularia reticulata*, *Zoopsis liukiuensis*) have been recorded only in glasshouses and have not been included in the species list. Some species (e.g., *Riccia crystallina*, *Sphaerocarpos michelii*, *S. texanus*, *Tortula freibergii*, Rumsey 1992), for which indications of being alien in parts of Europe exist (Hill et al. 2005), have been excluded on the grounds of too much uncertainty in the assessment. In the case of the Macaronesian islands, there are some mainly southern hemisphere species (e.g., *Campylopus incrassatus*, *Ditrichum punctulatum*), for which it is unclear if they are alien (JP Frahm, personal communication 2007). Taxonomical and nomenclatural changes are rather frequent in some recently discovered alien bryophytes (e.g., Perold 1997; Pfieffer et al. 2000).

Eight of the aliens are native to Europe as a whole but alien/cryptogenic to some countries, whereas the reminder are alien to the entire continent. Only 11 species have been recorded as alien or possibly alien from more than three countries (Table 3.1). Very few of these (especially heath star moss **Campylopus introflexus** and *Orthodontium lineare*) are widespread, although the ruderal thalloid liverwort *Lunularia cruciata*, which is certainly native to the Mediterranean region, has greatly expanded its range northward in recent decades and is considered to be an alien in much of northern Europe.

Table 3.1 Alien bryophytes in Europe, ranked by decreasing number of invaded countries/regions. Only species invading >3 countries/regions are shown

	Subphylum	Family	Species	No. countries/regions
Alien	Bryophyta	Dicraniaceae	*Campylopus introflexus*	21
	Bryophyta	Orthodontiaceae	*Orthodontium lineare*	15
	Bryophyta	Pottiaceae	*Didymodon australasiae*	11
	Hepaticopsida	Ricciaceae	*Ricciocarpos natans*[a]	8
	Bryophyta	Pottiaceae	*Leptophascum leptophyllum*	6
	Bryophyta	Pottiaceae	*Hennediella stanfordensis*	4
	Bryophyta	Pottiaceae	*Tortula bolanderi*	4
Cryptogenic	Hepaticopsida	Lunulariaceae	*Lunularia cruciata*[a]	12
	Hepaticopsida	Ricciaceae	*Riccia rhenana*[a]	12
	Bryophyta	Pottiaceae	*Scopelophila cataractae*	7
	Bryophyta	Rhabdoweisiaceae	*Dicranoweisia cirrata*[a]	4

[a]Native in some parts of Europe

In the whole of Europe, Hill et al. (2006) record 1,292 species of mosses and Söderström et al. (2002) record 474 species of liverworts (excluding subspecies and varieties). On this basis, only 1.8% of all European species are certainly alien; if cryptogenic species are included, the estimate rises to 2.5% of both mosses and of liverworts. This contrasts strongly with the much higher proportion of aliens amongst vascular plants.

One might therefore ask why there are so few alien bryophyte and lichen species in Europe? A key reason is that due to the lack of distribution data and historical knowledge, some alien bryophytes (especially inconspicuous species) might well have been overlooked and are wrongly considered to be native (Hill et al. 2006); this especially applies to older introductions (*Bryum gemmiferum* is one possibility), whose status is usually impossible to determine. The same is true for lichens (Wirth 1997; Breuss 2002). However, spores of both bryophytes and lichens are very efficient at long distance dispersal (Philippi 1976), which means that human activities play a much less important role in overcoming geographic barriers than in vascular plants. Partly for this reason, anomalous distributions are less reliable as indicators of human introduction than in less mobile taxa. *Scopelophila cataractae*, which is confined to mine spoil heaps contaminated by heavy metals, is a famous example. It was first recorded in 1967 in Wales, and has few European localities, all very distant from each other (Smith 2004). Its main range is in the southern hemisphere, and it is difficult to imagine how it reached these European outposts, whether by human or natural means. We include this species as cryptogenic.

No lichens are unequivocally known to be alien in Europe, although a few may be considered cryptogenic. The best known of these is *Lecanora conizaeoides*, which has expanded to become one of the commonest species encrusting trees with acid bark throughout Europe in recent decades. It is strongly pollution-tolerant and may have spread naturally from small relict populations following increasing acidification of rainwater by industrial emissions. Three species (*Anisomeridium nyssaegenum*, *Lecanora conizaeoides*, *Phaeophyscia rubropulchra*) are suspected to be alien in Austria (Poelt and Türk 1994; Breuss 2002). In the UK, five species (just 0.3% of the total lichen flora) are considered to be alien (Gilbert 2002). Three epiphytic *Parmelia* species, *P. elegantula*, *P. exasperatula* and *P. laciniatula* are rare but display similar distribution patterns to *Lecanora conizaeoides*. Another species, *Parmelia submontana*, has been discovered recently in Scotland. Wittenberg (2006) considered that no lichens are alien in Switzerland.

3.4 Temporal Trends

No alien bryophyte species is known to have arrived to Europe before 1800, but this is not a surprise in view of the late development of bryology as a discipline (few common native species had received their modern scientific names by this date). Most early records were made in Great Britain, but some from continental European countries date back to the 19th century. The first species to be recognised as introductions were *Lunularia cruciata* (1828, Karlsruhe, Germany; Frahm 1973) and *Atrichum crispum* (1848, Rochdale, England; Smith 2004). The cumulative number of alien bryophytes in Europe has increased exponentially (Fig. 3.1). Five species probably arrived in the last 20 years.

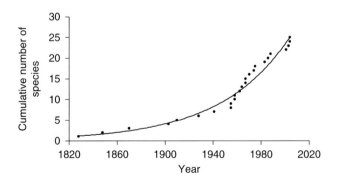

Fig. 3.1 Temporal trends in invasion of 25 alien and cryptogenic bryophytes in Europe, for which data on introduction dates are available. Shown is the cumulative number of recorded species ($R^2 = 0.98$; $y = 1.06e^{(x-1848)}$)

Due to its rapidity, the colonisation of Europe by *Campylopus introflexus* represents one of the best documented of all plant invasions (Hassel and Söderström 2005). It was first recorded in England in 1941, and by 1950, approximately 60 localities were known in the U.K. The first record from continental Europe was made in 1954 on the Channel coast of Bretagne (Finistère) (Richards 1963; Hassel and Söderström 2005). Within a year it was found near Paris (Fontainebleau) (Hahn 2006), and subsequently spread rapidly to neighbouring countries. Today, it is known as far east as Russia and as far south as the fringes of the Mediterranean. This example shows that rapid dispersal is possible in alien bryophytes and allows them to expand across their full potential range within a few decades, the distribution eventually becoming as stable as that of any native species. *Orthodontium lineare* is currently advancing from west to east at a rate similar to *Campylopus introflexus* (Hassel and Söderström 2005).

3.5 Biogeographic Pattern and Spatial Distribution

The countries with the most alien bryophytes are the UK, followed by France, Ireland, the Canary Islands and Spain (Table 3.2). This illustrates a biogeographic trend: countries and regions with a humid and cool climate are most invaded, whereas countries with drier and warmer climates are poor in alien bryophytes. So (north-)west European countries and the Macaronesian island groups (with humid climate in the mountainous interior) display the highest species numbers, and to a lesser extent, this also holds for Central European countries. Further, differing research intensity between European countries may also partly contribute to the observed distribution pattern, as most west European countries have exceptionally good research traditions in bryology.

3.6 Regions of Origin

The majority of European alien bryophytes are native to four main continental regions, each accounting for 13–19% of introductions to Europe: in order of decreasing importance, these are South America, Australasia, North America, and Africa. Another 7% of the species are native to oceanic islands. Only 7% and 8% of the species are native either to parts of Europe and temperate Asia, respectively.

Compared to other taxonomic groups, the important contribution of distant regions (especially from the southern hemisphere) to the alien bryophyte flora of Europe is remarkable. We argue that this is linked to the excellent dispersal capacities of bryophytes, which makes it easier to overcome geographic barriers without the assistance of humans. However, for southern hemisphere species, crossing the equator by natural means presents a barrier because of the prevailing

Table 3.2 Number of alien bryophytes in European countries and regions. Only countries/regions with >2 alien and cryptogenic bryophytes are included. Note that oceanic areas in the western part of Europe are consistently more invaded than the continental eastern and Mediterranean areas. Spain is classified in the oceanic group because most alien species have been exclusively recorded in the Pyrenees and north-west rather than the Mediterranean region

Region	Country	Area (km^2)	Naturalised	Casual	Cryptogenic	Total
Oceanic/western	UK	244,880	14	1	4	19
	France	551,500	3	2	5	10
	Ireland	70,283	6	2	1	9
	Canary Islands	7,447	2	2	4	8
	Spain	504,782	1	2	5	8
	Madeira	828	2	3	2	7
	Azores	2,355	2	3	1	6
	Belgium	30,520	3	0	3	6
	The Netherlands	41,253	3	0	2	5
	Denmark	43,090	3	0	2	5
	Sweden	449,960	3	0	2	5
	Norway	324,220	2	1	1	4
	Portugal	92,390	2	0	2	4
	Subtotal		**22**	**7**	**2**	**31**
Continental/eastern	Germany	356,910	3	0	3	6
	Czech Republic	78,864	2	0	3	5
	Poland	312,685	1	1	2	4
	Russia	2,500,000	1	1	2	4
	Italy	301,270	2	1	1	4
	Austria	49,035	2	1	0	3
	Finland	338,130	2	0	1	3
	Hungary	93,030	1	0	2	3
	Latvia	64,634	1	0	2	3
	Slovakia	49,035	0	1	2	3
	Switzerland	83,850	1	1	1	3
	Subtotal		**8**	**1**	**4**	**13**

wind directions in the inter-tropical convergence zone. For the few lichens which are assumed to be alien in Europe, the main regions of origin are considered to be North America and other parts of Europe (Wirth 1997; Breuss 2002).

In order to place the distribution of alien bryophytes into context, some comparison with the native flora is helpful. In the well-studied UK flora, 41% of native bryophytes occur in four or more of the continental regions listed in Fig. 3.2. Compared with vascular plants, this total is astonishingly high, and underlines both the efficacy of natural dispersal and the difficulty of identifying which parts of the range are natural. However, there are significant differences in the distributional trends of European aliens and natives (Fig. 3.2a). A much higher proportion of natives are found in the northern hemisphere regions, whereas alien species tend to occur more

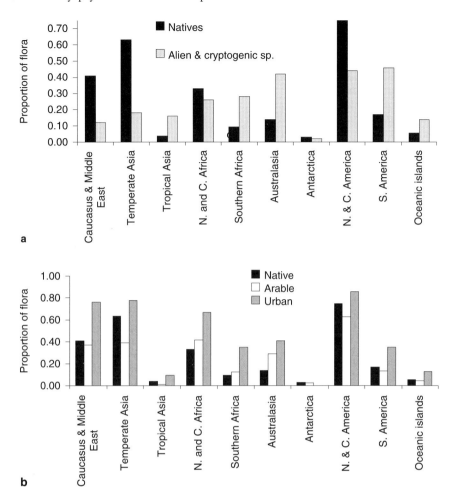

Fig. 3.2 Global distribution patterns of the complete UK bryophyte flora, from Smith (2004) and Paton (1999). (a) Comparison between native species (n = 1,027) and alien/cryptogenic species. (b) Comparison between all native species and those characteristic of arable or urban habitats (n = 1,027, 89, 31)

often in South Africa, Australia, New Zealand and South America. It may be interesting to examine whether other ecological groups of bryophytes, which we might suspect of containing yet unrecognised alien species, also possess a tendency towards a southern hemisphere distribution. Figure 3.2b shows the evidence for two such groups, again based on the UK flora. Species characteristic of arable land show a predominantly northern hemisphere pattern and are likely to be composed largely of genuine natives. The strong presence in North Africa and the Middle East suggests that a number of them may have adapted to such habitats early in the history of agriculture in the Mediterranean and surrounding area, as is thought to be the case with many agricultural vascular plant species (Chytrý et al. 2005). In contrast, urban bryophytes tend to be very widespread, with a high likelihood of occurrence in all global regions. It is likely that human-aided dispersal, and the novelty and homogeneity of the urban environment across the world have been important factors in shaping this distribution pattern, and many of these species are likely to be alien in parts of their range.

3.7 Main Introduction Pathways to Europe

The most important introduction pathway is with ornamental plants (15 species), as an epiphyte or weed. Some literature sources identify certain species as being introduced "with soil" (Paton 1999), but this may often also refer to soil in plant pots. The only other significant introduction pathway is unspecified accidental import as a stowaway on ships and planes, perhaps on clothing or goods. Only one species (*Ricciocarpos natans*), which floats on the surface of water and is used as an ornamental in garden ponds and aquaria is known to be introduced deliberately in northern Europe. For 45% of the alien bryophytes their mode of introduction is unknown. The negligible contribution of deliberate introductions contrasts strongly with vascular plants.

Secondary spread often mainly depends on the dispersal capacity of the species. However, human activity may enhance it. Spread may be favoured by anthropogenic changes to existing habitats (e.g., by air borne pollutants), by creating new habitats or by unintentional transport. Some alien bryophytes are known to have expanded in western and central Europe due to acid rain (e.g., **Campylopus introflexus**, *Orthodontium lineare*, JP Frahm, personal communication 2007). It is known (Ketner-Oostra and Sýkora 2000) that the spread of **Campylopus introflexus** in Europe is fostered by mechanical disturbance (e.g., caused by rabbits or trampling by people), which breaks up the moss mat and subsequently disperses propagules to initiate new populations. The formation of specialised structures for vegetative reproduction is a potential mechanism to aid invasion, via vegetative fragments, gemmae (simple asexual buds shed from the leaf tips or other organs), bulbils or tubers (larger gemmae-like structures borne respectively in the leaf axils or on rhizoids), although most species are probably only moved short distances by such means. Fifty four percent of UK alien bryophytes possess these structures, compared with only 28% of natives (PW Lambdon, unpublished 2007). However, many bryophytes are dioecious, and since there is often only one sex in the founder population (e.g., *Leptophascum leptophyllum*, *Lunularia cruciata* in its alien European range), they can only survive via vegetative propagation. This feature is therefore likely to be a consequence of their colonizing history. For lichens, introduction pathways are poorly understood; however, it is known that *Lecanora conizaeoides* has increased as a result of acidification of precipitation (Purvis et al. 1992), and *Parmelia submontana* seems to be associated with alien trees and may thus be an accidental garden import (Coppins 2002).

Another possible route of entry is via the spontaneous formation of chromosomal mutants in the introduced range. *Hennediella macrophylla*, for example, could have arisen several times throughout its range by haploidy from the more common *H. stanfordensis*, and this is one possibility of its appearance in the UK (Smith 2004).

3.8 Most Invaded Ecosystems

Alien bryophyte species in Europe display a strong affinity for habitats with a strong anthropogenic disturbance regime, especially gardens, roadsides, and walls (Table 3.3). The selection of micro-habitats underlines this pattern. Exposed soil created by mechanical disturbance is most often invaded (18 species). Therefore, most aliens are early-successional colonists, while natural habitats (e.g., dunes, rocks, broadleaved woodland) are rarely invaded, and several habitat types are not regularly invaded at all (e.g., dry grassland, alpine meadows, screes). This is in strong contrast to the patterns displayed amongst native bryophytes. Of 1,027 native species in the UK, only 89 are strongly adapted to agricultural habitats and only 31 have their core strongholds in urban habitats (Gilbert 1971; Paton 1999; Smith 2004).

So far, few bryophytes have naturalised widely in near natural vegetation (Söderström 1992; Stieperaere and Jacques 1995). *Campylopus introflexus* colonises mainly coastal dunes, heathlands and bogs, but away from dunes is often associated with burned areas, peat cuttings or roadside banks where there has been obvious human interference. *Orthodontium lineare* occurs on decaying wood in

Table 3.3 Habitats invaded by a total of 44 alien bryophytes in Europe. Given are EUNIS habitats and additional habitat categories more specific for the habitat preferences of bryophytes. Note that species restricted to glasshouses have been excluded and that species may occur in several habitats

EUNIS habitats	Bryophyte habitats	% of species
I	Gardens	36
H	Roadsides	20
J	Walls, buildings	20
C	Freshwater (incl. littoral)	18
J	Ruderal habitats	13
H	Rocks	13
J	Greenhouses	11
D	Mires, bogs, fens	11
G	Broadleaved woodlands	11
I	Arable land	9
G	Conifer woodlands	9
A	Dunes, coastal habitats	4
C	Freshwater	2
J	Urban (unspecified)	2
–	Unknown	9

woodlands (Hedenäs et al. 1989; Hill et al. 1992; Hassel and Söderström 2005). Besides anthropogenic habitats, *Lunularia cruciata* invades moist, shaded soils, rocks and walls. The few cryptogenic lichens mostly colonise acidic bark of trees (Wirth 1997; Breuss 2002).

3.9 Ecological and Economic Impacts

The impacts of alien bryophytes are less obvious than in vascular plants. They may compete with native bryophytes and lichens, or with germinating seedlings of vascular plants for light, nutrients or space. ***Campylopus introflexus*** is the only species known to cause strong impacts of this type. It forms dense mats, significantly reducing the species diversity of lower plants and lichen communities (Hahn 2006). Where ***Campylopus introflexus*** has colonised thatched roofs in southern England, there is a concern that the red data book moss *Leptodontium gemmascens* may also be at risk (R Porley, personal communication 2007). A few threats posed by other alien bryophytes have been documented, but so far the implications have not been well studied. In the UK, *Orthodontium lineare* competes with a rare relative, *O. gracile*, on sandstone rocks, and has caused the loss of this species from many localities (Porley and Hodgetts 2005). The moss *Hennediela stanfordensis* and the liverwort *Lophocolea semiteres* are both showing signs of rapid spread in western Europe and can be locally dominant. *L. semiteres* forms thick, monoclonal mats, for example in grazing marshes in the Thames basin, England, and in the Netherlands may be ousting the native *Lophocolea heterophylla* (Porley and Hodgetts 2005). Lichens are well-known for their slow growth, but this mainly applies to native saxicolous species. As the only widespread species among the potential aliens, *Lecanora conizaeoides* is a relatively rapid colonist and its continuous crustose colonies can dominate considerable areas of bark. Whether there is a threat from the exclusion of competing natives has yet to be demonstrated (Gilbert 2002).

Aside from competition, it is also likely that abundant alien bryophytes and lichens can alter ecosystem functioning occasionally by stabilizing soils, binding leaf litter, altering decay rates, and creating humid microhabitats which affect the composition of microfaunal communities. No such species is yet widespread enough to have a far-reaching impact, but the potential consequences of invasions at this micro-environmental level remain almost unexplored.

3.10 Management Options and Their Feasibility

Deriving feasible management strategies for bryophytes is very difficult for several reasons. Introduction is difficult to control, identification of species needs expert knowledge, long distance dispersal and thus reimmigration is likely to be frequent, and due to the small size of bryophytes, most management measures are difficult and costly to apply. However, since ***Campylopus introflexus*** is still the only species perceived as a widespread threat, few management options have been tested. These include liming of dunes, burning, herbicide treatment and introduction of grazing animals to trample the mats. They have generally met with only limited success and in most cases such direct control measures damage native vegetation more than they affect the invader (Ketner-Oostra and Sýkora 2002; Weidema 2006).

3.11 Future Expected Trends

Creation of new habitats is one of the most relevant drivers for the spread of alien bryophytes in Europe. So, the continuing increase of urban habitats (6% from 1990 to 2000 in Europe, EEA 2007), will favour urbanophilic species.

Climate change and increasing temperatures may in future foster range expansions of alien bryophytes (Frahm and Klaus 2000, 2001). There are some (sub)tropical bryophyte species restricted to glasshouses in Europe, and in recent years, some have started to naturalise outdoors (e.g., *Didymodon australasiae*, Müller 2002), which might reflect the recent warming trend.

Air borne acidification has been strongly reduced in the last few decades, so it is likely that this factor will lose relevance in future. However, nitrogen deposition in large parts of Europe exceeds critical loads and is still increasing (Posch et al. 2005). As several alien bryophytes can take advantage of air borne nitrogen (Söderström 1992), it is expected that some species will gain from this trend.

As inter-continental trade increases further, and especially with the increasing popularity of alien ornamental plants, there is great potential for increased movement of alien bryophyte species across the world. So we can probably expect that numbers, abundance and associated impacts of alien bryophytes will increase in the future considerably.

Acknowledgements The support of this study by the European Commission's Sixth Framework Programme project DAISIE (Delivering alien invasive species inventories for Europe, contract SSPI-CT-2003-511202) is gratefully acknowledged. We are also very thankful to Jan-Peter Frahm, who willingly shared his knowledge and checked and improved the species list and commented on the text. We are indebted to Christian Schröck, Mark Hill, Adam Stebel, Juana María González-Mancebo, Petr Pyšek, Ron Porley for critical checking of the species list of bryophytes. For comments and discussion we are obliged to Marie Desprez-Lousteau, and for comments on a previous version of the manuscript we are indebted to Phil Hulme, Wolfgang Nentwig and Petr Pyšek.

References

Ahmadjian V (1993) The lichen symbiosis. Wiley, New York
Breuss O (2002) Flechten. In: Essl F, Rabitsch W (eds) Neobiota in Österreich. Umweltbundesamt, Wien. 178–179
Chytrý M, Pyšek P, Tichy L, Knollova I, Danihelka J (2005) Invasions by alien plants in the Czech Republic: A quantitative assessment across habitats. Preslia 77:339–354

Coppins BJ (2002) Checklist of lichens of Great Britain and Ireland. British Lichen Society, London
EEA (2007) EIONET – European topic center on land use and spatial information http://terrestrial.eionet.europa.eu/CLC2000/changes/DATA.xls. Cited 30 Oct 2007
Frahm JP (1973) Über Vorkommen und Vebreitung von **Lunularia Cruciata** (L.) Dum. in Deutschland. Herzogia 2:396–409
Frahm JP, Klaus D (2000) Moose als Indikatoren von rezenten und früheren Klimafluktuationen in Mitteleuropa. NNA-Berichte 2/2000:69–75
Frahm JP, Klaus D (2001) Bryophytes as indicators of recent climate fluctuations in Central Europe. Lindbergia 26:97–104
Gilbert OL (1971) Urban bryophyte communities in north-east England. Trans Brit Bryol Soc 6:306–316
Gilbert OL (2002) Lichens. Collins New Naturalist Series, HarperCollins, London
Hahn D (2006) Neophyten der ostfriesischen Inseln. Schr-R Nationalpark Niedersächs Wattenmeer 9:1–175
Hassel K, Söderström L (2005) The expansion of the alien mosses *Orthodontium lineare* and *Campylopus introflexus* in Britain and continental Europe. J Hattori Bot Lab 97:183–193
Hedenäs L, Herben T, Rydin H, Söderström L (1989) Ecology of the invading moss *Orthodontium lineare* in Sweden: Spatial distribution and population structure. Holarct Ecol 12:163–172
Hill MO, Preston CD, Smith AJE (1992) Atlas of the bryophytes of Britain and Ireland. Harley Books, Colchester
Hill MO, Baker R, Broad G, Chandler PJ, Copp GH, Ellis J, Jones D, Hoyland C, Laing I, Longshaw M, Moore N, Parrott D, Pearman D, Preston C, Smith RM, Waters R (2005) Audit of non-native species in England. English Nature Research Reports 662:1–81
Hill MO, Bell N, Bruggeman-Nannenga MA, Brugués M, Cano MJ, Enroth J, Flatberg KI, Frahm JP, Gallego MT, Garilleti R, Guerra J, Hedenäs L, Holyoak DT, Hyvönen J, Ignatov MS, Lara F, Mazimpaka V, Muñoz J, Söderström L (2006) An annotated checklist of the mosses of Europe and Macaronesia. J Bryol 28:198–267
Ketner-Oostra R, Sýkora KV (2000) Vegetation succession and lichen diversity on dry coastal calcium-poor dunes and the impact of management experiments. J Coastal Cons 6:191–206
Müller F (2002) Ein Freilandnachweis von *Didymodon australasiae* var. *umbrosus* in Deutschland. Herzogia 15:87–190
Paton JA (1999) The liverwort flora of the British Isles. Harley Books, Colchester
Perold SM (1997) Studies in the liverwort genus *Fossombronia* (Metzgeriales) from southern Africa. 4. A re-examination of *F. crispa*, *F. leucoxantha* and *F. tumida*. Bothalia 27:105–115
Pfieffer T, Kruijer HJD, Frey W, Stech M (2000) Systematics of the *Hypopterygium tamarisci* complex (Hypopterygiaceae, Bryopsida). Implications of molecular and morphological data. J Hattori Bot Lab 89:55–70
Philippi G (1976) Einfluß des Menschen auf die Moosflora in der Bundesrepublik Deutschland. Schr-R Vegetationskde 10:163–168
Poelt J, Türk R (1994) *Anisomeridium nyssaegenum*, ein Neophyt unter den Flechten, in Österreich und Süddeutschland. Herzogia 10:75–81
Porley R, Hodgetts N (2005) Mosses and liverworts. New Naturalist Series. HarperCollins, London
Posch M, Slootweg J, Hettelingh JP (2005) European critical loads and dynamic modelling. www.mnp.nl/cce/publ/SR2005.jsp. Cited 30 Sept 2007
Purvis OW, Coppins BJ, Hawksworth DL, James PW, Moore DM (1992) The lichen flora of Great Britain and Ireland. Natural History Museum & British Lichen Society, London
Richards PW (1963) *Campylopus introflexus* (Hedw.) Brid. and *C. polytrichoides* De Not. in the British Isles; a preliminary account. Trans Brit Bryol Soc 4:404–417
Rumsey FJ (1992) The status of *Tortula freibergii* in the British Isles. J Bryol 17:371–373
Smith AJE (2004) The moss flora of the British Isles. Cambridge University Press, Cambridge
Söderström L (1992) Invasions and range expansions and contractions of bryophytes. In: Bates JW, Farmer AM (eds) Bryophytes and lichens in a changing environment. Clarendon, Oxford. 131–158

Söderström L, Urmi E, Váňa J (2002) Distribution of Hepaticae and Anthocerotae in Europe and Macaronesia. Lindbergia 27:3–47

Stieperaere H, Jacques A (1995) The spread of *Orthodontium lineare* and *Campylopus introflexus* in Belgium. Belg J Bot 128:117–123

Weidema I (2006) NOBANIS: Invasive alien species fact sheet *Campylopus introflexus*. www.nobanis.org. Cited 30 Sept 2007

Wirth V (1997) Einheimisch oder eingewandert? Über die Einschätzung von Neufunden von Flechten. Bibl Lichenol 67:277–288

Wittenberg R (ed) (2006) Invasive alien species in Switzerland. An inventory of alien species and their threat to biodiversity and economy in Switzerland. Environmental Studies 29. Federal Office for the Environment, Bern

Chapter 4
Alien Vascular Plants of Europe

Petr Pyšek, Philip W. Lambdon, Margarita Arianoutsou, Ingolf Kühn, Joan Pino, and Marten Winter

4.1 Introduction

In terms of invasion biology, vascular plants are the most intensively researched taxonomic group; at least 395 plant invaders have been addressed in detailed case studies globally, accounting for 44% of all invasive taxa studied; after North America, Europe is the continent enjoying the most intensive study with at least 80 invasive plant species having been addressed (Pyšek et al. 2008). However, although there is a considerable body of information on major plant invaders in Europe (see also Weber 2003), the situation is much less satisfactory as far as complete national inventories of alien plants are concerned. Prior to the DAISIE project (www.europe-aliens.org), only few countries had a sound information on the composition of their alien floras, available in specialised checklists, notably Austria (Essl and Rabitsch 2002), the Czech Republic (Pyšek et al. 2002), Germany (Klotz et al. 2002; Kühn and Klotz 2003), Ireland (Reynolds 2002) and the UK (Clement and Foster 1994; Preston et al. 2002, 2004). This situation directly translated into poor knowledge across the European continent. The only available continental analysis of plant invasion patterns in Europe (Weber 1997) was based on data from Flora Europaea (Tutin et al. 1964–1980), the only synthetic source of information on floras of particular countries, including alien species. This source is, however, nowadays outdated and contains numerous inaccuracies in data for individual countries (Pyšek 2003). In general, information on the presence and distribution of alien plant species for most European countries was scattered in a variety of published and unpublished accounts and databases; this is the case in other continents too (Meyerson and Mooney 2007). On the plant side, DAISIE was thus a major challenge of collating and assessing existing data on the most numerous group of European aliens and concentrating this information in an authoritative continental inventory.

The European area covered (Fig. 4.1) by the plant team of DAISIE was partly determined by the geographical coverage of source floras, but it was broadly attempted to use the limits set by Flora Europaea (Tutin et al. 1964–1980) for the north and central continental boundaries (i.e., as far east as the Urals, to the border of the Black Sea but excluding the Caucasus). In the south-east, Cyprus was the boundary of the area included, and Turkey was also considered. In total, 49 countries/regions were included (Fig. 4.1). For each national region, a data set was compiled from the most comprehensive literature sources available, and taxonomic treatment was standardised across all national checklists (see Lambdon et al. 2008 for data sources and details of the procedure). Naturalised and casual alien species were distinguished with respect to invasion status, and archaeophytes and neophytes with respect to the residence time, following criteria suggested by Richardson et al. (2000) and Pyšek et al. (2004). However, in some cases there was insufficient information to determine the exact category (archaeophyte vs. neophyte, naturalised vs. casual) and the taxon was recorded as alien without further specification. This designation occurs most commonly in poorly-recorded countries (e.g., Belarus, Bulgaria, Moldova), but such data sets generally only include prominent invaders which are mostly likely to be naturalised.

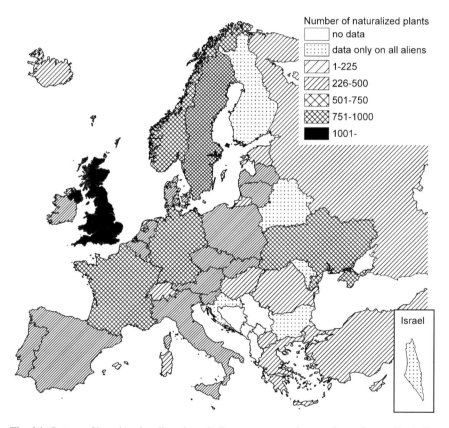

Fig. 4.1 Pattern of invasions by alien plants in Europe, expressed as numbers of naturalised alien plants in countries/regions; dotted areas indicate poorly researched regions for which only the information on the total number of aliens is available (Based on data from Lambdon et al. 2008); see this source for information on the numbers of species and assessment of the quality of data in particular countries/regions

Similarly, since the focus of DAISIE was primarily on neophytes, it may be safely assumed that the majority of all aliens, where reported, refer to naturalised neophytes. We distinguished between aliens in Europe, which group includes also species that are native in a part of Europe but alien to another part, and aliens to Europe, including species with native distribution area outside Europe; the latter is a subgroup of the former.

This chapter summarises the basic information on the structure of alien flora of Europe, presents the most common naturalised species, and describes robust large-scale geographical patterns in the level of invasion (in terms of Hierro et al. 2005; Chytrý et al. 2008b) across Europe and in the composition of regional alien floras. Finally, it points to current research gaps and outlines avenues for further research. Complete and more detailed information can be found in Lambdon et al. (2008).

4.2 Diversity of Alien Plants in Europe

The DAISIE database contains records of 5,789 alien plant species in Europe, of which 2,843 are alien to Europe, i.e. of extra-European origin (275 species were not assigned an origin status due to ambiguities over their native range). Of these aliens in and to Europe, 1,507 and 872, respectively, are casual in all regions where they occur. There are in total 3,749 naturalised alien plant species recorded in Europe, of which 1,780 are alien to Europe. We do not attempt to derive the total number of naturalised neophytes since it would have to be based on a limited subset of only 19 countries for which invasion and residence time status were designated, which would necessarily lead to an underestimation of the number of naturalised neophytes currently present in Europe (Lambdon et al. 2008).

The 11 years old overview of the alien flora of Europe (Weber 1997) reported 1568 naturalised species in Europe; this is much less than recorded by DAISIE and to explain this difference, two aspects need to be considered. The overview of Weber (1997) was based on Flora Europaea, which relied on data from 1960s–1970s (Tutin et al. 1964–1980). Since this period, there has been a continual influx of alien species to individual countries (Pyšek et al. 2003). When the publication of Flora of Europaea was completed, many more alien species than included in that work must have been present in Europe (Fig. 4.2). Taking into account that 65% of species in DAISIE database are naturalised, it can be estimated that there were 2,175 naturalised aliens in Europe in 1980, when the publication of the first edition of Flora Europaea was completed, i.e. much higher number of species than reported by Tutin et al. (1964–1980). Another principle reason for the low alien species number reported in Flora Europaea was rather low level of detail adopted for screening aliens. Raised awareness of the issue of alien species and increasing research intensity in the last decades (Pyšek et al. 2006), yielded dramatically higher numbers of species recorded in the DAISIE database, which is much closer to the reality.

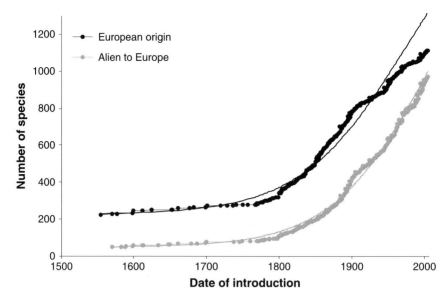

Fig. 4.2 Increase in numbers of alien neophytes introduced to Europe over the last 500 years. Cumulative data are shown separately for species with native distribution area outside Europe (n = 929) and those with European origin, but occurring as alien in other parts of the continent (n = 954); for these species introduction relates to their first record as alien outside their native range. Introduction dates are estimated from the minimum residence time, and species where this could not be evaluated with a reasonable degree of accuracy were excluded (Taken from Lambdon et al. 2008, published by courtesy of the Czech Botanical Society). Both relationships are approximately hyperbolic, and the following semi-logit transformation was most appropriate: $T(p) = -\ln(p/(2-p))$, where p is the proportion of the total number of species introduced, at a given time T, since AD 1500. Alien to Europe: $N = 0.0134p - 26.49$ ($r^2 = 0.97$); Aliens of European origin: $N = 0.0113p - 22.40$ ($r^2 = 0.95$)

4.3 Areas of Origin

For aliens in Europe, other parts of the continent are the main donor area. As much as 29% of all introductions (attributing species that originate from more than one region to each of these regions, Fig. 4.3) recruit from some European countries and invade in others. Combined with aliens of Asian origin (31%) illustrates the major contribution (60%) of the Eurasian super-continent to alien species richness in Europe. North and South America together account for 19% introduction of alien species in Europe. Considering species of extra-European origin separately yields a different picture (Fig. 4.3). Among aliens to Europe, 34% of introductions are of Asian origin (with temperate Asia providing more species than tropical), 23% and 22% originate from North and South America, respectively, and 17% from Africa. These figures are fairly consistent with distribution of origins in national floras (Kühn and Klotz 2003; Pyšek et al. 2002) and confirm patterns reported by Weber (1997) for Europe.

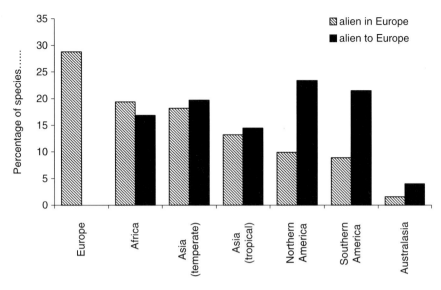

Fig. 4.3 Donor regions of the alien flora of Europe. Based on 2,094 naturalised aliens in Europe for which the region of origin was classified (hatched bars), 1,168 of these are alien to Europe, i.e. species that arrived to Europe from other continents (black bars). Note that the figure reflects the numbers of introductions from donor regions; species originating from more than one region were attributed to each of the regions of origins, so that the totals in both groups of aliens equal to 100% (Based on data from Lambdon et al. 2008)

4.4 Taxonomy

The European alien flora is dominated by large global plant families (Table 4.1); the highest numbers are found in the Asteraceae (692 alien representatives), Poaceae (597), Rosaceae (363), Fabaceae (subfamily Faboideae, 323) and Brassicaceae (247). These families have a weedy tendency (Daehler 1998; Pyšek 1998) and have undergone major radiations in temperate regions, with the exception of Rosaceae, where the majority of species introduced to Europe are boreo-temperate woody shrubs and trees. The only other predominantly woody, family highly represented is the Pinaceae (53 alien species). In total, alien species are from 213 families (Lambdon et al. 2008), almost twice as many as reported by Weber (1997). In some cases success may to be linked to the frequency of introductions, as some family characteristics make the species valuable for human uses (e.g., the Rosaceae as fruit crops, Pinaceae for timber and Lamiaceae as herbs and ornamental plants). Those families, which have diversified in Europe, sometimes have correspondingly greater numbers which are alien in Europe (Table 4.1). At higher taxonomic levels, families with a high representation of alien species cluster in the orders Asparagales, Ranunculales, Caryophyllales, Lamiales and Solanales (Lambdon et al. 2008). Obviously, the high diversity of European aliens in some families is largely determined by their high global species pools

Table 4.1 Most represented families in the alien flora of Europe (with at least 100 alien taxa), classified according to the Angiosperm Phylogeny Group (Stevens 2001 onwards) and Mabberley (1997). The total number of alien species recorded, the total number of naturalised aliens (Natur), and the number of naturalised neophytes (Neo) is given for aliens to Europe, aliens of European origin, and aliens in Europe. World species numbers were taken from Mabberley (1997). Species are ranked according to the total number of alien taxa in Europe (shown in bold) (See Lambdon et al. 2008 for a complete account on families)

	Aliens to Europe			Aliens of as in Table 4.2 European origin			Aliens in Europe			Total aliens in Europe as % of world number
	Total	Natur	Neo	Total	Natur	Neo	Total	Natur	Neo	
Asteraceae	334	193	138	334	225	144	**692**	424	283	3.0
Poaceae	340	136	93	257	159	99	**597**	295	192	6.3
Rosaceae	212	176	76	134	93	44	**363**	247	120	12.8
Fabaceae (Faboideae)	82	40	22	233	138	101	**323**	181	124	2.7
Brassicaceae	50	17	14	174	127	87	**247**	146	102	7.6
Amaranthaceae	128	51	45	56	40	27	**185**	91	72	9.0
Lamiaceae	55	29	17	97	69	34	**165**	102	52	2.5
Caryophyllaceae	13	5	2	141	74	37	**156**	80	39	6.8
Apiaceae	31	20	10	103	64	33	**143**	87	44	4.0
Plantaginaceae	38	21	7	82	58	23	**132**	84	31	2.4
Onagraceae	80	45	36	10	8	4	**112**	68	43	17.2
Solanaceae	88	53	39	12	11	6	**107**	66	45	3.6
Polygonaceae	45	30	18	46	28	16	**106**	63	36	9.6
Boraginaceae	26	18	10	63	47	30	**105**	69	41	4.6

(Asteraceae, Faboidae, Poaceae, Lamiaceae), but the proportions of the total numbers of world representatives present as aliens in Europe indicate that some families (notably Onagraceae with 17%, Salicaceae 15%, Rosaceae 13%, Geraniaceae 10%, Polygonaceae 10% and Amarathaceae 9% of the world species pool occurring as aliens in Europe) are more predisposed to invade than others (Daehler 1998; Pyšek 1998).

There are 1,567 genera with at least one alien representative in Europe (Table 4.2). The commonest genera are those with evolutionary centres in Europe or Eurasia, with a high native diversity (*Centaurea*, *Silene*, *Euphorbia*, *Rumex*) or those extensively used by humans (*Trifolium*, *Vicia*, *Rosa*). The pattern is very different for alien species with extra-European origin; the commonest genera among aliens to Europe are globally-diverse ones comprising mainly urban and agricultural weeds (*Amaranthus*, *Chenopodium* and *Solanum*), but also frequently cultivated (the largest *Cotoneaster* with 70 species alien to Europe comprises almost exclusively introductions for ornamental purposes, or other ornamentals such as those in genera *Sedum* or *Narcissus*). One special case is the genus *Oenothera*, where hybrid swarms tend to become true breeding after a few generations of isolation. Only a few large genera which have successfully invaded (e.g., *Oxalis*, *Panicum*, *Helianthus*) are predominantly extra-European.

Table 4.2 Most represented genera in the alien flora of Europe (with at least 35 alien taxa), classified according to the Angiosperm Phylogeny Group (Stevens 2001 onwards) and Mabberley (1997). The total number of alien species recorded, the total number of naturalised aliens (Natur), and the number of naturalised neophytes (Neo) is given for aliens to Europe, aliens of European origin, and aliens in Europe. Species are ranked according to the total number of alien taxa *in* Europe (shown in bold) (Adapted from Lambdon et al. 2008, where a more complete account on genera can be found)

Genus	Family	Aliens to Europe			Aliens of European origin			Aliens in Europe		
		Total	Natur	Neo	Total	Natur	Neo	Total	Natur	Neo
Cotoneater	Rosaceae	70	62	16	4	3	1	**75**	65	17
Oenothera	Onagraceae	43	32	28	3	2	2	**64**	49	33
Chenopodium	Amaranthaceae	29	10	10	22	17	9	**52**	27	19
Centaurea	Asteraceae	7	3	2	42	22	18	**51**	25	20
Rumex	Polygonaceae	11	3	2	26	16	12	**45**	20	14
Trifolium	Fabaceae	7	4	2	42	25	15	**49**	29	17
Euphorbia	Euphorbiaceae	10	7	2	29	23	14	**47**	33	17
Silene	Caryophyllaceae	5	2	1	40	18	18	**47**	21	9
Solanum	Solanaceae	35	25	19	3	2	1	**45**	29	20
Eragrostis	Poaceae	38	5	4	7	5	5	**45**	10	9
Senecio	Asteraceae	22	14	7	15	11	8	**44**	27	16
Amaranthus	Amaranthaceae	37	19	19	2	2	2	**39**	21	21
Bromus	Poaceae	13	7	3	24	15	18	**37**	22	11
Sedum	Crassulaceae	15	10	6	18	18	12	**36**	29	18
Veronica	Plantaginaceae	8	5	4	25	19	5	**35**	25	9
Rubus	Rosaceae	21	18	9	10	7	2	**35**	27	11
Cyperus	Cyperaceae	26	16	6	9	4	3	**35**	20	9

There are 128 species recorded from more than a half of the countries considered (Lambdon et al. 2008). The most common European alien species is Canadian fleabane *Conyza canadensis*, native to North America, occurring in 47 countries/regions (94%). Other species occurring in more than 80% of the regions studied include (Table 4.3): thorn-apple *Datura stramonium*, Jerusalem artichoke *Helianthus tuberosus*, black locust **Robinia pseudoacacia** (all native to North America), common amaranth *Amaranthus retroflexus*, least pepperwort *Lepidium virginicum* (North and Central America), shaggy-soldier *Galinsoga quadriradiata* (Central and South America), gallant-soldier *Galinsoga parviflora*, pineapple-weed *Matricaria discoidea* (South America), rough cocklebur *Xanthium strumarium* (Eurasia), common millet *Panicum miliaceum*, common field-speedwell *Veronica persica* (Asia) and common evening-primrose *Oenothera biennis* (this species probably originated in Europe but is considered alien in most countries). Since the widely distributed aliens are naturalised in the vast majority of regions from which the information on invasion status is available, it is reasonable to assume that the same is likely to be true for those where such assessment is missing (unspecified occurrences in Table 4.3). Notably, all of them are aliens to Europe, mostly originating from North America.

Table 4.3 The most widespread alien plant species in Europe. Their status as naturalised or casual is shown. Unspecified status refers to regions where the species is definitely alien but classification as to whether it is casual or naturalised is not available or it is impossible to decide about the status with certainty. Occurrence as neophyte or archaeophyte is not distinguished but given the focus of DAISIE, the majority of taxa are neophytes. Species are ranked according to the decreasing number of total occurrences in Europe as aliens (Adapted from Lambdon et al. 2008)

Species	Family	Naturalised	Casual	Unspecified	Total
Conyza canadensis	Asteraceae	33	1	13	47
Datura stramonium	Solanaceae	25	7	13	45
Amaranthus retroflexus	Amaranthaceae	30	4	10	44
Galinsoga parviflora	Asteraceae	27	2	15	44
Helianthus tuberosus	Asteraceae	26	5	12	43
Xanthium strumarium	Asteraceae	22	5	16	43
Lepidium virginicum	Brassicaceae	16	11	15	42
Oenothera biennis	Onagraceae	28	2	12	42
Robinia pseudoacacia	Fabaceae	32	2	8	42
Galinsoga quadriradiata	Asteraceae	25	1	15	41
Matricaria discoidea	Asteraceae	23	3	15	41
Panicum miliaceum	Poaceae	16	20	5	41
Veronica persica	Plantaginaceae	27	0	14	41
Ailanthus altissima	Simaroubaceae	30	1	9	40
Amaranthus albus	Amaranthaceae	24	5	11	40
Erigeron annuus	Asteraceae	27	3	10	40
Fallopia japonica	Polygonaceae	29	1	10	40
Medicago sativa	Fabaceae	23	4	13	40
Amaranthus blitoides	Amaranthaceae	24	6	9	39
Lepidium sativum	Brassicaceae	10	21	8	39
Papaver somniferum	Papaveraceae	12	17	10	39
Solidago canadensis	Asteraceae	28	0	11	39
Acer negundo	Sapindaceae	26	3	9	38
Chenopodium ambrosioides	Amaranthaceae	22	6	10	38
Elodea canadensis	Hydrocharitaceae	26	0	12	38
Juncus tenuis	Juncaceae	26	0	12	38
Panicum capillare	Poaceae	17	16	5	38
Phalaris canariensis	Poaceae	10	15	13	38
Vicia sativa	Fabaceae	15	13	10	38

However, the widely distributed species are not necessarily the most invasive in terms of impact and human perception. Many taxa listed in Table 4.3 are agricultural weeds or ruderal species of urban habitats; only few of the species included in the 100 worst alien species in this book are also among the most widely-distributed aliens, such as woody invaders black locust ***Robinia pseudoacacia*** (41 countries/regions) and tree of heaven ***Ailanthus altissima*** (39), or noxious weeds as Japanese knotweed ***Fallopia japonica*** (39), annual ragweed ***Ambrosia artemisiifolia*** (35), Himalayan balsam ***Impatiens glandulifera*** (34), Japanese rose ***Rosa rugosa*** (34), giant hogweed ***Heracleum mantegazzianum*** (27) and iceplant ***Carpobrotus edulis*** (24). Some of the serious invaders are actually quite localised to the regions concerned, e.g. wild ginger ***Hedychium gardnerianum*** (3) or giant rhubarb *Gunnera tinctoria* (4).

4.5 Temporal Trends of Invasion

There was a steady increase in the number of neophyte species over the last two centuries, both in terms of the rate at which new species were being imported to Europe and the rate of increase of the total number of neophytes in Europe (Fig. 4.2). Visual inspection of the figure suggests that arrivals from regions outside of Europe started to increase exponentially approximately at the middle of the 19th century. Of the nowadays naturalised neophytes alien to Europe, 50% arrived after 1899, 25% arrived after 1962 and 10% arrived after 1989. At present, 6.2 new alien species, capable of naturalisation, are arriving to Europe each year. The slope is marginally less steep for naturalised neophytes of European origin, which tended to start their spread historically earlier. In this case, 50% had first been detected as alien in a European country by 1876, and the most recent 10% had started to appear by 1969. Today, approximately 4.4 European species capable of naturalisation are newly found in parts of the continent outside their native range each year (Lambdon et al. 2008). Overall, it seems that the rate of new introductions has increased sharply throughout the two past centuries and is showing little sign of slowing down.

4.6 Main Pathways to Europe

Among 2,024 naturalised plant taxa alien to Europe with information on the pathway of introduction, intentional introductions account for 63% and unintentional for 37% (Fig. 4.4). Escapes of species cultivated for ornament and horticulture account for the highest number of species, 58% of the total. Only about 11 species can with certainty be attributed to intentional releases in the wild; this group is in many cases difficult to distinguish from "amenity" species (planted in semi-wild situations for practical purposes such as landscaping, e.g. iceplant **Carpobrotus edulis** or black locust **Robinia pseudoacacia**, often used for stabilisation of soil). Examples of deliberate releases include purple pitcher plant *Sarracenia purpurea*, which was deliberately introduced to bogs in the UK and Ireland by botanists, and smooth cordgrass *Spartina alterniflora*, introduced to salt-marshes, although arguably the latter can be also considered rather an amenity use.

Contaminants of seed, mineral materials and other commodities are responsible for 403 introductions to Europe (17% of all species) and 235 species are assumed to have arrived as stowaways (directly associated with human transport but arriving independently of commodity, see Hulme et al. 2008a) (Fig. 4.4). However, the number of stowaways is almost certainly underestimated due to technically difficult systematic recording of this pathway, e.g. seed admixtures (Mack 2000). This underestimation is likely to be even more pronounced in unaided species, which are assumed to arrive by means independent of humans from a neighbouring region where they are not native (Hulme et al. 2008a). Forty one aliens of extra-European

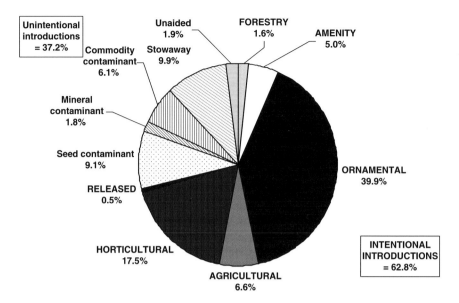

Fig. 4.4 Relative contribution of pathways of introduction shown for naturalised aliens to Europe, i.e. species with the area of origin outside Europe. Pathways of intentional introductions are in upper case letters, unintentional in lower case (Based on 1,983 naturalised aliens. Data from Lambdon et al. 2008)

origin (2% of species alien to Europe) are a product of spontaneous hybridisation involving one or both alien parents (Fig. 4.4). The spectrum of pathways is very similar for the complete European alien flora, as both groups of aliens, in Europe and to Europe, do not substantially differ in the proportional contribution of individual pathways (Lambdon et al. 2008).

4.7 Biogeographical Patterns

The pattern in the level of invasion of European regions by naturalised aliens (a measure based on species numbers, which does not necessarily reflect the invasibility of the region; Lonsdale 1999; Richardson and Pyšek 2006; Chytrý et al. 2008a) is summarised in Fig. 4.1. From the continental perspective, the highest richness of alien species is associated with large industrialised north-western countries with a tradition of good botanical recording or intensive recent research. The highest number of all alien species, regardless of status, is reported from Belgium (1,969), United Kingdom (1,779), Czech Republic (1,378), France (1,258), Sweden (1,201) and Austria (1,086); this is due to detailed inclusion of casuals. Surprisingly, only 371 alien species are reported from Russia, 149 of them naturalised (Lambdon et al. 2008). In terms of naturalised neophytes, United Kingdom (857), Germany

(450), Belgium (447), Italy (440), Ukraine (297), Austria (276), Poland (259), Lithuania (256), Portugal (250) and Czech Republic (229) harbour more than 200 species (Fig. 4.1).

The species-area relationship for naturalised alien plants in Europe indicates a diminishing increase of species numbers with increasing area (Fig. 4.5), the same as found for mammals. Plotting a subgroup of naturalised neophytes from countries where classification of species status was possible yields a steeper slope (Fig. 4.5).

Lambdon et al. (2008) used ordination analysis on assemblages of alien species in European countries/regions and identified five major distribution types: (1) north-western, comprising Scandinavia and the UK; (2) west-central, extending from Belgium and the Netherlands to Germany and Switzerland; (3) Baltic, including only the former Soviet Baltic states; (4) east-central, comprising the remainder of central and Eastern Europe; (5) southern, covering the entire Mediterranean region. Some prominent European alien invaders represent these biogeographical zones, e.g. **Rhododendron ponticum** the north-western, *Heracleum sosnowskyi* the Baltic, wild cucumber **Echinocystis lobata** the east-central and Indian fig **Opuntia ficus-indica** the southern distribution type (see the 100 worst alien species for distribution maps of these species, except for *H. sosnowskyi* see Lambdon et al. 2008).

Although it cannot be excluded that the pattern observed arises partly due to regional differences in the approach to botanical recording, there are almost certainly strong cultural and climatic influences. Gross Domestic Product, and mean annual rainfall and air

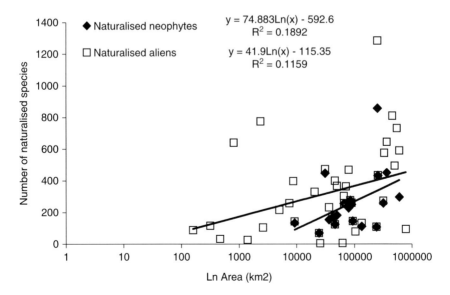

Fig. 4.5 Species-area relationship for alien plants in Europe. Based on the total number of naturalised aliens in 41 countries/regions. Trend is also shown for a subset of countries/regions where the information on the number of naturalised neophytes is available (n = 19). Note the semilog scale (Based on data from Lambdon et al. 2008)

temperature seem to play a role in the differentiation of alien floras, but are difficult to interpret since these factors are highly confounded with either latitudinal or longitudinal gradients. Country area and human population were poor explanatory variables, suggesting that factors associated with population density (e.g. urbanisation) are minor determinants of floristic composition. Therefore, it seems that bioclimatic constraints, dictating the suitability of species to the physical environment, are of primary importance and play major role in shaping the assemblages of alien species in Europe. Cultural factors such as regional trade links and traditional local preferences for crop, forestry and ornamental species, may also be important in influencing the introduced species pool (Lambdon et al. 2008). Kühn et al. (2003) showed that at least for Germany, alien species largely follow biogeographic patterns of native species.

Despite regional differences (Chytrý et al. 2008b), there is a high level of uniformity across the alien floras of the five biogeographical zones at the continent – there are only few distinctions between their alien species assemblages, as to be expected from the pattern reported for a continental scale elsewhere (e.g., McKinney 2004). For all five distribution types, the dominant families and geographical regions of origin were not substantially different from those displayed for the whole of Europe, and the dominant genera were also similar across the zones. However, the southern assemblage was most distinct, contained lower numbers of species in the temperate weedy genera such as *Chenopodium* and a stronger representation of genera with a tropical or New World bias (e.g., *Acacia*, *Opuntia*). That this assemblage coincides with the Mediterranean region, which has a particularly contrasting climate compared with the rest of Europe, confirms the important role of climate in shaping these alien species assemblages. Many of the common agricultural weeds alien in Europe are also native to the Mediterranean (Pyšek et al. 2004) and therefore excluded from the alien assemblage. In addition, countries in the Mediterranean region are positioned at the crossroads of three continents, which makes them accessible to biotic elements originating from a variety of sources; this may also contribute to the floristic distinctness of the southern alien assemblage. The east-central zone was most influenced by temperate genera (e.g., *Cotoneaster*, *Salix*), whilst the remaining three assemblages were very similar. The distribution patterns of alien species in Europe can be formally classified into four groups: (1) widespread: naturalised across much of the continent (448 species, with 76 occurring in all biogeographical zones); (2) regionally-common: naturalised consistently across a major biogeographical zone (196 species); (3) sporadic: occurring rarely and inconsistently across several biogeographic zones; (4) local: naturalised in only a small part of Europe; the latter two categories comprise the remaining vast majority of species (Lambdon et al. 2008).

4.8 Most Invaded Habitats

Since habitat descriptions in most floras are relatively coarse (Chytrý et al. 2008a, b), the recording in DAISIE database is only down to EUNIS Level 2 (Davies and Moss 2003), although in many cases only Level 1 was possible, either through low resolution or ambiguities in the source literature (Lambdon et al. 2008). Information on habitat affinities is available for 30 countries/regions. For Europe as a whole, 57% of recorded naturalised aliens in Europe (2,122 species) and 58% to Europe (1,059 species) were classified with respect to the occurrence in habitats (Table 4.4). Recent research on the representation of alien species in plant communities showed that habitat identity is the major determinant of not only the level of invasions (in the sense of Hierro et al. 2005) but also of invasibility and that it is more important than climate and factors related to propagule pressure (Chytrý et al. 2008a). The information on the level of invasions of individual habitats is therefore crucial to the management of alien species in Europe.

Human made habitats (industrial habitats and arable land, parks and gardens) harbour most alien species (Chytrý et al. 2005, 2008a, b). Of all naturalised aliens present in Europe, 64% occur in industrial habitats and 58% on arable land and in parks and gardens. Grasslands and woodlands are also highly invaded, with 37% and 32%, respectively, of all naturalised aliens in Europe present in these habitats. Mires, bogs and fens are least invaded of terrestrial habitats; approximately 10% of aliens in Europe occur there. In marine habitats, only 12 vascular plant species were recorded (7 of them alien to Europe), representing 0.6% of all species (Table 4.4).

Aliens in Europe on average occur in more habitat types than aliens to Europe, as indicated by a tendency to higher proportional values for the former group in Table 4.4 (see Lambdon et al. 2008 for statistical analysis). This can be interpreted in terms of better preadaptation of aliens originating in other parts of Europe to a

Table 4.4 Level of invasion of European EUNIS habitats by naturalised aliens, shown separately for aliens to Europe (n = 1,059) and aliens in Europe (n = 2,122) The sums of percentages across habitats do not equal 100% because many species occur in more than one habitat type. Habitat types were classified according to the EUNIS system (Davies and Moss 2003) (Based on data from Lambdon et al. 2008)

Category	Number of species		Percentage of the total number of classified species	
	Aliens in Europe	Aliens to Europe	Aliens in Europe	Aliens to Europe
Number of species classified	2,122	1,059	–	–
% classified of the total	56.6	57.7	–	–
A. Marine habitats	12	7	0.6	0.7
D. Mires, bogs and fens	220	118	10.4	11.1
B. Coastal habitats	343	170	16.2	16.1
C. Inland surface waters	444	260	20.9	24.6
F. Heathland and scrub	462	206	21.8	19.5
H. Inland sparsely vegetated habitats	497	211	23.4	19.9
G. Woodland and forest	668	310	31.5	29.3
E. Grasslands	793	276	37.4	26.1
I. Arable land, gardens and parks	1,240	533	58.4	50.3
J. Industrial habitats	1,360	658	64.1	62.1

wider range of European habitats – these species seem to profit from a better habitat match compared to extra-European aliens, which need to adapt to the character of European habitats during the invasion process (Lambdon et al. 2008). Another reason may be longer residence times of aliens with European native range (Pyšek et al. 2003), providing them with more time to colonise a wider range of habitats.

4.9 Ecological and Economic Impacts

Comprehensive data exist in the DAISIE database only for few countries (Latvia, Lithuania, and UK), where about 20% of naturalised species are considered to have an impact. The best data set, in terms of evaluation of the impact, refers to Switzerland; of 97 naturalised alien plant species, 50 are documented as having impact (see synthesis in Wittenberg 2006). However, some insights into a variety of impacts caused by alien plant species in Europe can be obtained from the inspection of species included in the 100 worst alien species; these taxa were selected so as to provide a representative sample of diverse impacts known to occur in Europe. Of the 18 plant taxa included, 17 are known to reduce the habitat of native species, and 8 are reported to cause disruption of the community assemblages. Iceplant *Carpobrotus edulis* (Vilà et al. 2006) and giant hogweed *Heracleum mantegazzianum* (Pyšek et al. 2007) are examples of alien species causing serious decrease in species richness of invaded communities. This is, however, not always the case as documented for Himalayan balsam *Impatiens glandulifera*; under some circumstances the invasion of this species does not necessarily result in loss of the diversity of invaded communities, only in shifts in species composition towards ruderal, nitrogen demanding species (Hejda and Pyšek 2006, but see Hulme and Bremner 2006).

Alien plant species exert ecological and economic impacts, both direct and indirect, at multiple levels. Of the 22 impact types defined by Binimelis et al. (2007), plants included in the 100 worst alien species are attributed, on average, with more than four types of impacts per species, which makes them the group with the second most diverse impact, following terrestrial mammals. Regarding economic impacts, Bermuda buttercup *Oxalis pes-caprae*, *Opuntia maxima*, knotgrass *Paspalum paspalodes* and *Rhododendron ponticum* are known to negatively affect commercial production and yield of agricultural and forest products. Black locust *Robinia pseudoacacia*, tree of heaven *Ailanthus altissima* and pampas grass *Cortaderia selloana* are typical examples of aliens to Europe causing serious damage to infrastructures and utilities.

The unique genetic nature of native or even endemic species of special conservation value can be lost through introgression with widespread aliens (Vilà et al. 2000). In the Czech flora (Pyšek et al. 2002), hybrids contribute 13% to the total number of aliens, and the hybridisation is more frequent in archaeophytes (19%) than in neophytes (12%). Probably the best-known European example of the recent evolution of an invader relates to a North-American smooth cordgrass *Spartina alterniflora*, which hybridised with European *S. maritima* in England and France to produce a sterile *S.* × *townsendii*. The hybrid originated only at two places in

Europe and no invasion occurred until an allotetraploid form, common cordgrass *S. anglica* arose in 1890s. This form is able to grow in a wider range of conditions and became an aggressive invader in Europe and elsewhere (Williamson 1996). In Germany, the hybridisation of alien Austrian yellowcress *Rorippa austrica* with native *R. sylvestris* produces complex hybrid forms; those with ploidy levels 3x–5x reproduce sexually and spread to areas where parents do not grow (Bleeker 2003).

Alien plants are reported to reduce availability of pollinators to native species as documented for Himalayan balsam **Impatiens glandulifera** (Chittka and Schürkens 2001). In the Balearic Islands, native *Lotus cytisoides* receives less pollinator visits in communities invaded by the South African iceplant *Carpobrotus* spp. than in uninvaded communities (Traveset and Moragues 2004). Plant invaders can also modify community structure at higher trophic levels; the grass *Elymus athericus* was shown to affect spider population dynamics in salt marshes in France (Petillon et al. 2005).

That invading species alter ecosystem functioning (transformers sensu Richardson et al. 2000) was documented in Europe for example for black locust **Robinia pseudoacacia**, which is reported to interfere with nitrogen cycle. Other species affect fire regimes; examples include pampas grass **Cortaderia selloana** and Mauritania vine reed *Ampelodesmos mauritanica*, which increase the risk of fire in Mediterranean scrub by higher fuel loads and slow decomposition resulting in formation of thick litter layers in invaded stands (Vilà et al. 2001; Grigulis et al. 2005). Finally, several alien terrestrial plants seem to affect cultural aspects of established human civilisations by altering their perception of natural landscapes (e.g., *Opuntia maxima*, Japanese rose **Rosa rugosa**, Bermuda buttercup **Oxalis pes-caprae**), reducing the area of recreational natural sites (e.g., giant hogweed **Heracleum mantegazzianum**) or becoming abundant over ancient ruins (e.g., tree of heaven **Ailanthus altissima**).

4.10 Expected Future Trends, Management Options and Their Feasibility

The data collated by DAISIE strongly suggests that up to now, the number of naturalised alien plant species in Europe tended to be underestimated and the outputs from DAISIE database represent the first European-wide detailed account of alien plant at this continent. The rate of import of new alien plant species to Europe shows very little signs of slowing down (Fig. 4.2). Moreover, even if introductions ceased, the number of naturalised plant species would increase due to the lag phase (Kowarik 1995; Richardson and Pyšek 2006). Since there is a close correlation between the total number of naturalised species and that of pests, more species mean more impact (Rejmánek and Randall 2004). As in other parts of the world, alien plant species are likely to increase in numbers (Levine and D'Antonio 2003), and the threat from plant invasions is unlikely to diminish in Europe in the near future.

Despite constant conservation efforts, biodiversity loss is estimated to be occurring at 100–10,000 times the background rate of the fossil record for the Cenozoic era (May et al. 1995) and alien species invasions are, along with loss of habitats through land-use change, direct exploitation, pollution and climate change, one of its major causes (Sala et al. 2000). In Europe, the Mediterranean region with highly diverse plant communities, is among the most endangered (Hulme et al. 2008b). This region is also likely to suffer from higher air temperatures and increasing drought, which are expected to change future fire regimes (Piñol et al. 1998; Lavorel et al. 1998). Increased fire frequencies have been reported to favour the establishment of alien species (Vilà et al. 2001). In northernly located regions, warmer conditions, likely to be brought about by climate changes, can be assumed to favour invasions by alien species as well (Walther et al. 2007), since many alien species originating from warmer regions are currently restricted from achieving wider distribution by not being able to complete their life cycle in cooler invaded areas (Pyšek et al. 2003).

The problem of alien plants needs to be addressed at the European scale. Dispersed and disconnected knowledge cannot easily be marshalled to deliver the information to politicians, but improving information exchange can build regional capacity to identify and manage invasive alien species threats. This implies that coordination of action against invasive species is crucial; a cross-European regulatory framework is needed. This holds true for plants in particular, as plants spread very easily and are more difficult to monitor and control, compared to some other taxa, such as vertebrates where substantial proportion of introductions is due to intentional releases. The spread of plants is highly correlated with human transport (Levine and D'Antonio 2003), which calls for cooperation between different sectors, including conservation agencies and transport companies with international scope (Meyerson and Mooney 2007).

Acknowledgements The support of this study by the European Commission's Sixth Framework Programme project DAISIE (Delivering alien invasive species inventories for Europe, contract SSPI-CT-2003–511202) is gratefully acknowledged. We also thank the following colleagues for contributing to national checklists and development of the plant part of DAISIE database: Paulina Anastasiu, Pavlos Andriopoulos, Corina Basnou, Ioannis Bazos, François Bretagnolle, Giuseppe Brundu, Emanuela Carli, Vitor Carvalho, Laura Celesti-Grapow, Philippe Chassot, Terry Chambers, Charalambos S. Christodoulou, Damian Chmura, Avinoam Danin, Pinelopi Delipetrou, Viktoras Didžiulis, Franz Essl, Helena Freitas, Kyriakos Georghiou, Zigmantas Gudžinskas, George Hadjikyriakou, Martin Hejda, Stephan Hennekens, Phil Hulme, Nejc Jogan, Melanie Josefsson, Nora Kabuce, Salit Kark, Mark Kenis, Stefan Klotz, Yannis Kokkoris, Mitko Kostadinovski, Svetlana Kuzmenkova, J. Rita Larrucea, José Maia, Elizabete Marchante, Hélia Marchante, Dan Minchin, Eva Moragues Botey, Josep Maria Ninot, Jorge Paiva, Eirini Papacharalampous, Francesca Pretto, Jan Pergl, Irena Perglová, David Roy, Hanno Schäfer, Christian Schröck, Rui Manuel da Silva Vieira, Culita Sîrbu, Wojtek Solarz, Walter Starmühler, Tom Staton, Oliver Stöhr, Rosenberg Tali, Barbara Tokarska-Guzik, Vladimir Vladimirov, Johannes Walter, Artemios Yannitsaros and Andreas Zikos. P.P. was supported by grants no. AV0Z60050516 from the Academy of Sciences of the Czech Republic, and nos. 0021620828 and LC06073 from MŠMT CR.

References

Binimelis R, Born W, Monterroso I, Rodriguez-Lapajos B (2007) Socio-economic impacts and assessment of biological invasions. In: Nentwig W (ed) Biological invasions. Ecological Studies 193, Springer, Berlin, pp 331–347

Bleeker W (2003) Hybridization and *Rorippa austriaca* invasion in Germany. Mol Ecol 12:1831–1841

Chittka L, Schürkens S (2001) Succesful invasion of a floral market: an exotic plant has moved in on Europe's river banks by bribing pollinators. Nature 411:653

Chytrý M, Pyšek P, Tichý L, Knollová I, Danihelka J (2005) Invasions by alien plants in the Czech Republic: a quantitative assessment across habitats. Preslia 77:339–354

Chytrý M, Jarošík V, Pyšek P, Hájek O, Knollová I, Tichý L, Danihelka J (2008a) Separating habitat invasibility by alien plants from the actual level of invasion. Ecology 89:1541–1553

Chytrý M, Maskell L, Pino J, Pyšek P, Vilà M, Font X, Smart S (2008b) Habitat invasions by alien plants: a quantitative comparison between Mediterranean, subcontinental and oceanic regions of Europe. J Appl Ecol 45:448–458

Clement EJ, Foster MC (1994) Alien plants of the British Isles. A provisional catalogue of vascular plants (excluding grasses). Botanical Society of the British Isles, London

Daehler CC (1998) The taxonomic distribution of invasive angiosperm plants: ecological insights and comparison to agricultural weeds. Biol Conserv 84:167–180

Davies CE, Moss D (2003) EUNIS habitat classification. European Topic Centre on Nature Protection and Biodiversity, Paris

Essl F, Rabitsch W (eds) (2002) Neobiota in Österreich. Umweltbundesamt, Wien

Grigulis K, Lavorel S, Davies ID, Dossantos A, Lloret F, Vilà M (2005) Landscape positive feedbacks between fire and expansion of a tussock grass in Mediterranean shrublands. Glob Change Biol 11:1042–1053

Hejda M, Pyšek P (2006) What is the impact of *Impatiens glandulifera* on species diversity of invaded riparian vegetation? Biol Conserv 132:143–152

Hierro JL, Maron JL, Callaway RM (2005) A biogeographical approach to plant invasions: the importance of studying exotics in their introduced and native range. J Ecol 93:5–15

Hulme PE, Bremner ET (2006) Assessing the impact of *Impatiens glandulifera* on riparian habitats: partitioning diversity components following species removal. J Appl Ecol 43:43–50

Hulme PE, Bacher S, Kenis M, Klotz S, Kühn I, Minchin D, Nentwig W, Olenin S, Panov V, Pergl J, Pyšek P, Roque A, Sol D, Solarz W, Vilà M (2008a) Grasping at the routes of biological invasions: a framework for integrating pathways into policy. J Appl Ecol 45:403–414

Hulme PE, Brundu G, Camarda I, Dalias P, Llambdon P, Lloret F, Medail F, Moragues E, Suehs C, Traveset A, Troumbis A, Vilà M (2008b) Assessing the risks of alien plant invasions on Mediterranean islands. In: Tokarska-Guzik B, Brundu G, Brock JH, Child LE, Pyšek P, Daehler C (eds) Plant invasions? Human perception, ecological impacts and management. Backhuys, Leiden, pp 39–56

Klotz S, Kühn I, Durka W (eds) (2002) BIOLFLOR: Eine Datenbank mit biologisch-ökologischen Merkmalen zur Flora von Deutschland. Schriftenr Vegetationsk 38:1–334

Kowarik I (1995) Time lags in biological invasions with regard to the success and failure of alien species. In: Pyšek P, Prach K, Rejmánek M, Wade M (eds) Plant invasions: general aspects and special problems. SPB Academic Publishers, Amsterdam, pp 15–38

Kühn I, Klotz S (2003) The alien flora of Germany: basics from a new German database. In: Child LE, Brock JH, Brundu G, Prach K, Pyšek P, Wade PM, Williamson M (eds) Plant invasions: ecological threats and management solutions. Backhuys, Leiden, pp 89–100

Kühn I, May R, Brandl R, Klotz S (2003) Plant distribution patterns in Germany: will aliens match natives? Feddes Repert 114:559–573

Lambdon PW, Pyšek P, Basnou C, Hejda M, Arianoutsou M, Essl F, Jarošík V, Pergl J, Winter M, Anastasiu P, Andriopoulos P, Bazos I, Brundu G, Celesti-Grapow L, Chassot P, Delipetrou P, Josefsson M, Kark S, Klotz S, Kokkoris Y, Kühn I, Marchante H, Perglová I, Pino J, Vilà M, Zikos A, Roy D, Hulme PE (2008) Alien flora of Europe: species diversity, temporal trends, geographical patterns and research needs. Preslia 80:101–149

Lavorel S, Canadell J, Rambal S, Terradas J (1998) Mediterranean terrestrial ecosystems: research priorities on global change effects. Global Ecol Biogeogr Lett 7:157–166

Levine JM, D'Antonio CM (2003) Forecasting biological invasions with increasing international trade. Conserv Biol 17:322–326

Lonsdale M (1999) Global patterns of plant invasions and the concept of invasibility. Ecology 80:1522–1536
Mabberley D (1997) The plant book. Cambridge University Press, Cambridge
Mack RN (2000) Cultivation fosters plant naturalization by reducing environmental stochasticity. Biol Invasions 2:111–122
May RM, Lawton JH, Stork NE (1995) Assessing extinction rates. In: Lawton H, May RM (eds) Extinction rates. Oxford University Press, Oxford, pp 1–24
McKinney ML (2004) Do exotics homogenize or differentiate communities? Roles of sampling and exotic species richness. Biol Invasions 6:495–504
Meyerson LA, Mooney HA (2007) Invasive alien species in an era of globalization. Front Ecol Environ 5:199–208
Petillon J, Ysnel F, Canard A, Lefeuvre JC (2005) Impact of an invasive plant (*Elymus athericus*) on the conservation value of tidal salt marshes in western France and implications for management: responses of spider populations. Biol Conserv 126:103–117
Piñol J, Terradas J, Lloret F (1998) Climate warming, wildfire hazard and wildfire occurrence in coastal eastern Spain. Climate Change 38:345–357
Preston CD, Pearman DA, Dines TD (2002) New atlas of the British and Irish flora. Oxford University Press, Oxford
Preston CD, Pearman DA, Hall AR (2004) Archaeophytes in Britain. Bot J Linn Soc 145:257–294
Pyšek P (1998) Is there a taxonomic pattern to plant invasions? Oikos 82:282–294
Pyšek P (2003) How reliable are data on alien species in Flora Europaea? Flora 198:499–507
Pyšek P, Sádlo J, Mandák B (2002) Catalogue of alien plants of the Czech Republic. Preslia 74:97–186
Pyšek P, Sádlo J, Mandák B, Jarošík V (2003) Czech alien flora and a historical pattern of its formation: what came first to Central Europe? Oecologia 135:122–130
Pyšek P, Richardson DM, Rejmánek M, Webster G, Williamson M, Kirschner J (2004) Alien plants in checklists and floras: towards better communication between taxonomists and ecologists. Taxon 53:131–143
Pyšek P, Richardson DM, Jarošík V (2006) Who cites who in the invasion zoo: insights from an analysis of the most highly cited papers in invasion ecology. Preslia 78:437–468
Pyšek P, Cock MJW, Nentwig W, Ravn HP (eds) (2007) Ecology and management of Giant Hogweed (*Heracleum mantegazzianum*). CAB International, Wallingford
Pyšek P, Richardson DM, Pergl J, Jarošík V, Sixtová Z, Weber E (2008) Geographical and taxonomic biases in invasion ecology. Trends Ecol Evolut 23:237–244
Rejmánek M, Randall JM (2004) The total number of naturalized species can be a reliable predictor of alien species pests. Diversity Distrib 10:367–369
Reynolds SCP (2002) A catalogue of alien plants in Ireland. National Botanic Gardens Glasnevin Occasional Papers 14:1–414
Richardson DM, Pyšek P (2006) Plant invasions: merging the concepts of species invasiveness and community invasibility. Progr Phys Geogr 30:409–431
Richardson DM, Pyšek P, Rejmánek M, Barbour MG, Panetta FD, West CJ (2000) Naturalization and invasion of alien plants: concepts and definitions. Diversity Distrib 6:93–107
Sala O, Chapin Stuart F, Armesto J, Berlow E, Bloomfield J, Dirzo R, Huber-Sanwald E, Huenneke L, Jackson R, Kinzig A, Leemans R, Lodge D, Mooney H, Oesterheld M, Poff N, Sykes M, Walker B, Walker M, Wall D (2000) Global biodiversity scenarios for the year 2100. Science 287:1770–1774
Stevens PF (2001 onwards) Angiosperm phylogeny website. Version 7, May 2006, URL: [http://www.mobot.org/MOBOT/research/APweb/]
Traveset A, Moragues E (2004) Effect of *Carpobrotus* spp. on the pollination success of native plant species of the Balearic Islands. Biol Conserv 122:611–619
Tutin TG, Heywood VH, Burges NA, Moore DM, Valentine DH, Walters SM, Webb DA (eds) (1964–1980) Flora Europaea. Vols. 1–5. Cambridge University Press, Cambridge

Vilà M, Weber E, D'Antonio CM (2000) Conservation implications of invasion by plant hybridization. Biol Invasions 2:207–217

Vilà M, Lloret F, Ogheri E, Terradas J (2001) Positive fire-grass feedback in Mediterranean Basin woodlands. Forest Ecol Manage 147:3–14

Vilà M, Tessier M, Suehs CM, Brundu G, Carta L, Galanidis A, Lambdon P, Manca M, Médail F, Moragues E, Traveset A, Troumbis AY, Hulme PE (2006) Local and regional assessment of the impacts of plant invaders on vegetation structure and soil properties of Mediterranean islands. J Biogeogr 33:853–861

Walther G-R, Gritti ES, Berger S, Hickler T, Tang Z, Sykes MT (2007) Palms tracking the climate change. Global Ecol Biogeogr 16:801–809

Weber E (2003) Invasive plant species of the world. A reference guide to environmental weeds. CABI International, Wallingford

Weber EF (1997) The alien flora of Europe: a taxonomic and biogeographic overview. J Veg Sci 8:565–572

Williamson M (1996) Biological invasions. Chapman & Hall, London

Wittenberg R (ed) (2006) An inventory of alien species and their threat to biodiversity and economy in Switzerland. Environmental Studies 29. Federal office for the Environment, Bern

Chapter 5
Alien Terrestrial Invertebrates of Europe

Alain Roques, Wolfgang Rabitsch, Jean-Yves Rasplus,
Carlos Lopez-Vaamonde, Wolfgang Nentwig, and Marc Kenis

5.1 Introduction

Unlike other groups of animals and plants, no checklist of alien terrestrial invertebrates was available in any of the European countries until recently. Since 2002, such checklists were successively provided by Austria (Essl and Rabitsch 2002), Germany (Geiter et al. 2002), the Czech Republic (Šefrová and Laštůvka 2005), Scandinavia (NOBANIS 2007), the United Kingdom (Hill et al. 2005), Switzerland (Wittenberg 2006) and Israel (Roll et al. 2007). However, most European regions remained uncovered and, furthermore, comparisons between the existing lists were inherently difficult because they used different definitions of alien. Thus, estimating the importance of terrestrial alien invertebrates at the European level remained impossible, mostly because of poor taxonomic knowledge existed for several groups. By gathering taxonomists and ecologists specialised on most invertebrate taxa together with collaborators working at the national level in 35 European countries, the DAISIE project intended to fill this gap. However, a lack of European expertise in some taxonomic groups did not allow coverage of all the terrestrial invertebrates with the same level of precision. Data on insects were more reliable than those of other taxa, and consequently the analyses presented below will mostly refer to this group.

5.2 Taxonomy

Alien terrestrial invertebrates represent one of the most numerous groups of introduced organisms in Europe. A total of 1,296 species originating from other continents have established so far, to which we add 221 cosmopolitan species of uncertain origin (cryptogenic). Both groups will hereafter be referred as alien species. Additionally, 964 species of European origin are considered to have been introduced from one to another European region. More than a half of these intra-European aliens (551 species) are species from continental Europe newly observed on islands, while a further significant proportion are Mediterranean species newly reported in northern and western areas of Europe. However, it was not possible to ascertain for a large part of these species of European origin if they were introduced through human activities or were naturally expanding, e.g., due to global warming. Therefore, we will essentially consider alien species of non-European origin in this chapter.

Arthropods, mostly insects, dominate and represent nearly 94% of the alien terrestrial invertebrates (Fig. 5.1). An obvious lack of knowledge probably led to an underestimate of the importance of other invertebrate phyla unless they constitute economic pests, phytosanitary threats or vectors of disease. Most of the 47 alien terrestrial nematodes consist of either serious pests of agriculture (e.g., *Globodera* and *Xiphinema* species affecting crops; Grubini et al. 2007) and forests (pine wood nematode **Bursaphelencus xylophilus** introduced to Portugal), or parasites of alien animals introduced to Europe (e.g., raccoon nematode *Baylisascaris procyonis*; Küchle et al. 1993). Only 13 alien species of Platyhelminthes (terrestrial flatworms) and 14 annelids (segmented worms) have yet been observed. Representatives of these groups include the American liver fluke *Fascioloides magna*, an important trematode parasite imported with game animals (Novobilský et al. 2007), the New Zealand flatworm *Arthurdendyus triangulata*, a predatory planarian species feeding on earthworms (Boag and Yeates 2001), and earthworms related to the degradation of organic wastes such as the Japanese red worm *Eisenia japonica* (Graff 1954) and the tropical *Eudrilus eugeniae* (Dominguez et al. 2001).

Terrestrial habitats have been little colonised by alien molluscs and only 16 species of gastropods, mostly slugs, are reported. They include a predatory Caucasian slug *Boettgerilla pallens* and several species of snails first restricted to greenhouses and then found outdoors (e.g., the orchid snail *Zonitoides arboreus*; Dvořák and Kupka 2007). Besides these truly alien species, a number of other molluscs have

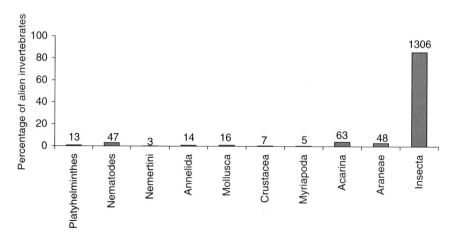

Fig. 5.1 Relative importance of higher taxonomic groups in the 1,522 alien invertebrate species established in Europe. The numbers above the bars correspond to the total number of species

been introduced from southern and western Europe towards northern and eastern countries. The Iberian slug ***Arion vulgaris*** *(= lusitanicus)*, several species of *Deroceras*; and snails such as *Milax gagates* and *Cryptomophalus aspersus* were unintentionally translocated within Europe (Wittenberg 2006).

Alien terrestrial crustaceans consist of only seven synanthropic isopods, mostly of subtropical origin, which are essentially present in warm man-made habitats (e.g., *Reductoniscus costulatus*, Kontschán 2004; *Trichorhina tomentosa* used as aquarium fish food) and one cosmopolitan species without known origin (*Porcellionides pruinosus*, Hornung et al. 2007). Similarly, there are only five alien species of myriapods established in Europe, two centipedes (chilopods) of the family Henicopiidae (*Lamyctes coeculus* and *L. emarginatus*; Bergersen et al. 2006) and three millipedes (diplopods), although more Mediterranean species occur regularly in anthropogenous habitats such as compost and glasshouses in northern and central Europe. Mites (Acari) are represented by a total of 63 species belonging to 14 different families but most belong to only 3 families: Tetranychiidae, spider mites (27 species, mostly *Oligonychus* and *Tetranychus* species; Migeon 2005), Eriophyiidae plant gallers (12 species; e.g., the fuchsia gall mite *Aculops fuchsiae*) and Amblyommidae ticks (7 species). In addition, the introduction of the honey bee ectoparasitic *Varroa destructor* is of important concern to bee keepers (Griffiths and Bowman 1981).

Among alien spiders (Araneae) 43 are of non-European origin and 44 expanded their range from Mediterranean and east Palaearctic origin to western and northern Europe. These 87 species belong to 25 spider families of which the most dominant families are Theridiidae (13 species), Salticidae (9 species), Pholcidae (9 species), and Linyphiidae and Oonopidae (8 species each) (Kobelt and Nentwig 2008).

The 1,306 alien insect species established in Europe belong to 16 different orders, all of which are already present in the native entomofauna. However, Coleoptera and Hemiptera largely dominate the aliens, representing 29% and 26% respectively, followed by Hymenoptera (15%), Lepidoptera (10%), Diptera (7%), Thysanoptera (4%), Psocoptera (3%), Phtiraptera (2%), and Blattodea (2%); the other orders (Orthoptera, Collembola, Siphonaptera, Phasmatodea, Dermaptera, Isoptera, Zygentoma) each accounting for less than 1%. The alien entomofauna is highly diverse with a total of 205 insect families involved but only one family (Castniidae, Lepidoptera) was not known from Europe before its introduction. Only 29 families contribute for more than 10 alien species, and three Hemiptera families in the suborder Sternorrhyncha contribute the most: Aphididae (aphids, 99 species), Diaspididae (diaspidid scales, 68 species), and Pseudococcidae (pseudococcid scales; 40 species).

At the order level, the taxonomic composition of the alien entomofauna significantly differs from that of the native European entomofauna which includes 25 orders (data from Fauna Europaea in Kenis et al., 2007; $\chi^2 = 568.50$; $P < 0.001$). Establishment patterns differ between orders. Hemiptera are more than three times better represented in the alien fauna than in the native fauna (26% vs. 8.0%). The alien entomofauna also includes significantly more Thysanoptera (4% vs. 0.6%), Psocoptera (3% vs. 0.3%) and Blattodea (2% vs. 0.2%) but many fewer Diptera (7% vs. 21%) and Hymenoptera (15% vs. 25%) than the native fauna. A similar proportion is observed in Coleoptera (29% vs. 30%) and Lepidoptera (10% both).

5.3 Temporal Trends

Fragments of insects found in Roman and Viking graves (e.g., *Sitophilus granarius*; Levinson and Levinson 1994; *Pulex irritans*, Beaucornu and Launay 1990) proved that some invertebrates species were introduced to Europe long ago. Similarly, parasites of early-domesticated alien mammals probably arrived with their host (e.g., cat flea *Ctenocephalides felis*). However, a clear identification of the archaeozoan invertebrates appeared to be more difficult than in other animal and plant groups. Therefore, we only qualified as aliens the neozoan species, i.e. those introduced after 1500. The few species identified as archaeozoans (e.g., *Blattella germanica, Stegobium paniceum, Cimex lectularis, Labia minor, Lepisma saccharina, Acheta domestica, Tenebrio molitor*) were thus excluded from the list of aliens.

The precise date of arrival in Europe is not known for most species because most introductions happened unintentionally (see Section 5.5). Even conspicuous species, such as the Asian long-horned beetle **Anoplophora glabripennis**, were reported with a delay of at least 3–5 years (Hérard et al. 2006). An analysis of the 995 alien species for which the date of the first record in Europe is known shows that the arrival of alien invertebrates has increased exponentially since the 15th century but a significant acceleration was observed during the second half of the 20th century (Fig. 5.2). As a probable result of globalisation, this trend is still increasing with an average of 19.1 alien species newly reported per year in Europe between 2000 and 2007; i.e., an average which is two times more than during the period 1950 to 1974 (10.2 species/year). The same trend was observed for all groups of invertebrates analyzed separately. An average of 17.5 new species of insects per year was recorded between 2000 and 2007, while this value was only 8.1 from 1950 to 1974. Between 1900 and 1950 1.5 alien spider species arrived per decade, between 1950 and 2000 2.4 species, and the most recent figures allow a prediction of one alien spider species arriving annually since 2000 (Kobelt and Nentwig 2008).

5.4 Biogeographic Patterns

A precise region of origin was ascertained for 79% (1,255 species) of the alien invertebrate species while 7% (102 species) were only known to be native of tropical or subtropical regions. The remaining cryptogenic invertebrates (14%) are mostly cosmopolitan species for which there is no agreement regarding their area of origin. This is particularly true for stored products pests and for some ectoparasites on cattle and pets that occur on other continents. A few other cryptogenic species appeared in Europe without having been described elsewhere. However, data on their phylogeography, population ecology, parasites and dispersal biology strongly suggest that they originate from another continent. The horse-chestnut leaf

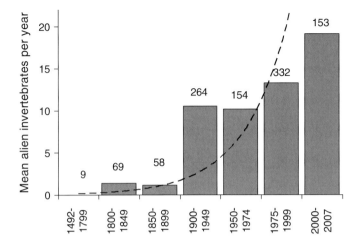

Fig. 5.2 Rate of established alien invertebrate species in Europe since 1492 as mean number of alien invertebrates recorded per year. Calculations made on 995 species for which the first record is precisely known. The numbers above the bars correspond to the number of new species recorded per period

miner *Cameraria ohridella*, is illustrative of the difficulty in identifying the native range of such species. Whereas this leaf miner was previously considered as an extra-European alien, recent genetic studies indicate that it originates from the Balkan (C. Lopez-Vaamonde, unpublished).

Asia has contributed the most alien invertebrates occurring in Europe (29% of the species of identified origin (Table 5.1), followed by North America (20%). The trends are similar for arthropods and insects when considered separately, but the contribution of both continents is higher for non-arthropods (32% and 23%, respectively).

Analysing insect data per time unit revealed that the relative contribution of Asia and North America was stable over time. During the periods 1950–1989 and 1990–2007, 29% and 21% of the established insects were of Asian and North American origin respectively. The contribution of tropical and subtropical areas is surprisingly important. The overall contribution of species from Australasia, Africa, Central and South America in combination with species of undefined tropical areas represent 37% of all alien insects in Europe. While we agree that insect species coming from these areas are not only native of tropical ecosystems, this proportion is nevertheless outstanding.

A comparison of the native range of alien insects from the different orders with that of all alien insects also revealed significant differences ($\chi_2 = 388.26$; $P = 0.00$). Insects in different orders came to Europe from different parts of the world. Most Hymenoptera (38%), Lepidoptera (35%) and Hemiptera (33%) have an Asian

Table 5.1 Origin of 1,517 alien invertebrate species established in Europe (% of the total are shown)

	Total invertebrates	Arthropods	Non-arthropods
Africa	12.3	12.9	3.2
North America	19.8	19.6	22.6
C and S America	10.8	10.9	9.7
Asia	29.4	29.3	32.3
Australasia	6.5	6.6	4.3
Tropical	6.7	7.1	1.1
Cryptogenic	14.5	13.7	26.9
Total	1,517	1,424	93

origin whilst Diptera arrived predominantly from North America (30%). Coleoptera came from various regions including a noticeable part from Australasia (11%) mostly linked to the introduction of *Eucalyptus* and *Acacia* spp. in the Mediterranean regions of Europe. Coleoptera also represent a large proportion of the cosmopolitan stored product pests that are predominantly of tropical or subtropical origin.

Large differences also exist between European countries in the number of alien insects recorded per country (Fig. 5.3). This is likely due to differences in sampling efforts and in local taxonomic expertise. The number of alien insects is significantly and positively correlated with the country surface area ($r = 0.52$; $P = 0.046$). The western countries and islands appear relatively more colonised. The number of alien species significantly decreases with the longitude of the countries centroids ($r = -0.699$; $P = 0.004$) whereas latitude did not seem to have a significant influence ($r = -0.39$; $P = 0.17$). Islands also host proportionally more alien species than continental countries relatively to their size (ANOVA on the number of alien species per square kilometre; $F_{1,55} = 4.53$; $P = 0.038$) but this is independent of the coast length ($r = 0.17$; $P = 0.38$). In continental countries, a direct access to the sea does not influence the number of alien insect species ($P = 0.64$).

Only 1% (13 out of 1,306) of the alien insect species are present in more than 40 countries, among them are the melon and cotton aphid **Aphis gossypii**, and several beetles associated with stored products especially seed bruchids (e.g., *Callosobruchus chinensis*). By contrast, most alien insects remain confined to one country (390 species) or two countries (180 species).

5.5 Main Pathways to Europe

The exact pathway of introduction is only known with confidence for the deliberately released biological control agents. Some were released in the field but others were first released in glasshouses, and then escaped and became established outdoors. This group includes several ladybeetles, among which the multicoloured Asian ladybeetle **Harmonia axyridis**, has now spread throughout western and central Europe, several parasitic wasps (e.g., the whitefly parasitoid *Encarsia*

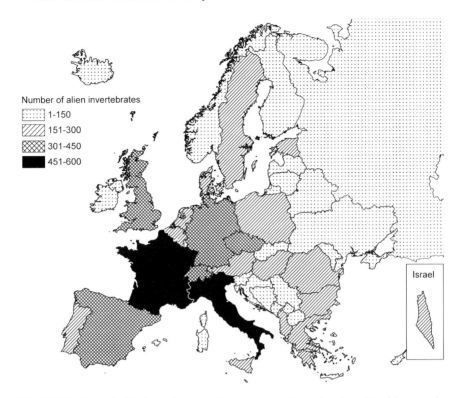

Fig. 5.3 Numbers of alien invertebrates in European countries and regions. The Macaronesian islands (not shown) have 163–203 alien species

formosa; Noyes 2007), predatory true bugs and several predatory mites (e.g., a phytoseiid predator of spider mites, *Amblyseius californicus*). A few other species were introduced for leisure or as pets, and then also escaped to the wild, e.g., some lepidopteran saturnids now found on urban trees (e.g., *Samia cynthia* living on *Ailanthus altissima*) and cockroaches used as food for reptile pets. However, intentional introductions play a minor role in invertebrate invasions, in contrast to mammals and plants. Only 131 (9%) and 7 (<1%) of the 1,517 alien terrestrial invertebrate species in Europe were introduced for biological control or leisure, respectively. Thus, ca. 90% of the alien invertebrates were introduced unintentionally through human activity.

Finding a likely vector for such accidental introductions is not easy, but for a number of species it could be inferred based on the species biology, plant or animal host. In insects, stowaway species, which are not directly related to the vector, represent about 15% of the total. Representatives include diverse species, from mosquitoes such as the Asian tiger mosquito *Aedes albopictus* whose eggs and larvae are introduced with second-hand tyres or inside bags watering lucky bamboos (*Dracaena sanderiana*), to staphylinid predators transported with compost

(Ødegaard and Tømmerås 2000) and 'tramp' ants of tropical and subtropical origin (e.g., the Argentine ant *Linephitema humile*). However, most alien insects, mites, and nematodes, probably arrived as contaminants of a commodity. For these species, the main pathway of introduction seems to be horticultural and ornamental trade (38%), which includes ornamental plants for planting, cut flowers, bonsais (a pathway of increasing importance; e.g., for the Asian long-horned beetle *Anoplophora chinensis* and some nematodes), seeds and aquarium plants. Then, in decreasing order insects arrived with stored products (18%), vegetables and fruits (12%), fresh and manufactured wood material (10%), and animal husbandry (3%), the pathway remaining unclear in 19% of the species.

Thus, it is not surprising that the number of alien invertebrate species per country is correlated with several macro-economic variables, more or less reflecting the trading activity of the country. For insects, strong positive correlations exist with both the size of the road network ($r = 0.72$; $P = 0.003$), the volume of recent merchandise imports (mean 2002–2005; $r = 0.78$; $P = 0.0007$), and of agriculture imports (mean 2002–2005; $r = 0.83$; $P = 0.00$) although independent increasing trends of both variables cannot be excluded. A significant correlation was also observed for the gross domestic product per capita (value 2002; $r = 0.53$; $P = 0.04$).

The introduction of alien spiders to Europe also appears clearly linked to global trade. The number of imported spider species fits best with the amount of manufactures (i.e. industrial goods shipped in containers) and less with mining products or agricultural products since spiders are, in contrast to most insects, not herbivorous (Kobelt and Nentwig 2008).

5.6 Most Invaded Ecosystems, Habitats and Environments

Approximately half (48%) of the alien invertebrates are phytophagous, whereas parasitoids and predators represent 22% and detrivores 21%. The 9% remaining species have unknown feeding habits. The proportion of phytophagous species is a bit higher in insects (53%) than in the other invertebrate groups. No global analysis of the relative importance of the different feeding niches has been done yet. Kenis et al. (2007) showed that the alien insect faunas of Switzerland and Austria differ significantly from the native European fauna in niche occupancy. The proportion of sap feeders (mostly Sternorrhyncha hemipterans and thrips) and detritivores (mostly represented by pests of stored products) is more important in the alien entomofauna of these countries. Other guilds that are proportionally more represented in the alien fauna than in the European fauna are wood borers, fruit borers, seed feeders/borers and omnivores. In contrast, external defoliators, root feeders/borers, stem borers, gall makers, predators and parasitoids seem to travel or establish less successfully. The analysis remains to be done at the European level but it is likely that sap feeders are similarly a major invading group because they represent 28% of all alien insects.

The main part of the alien invertebrates established in Europe is confined to man-made habitats. About 34% of these species live indoors in domestic, industrial, and other artificial habitats, e.g., in greenhouses (Table 5.2). Parks and gardens host another 24% of the species, frequently introduced with their native host plant, whilst 15% have colonised agricultural land. Thus, only 20% of the alien insects have established in a more or less natural environment, mostly in woodlands and forests (12%). In addition, no correlation was observed between the number of alien species per country and the surface of the forested area ($r = 0.28$; $P = 0.31$). Similarly, most alien spiders (71%) live synanthropically at least in a part of their invaded habitat, e.g., in houses or greenhouses, with Oecobiidae, Oonopidae, Pholcidae, and Theridiidae as dominant families. Alien Gnaphosidae, Linyphiidae, and Salticidae usually only occur in natural habitats (Kobelt and Nentwig 2008).

As Kenis et al. (2007) pointed out, human habitats may be more readily invaded by insects than natural habitats because insects linked to human environments and activities are more likely to be carried by human transport into a new region than insects living in natural areas. This hypothesis is confirmed by the interception data that invariably contain far more agricultural and domestic pests than insects exclusively linked to natural habitats (Roques and Auger-Rozenberg 2006). In addition, man-made habitats have been studied more thoroughly than natural habitats and economic pests and vectors are also better known than natural entomodiversity in Europe. However, the existence of a time-lag between introduction to human habitats and the subsequent spread to natural habitats was observed for several species (e.g., Ødegaard and Tømmerås 2000). The Asian long-horned beetle *Anoplophora glabripennis* which occurs in both subnatural forests and man-made plantations in its native Chinese range, is presently restricted to parks and tree edges in Europe. It should be especially surveyed in this context.

Table 5.2 Main habitats of 1,306 alien insect species established in Europe (note that a species may live in several habitats)

Habitat	Number of alien species	Percentage of alien species living in this habitat
Coastal areas	20	1.1
Inland surface waters	10	0.6
Mires and bogs	16	0.9
Grasslands	73	4.1
Heathlands, edges, scrubs and tundra	68	3.8
Woodlands and Forests	219	12.3
Miscellaneous without vegetation	12	0.7
Agricultural lands	273	15.3
Parks and gardens	423	23.7
Urban areas	399	22.3
Glasshouses	202	11.3
Unknown	72	4.0

5.7 Ecological and Economic Impacts

Alien invertebrates are clearly better known for their economic or sanitary impact (pests of agriculture, horticulture, stored products and forestry; vectors of human and animal diseases) than for their ecological impact. A negative economic or health impact was found for 45% of the alien invertebrates established in Europe, and for 49% of the insects, whereas the rate of native insects reaching pest status in temperate countries is probably below 5% (Kenis et al. 2007).

Consequently, the species factsheets for the "100 of the worst" are essentially presenting the economically most important species of invertebrates (aphids, whiteflies, agromyzid leaf miners, chrysomelid beetles, noctuids, slugs damaging crops and fruit flies, thrips, xylophagous beetles, lepidopteran leaf miners, and nematodes damaging trees). Other categories of economically important alien invertebrates not included in the factsheets are beetles feeding on stored products (e.g., a number of Dermestidae, Bruchidae, Nitidulidae, Tenebrionidae), bark beetles damaging trees and logs (e.g., the North American *Gnathotrichus materarius*, Faccoli 1998), phytophagous scales attacking vegetables, orchard trees and ornamentals (e.g., the San José scale *Diaspidiotus perniciosus*), leafhoppers transmitting viral diseases to crops (e.g., *Scaphoides titanus* vector of yellow speckle disease on grapevine; Della Giustina 1989), seed insects (e.g., *Megastigmus* seed chalcids; Roques and Skrzypczynska 2003), cynipid gall-makers (e.g., the Asian *Dryocosmus kuriphilus* on chestnut; Brussino et al. 2002), phytophagous true bugs (e.g., the North American conifer seed bug *Leptoglossus occidentalis*; Taylor et al. 2001), and the lace bugs *Corythucha ciliata* and *C. arcuata*; Servadei 1966; Bernardinelli and Zandigiacomo 2000), termites (e.g., the North American *Reticulitermes flavipes*; Austin et al. 2005), phytophagous mites (e.g., the red spider mite *Tetranychus evansi*; Migeon 2005), and nematodes attacking crops (e.g., the potato cyst nematodes *Globodera pallida* and *G. rostochiensis*; Grubini et al. 2007). Some species also represent social nuisance due to their presence in human buildings and gardens (e.g., *Leptoglossus occidentalis*, **Harmonia axyridis**, several tramp ants and cockroaches).

Nearly 100 (ca. 7%) of the alien invertebrate species affect human and animal health. Biting insects, which can potentially transmit diseases, include seven mosquitoes (Culicidae, e.g., the tiger mosquito, **Aedes albopictus**), and a number of ectoparasites among which six flea species (Siphonaptera such as the rat flea *Nosopsyllus fasciatus*; Beaucornu and Launay 1990), and 27 sucking louse species (Phthiraptera such as the south American *Gyropus ovalis* on guinea pigs; Stojcevic et al. 2004). In addition, 20 alien species of mites are susceptible of transmitting virus to humans (e.g., several Amblyommidae ticks of the genus *Hyalomma* transmitting *Rickettsia*; Parola 2004), or causing allergies and dermatitis (e.g., the tropical fowl mite *Ornithonyssus bursa*, Gjelstrup and Møller 1985). More than half (57%) of the 47 introduced nematodes are endoparasites of humans or cause zoonosis to cattle, game animals (e.g., *Ashworthius sidemi*; Drozdz et al. 2003), and poultry (*Ascaridia dissimilis*; Šnábel et al. 2001). Similarly, 7 of the 11 alien flatworms have a direct health impact on poultry and mammals (e.g., the American liver fluke *Fascioloides magna*; Novobilský et al. 2007). Among alien spiders some species of medical importance to humans live in or around buildings; e.g., two *Loxosceles* species from America and the black widow *Latrodectus hasselti* from Australia (Kobelt and Nentwig 2008).

Ecological impacts of alien invertebrate species have been much less investigated than their economic or human health impacts. A recent worldwide review by Kenis et al. (2008) listed several types of ecological impacts for alien insects, which also apply to other invertebrates. Alien invertebrates can affect the native biodiversity through direct interactions, e.g., hybridisation with native related species, feeding on native plants, preying or parasitizing native species. They can also affect native species and ecosystems indirectly, through cascading effects, or through various mechanisms, e.g., by competing for food or space, carrying diseases, or sharing natural enemies with native species. Most examples of ecological impact of alien insects have been observed in North America, Oceania and oceanic islands (Kenis et al., 2008). In Europe, only a handful of examples of terrestrial invertebrates showing a clear impact on native biodiversity or ecosystems have been identified so far.

Whereas hybridisation between introduced and native species is of major concern in vertebrates and plants only intraspecific introgression has been reported between subspecies of bees (*Apis mellifera*) and bumblebees (*Bombus terrestris*) following repeated introductions as pollinators in fields and glasshouses. Thus, introgression threatens native populations of the honeybee, *Apis mellifera*, in NW Europe (Goulson 2003) and in the Canary islands (De La Rùa et al. 2002). As few researches have yet been done in this field, it is to be expected that more studies will reveal more cases of genetic introgression in invertebrates.

Despite the importance of phytophagous species in the alien fauna, the direct impact of alien herbivores on the survival of native plant populations is poorly documented in Europe. Only a few studies analyzed competition between native and alien herbivores for the same resource. Some alien seed chalcids in the genus *Megastigmus* tend to displace the native species when exploiting fir seeds (Fabre et al. 2004).

More information is available on the ecological impact of alien predators. Widespread invasive ants such as the Argentine ant **Linepithema humile**, are considered to have dramatic effects by displacing native ants, other invertebrates or even vertebrates, either through resource competition or direct predation. By competing with native ants, **L. humile** also alters the seed dispersal process of myrmecochorous plants. However, climatic requirements may limit their impact in some areas (Wetterer et al. 2006). The ladybeetle **Harmonia axyridis**, introduced from eastern Asia, out-competes and displaces native aphidophagous ladybeetles by intra-guild predation and competition for food in North America, and is suspected to have a similar impact in Europe. A recently introduced Asian hornet, *Vespa velutina nigrithorax*, is rapidly spreading in SW France and threatens local honey bee populations by its preying activity (Haxaire et al. 2006). Introduced parasitoids may also displace native parasitoids by competition. Introduced *Cales noaki* (Aphelinidae) has replaced the native aphelinid *Encarsia margaritiventris* as the dominant parasitoid of the *Viburnum* whitefly *Aleurotuba jelineki* in Italy (Viggiani 1994). Among invertebrates other than insects, two species have an impact on native biodiversity or ecosystems: the New Zealand flatworm *Arthurdendyus triangulata*, and the Spanish slug **Arion vulgaris**. The first locally eliminates native earthworms in Great Britain, which strongly affects soil ecosystems (Boag and Yeates 2001), and the second competes with and sometimes preys on native slugs in Central Europe (Wittenberg 2006).

5.8 Management Options and Their Feasibility

Most introductions of alien invertebrates are unintentional and thus remain rather unpredictable. The identification of species traits which may favour introduction and establishment of alien species as well as the characteristics which may make habitats more prone to invasion may help to develop a prevention strategy. For example, the identification of sap-feeding insects as a major invading group associated with the trade of ornamental plants strongly suggests a need to intensify the quarantine surveys for this group and to strengthen the regulations of ornamental trade. The information on specific biological attributes such as host specificity, mode of reproduction and fecundity, and thermal requirements for survival and development may help to target potential invaders. Furthermore, phylogenetic information can also be predictive and of interest to identify potential future invaders.

However, is such an analysis sufficient to predict potential invertebrate invaders? Kenis et al. (2007) pointed out that the two most damaging alien insects in outdoor agriculture in Europe, the Colorado potato beetle **Leptinotarsa decemlineata** and the western corn rootworm **Diabrotica virgifera virgifera**, belong to the leaf beetle family, Chrysomelidae, which does not include many successful invaders in Europe (23 species, and mostly seed beetles). These two species also develop in trophic niches (external defoliators and root feeders) that are rarely exploited by the alien entomofauna. However, they are both associated with introduced crops and in that sense may have been predicted. More generally, less than 20% of the alien invertebrates established in Europe have been intercepted before their arrival. *D. virgifera* was intercepted once (Roques and Auger-Rozenberg 2006) whereas repeated introductions from North America are likely to have occurred (Miller et al. 2005). Moreover, several species have a limited impact, or more often no known impact, in their country of origin. An alternative approach to predict the potential impact of alien species not yet present in Europe could be, for phytophagous invertebrates at least, the development of a 'sentinel' plant strategy where European plants are grown in other continents and thus offered to local invertebrates for infestation. Such sentinel experiments have just been initiated with several species of European trees planted in China, which is the largest potential source for tree pest invaders (Roques and Auger-Rozenberg 2006).

These elements limit the usefulness of species-based risk assessments for most alien invertebrates except for intentionally introduced species, such as biological control agents or species used for leisure, for which no introduction should be carried out without a proper risk assessment (van Lenteren et al. 2006).

Eradication of established invertebrate invaders is not an easy task. Because of the latency between arrival and first record, eradication is rarely successful except in confined areas such as islands or at the very early stages of establishment. In most cases, founder populations have time to develop and spread before being even noticed and any attempt at eradication is possible. Several attempts have been undertaken in Austria and France to eradicate the Asian long-horned beetle *Anoplophora glabripennis*, in the few places it had invaded but new damage was recently observed (Tomiczek and Hoyer-Tomiczek 2007). Similarly, the populations of another Asian long-horned beetle, *A. chinensis*, are now widespread in northern Italy although an eradication programme is still ongoing (Maspero et al. 2007). Several options for what needs to be done to manage alien pests are available (e.g., Hulme 2006; Nentwig 2007). However, some are more realistic than others, particularly when it concerns invertebrates. We believe that the following two general strategies are of particular relevance to deal with alien invertebrate species in Europe:

1. Tackle pathways: this approach should consist of identifying and comparing the key pathways and commodities, estimating the risks associated with these, and restricting the riskiest (Simberloff 2005). Work et al. (2005) estimated the arrival rates of alien insects in the USA via four cargo pathways, viz. air cargo, refrigerated and non-refrigerated maritime cargo and US-Mexico border cargo. They determined the risks of invasion associated with each of these pathways and were able to evaluate the effectiveness of current efforts to monitor the arrival of alien species. We strongly recommend similar analyses on pathways of introduction in Europe (Hulme et al. 2008). The loss of border controls within the European Union mimics the situation in the United States. If a species enters a country, its further spread within the EU is uncontrolled. Although it is neither realistic to call for more border control measures in the public transport (although successfully employed in other parts of the world) nor desirable from an economic point of view, it may become necessary to restrict some sectors of international trade or at least to install more sophisticated quarantine or control measures at invasion hubs (e.g., ornamental plant trade, wood import, wood packages). Existing plant protection instruments and quarantine services do quite well, but obviously cannot cope with the vast amount of transported goods each and every day throughout Europe. For example, in 2001, the South African small hive beetle *Aethina tumida* (Nitidulidae) was detected in quarantine in Portugal and immediately eradicated. Another problem is that most invertebrate invaders arrive as eggs and larvae, which are usually not easily identifiable at species level. The development of new diagnostic tools based on barcode molecular sequences should also allow a faster and more accurate detection and identification of the aliens at the very early stage of invasion. Sharing of such tools with North American, Asiatic and peri-Mediterranean countries could also help to develop a global surveillance.

2. Reduce the taxonomic impediment and strengthen ecological research: one of the conclusions of DAISIE is the paucity of taxonomic expertise on terrestrial invertebrates in Europe. To reduce this taxonomic impediment is a prerequisite to deal with biological invasions. Although much data has been accumulated recently on biological invasions, we clearly need more information on the ecological effects of single species, particularly invertebrates. We know next to nothing about their effects on native invertebrate communities.

5.9 Future Expected Trends

The exponential increase in the establishment of alien invertebrates observed in Europe will probably continue in the next decades due to expanding markets, globalisation, increasing amount of transported goods and transporting agents. This is equally true for alien species from outside Europe as well as within Europe. Even more remote areas will become easily reachable in a near future. Commodities presently considered of less importance may provide new pathways in relation with new consumer behaviour and the facilitation of transcontinental human migration. A recent study on interception data showed that bonsais carry a more diverse alien fauna, including large xylophagous beetles, nematodes, aphids and scales, than timber (Roques and Auger-Rozenberg 2006). Pathways as diverse as the trade in aquarium plants and in reptile pets (González-Acuña et al. 2005) or fruits taken back home during international flights (Liebhold et al. 2006) are other opportunities of translocation for alien invertebrates.

Global warming is also likely to promote the arrival and survival of alien species, allowing more invertebrates from subtropical and possibly tropical areas to survive under European winter conditions at least locally, e.g., along the Mediterranean coast. The recent arrival and establishment of several tropical species associated with palms is illustrative of this process. Since 1993, 31 palm pests were recorded, among them a Castniidae moth from South America, *Paysandisia archon* (Montagud Alario and Rodrigo Coll 2004), and the red palm weevil *Rhynchophorus ferrugineus* from Melanesia, which successfully colonised southern France, Corsica, Italy, continental Greece, Crete and Cyprus from 2004 to 2006 (Rochat et al. 2006).

Finally, there is an urgent need for studies to assess the ecological impact of alien invertebrates in order to define appropriate conservation measures to preserve natural ecosystems. Most of these species appear to live in man-made habitats, and only a small proportion of them seem to have colonised natural ecosystems yet. However, whether our data reflect the reality or just a paucity of investigations in natural habitats compared to man-made habitats remains unknown.

Acknowledgements The support of this study by the European Commission's Sixth Framework Programme projects DAISIE (Delivering alien invasive species inventories for Europe, contract SSPI-CT-2003-511202) and ALARM (Assessing large scale environmental risks for biodiversity with tested methods; integrated project GOCE-CT-2003-506675) is gratefully acknowledged. We also thank very much the numerous collaborators having supplied data at taxonomic and national level:

C. Cocquempot, M.Martinez, A. Cœur d'Acier, F. Dorkeld, P.O. Cochard, A. Migeon, S. Augustin, D. Sauvard, M. Navajas, P. Zagatti, A. Yart, C. Villemant, P. Reynaud, J.C. Streito, J.F. Germain, D. Morin, A. Bertrand, E. Sechet, E. Çota, A.C. Mailleux, R. Tomov, K. Trencheva, D. Pilarska, P. Stoev, V. Skuhravy, M. Skuhravá, M. Kalapanida, G. Pellizzari, M. Faccoli, F. Lakatos, O. Dumčius, D. Mifsud, B. Ivanov, S. Naceski, I. Papazova-Anakieva, B. Økland, D. Roganovič, W. Solarz, M.R. Paiva, M. Branco Simoes, C. Tavares, N. Olenici, M. Glavendekić, O. Petrović- Obradović, S. Gomboć, G. Seljak, N. Pérez, A. Estrada-Peña, Å. Lindelöw, L.G. Moraal, P. Lambdon.

References

Austin JW, Szalanski AL, Scheffrahn RH, Messenger MT, Dronnet S, Bagnères AG (2005) Genetic evidence for the synonymy of two *Reticulitermes* species: *Reticulitermes flavipes* (Kollar) and *Reticulitermes santonensis* (Feytaud). Ann Entomol Soc Am 98:395–401

Beaucornu JC, Launay H (1990) Les puces de France et du Bassin Méditerranéen Occidental. Faune de France 76, Lechevallier, Paris

Bergersen R, Olsen KM, Djursvoll P, Nilssen AC (2006) Centipedes (Chilopoda) and millipedes (Diplopoda) in North Norway. Norw J Ent 53:23–38

Bernardinelli I, Zandigiacomo P (2000) Prima segnalazione di *Corythucha arcuata* (Say) (Heteroptera, Tingidae) in Europa. Inform Fitopat 50:47–49

Boag B, Yeates GW (2001) The potential impact of the New Zealand flatworm, a predator of earthworms, in western Europe. Ecol Appl 11:1276–1286

Brussino G, Bosio G, Baudino M, Giordano R, Ramello F, Melika G (2002) Pericoloso insetto esotico per il castagno europeo. Inform Agrar 58:59–61

De La Rùa P, Serrano J, Galian J (2002) Biodiversity of *Apis mellifera* populations from Tenerife (Canary Islands) and hybridisation with east European races. Biol Conserv 11:59–67

Della Giustina W (1989) Homoptères Cicadellidae. Vol 3, Compl Faune de France 73, Lechevallier, Paris

Dominguez J, Edwards CA, Dominguez J (2001) The biology and population dynamics of *Eudrilus eugeniae* (Kinberg) (Oligochaeta) in cattle waste solids. Pedobiologia 45:341–353

Drozdz J, Demiaszkiewicz AW, Lachowicz J (2003) Expansion of the Asiatic parasite *Ashworthius sidemi* (Nematoda, Trichostrongylidae) in wild ruminants in Polish territory. Paras Res 89:94–97

Dvořák L, Kupka J (2007) The first outdoor find of an American snail *Zonitoides arboreus* (Say, 1816) from the Czech Republic. Malacol Bohem 6:1–2

Essl F, Rabitsch W (eds) (2002) Neobiota in Österreich. Umweltbundesamt, Wien

Fabre JP, Auger-Rozenberg MA, Chalon A, Boivin S, Roques A (2004) Competition between exotic and native insects for seed resources in trees of a Mediterranean forest ecosystem. Biol Invasions 6:11–22

Faccoli M (1998) The North American *Gnathotrichus materiarius* (Fitch) (Coleoptera Scolytidae): an ambrosia beetle new to Italy. Redia 81:151–154

Geiter O, Homma S, Kinzelbach R (2002) Bestandaufnahme und Bewertung von Neozoen in Deutschland. Forschungsbericht 296, Umweltbundesamt, Berlin

Gjelstrup P, Møller AP (1985) A tropical mite, *Ornithonyssus bursa* (Berlese, 1888) (Macronyssidae, Gamasida) in Danish swallow (*Hirundo rustica*) nests, with a review of mites and ticks from Danish birds. Ent Meddel 53:119–125

González-Acuña D, Beldoménico PM, Venzal JM, Fabry M, Keirans JE, Guglielmone AA (2005) Reptile trade and the risk of exotic tick introductions into southern South American countries. Exp Appl Acar 35:335–339

Goulson D (2003) Effects of introduced bees on native ecosystems. Annu Rev Ecol Syst 34:1–26

Graff O (1954) Die Regenwurmfauna im östlichen Niedersachsen und in Schleswig-Holstein. Beitr Naturk Niedersachs 7:48–56

Griffiths DA, Bowman CE (1981) World distribution of the mite *Varroa jacobsoni*, a parasite of honeybees. Bee World 62:154–163

Grubini TD, Ontrec L, Buljak TG, Blümel S (2007) The occurrence and distribution of potato cyst nematodes in Croatia. J Pest Sci 80:21–27

Haxaire J, Bouguet JP, Tamisier JP (2006) *Vespa velutina* Lepeletier, 1836, une redoutable nouveauté pour la Faune de France. Bull Soc Ent Fr 111:194

Hérard F, Ciampitti M, Maspero M, Krehan H, Benker U, Boegel C, Schrage R, Bouhot-Delduc L, Bialooki P (2006) *Anoplophora* spp. in Europe: infestations and management process. EPPO Bull 36:470–474

Hill M, Baker R, Broad G, Chandler PJ, Copp GH, Ellis J, Jones D, Hoyland C, Laing I, Longshaw M, Moore N, Parrott D, Pearman D, Preston C, Smith RM, Waters R (2005) Audit of non-native species in England. Research Report 662, English Nature, Peterborough

Hornung E, Tothmérész B, Magura T, Vilisics F (2007) Changes of isopod assemblages along an urban- suburban- rural gradient in Hungary. Eur J Soil Biol 43:158–165

Hulme PE (2006) Beyond control: wider implications for the management of biological invasions. J Appl Ecol 43:835–847

Hulme PE, Bacher S, Kenis M, Klotz S, Kühn I, Minchin D, Nentwig W, Olenin S, Panov V, Pergl J, Pyšek P, Roque A, Sol D, Solarz W, Vilà M (2008) Grasping at the routes of biological invasions: a framework for integrating pathways into policy. J Appl Ecol 45:403–414

Kenis M, Rabitsch W, Auger-Rozenberg MA, Roques A (2007) How can alien species inventories and interception data help us prevent insect invasions? Bull Ent Res 97:489–502

Kenis M, Auger-Rozenberg MA, Roques A, Timms L, Péré C, Cock MJW, Settele J, Augustin S, Lopez-Vaamonde C (2008) Ecological impact of invasive alien insects – a world review. Biol Invasions DOI 10.1007/s 10530-008-9318-y

Kobelt M, Nentwig W (2008) Alien spider introductions to Europe supported by global trade. Diversity Distrib 14:273–280

Kontschán J (2004) *Reductoniscus costulatus* Kesselyák, 1930, a new isopod species from Hungary. Folia Hist Nat Mus Matraensis 28:89–90

Küchle M, Knorr HLJ, Medenblik-Frysch S, Weber A, Bauer C, Naumann GOH (1993) Diffuse unilateral subacute neuroretinitis syndrome in a German most likely caused by the raccoon roundworm, *Baylisascaris procyonis*. Graefe's Arch Clin Exp Ophthalmol 231:48–51

Levinson H, Levinson A (1994) Origin of grain storage and insect species consuming desiccated food. J Pest Sci 67:47–60

Liebhold AM, Work TT, McCullough DG, Cavey JF (2006) Airline baggage as a pathway for alien insect species invading the United States. Am Ent 52:48–54

Maspero M, Cavalieri G, D'Angelo G, Jucker C, Valentini M, Colombo M, Herard F, Lopez, Ramualde N, Ciampitti M, Caremi G, Cavagna B (2007) A*noplophora chinensis* eradication programme in Lombardia. www.eppo.org/QUARANTINE/anoplophora_chinensis/chinensis_IT_2007.htm. Cited Sept 2007

Migeon A (2005) A new spider mite pest in France, *Tetranychus evansi* Baker and Pritchard. Phytoma 579:38–43

Miller N, Estoup A, Toepfer S, Bourguet D, Lapchin L, Derridj S, Kim KS, Reynaud P, Furlan F, Guillemaud T (2005) Multiple transatlantic introductions of the western corn rootworm. Science 310:992

Montagud Alario S, Rodrigo Coll I (2004) *Paysandisia archon* (Burmeister, 1880) (Lepidoptera, Castniidae): nueva plaga de palmáceas en expansión. Phytoma España 157:40–53

Nentwig W (ed) (2007) Biological invasions. Ecological Studies 193, Springer, Berlin

NOBANIS (2007) North European and Baltic network on invasive alien species. www.nobanis.org. Cited Sept 2007

Novobilský A, Horáčková E, Hirtová L, Modrý D, Koudela B (2007) The giant liver fluke *Fascioloides magna* (Bassi 1875) in cervids in the Czech Republic and potential of its spreading to Germany. Parasit Res 100:549–553

Noyes J (2007) Universal Chalcidoidea database. www.nhm.ac.uk/researchcuration/projects/chalcidoids/. Cited Sept 2007

Ødegaard F, Tømmerås BA (2000) Compost heaps - refuges and stepping-stones for alien arthropod species in northern Europe. Diversity Distrib 6:45–59

Parola P (2004) Tick-borne rickettsial diseases: emerging risks in Europe. Comp Immun Microb Infect Dis 27:297–304

Rochat D, Chapin E, Ferry M, Avand-Faghih A, Brun L (2006) Le charançon rouge du palmier dans le bassin méditerranéen. Phytoma 595:20–24

Roques A, Auger-Rozenberg MA (2006) Tentative analysis of the interceptions of nonindigenous organisms in Europe during 1995–2004. EPPO Bull 36:490–496

Roll E, Dayan T, Simberloff D (2007) Non-indigenous species in Israel and adjacent areas. Biol Invasions 9:629–643

Roques A, Skrzypczynska M (2003) Seed-infesting chalcids of the genus *Megastigmus* Dalman (Hymenoptera: Torymidae) native and introduced to Europe: taxonomy, host specificity and distribution. J Nat Hist 37:127–238

Šefrová H, Laštůvka Z (2005) Catalogue of alien animal species in the Czech Republic. Acta Univ Agric Silv Mendelianae Brunensia 53:151–170

Servadei A (1966) Un tingide neartico comparso in Italia (*Corythuca ciliata* Say). Boll Soc Ent Ital 96:94–96

Simberloff D (2005) The politics of assessing risk for biological invasions: the USA as a case study. Trends Ecol Evol 20:216–222

Šnábel V, Permin A, Magwisha HB, et al. (2001) On the species identity of *Ascaridia galli* (Schrank, 1788) and *Ascaridia dissimilis* (Perez Vigueras, 1931): a comparative genetic study. Helminthol 38:221–224

Stojcevic D, Mihaljevic Z, Marinculic A (2004) Parasitological survey of rats in rural regions of Croatia. Vet Med Czech 49:70–74

Taylor SJ, Tescari G, Villa M (2001) A Nearctic pest of Pinaceae accidentally introduced into Europe: *Leptoglossus occidentalis* (Heteroptera: Coreidae) in northern Italy. Ent News 112:101–103

Tomiczek C, Hoyer-Tomiczek U (2007) Der asiatische Laubholzbockkäfer (*Anoplophora glabripennis*) und der Citrusbock (*Anoplophora chinensis*) in Europa – ein Situationsbericht. Forstschutz Akt 38:2–5

Van Lenteren JC, Bale J, Bigler F, Hokkanen HMT, Loomans AJM (2006) Assessing risks of releasing exotic biological control agents of arthropod pests. Annu Rev Ent 51:609–634

Viggiani G (1994) Recent cases of interspecific competition between parasitoids of the family Aphelinidae (Hymenoptera: Chalcidoidea). Norw J Agric Sci Suppl 16:353–359

Wetterer JK, Espalader X, Wetterer AL, et al. (2006) Long-term impact of exotic ants on the native ants of Madeira. Ecol Ent 31:358–368

Wittenberg R (2006) Invasive alien species in Switzerland. An inventory of alien species and their threat to biodiversity and economy in Switzerland. Environmental Studies 29, Federal Office for the Environment, Bern

Work TT, McCullough DG, Cavey JF, Komsa R (2005) Arrival rate of non-indigenous insect species into the United States through foreign trade. Biol Invasions 7:323–332

Chapter 6
Alien Invertebrates and Fish in European Inland Waters

Francesca Gherardi, Stephan Gollasch, Dan Minchin, Sergej Olenin, and Vadim E. Panov

6.1 The Vulnerability of Inland Waters to Alien Species

It seems axiomatic that rivers, lakes, freshwater marshes, and other inland wetlands have an infinite value to humankind. They contribute for 20% (about US$6.6 trillion) to the estimated annual global value of the entire biosphere (Costanza et al. 1997). High-quality water has also become a strategic factor that allows for the viability and development of an increasing number of countries affected by both climate change and rising water-demand. All this justifies the current concern about the degradation of freshwater systems leading to rapid extinctions of organisms – in some cases even matching the declines recorded in tropical forests (Ricciardi and Rasmussen 1999).

There is general consensus today that some alien species will continue to be major drivers of degradation and loss of aquatic systems (Sala et al. 2000; Millennium Ecosystem Assessment 2005). The vulnerability of inland waters to bioinvasions is due to several factors (Gherardi 2007a), including the higher intrinsic dispersal ability of freshwater species compared with terrestrial organisms (Beisel 2001) and the strong impact of both human disturbance (Ross et al. 2001) and altered seasonal temperature regimes (Eaton and Scheller 1996). Species introduction into inland waters is associated with the intensity with which humans utilise these systems for recreation, food sources, and commerce (Rahel 2000), being a direct consequence of economic activity and trade globalisation that benefit millions worldwide (Lodge and Shrader-Frechette 2003). This situation has generated a conflict between two often competing goals – increasing economic activity and protecting the environment, which makes it difficult to decision-makers to develop policies aimed at containing the spread of aliens and mitigating the ecological risks they pose (Gherardi 2007a).

6.2 Diversity of Animal Alien Species in European Inland Waters

European inland waters have been affected since centuries by a high rate of colonisation by animal alien species. From the DAISIE database, integrated with the data collected by the IMPASSE consortium (www.hull.ac.uk/hifi/IMPASSE/), we extracted a total of 296 and 136 invertebrate and fish species alien to the entire Europe and alien to at least one European country but native to others, respectively (Fig. 6.1). Species of unknown origin which cannot be ascribed as being native or alien have been excluded from the analysis. A large component comprises chordates (mainly bony fishes) and crustaceans, whereas molluscs, flatworms, and segmented worms are less well represented. These numbers certainly underestimate the effective diversity of animal aliens in European inland waters. In fact, notwithstanding the increased scientific interest for biological invasions in the last decade and the surge of research focused on the identification of alien species in fresh waters (Gherardi 2007b), there is still a gap in knowledge regarding invertebrate taxa. On the other hand some taxa, such as fishes, have attracted the most scientific attention due to their higher visibility and greater economic importance. Additionally, the area of origin of some species is still unknown. Finally, human-aided movement of species is a more widespread phenomenon if we count the many translocation events between geographically distinct drainage basins within each country. For instance, through human intervention several species endemic to northern Italy (e.g., *Alburnus arborella*, *Cobitis taenia*, and *Chondrostoma genei*) have achieved a nearly pan-Italian distribution (Bianco 1998).

The majority of animal alien species originates from Asia, the others from the Americas (in particular North America), Africa, and Oceania (Fig. 6.2). Of the 36

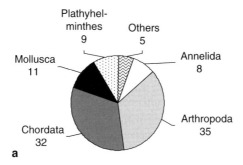

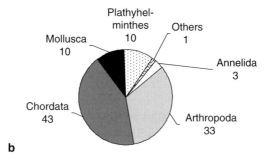

Fig. 6.1 Percentage per taxon of animal alien species recorded in European inland waters, distinguished between species (n = 296) alien to entire Europe (**a**) and species (n = 136) alien to at least one European country but native to others (**b**)

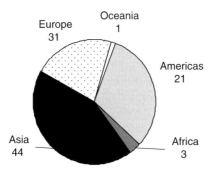

Fig. 6.2 Percentage distribution of animal alien species recorded in European inland waters per donor continent (n = 434). Species whose native range includes two or more continents were tallied more than once

European countries here examined, 19 bony fish species were recorded from 10 or more countries, such as the rainbow trout *Oncorhynchus mykiss* (31 countries), the grass carp *Ctenopharyngodon idella* (27 countries), and the brook trout **Salvelinus fontinalis** (26 countries). Among the other taxa, the signal crayfish *Pacifastacus leniusculus* was recorded from 25 countries, the zebra mussel ***Dreissena polymorpha*** from 22, the gastropod *Potamopyrgus antipodarum* from 18, the nematode ***Anguillicola crassus*** from 17, the flatworm *Bdellocephala punctata* from 13, and the jellyfish *Craspedacusta sowerbyi* from 12. With two notable exceptions (***D. polymorpha*** and *C. sowerbyi*), the pan-European success of these species seems to be related to frequent "inoculations" due to angling or aquaculture interest or to their association as non-targets with some "desired" species. Their spread, thus, seems to be a function of the intensity and diversity of imports as a consequence of "propagule pressure" (Williamson 1996; Colautti and MacIsaac 2004).

No European country is immune to the colonisation of animal alien species but their diversity varies widely across Europe, ranging from four in Iceland to 129 in the Ukraine. On the one hand, this large variance can be an artifact due to the diverse level of knowledge and of field survey efforts across countries. On the other hand, countries may differ in the extension of their basins (e.g., Malta vs. Germany) and in their exposure to species introductions due to their reciprocal connectivity/isolation, the presence/absence of harbours subject to intercontinental traffic, or the intensity of voluntary introductions of species due to sport and commercial fisheries or aquaculture (e.g., Italy, the Iberian Peninsula, and British Isles).

6.3 The Long History of Species Introduction

Human modifications of freshwater systems through species introduction have a long history in Europe, starting well before the 20th century. Not all of them have been recognised as causing harm, but some introduced species progressively integrated to become keystone species and are now protected under law. Notable examples include various species of crayfish, such as *Astacus astacus*: as reported by Linnaeus (1746: 358) and confirmed by Pontoppidan (1775: 175), the noble crayfish *Astacus astacus* seems to have been imported into Sweden from Russia by the Swedish King John III after 1568 and later into Finland (Westman 1973).

The introduction of freshwater animal alien species, mainly fishes, intensified after the mid 19th century under the promotion in Europe by some "acclimatisation societies" (Copp et al. 2005). In Russia, for instance, almost 250 introductions that included 35 fish and 13 invertebrate species were annually conducted by the Russian Society of Acclimatisation since 1857 (see references in Alexandrov et al. 2007). Similarly, the Society for Acclimatisation of Animals, Birds, Fishes, Insects and Vegetables, established in 1860 in Britain, was responsible for the majority of the introductions from continental Europe. These included pikeperch *Sander lucioperca* and the Danube catfish *Silurus glanis* (Lever 1977). In Germany, fish importations, pioneered by Max von dem Borne (1826–1894), commenced in 1882 with the introduction of North American species (e.g., **Salvelinus fontinalis**, *Ameiurus nebulosus*, and *Micropterus dolomieu*). In Italy, the introduction of alien fishes, in particular North American species (e.g., *M. dolomieu*, *Ameiurus melas*, *Ictalurus punctatus*, and *Salvelinus namaycush*), began in the mid-19th century with most being introduced during 1897–1898 by two centers of ichthyology based in Brescia and Rome (Bianco 1998). The main motivation for these introductions was to increase the supply and diversity of aquatic food and game resources, both through natural production and in the then recently developed field of fish husbandry (Copp et al. 2005) adopting new fish hatchery techniques (Wilkins 1989). The novelty and ornamental value of the new species also played a role, as explicitly stated in the aims of the UK Acclimatisation Society ("The introduction, acclimatisation, and domestication of all animals, birds, fishes, insects, and vegetables, whether useful or ornamental"; Lever 1977).

During the 20th century, the overall number of animal alien species arriving in European inland waters greatly increased (Fig. 6.3). This "explosion" (in the sense of Elton 1958) of alien crustaceans and fishes, particularly evident after the 1940s, was likely fostered by the greater mobility and trade following the Second World War.

6.4 Pathways of Introductions

A wide range of human activities became responsible for animal alien species expansion arising from extensive fish culture and sport fishing (30%) and intensive aquaculture (27%), followed by the passive transportation by vessels (25%), ornamental use (with introduction to lakes on private estates, small garden ponds, and indoor aquaria; 9%) and dispersal through canals (8%) (Fig. 6.4a). The only species to be imported for biological control were *Gambusia* spp. (1%); the eastern mosquitofish *G. holbrooki*, for instance, was introduced to Spain in 1920, Germany in

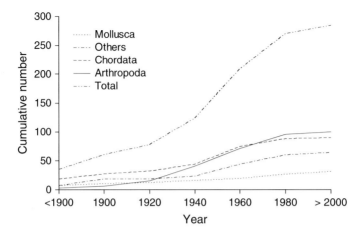

Fig. 6.3 Increase with time in the frequency of the animal alien species in European inland waters. Dates refer to the exact or approximate year of introduction into the wild or, when this datum is absent, to the year of the first record in the published literature. The date is missing for 22 species

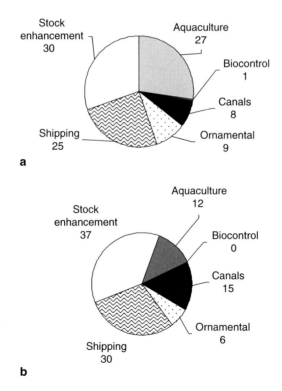

Fig. 6.4 Percentage of pathways of introduction of animal alien species in European inland waters, distinguished between species (n = 262) alien to the entire Europe (a) and species (105) alien to at least one European country but native to others (b). Species that show two or more pathways of introduction have been tallied more than once

1921, Italy in 1922 (imported from Spain) and has since then spread to many warm-water systems (Copp et al. 2005).

The role that the diverse pathways have had in introducing species from one European country to the others seems to be different (Fig. 6.4b). The importance of transportation through vessels and dispersal through canals is obviously higher (30% and 15%, respectively), whereas the frequency of escapes from aquaculture (12%) decreases and the effect of stock enhancement and ornamental use remains unchanged (37% and 6%, respectively). However, a realistic assessment of the diverse pathways of species introduction should be made on a case-by-case basis. For instance, in Germany the most frequent pathways involve the gradual incremental spread of generations through artificial canals that act as corridors (Gollasch and Nehring 2006). More direct arrivals took place as stowaways on or in the hulls of ships (Gollasch 2007), which seems to be less important in Italy (the Alps impede the construction of canals and navigation from the ports in the Mediterranean to inland waters is scarce). Motives may also vary between countries for each single species. For instance, as reported by Copp et al. (2005), the introduction of pumpkinseed (*Lepomis gibbosus*) in the late 19th and early 20th century resulted from deliberate releases for angling in France (e.g., Künstler 1908) but as escapes of ornamental fish in England (Copp et al. 2002), Slovenia (Povž and Šumer 2005), and Spain (García-Berthou and Moreno-Amich 2000).

Many introductions of animal alien species were intended for the development of fisheries and for angling via stock enhancement. The stocking of new species was a common practice in Europe in the 1960s and 1970s. This was aimed at promoting fishery diversity to counterbalance the perceived decline in the status of many fisheries (Welcomme 1988) and was accompanied by the introduction of large quantities of crustaceans, in particular amphipods (e.g., *Gammarus tigrinus*, *G. pulex*, *Pontogammarus robustoides*, *Obesogammarus crassus*, and *Echinogammarus ischnus*; Jazdzewski 1980), to increase commercial fish production. Several introduced species were contaminants of licensed fish consignments (e.g., top-mouth gudgeon, accidentally imported from Asia in 1960 to a pond in Romania along with a deliberate introduction of young of Chinese carp; Bănărescu 1964) or as parasites of the target species (e.g., the Platyhelminth **Gyrodactylus salaris**). In the European part of the former USSR (presently Belarus, Estonia, Latvia, Lithuania, Moldova, Russia, and Ukraine), fish release (under the historical term of "acclimatisation") were even "tasks from government" in annual and 5-year Soviet State Plans that had the character of state laws; large quantities of crustaceans from the Ponto-Caspian region were also moved during 1940–1970, as live food for commercial fish species (nine species of Mysidacea, seven species of Cumacea, and 17 species of Amphipoda; Karpevich 1975). Since those times, several fish species have been illegally introduced to increase the diversity of the target species for anglers (e.g., large fish species, such as *Silurus glanis*) and others have been released as bait and forage fishes (e.g., the bleak *Alburnus alburnus* in the Iberian Peninsula; Elvira and Almodóvar 2001). Most planned introductions were conducted without any scientific basis. In the Iberian Peninsula, apparent "vacant niches" were filled by stocking piscivorous fish, such as *Esox lucius*, into newly constructed reservoirs (Godinho et al. 1998). Some introductions were aimed at alleviating poverty in lowly developed areas, e.g., the red swamp crayfish, *Procambarus clarkii*, released in the rice fields of Andalusia, Spain (Gherardi 2006). Unfortunately, several of these attempts to reduce societal problems have had unexpected negative consequences, ultimately causing more problems than they have solved (the "Frankenstein effect" of Moyle et al. 1986).

The diversification of the ornamental fish trade has been responsible for the appearance in the wild of an increasing number of aliens, either fishes (e.g., the North American fathead minnow *Pimephales promelas*, first recorded to reproduce in small enclosed private waters of the UK; Copp et al. 2007) or some molluscs used to clean aquaria by the scraping of their radula (e.g., *Melanoides tuberculata*; Cianfanelli et al. 2007). The impact of ornamental species on freshwater communities is confirmed by recent reports of the parthenogenetic marbled crayfish (*Procambarus* sp.) in the wild in Germany and the Netherlands (Souty-Grosset et al. 2006), and of *Ameiurus catus* (Britton and Davies 2006) and *Aristichthys nobilis* (Britton and Davies 2007) in Britain.

Finally, the interconnection of river basins by means of canals has facilitated the range expansion of many species within Europe either or both aided by active movement or by ship transport (e.g., Jażdżewski 1980; Bij de Vaate et al. 2002; Galil et al. 2007; Panov et al. 2007a). Numerous canals have been constructed during the last two centuries in Europe to promote trade, forming complex European networks of inland waterways which connect 37 countries in Europe and beyond (Galil et al. 2007): in Germany alone there are 1,770 km of inland waterways (Tittizer 1996). Several studies showed the penetration of Ponto-Caspian species via three important canal corridors spreading either actively or passively throughout Europe (Bij de Vaate et al. 2002; Ketelaars 2004; Galil et al. 2007). Transmission via the northern corridor involved the zebra mussel **Dreissena polymorpha** and the cladocerans **Cercopagis pengoi**, *Evadne anonyx*, and *Cornigerius maeoticus* (Panov et al. 2007b), the central corridor the amphipod *Chelicorophium curvispinum*, and the southern corridor the amphipod **Dikerogammarus villosus**, isopod *Jaera istri*, mysid *Limnomysis benedeni*, and the polychaete *Hypania invalida* (Bij de Vaate et al. 2002). The arrival of Ponto-Caspian species such as **Dreissena polymorpha** and **Cercopagis pengoi** at the main harbours of the North and the Baltic Seas was followed by their "jump" in ballast to the Laurentian Great Lakes (MacIsaac et al. 2001; Panov et al. 2007b).

6.5 The Multilevel Impact of Freshwater Animal Alien Species

Not all alien species can be considered to have negative impacts; in fact some of them are universally recognised as being of benefit (Ewel et al. 1999). Also among those species that have been introduced by humans and were able to form self-sustaining populations, as many as 80–90% (Williamson 1996) – or less than 75%, at least for some taxa (Jerscke and Strayer 2005) – may actually have minimal detectable effects on the environment. So, the fraction of aliens that yield problems may be small, but although few species they have had catastrophic impacts on the environment and human economy. For some (but not all) species recorded, a multilevel impact (Parker et al. 1999) on the recipient communities and ecosystems is known, even if seldom quantified (Gherardi 2007b), and includes (1) competitive superiority over native species, possibly leading to extinction, e.g., **Dikerogammarus villosus** (Dick and Platvoet 2000); (2) hybridisation with native species with the consequent reduction of genetic diversity, e.g., *Carassius auratus* and *C. carpio* (Hänfling et al. 2005); (3) disruption of the pristine interactions between species and of the existing food web links, e.g., rainbow trout (*O. mykiss*) (Nyström et al. 2001) and *Pacifastacus leniusculus* (Nyström 1999); (4) habitat modification and changes to ecosystem functioning, e.g., **Procambarus clarkii** (Gherardi 2007c) and *Chelicorophium curvispinum* (Devin et al. 2005); (5) transmission of parasites and diseases, e.g., **Anguillicola crassus** (Kirk 2003) and **Pseudorasbora parva** (Gozlan et al. 2005); and (6) damages to socio-economics, recreation, human health, and well-being, e.g., **Dreissena polymorpha** (Karatayev et al. 1997). Conversely, we are still ignorant about the long-term ecological and evolutionary feedbacks between invasive species and the invaded communities and ecosystems (Strayer et al. 2006).

At the river basin level, some animal aliens have dominated aquatic communities, as in the case of the red swamp crayfish in southern Europe (Gherardi 2006). Large catchments may function as hotspots of alien diversity. For instance, in the River Rhine, more than 95% of macroinvertebrates consists of aliens (Bij de Vaate et al. 2002). These will have originated from different biogeographic regions (North America, e.g., *Gammarus tigrinus* and *Orconectes limosus*, Mediterranean, the freshwater shrimp *Atyaephyra desmaresti*, and Ponto-Caspian region, e.g., *Gammarus roeseli* and **Dikerogammarus villosus**; Beisel 2001). Biotic homogenisation among basins, defined as the ecological process by which formerly disparate biota lose biological distinctiveness at any level of organisation (McKinney and Lockwood 1999), has evolved as the results of global alien species spread and the extinction of endemic species. The fish fauna in the Iberian Peninsula, for instance, showed in 2001 an increased similarity of over 17% from the original pristine situation (Clavero and Garcia-Berthou 2006). Similar processes are taking place elsewhere in Europe, although their quantification is still missing.

6.6 Concluding Remarks

Several animal aliens are today affecting freshwater communities, imperiling native species, altering ecosystem processes, and causing damage to human endeavors. Recognizing these threats certainly lays the strongest and possibly the only ethical basis for the concern that scientists, laypeople, and institutions have today about the problem of introduced species (Simberloff 2003). This general awareness of the detrimental effects of aliens is in the process of being translated into implemented policies aimed at preventing new undesirable introductions (e.g., Council Regulation No 708/2007 concerning use of alien and locally absent species in aquaculture), responding quickly to newly discovered alien species, and controlling the most damaging established aliens (e.g., European Strategy on Invasive Alien Species, CEC 2002). However, despite the progress made in the last decade in the comprehension of biological invasions, current efforts on this front still suffer from a lack of basic scientific information about the extent and distribution of alien diversity, particularly in inland water systems. A much greater and more urgently applied investment to address these deficiencies is thus warranted.

Acknowledgements This study has been supported by the European Commission Sixth Framework Programme projects DAISIE: Delivering alien invasive species inventories for Europe, contract SSPI-CT-2003-511202 (VP, SG, SO, DM), ALARM: Assessing large scale environmental risks for biodiversity with tested methods, contract GOCE-CT-2003-506675 (VP, SO, DM), and IMPASSE: Environmental impacts of alien species in aquaculture, contract SSP-2005-5A (FG, SG, SO).

References

Alexandrov B, Boltachev A, Kharchenko T, Lyashenko A, Son M, Tsarenko P, Zhukinsky V (2007) Trends of aquatic species invasions in Ukraine. Aquat Invasions 2:215–242

Bănărescu P (1964) Pisces – Osteichthyes. Fauna Republicii Populare Romîne. Vol. XIII. Ed Acad Rep Pop Rom, Bucureti

Beisel J-N (2001) The elusive model of a biological invasion process: time to take differences among aquatic and terrestrial ecosystems into account? Ethol Ecol Evol 13:193–195

Bianco PG (1998) Freshwater fish transfers in Italy: history, local modification of fish composition, and a prediction on the future of native populations. In: Cowx IJ (ed) Stocking and introductions of fishes. Blackwell, Oxford. 165–197

Bij de Vaate A, Jażdżewski K, Ketelaars HAM, Gollasch S, Van der Velde G (2002) Geographical patterns in range extension of Ponto-Caspian macroinvertebrate species in Europe. Can J Fish Aquat Sci 59:1159–1174

Britton JR, Davies JM (2006) First record of the white catfish *Ameiurus catus* in Great Britain. J Fish Biol 69:1236–1238

Britton JR, Davies JM (2007) First U.K. recording in the wild of the bighead carp *Hypophthalmichthys nobilis*. J Fish Biol 70:1280–1282

Cianfanelli S, Lori E, Bodon M (2007) Alien freshwater molluscs in Italy and their distribution. In: Gherardi F (ed) Biological invaders in inland waters: profiles, distribution, and threats. Springer, Dordrecht. 103–121

Clavero M, García-Berthou E (2006) Homogenization dynamics and introduction routes of invasive freshwater fish in the Iberian Peninsula. Ecol Appl 16:2313–2324

Colautti RI, MacIsaac HJ (2004) A neutral terminology to define 'invasive' species. Diversity Distrib 10:135–141

Copp GH, Fox MG, Kováč V (2002) Growth, morphology and life history traits of a coolwater European population of pumpkinseed *Lepomis gibbosus*. Arch Hydrobiol 155:585–614

Copp GH, Bianco PG, Bogutskaya NG, Erós T, Falka I, Ferreira MT, Fox MG, Freyhof J, Gozlan RE, Grabowska J, Kováč V, Moreno-Amich R, Naseka AM, Peňáz M, Povž M, Przybylski M, Robillard M, Russell IC, Stakėnas S, Šumer S, Vila-Gispert A, Wiesner C (2005) To be, or not to be, a non-native freshwater fish? J Appl Ichthyol 21:242–262

Copp GH, Templeton M, Gozlan RE (2007) Propagule pressure and the invasion risks of non-native freshwater fishes: a case study in England. J Fish Biol 71 (Suppl D):1–12

Costanza R, d'Arge R, de Groot R, Farber S, Grasso M, Hannon B, Limburg K, Naeem S, O'Neill RV, Paruelo J, Raskin RG, Sutton P, van den Belt M (1997) The value of the world's ecosystem services and natural capital. Nature 387:253–260

Devin S, Bollache L, Noël PN, Beisel J-N (2005) Patterns of biological invasions in French freshwater systems by non-indigenous macroinvertebrates. Hydrobiologia 551:137–146

Dick JTA, Platvoet D (2000) Invading predatory crustacean *Dikerogammarus villosus* eliminates both native and exotic species. Proc R Soc Lond 267:977–983

Eaton JG, Scheller RM (1996) Effects of climatic warming on fish thermal habitat in streams of the United States. Limnol Ocean 41:110–1115

Elton CS (1958) The ecology of invasions by animals and plants. Methuen, London

Elvira B, Almodóvar A (2001) Freshwater fish introductions in Spain: facts and figures at the beginning of the 21st century. J Fish Biol 59:323–331

Ewel JJ, O'Dowd DJ, Bergelson J, Daehler CC, D'Antonio CM, Gomez D, Gordon DR, Hobbs RJ, Holt A, Hopper KR, Hughes CE, Lahart M, Leakey RRB, Lee WG, Loope LL, Lorence DH, Louda SM, Lugo AE, Mcevoy PB, Richardson DM, Vitousek PM (1999) Deliberate introductions of species: research needs. Bioscience 49:619–630

Galil BS, Nehring S, Panov VE (2007) Waterways as invasion highways: impact of climate change and globalization. In: Nentwig W (ed) Biological invasions. Ecological Studies 193, Springer, Berlin. 59–74

García-Berthou E, Moreno-Amich R (2000) Introduction of exotic fish into a Mediterranean lake over a 90-year period. Arch Hydrobiol 149:271–284

Gherardi F (2006) Crayfish invading Europe: the case study of *Procambarus clarkii*. Mar Fresh Behav Physiol 39:175–191

Gherardi F (2007a) The impact of freshwater NIS: what are we missing? In: Gherardi F (ed) Biological invaders in inland waters: profiles, distribution, and threats. Springer, Dordrecht. 437–462

Gherardi F (2007b) Biological invasions in inland waters: an overview. In: Gherardi F (ed) Biological invaders in inland waters: profiles, distribution, and threats. Springer, Dordrecht. 3–25

Gherardi F (2007c) Understanding the impact of invasive crayfish. In: Gherardi F (ed) Biological invaders in inland waters: profiles, distribution, and threats. Springer, Dordrecht. 507–542

Godinho FN, Ferreira MT, Portugal e Castro MI (1998) Fish assemblage composition in relation to environmental gradients in Portuguese reservoirs. Aquat Living Resour 11:325–334

Gollasch S (2007) Marine vs. freshwater invaders: Is shipping the key vector for species introductions to Europe? In: Gherardi F (ed) Biological invaders in inland waters: profiles, distribution, and threats. Springer, Dordrecht. 339–345

Gollasch S, Nehring S (2006) National checklist for aquatic alien species in Germany. Aquat Invasions 1:245–269

Gozlan RE, St Hilaire S, Feist SW, Martin P, Kent ML (2005) Disease threat to European fish. Nature 435:1046

Hänfling B, Bolton P, Harley M, Carvalho GR (2005) A molecular approach to detect hybridisation between crucian carp (*Carassius carassius*) and non-indigenous carp species (*Carassius* spp. and *Cyprinus carpio*). Freshwater Biol 50:403–417

Jażdżewski K (1980) Range extensions of some gammaridean species in European inland waters caused by human activity. Crustaceana Suppl 6:84–106

Jerscke JM, Strayer DL (2005) Invasion success of vertebrates in Europe and North America. Proc Natl Acad Sci USA 102:7198–7202

Karatayev AY, Burlakova LE, Padilla DK (1997) The effects of *Dreissena polymorpha* (Pallas) invasion on aquatic communities in eastern Europe. J Shellfish Res 16:187–203

Karpevich AF (1975) Teorija i praktika akklimatizacii vodnykh vodoemov. Pishchevaya Promyshlennost, Moskva

Ketelaars HAM (2004) Range extensions of Ponto-Caspian aquatic invertebrates in continental Europe. In: Dumont HJ, Shiganova TA, Niermann U (eds) Aquatic invasions in the Black, Caspian, and Mediterranean seas. Kluwer, Dordrecht. 209–236

Kirk RS (2003) The impact of *Anguillicola crassus* on European eels. Fish Manage Ecol 10:385–394

Künstler J (1908) *Ameiurus nebulosus* et *Eupomotis gibbosus*. Bull Soc Acclimatation 238–244

Lever C (1977) The naturalised animals of the British Isles. Hutchinson, London

Linnaeus C (1746) Fauna Suecica. Salvius, Stockholm

Lodge DM, Shrader-Frechette K (2003) Nonindigenous species: ecological explanation, environmental ethics, and public policy. Conserv Biol 17:31–37

MacIsaac HJ, Grigorovich IA, Ricciardi A (2001) Reassessment of species invasion concepts: the Great Lakes basin as a model. Biol Invasions 3:405–416

McKinney ML, Lockwood JL (1999) Biotic homogenization: a few winners replacing many losers in the next mass extinction. Trends Ecol Evol 14:450–453

Millennium Ecosystem Assessment (2005) Ecosystem and human well-being: biodiversity synthesis. World Resources Institute, Washington, DC

Moyle PB, Li HW, Barton BA (1986) The Frankenstein effect: impact of introduced fishes on native fishes in North America. In: Strond RH (ed) Fish culture in fisheries management. American Fisheries Societies, Bethesda

Nyström P (1999) Ecological impact of introduced and native crayfish on freshwater communities: European perspectives. In: Gherardi F, Holdich DM (eds) Crayfish in Europe as alien species: How to make the best of a bad situation? Balkema, Rotterdam. 63–84

Nyström P, Svensson O, Lardner B, Brönmark C, Granéli W (2001) The influence of multiple introduced predators on a littoral pond community. Ecology 82:1023–1039

Panov VE, Dgebuadze YY, Shiganova TA, Filippov AA, Minchin D (2007a) A risk assessment of biological invasions in the inland waterways of Europe: the northern invasion corridor case study. In: Gherardi F (ed) Biological invaders in inland waters: profiles, distribution, and threats. Springer, Dordrecht. 639–659

Panov VE, Rodionova NV, Bolshagin PV, Bychek EA (2007b) Invasion biology of Ponto-Caspian onychopod cladocerans (Crustacea: Cladocera: Onychopoda). Hydrobiologia 590:3–14

Parker IM, Simberloff D, Lonsdale WM, Goodell K, Wonham M, Kareiva PM, Williamson MH, Von Holle B, Moyle PB, Byers JE, Goldwasser L (1999) Impact: toward a framework for understanding the ecological effects of invaders. Biol Invasions 1:3–19

Pontoppidan E (1775) The natural history of Norway. Linde, London

Povž M, Šumer S (2005) A brief review of non-native freshwater fishes in Slovenia. J Appl Ichthyol 21:316–318

Rahel FJ (2000) Homogenization of fish faunas across the United States. Science 288:854–856

Ricciardi A, Rasmussen JB (1999) Extinction rates of North American freshwater fauna. Conserv Biol 13:1220–1222

Ross RM, Lellis WA, Bennett RM, Johnson CS (2001) Landscape determinants of nonindigenous fish invasions. Biol Invasions 3:347–361

Sala OE, Chapin FS III, Armesto JJ, Berlow E, Bloomfield J, Dirzo R, Huber-Sannwald E, Huenneke L, Jackson RB, Kinzig A, Leemans R, Lodge DM, Mooney HA, Oesterheld M, Poff NL, Sykes MT, Walker BH, Walker M, Wall DH (2000) Biodiversity scenarios for the year 2100. Science 287:1770–1774

Simberloff D (2003) How much information on population biology is needed to manage introduced species? Conserv Biol 17:83–92

Souty-Grosset C, Holdich DM, Noël PY, Reynolds JD, Haffner P (2006) Atlas of crayfish in Europe. Muséum national d'Histoire naturelle, Paris

Strayer DL, Eviner VT, Jeschke JM, Pace ML (2006) Understanding the long-term effects of species invasions. Trends Ecol Evol 21:645–651

Tittizer T (1996) Vorkommen und Ausbreitung aquatischer Neozoen (Makrozoobenthos) in den Bundeswasserstrassen. In Gebietsfremde Tierarten. Auswirkungen auf einheimischen Arten, Lebensgemeinschaften und Biotope. Situationsanalyse. In: Gebhardt H, Kinzelbach R,

Schmidt-Fischer S (eds) Umweltministerium Baden Württemberg. Ecomed, Landsberg. 49–86

Welcomme RL (1988) International introductions of inland aquatic species. FAO Fisheries Technical Paper 294. FAO, Rome

Westman K (1973) The population of the crayfish *Astacus astacus* in Finland and the introduction of the American crayfish *Pacifastacus leniusculus* Dana. Freshwater Crayfish 1:41–55

Wilkins NP (1989) Ponds, passes and parcs: aquaculture in Victorian Ireland. Glendale Publishers, Northdale

Williamson M (1996) Biological invasions. Chapman & Hall, London

Chapter 7
Alien Marine Biota of Europe

Bella S. Galil, Stephan Gollasch, Dan Minchin, and Sergej Olenin

7.1 In the Beginning

The recognition that marine species in European coastal waters originate from other parts of the world has lagged, in some cases centuries, behind their postulated arrival. The NW Atlantic soft shell clam *Mya arenaria* is considered one of the earliest introductions: it was proposed that it had been brought in the 16th century intentionally as bait or food, unintentionally with in solid ships' ballast, with oysters (Hessland 1946). But it may have been brought earlier still: a find of *Mya* shells from the Kattegat, Denmark that was dated to 1245–1295 excited a debate on possible transport in Vikings' vessels (Petersen et al. 1992). Vikings aside, the history of marine biological invasions most likely began with the development of "global" maritime trade routes in the 16th century. Some of the first observations of vessel-transported alien species date back to the 17th and 18th century: live specimens of *Balanus tintinnabulum* were described and illustrated from a vessel coming from West Africa and wrecked off the Dutch coast in 1764 (Holthuis and Heerebout 1972). The first record of the east Pacific Megabalanus coccopoma in European waters dates to 1851, from a vessel in Le Havre, France (Kerckhof and Cattrijsse 2001). Indeed, Darwin (1854) himself suggested that barnacles had been transported as foulants on ship hulls, and he was the first to record the Indo West-Pacific *Balanus amphitrite* from the Mediterranean and the Portuguese coast. But it was the excitement attending the excavation of the Suez Canal that focused attention on possible mass incursion of alien marine biota from one sea to another. On the eve of the opening of the Canal Vaillant (1865: 97) argued that the cutting through the Isthmus of Suez offered an opportunity to examine the immigration of species and the mix of faunas from both seas. Indeed, within 20 years two Red Sea molluscs, the Gulf pearl oyster **Pinctada radiata** and *Cerithium scabrum*, were collected in Alexandria and Port Said (Monterosato 1878; Keller 1883).

7.2 Temporal Trends Since 1900

In total, 737 alien multicellular species have been recorded from European seas, the great majority, 569 species, were found in the Mediterranean, 200 from the Atlantic coast (Norway to the Azores, including Britain and Ireland), and 62 from the Baltic

Sea. The numbers of aliens recorded in European coastal waters per decade over the past century are given in Fig. 7.1.

Mediterranean Sea: Records of the 569 aliens reflect political crises, economic development and scientific interest in recording marine alien species. The few alien species recorded prior to the 20th century reflect most probably ignorance of the phenomenon coupled with lack of detailed marine biological surveys. The gap in the 1910s indicates the First World War and the political upheavals attending the breakup of the Ottoman Empire, whereas the dip in the 1940s relates to the devastation of the Second World War. The rather sharp decline in the number of records in the 1980s may be due to the closure of the Suez Canal and the impact of the Arab Oil Embargo that limiting the number of vessels transiting the Mediterranean. The increasing role of the Mediterranean as a hub of international commercial shipping, a surge in development of marine aquaculture farming over the last 25 years and the continuous enlargement of the Suez Canal, all contribute to increase of aliens since the 1950s. The greater number of records in the 1920s and 1970s reflects the publication of the results of 'The Cambridge Expedition to the Suez Canal', and the joint program by the Smithsonian Institution, the Hebrew University of Jerusalem, and the Sea Fisheries Research Station, Haifa, respectively. Though no concerted regional research effort was undertaken since, there seem to be slightly more introductions recorded in the first years of the 21st century (83 species) than in the 1990s (101 species).

Atlantic coast of Europe: The 200 aquatic alien species known from the Atlantic coasts of Europe also reflect political events with the lowest number of new alien records occurring during the Second World War, a time of few marine investigations. Since the 1950s the number of first records of new species is increasing, possibly due to increased shipping, further development of aquaculture, improved surveying techniques and sampling methods including diving.

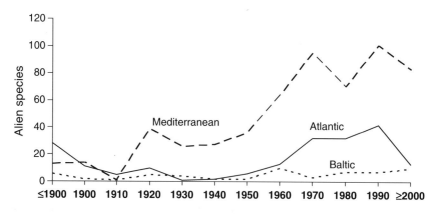

Fig. 7.1 The numbers of multicellular alien species recorded in European Seas for each decade over the past century

Baltic Sea: The 62 marine alien species established in the Baltic Sea reflect it smaller size, its brackish waters (Johannesson and André 2006), the less intensive transoceanic shipping activities, and fewer species used for aquaculture. The Baltic Sea receives secondary introductions from both the North Sea and from adjacent inland waters. Some brackish tolerant species were widely distributed as a forage food in the 1960–1970s, especially in former Soviet Union republics. The increase in the numbers of new introductions in the past two decades may also reflect a greater knowledge of the area (Leppäkoski 2005).

7.3 Biogeographic Patterns

The native range of alien species is often difficult to assess accurately. The true donor region may be obscured by the choice of vector, by repeated primary introductions from different populations within the species native range, or by succession of secondary introductions (the dispersal of an alien beyond its primary location of introduction). With few notable exceptions, the source populations of alien species in European seas have not been ascertained by molecular means (Jousson et al. 1998; Meusnier et al. 2004; Terranova et al. 2006; Andreakis et al. 2004, 2007). However, intentional introductions come from known areas. One case is the Pacific cupped oyster ***Crassostrea gigas***: a strain from Taiwan was probably introduced to Portugal in vessel fouling between the 16th and 18th centuries, and was in fact known as the 'Portuguese' Oyster until genetic studies clarified its taxonomy and origin (Boudry et al. 1998; O'Foighil et al. 1998). In 1866 the 'Portuguese' Oyster (now recognised as ***C. gigas***) was intentionally introduced to the Bay of Arcachon, France. During the 19th and 20th century it was repeatedly imported from Portugal to Ireland and southern North Sea coasts (Wolff and Reise 2002). In the 1960s ***C. gigas*** spat produced in British Columbia from Japanese stock, was imported to Britain under quarantine and later distributed to replenish native stocks in the Netherlands and elsewhere (Wolff and Reise 2002). There have been several consignments of molluscs to Europe from the North Pacific that will have resulted in several associated species becoming established (Minchin 1996) and some of these imports will have been unauthorised (Goulletquer et al. 2002). Many of the unwanted biota will have been spread as a result of shellfish movements either in trade or farming practices.

Given the long history of maritime traffic between Europe and the Americas, the expected origin for many alien species would be the NW Atlantic Ocean but the NW Pacific Ocean dominate as a source of origin with 44 species along the Atlantic coast of Europe and 10 in the Baltic Sea (Table 7.1). Eighteen aliens in the Baltic Sea originate from the NW Atlantic Ocean and 17 from the Ponto-Caspian region. Most of the alien species in the Mediterranean are thermophilic, originating in tropical waters: the Indo-Pacific Ocean (226), Indian Ocean (101), the Red Sea (63), and pan-tropical (56).

Table 7.1 The native ranges of the multicellular alien taxa recorded in European Seas

Native range	Mediterranean Sea	Atlantic coast of Europe	Baltic Sea
Indo-Pacific	226	8	
Indian Ocean	101	3	
Red Sea	63		
Pantropical	56	9	2
Pacific	33	41	4
Pacific NW	24	44	10
Atlantic NW		37	18
Atlantic W	22	12	2
Atlantic S		9	
Atlantic E	11		
Cosmopolitan	15		
Circumboreal	5		
Ponto-Caspian		7	17
Mediterranean		6	
North Sea			2
Other	13	24	7
Total	569	200	62

7.4 Main Pathways to Europe

It is often difficult to identify a pathway for a specific introduction. This is because pathways may differ between regions or multiple pathways may overlap (Minchin 2007). Therefore, most pathways cannot be positively identified, except where there has been an intentional introduction, as in the case of stock imports of cultured species. Even when a pathway of an alien species in one site is known, its introduction to other regions may utilise different means including natural dispersal processes. Some areas with concentration of maritime activities such as ports, marinas, and aquaculture farming act as nodes and crossover points, aiding to further dispersal of alien species along coastal zones. The vicinity of invasion corridors enhances such opportunities as happened following the opening of the Suez Canal that allowed entry of Indo-Pacific and Erythrean biota to the Mediterranean (Galil 2000).

European coastal waters are particularly susceptible to biological invasions because of the region's busy maritime traffic, and estuaries, lagoons and bays some of which are crowded with marine aquaculture farms and the presence of the Suez Canal. The main pathways of the alien taxa (Table 7.2) recorded off the Atlantic coast of Europe are the vessels (hull fouling/ballast) (47%), aquaculture (24%), and aquaculture/vessels (primary/secondary introductions) (13%). The majority of aliens in the Baltic Sea were introduced by vessels (45%) and aquaculture (18%). Most of the alien species in the Mediterranean have entered through the Suez Canal (54%), vessels (21%), aquaculture (11%), and canals/vessels (primary/secondary introductions) (10%).

It has been assumed that Erythrean aliens (Indo-West-Pacific organisms that were introduced into the Mediterranean through the Suez Canal) progress through

Table 7.2 The main pathways of the multicellular alien taxa recorded in European Seas

Pathway	Mediterranean Sea	Atlantic coast of Europe	Baltic Sea
Canals	307		
Vessels	117	93	28
Aquaculture	61	48	11
Canals, vessels	56		4
Aquaculture, vessels	15	28	4
Habitat management		4	2
Canals, fisheries			2
Unknown	8	17	6
Other	5	10	5
Total	569	200	62

the Suez Canal and along the coasts of the Levant as a result of "natural" dispersal, by autochthonous active or passive larval or adult movements, unaided further by human activity. Indeed, a temporal succession of directional ("stepping stones") records from the Red Sea, the Suez Canal, and along the coasts of the Levant may confirm a species status as a naturally dispersing Erythrean alien. However, dispersal could also result from anthropogenic translocation. Already Fox (1926: 20) wrote "*It is, of course, well known that ships have in more than one instance dispersed marine organisms from one part of the world to another. This must apply equally to transport through the Suez Canal. .... Possibly tugs and barges permanently employed in the Canal may take a larger share than other vessels in this transport from one end of the Canal to the other*". Even where records are consistent with long-shore autochthonous dispersal, there might be a degree of uncertainty where fouling organisms (such as serpulid polychaetes or mussels) are concerned, as they are more susceptible to shipping-mediated transfer. In some cases we suspect simultaneous mechanisms of transport (Galil 2006; Gollasch et al. 2006).

Though it is seldom possible to ascertain the precise means of transmission, yet it is inferred that port and port-proximate aliens are primarily dispersed by vessels (but see Mineur et al. 2007). The transport of boring, fouling, crevicolous or adherent species on the hulls of ships has been long recognised as the most ancient vector of aquatic species introduction: in 1873 the tri-masted *Karikal* arrived at the port of Marseille from India carrying on its flanks a "*a small forest populated by crustaceans*" (Catta 1876: 4). Fouling generally concerns small-sized sedentary, burrow-dwelling or clinging species, though large species whose life history includes an appropriate life stage may be disseminated as well (Zibrowius 1979; Gollasch 2002). Ballast (formerly solid, but for the past 130 years aqueous) is usually taken into dedicated tanks or into empty cargo holds when offloading cargo, and discharged when loading cargo or bunkering (fuelling), to increase vessel stability in heavy weather and to submerge the propeller when empty. Ballast water therefore have originated mostly in port or near port waters until the recent recommendation that exchange of ballast water be carried out in the open ocean (IMO Ballast Water Management Convention 2004). Water and sediment carried in ballast tanks, even after voyages of several weeks' duration, have been found to contain viable organisms (e.g., Gollasch et al. 2000; Olenin et al. 2000; Wonham et al. 2001; Drake et al. 2002). Indeed, the global maritime trade connections of European ports supply us with notorious examples for transoceanic ballast dispersal: the North American comb jelly ***Mnemiopsis leidyi*** reached the Black Sea in ballast water in the 1980s substantially impacting its ecosystem (e.g., Kideys 2002), and the Erythrean alien portunid crab *Charybdis hellerii* was transported in ballast water from the eastern Mediterranean to the Caribbean Sea, and has since rapidly spread along the Atlantic coast from Florida to Brazil (Mantelatto and Dias 1999).

Market-driven demands for alien fish and shellfish are on the rise with the increasing affluence of European countries. This, coupled with the crisis in wild fisheries, has created a surge in the development in mariculture industry over the past 25 years. Transport and transplantation of commercially important alien shellfish has resulted in numerous unintentional introductions of pathogens, parasites and pest species. Shellfish farms have served as gateways into Mediterranean coastal waters for the spread of unintentional associated alien organisms. Many aliens from a variety of taxa were introduced with oyster stock from east Asia: the brown algae *Sargassum muticum,* the Japanese kelp ***Undaria pinnatifida***, and the predatory snail *Ocinebrellus inornatus*. Two oyster-parasites, *Mytilicola orientalis* and *Myicola ostreae*, also arrived with their host.

7.5 Most Invaded Ecosystems

Within the marine regions certain areas receive more aliens on account of the greater number and frequency of transmission modes and also because some of these environments may be more susceptible to invasion on account of containment of water within reduced tidal movements, "empty niches" or provided settlement surfaces. In all European marine regions the aliens generally occur in the shallow coastal zone. For example, regions within the Baltic Sea at high risk include the Gulf of Finland, the Gulf of Riga and the coastal lagoons (Leppäkoski et al. 2002; Panov et al. 2003, 2007b). Such areas are known for their diversity of alien species, and act as bridgeheads for secondary introductions (Gruszka 1999; Leppäkoski et al. 2002). Hundreds of alien macrophytes, invertebrates and fish species of Indo-Pacific and Erythrean origin occur in most coastal habitats of the eastern Mediterranean as the result of introduction through the Suez Canal (Galil 2000).

In European brackish water areas (Baltic, Black, Azov and Caspian Seas) the diversity of alien species is greater and may be due to the poor native species richness at intermediate brackish salinities (Paavola et al. 2005). In such salinities with lowest native species richness aliens may be capable of penetrating niches and becoming established without suppression by native biota (Wolff 1999; Nehring 2006).

Specific substrate types are also of importance according to the habit of an alien species and infauna are particularly able to colonise coastal inlets and lagoons. The availability of firm substrata are also of importance particularly where fouling species can colonise adjacent man-made structures. At the Atlantic coast, the Oosterschelde estuary (The Netherlands) is a 'hot spot' for alien species: here about 30% of all Dutch introduced marine and estuarine species occur only within this estuary, and ~50% of all species occur only in this estuary and neighboring waters (Wolff and Reise 2002). In the Mediterranean Sea, the Thau Lagoon (France) is considered a 'hot spot' of aquaculture introductions (Verlaque 2001). In the Baltic Sea, the Curonian Lagoon (Lithuania/Russia) hosts a significantly higher number of established alien benthic species when compared with the adjacent open coast area (Zaiko et al. 2007).

7.6 Ecological and Economic Impacts

Alien species are prominent in most coastal habitats in European seas with few exceptions, their ecological impacts on the native biota is poorly known, although invasive species can cause major shifts in community composition. Little is known about the inter-relationships of native and alien biota, yet sudden concurrent changes in abundance of native biota occur as a result of displacement and competition. This may be aided by anthropogenic alteration of the marine ecosystem through habitat destruction, pollution, and rising sea-water temperature (Galil 2007). The reasons for these changes are not always easily determined. While no extinction of a native species is known, sudden declines in abundance, and even local extirpations, concurrent with proliferation of aliens, had been recorded, for example, the lower temperature tolerant strain of the killer alga ***Caulerpa taxifolia*** released to the Mediterranean (Longepierre et al. 2005), and the less well known but not less spectacular replacements and displacements of native species by several Erythrean aliens, in particular ***Brachidontes pharaonis***, and the kuruma prawn ***Marsupenaeus japonicus*** (Galil 2007). Population declines and niche contraction of native species may reduce genetic diversity, or otherwise alter ecosystem function and changes in habitat structure, leading to reduced species richness (Galil 2007).

Some economies have a dependency on aliens for food production (Black 2001) but alien pests, parasites and diseases can have serious impacts on production in aquaculture or impact on fisheries. For example the halposporidian *Bonamia ostreae* resulted in serious mortalities of the native oyster *Ostrea edulis* in northern Europe that led to imports in the 1970s of ***Crassostrea gigas***, flown from Japan to France, to replace production. These consignments brought pests, parasites and diseases which will have reduced some of this production (Hèral 1990) although in some areas it will have become a fouling organism. Yet French landings yielded 105,000 t in 2003. Imports of an oriental eel for culture trials resulted in the nematode parasite ***Anguillicola crassus*** harming the internal organs of the European eel *Anguilla anguilla*, in particular the air bladder thereby reducing swimming ability making oceanic spawning migrations less likely with a possible affect on recruitment. Aliens have been used in the development of fisheries, the stockings of the red king crab *Paralithodes camtschaticus* released over decades to the Barents Sea evolved a population now extending into Norwegian waters. Although it is incompatible with non-trap fishing methods, due to entanglement in gill-nets and from eating long-line baits, it is worth 9 million Euros a year. While in the eastern Mediterranean, the large numbers of the Erythrean alien crab *Charybdis longicollis* entangle the nets and damage the catch, and the scyphozoan **Rhopilema nomadica** not only makes trawling impractical but its stings deter humans from bathing and thereby has an impact on tourism (Galil 2000).

Fouling on vessel hulls adds to fuel costs and interferes with water abstraction by industry requiring high levels of general maintenance (Jenner et al. 1998) and man-made structures in port regions can provide significant firm surfaces for fixed alien species (Minchin 2006) such as barnacles *Balanus* spp. and tube worms such as **Ficopomatus enigmaticus** that can form significant growths.

Overall aliens have a wide range of impacts and there will be different levels of dependency upon them in different local economies. They are likely to become more widely spread with open trade agreements, improved transport vectors, new species entering cultivation, and with alterations in climate.

7.7 Management Options and Their Feasibility

The main pathways for the entry of alien species to European seas are canals, shipping and aquaculture. Measures to manage the spread with the most prominent pathways makes good sense. In relation to shipping it is generally accepted that fouling remains as a potent means of spreading species and is as important as the management of ballast water. To address hull fouling, the International Maritime Organization, the United Nations body which deals with shipping, developed the *International Convention on the Control of Harmful Anti-fouling Systems on Ships* (IMO 2001). Its aim, however, was not the reduction of unintentional species movements, but to reduce the toxic effects caused by the widely used paint additives such as tri-butyl-tin (TBT) used on ship hulls to discourage settlement and growth of fouling organisms. The paint acts as biocide against a broad range of organisms. While this convention is not yet in force, the EC Council Regulation No 782/2003 involves the phasing-out of the usage of TBT antifoulant paints. There are concerns that alternative paints may not be as efficient as TBT-containing paints and that port regions will as a result become less toxic leading to an increase in alien invasions (Minchin and Gollasch 2003).

To reduce species introductions in ballast water, the IMO agreed upon the *International Convention for the Control and Management of Ship' Ballast Water and Sediments* (IMO 2004). This convention recommends ballast water exchange at sea as an intermediate management option until more stringent standards are required. This standard may only be reached with ballast water treatment systems which are currently being developed. However, as above, this convention has not yet entered into force.

In aquaculture, quarantine and other measures to avoid unintentional species introductions with target species are available. Some of these guidelines are voluntary (e.g., ICES 2005; Hewitt et al. 2006), others have a stronger mandatory character (EC Council Regulation No 708/2007). These instruments include provisions to assess the risk which target species may pose in case they escape into the wild.

The invasion pathways contributing the most alien species in European Seas are shipping canals (e.g., Suez Canal or Baltic-Volga-Caspian waterway) (Gollasch et al. 2006; Galil et al. 2007; Panov et al. 2007a). Here avoidance mechanisms are more difficult to apply. The methods to reduce passage of species from one end to the other may include the insertion of environmental barrier(s) to reduce free movements of aliens by using a saline lock, bubble curtains or electrical fields. Presently no canal in Europe, open to international shipping, is equipped with such measures. Installation of some environmental or mechanical barriers may also reduce the risk of spread.

7.8 Future Expected Trends

Biological invasions continue to take place and there have been particular increases since the 1950s in European seas. There have been many case histories that have shown harmful effects on the environment and dependent industries and users. It is widely recognised that the global change-induced effects in the oceans favor the advent of thermophilic alien species, thus affecting the structural and functional biodiversity of the coastal zones, and that this impact may be irreversible.

Unless efforts are undertaken to implement the instruments listed above in control of alien species range expansions further new species introductions, some unexpected, will take place with unwanted impacts. It is in the best interest of the environment and for society to reduce the number and extent of invasions in order to preserve biodiversity and maintain water quality.

Acknowledgements The support of this study by the European Commission's Sixth Framework Programme project DAISIE (Delivering alien invasive species inventories for Europe, contract SSPI-CT-2003-511202) is gratefully acknowledged. We also thank our colleague Vadim Panov for his contributions to this work.

References

Andreakis N, Procaccini G, Kooistra WHCF (2004) *Asparagopsis taxiformis* and *Asparagopsis armata* (Bonnemaisoniales, Rhodophyta): genetic and morphological identification of Mediterranean populations. Eur J Phycol 39:273–283

Andreakis N, Procaccini G, Maggs C, Kooistra WHCF (2007) Phylogeography of the invasive seaweed *Aspargopsis* (Bonnemaisoniales, Rhodophyta) reveals cryptic distribution. Mol Ecol 16:2285–2299

Black KD (ed) (2001) Environmental impacts of aquaculture. Sheffield Academic Press, Sheffield

Boudry P, Heurtebise S, Collet B, Comette F, Gerard A (1998) Differentiation between populations of the Portuguese oyster, *Crassostrea angulata* (Lamarck) and the Pacific oyster, *Crassostrea gigas* (Thunberg), revealed by mtDNA RFLP analysis. J Exp Mar Biol Ecol 226:279–291

Catta JD (1876) Note sur quelques crustacés erratiques. Ann Sci Nat Zool 3:1–33

Darwin C (1854) A monograph on the subclass Cirripedia, with figures of all the species. The Balanidae (or sessile cirripedes); the Verrucidae. Ray Society, London. 684

Drake LA, Ruiz GM, Galil BS, Mullady TL, Friedmann DO, Dobbs FC (2002) Microbial ecology of ballast water during a trans-oceanic voyage. Mar Ecol Progr Ser 233:13–20

Fox HM (1926) General part. Zoological results of the Cambridge expedition to the Suez Canal, 1924. 1. Trans Zool Soc Lond 22:1–64

Galil BS (2000) A sea under siege – alien species in the Mediterranean. Biol Invasions 2:177–186

Galil BS (2006) Shipwrecked: shipping impacts on the biota of the Mediterranean Sea. In: Davenport JL, Davenport J (eds) The ecology of transportation: managing mobility for the environment. Environmental pollution 10. Springer, Dordrecht. 39–69

Galil BS (2007) Loss or gain? Invasive aliens and biodiversity in the Mediterranean Sea. Mar Poll Bull 55:314–322

Galil BS, Nehring S, Panov VE (2007) Waterways as invasion highways – Impact of climate change and globalization. In: Nentwig W (ed) Biological invasions. Ecological studies 193, Springer, Berlin. 59–74

Gollasch S (2002) The importance of ship hull fouling as a vector of species introductions into the North Sea. Biofouling 18:105–121

Gollasch S, Lenz J, Dammer M, Andres HG (2000) Survival of tropical ballast water organisms during a cruise from the Indian Ocean to the North Sea. J Plankton Res 22:923–937

Gollasch S, Galil BS, Cohen AN (2006) Bridging divides: maritime canals as invasion corridors. Monograph biology 83. Springer, Dordrecht. 315

Goulletquer P, Bachelet G, Sauriau PG, Noel P (2002) Open Atlantic coast of Europe: a century of introduced species into French waters. In: Leppäkoski E, Gollasch S, Olenin S (eds) Invasive aquatic species of Europe. Distribution, impact and management, Kluwer, Dordrecht. 276–290

Gruszka P (1999) The River Odra estuary as a gateway for alien species immigration to the Baltic Sea basin. Acta Hydrochim Hydrobiol 27:374–382

Héral M (1990) Traditional oyster culture in France. In: Barnabé G (ed) Aquaculture. Ellis Horwood, London. 342–387

Hessland I (1946) On the Quaternary *Mya* period in Europe. Arkiv Zool 37A:1–51

Hewitt C, Minchin D, Olenin S, Gollasch S (2006) Canals, invasion corridors and Introductions. In: Gollasch S, Galil BS, Cohen AN (eds) Bridging divides: maritime canals as invasion corridors. Kluwer, Dordrecht. 301–306

Holthuis LB, Heerebout GR (1972) Vondsten van de zeepok *Balanus tintinnabulum* (Linnaeus, 1758) in Nederland. Zool Bijdr 13:24–31

Jenner HA, Whitehouse JW, Taylor CJL, Khalanski M (1998) Coolong water management in European power stations and control of fouling. Hydroecol Appl 10:1–225

Johannesson K, André C (2006) Life on the margin-genetic isolation and loss of variation in a peripheral marine ecosystem. Mol Ecol 15:2013–2030

Jousson O, Pawlowski J, Zaninetti L, Meinesz A, Boudouresque CF (1998) Molecular evidence for the aquarium origin of the green alga *Caulerpa taxifolia* introduced to the Mediterranean Sea. Mar Ecol Progr Ser 172:275–280

Keller C (1883) Die Fauna im Suez-Kanal und die Diffusion der mediterranen und erythräischen Tierwelt. N Denkschr Allg Schweiz Ges Gesamt Naturwiss, ser 3, 28:1–39

Kerckhof F, Cattrijsse A (2001) Exotic Cirripedia (Balanomorpha) from buoys off the Belgian coast. Senckenberg Mar 31:245–254

Kideys A (2002) The comb jelly *Mnemiopsis leidyi* in the Black Sea. In: Leppäkoski E, Gollasch S, Olenin S (eds) Invasive aquatic species of Europe: distribution, impact and management. Kluwer, Dordrecht. 56–61

Leppäkoski E (2005) The first twenty years of invasion biology in the Baltic Sea area. Oceanolog Hydrobiol Stud 34:5–17

Leppäkoski E, Olenin S, Gollasch S (2002) The Baltic Sea: a field laboratory for invasion biology. In: Leppäkoski E, Gollasch S, Olenin S (eds) Invasive aquatic species of Europe: distribution, impacts and management. Kluwer, Dordrecht. 253–259

Longepierre S, Robert A, Levi F, Francour P (2005) How an invasive alga species (*Caulerpa taxifolia*) induces changes in foraging strategies of the benthivorous fish *Mullus surmuletus* in coastal Mediterranean ecosystems. Biodivers Conserv 14:365–376

Mantelatto FLM, Dias LL (1999) Extension of the known distribution of *Charybdis hellerii* (A. Milne-Edwards, 1867) (Decapoda, Portunidae) along the western tropical South Atlantic. Crustaceanana 72:617–620

Meusnier I, Valero M, Olsen JL, Stam WT (2004) Analysis of rDNA ITS1 indels in *Caulerpa taxifolia* (Chlorophyta) supports a derived, incipient species status for the invasive strain. Eur J Phycol 39:83–92

Minchin D (1996) Management of the introduction and transfer of marine molluscs. Aquat Conserv Mar Freshwater Ecosyst 6:229–244

Minchin D (2006) The transport and the spread of living aquatic species. In: Davenport J, Davenport JL (eds) The ecology of transportation: managing mobility for the environment. Springer, Dordrecht. 77–97

Minchin D (2007) Aquaculture and transport in a changing environment: overlap and links in the spread of alien biota. Mar Poll Bull 55:302–313

Minchin D, Gollasch S (2003) Fouling and ships' hulls: how changing circumstances and spawning events may result in the spread of exotic species. Biofouling Suppl 19:111–122

Mineur F, Johnson MP, Maggs CA, Stegenga H (2007) Hull fouling on commercial ships as a vector of macroalgal introduction. Mar Biol 151:1299–1307

Monterosato di TA (1878) Enumerazione e sinonimia delle conchiglie mediterranee. Giornale Sc Nat Econ Palermo 13:61–115

Nehring S (2006) Four arguments why so many alien species settle into estuaries, with special reference to the German river Elbe. Helgoland Mar Res 60:127–134

O'Foighill D, Gaffney PM, Wilbur AE, Hilbish TJ (1998) Mitochondrial cytochrome oxidase I gene sequences support an Asian origin for the Portuguese oyster, *Crassostrea angulata*. Mar Biol 131:497–503

Olenin S, Gollasch S, Jonusas S, Rimkutė I (2000) En-route investigations of plankton in ballast water on a ships' voyage from the Baltic Sea to the open Atlantic coast of Europe. Int Rev Hydrobiol 85:577–596

Paavola M, Olenin S, Leppäkoski E (2005) Are invasive species most successful in habitats of low native species richness across European brackish water seas? Estuarine Coast Shelf Sci 64:738–750

Panov VE, Bychenkov DE, Berezina NA, Maximov AA (2003) Alien species introductions in the eastern Gulf of Finland: current state and possible management options. Proc Estonian Acad Sci Biol Ecol 52:254–267

Panov VE, Rodionova NV, Bolshagin PV, Bychek EA (2007a) Invasion biology of Ponto-Caspian onychopod cladocerans (Crustacea: Cladocera: Onychopoda). Hydrobiologia 590:3–14

Panov V, Dgebuadze Y, Shiganova T, Filippov A, Minchin D (2007b) A risk assessment of biological invasions: inland waterways of Europe – the northern invasion corridor case study. In: Gherardi F (ed) Biological invaders in inland waters: profiles, distribution and threats. Springer, Dordrecht. 639–656

Petersen KS, Rasmussen KL, Heinemeier J, Rud N (1992) Clams before Columbus? Nature 359:679

Terranova MS, Lo Brutto S, Arculeo M, Mitton JB (2006) Population structure of *Brachidontes pharaonis* (Fischer P, 1870) (Bivalvia, Mytilidae) in the Mediterranean Sea, and evolution of a novel mtDNA polymorphism. Mar Biol 150:89–101

Vaillant L (1865) Recherches sur la faune malacologique de la baie de Suez. J Conchyl 13:97–127

Verlaque M (2001) Checklist of the macroalgae of Thau Lagoon (Hérault, France), a hot spot of marine species introduction in Europe. Oceanol Acta 24:29–49

Wolff WJ (1999) Exotic invaders of the meso-oligohaline zone of estuaries in the Netherlands: why are there so many? Helgoland Meeresunters 52:393–400

Wolff WJ, Reise K (2002) Oyster imports as a vector for the introduction of alien species into northern and western European coastal waters. In: Leppäkoski E, Gollasch S, Olenin S (eds) Invasive aquatic species of Europe: distribution, impacts and management. Kluwer, Dordrecht. 193–205

Wonham MJ, Walton WC, Ruiz GM, Frese AM, Galil BS (2001) Going to the source: role of the invasion pathway in determining potential invaders. Mar Ecol Progr Ser 215:1–12

Zaiko A, Olenin S, Daunys D, Nalepa T (2007) Vulnerability of benthic habitats to the aquatic invasive species. Biol Invasions 9:703–714

Zibrowius H (1979) Serpulidae (Annelida Polychaeta) de l'Océan Indien arrives sur des coques de bateaux à Toulon (France, Méditeranée). Com Int Exploit Sci Mer Méd 25/26:133–134

Chapter 8
Alien Birds, Amphibians and Reptiles of Europe

Salit Kark, Wojciech Solarz, François Chiron, Philippe Clergeau, and Susan Shirley

8.1 Introduction

DAISIE aims to integrate information on current invasions across Europe through an online freely available database of alien species (www.europe-aliens.org, Shirley and Kark 2006). Overall, the DAISIE database includes 55 islands or countries in Europe (including European Russia), Israel and the Macaronesian islands (hereby referred to as Europe). Patterns of alien introductions, their impacts and management tools differ for birds vs. reptiles and amphibians in various ways. Birds are one of the best recorded groups and much better data exists on their introductions. Reptiles and amphibians have smaller numbers of recorded alien species, and information is less detailed in many cases. However, some of the issues concerning aliens of these groups in Europe are similar. For example, prevention seems to be the best strategy to reduce the long-term impacts and costs of dealing with most species. Also, both groups show increase in the number of introduction events during the 20th century, which could be related to rise in human immigration into Europe (Jeschke and Strayer 2006) and in the international trade, both legal and illegal during this period, leading to the deliberate and non-deliberate release of alien birds in the wild (Jenkins 1999). Here we provide information and discussion on each of the two groups and comparisons where relevant. Finally, we provide joint discussion on management options and on future trends.

8.2 Alien Birds

We built a database of bird introductions to Europe. Our main sources were Long (1981) and references therein, Lever (1987) and Hagemeijer and Blair (1997), and these were supplemented with grey literature, country-based reports, publications and information from local experts. Countries that include both mainland and island areas were separated into two regions.

Overall, we recorded 193 bird species belonging to 37 families introduced in or to Europe since 1850, corresponding to 1883 introduction events. Of these, 140 alien bird species occur in 2007 in at least one of the 55 European regions. Out of these 140, only 77 species have established breeding populations and are termed established species.

8.2.1 Temporal Trends Since 1900

Although bird introductions by humans began thousands of years ago (Long 1981), 85% of the known events of bird introductions in Europe occurred after 1850. When considering temporal trends of introductions, two main factors should be taken into account. The first is the number of species introduced and the second is the number of introduction events (a species can be introduced multiple times to similar or different locations). The latter are known to positively affect the potential of a species to establish successfully (Duncan et al. 2003; Cassey et al. 2004; Jeschke and Stayer 2006). The number of alien bird species introduced into Europe has increased between 1900 and 2000 (Fig. 8.1). This includes both successful and failed introductions, in which the alien species failed to establish populations. While the rate of new introduction events was relatively stable in the first decades of the 20th century, the major increase in the number of introduction events in Europe occurred during its last decades. Introductions occurring after 2000 are not included in this paper. This is because we do not yet know the fate of these introductions, which makes it hard to compare them with older introductions (Kark and Sol 2005).

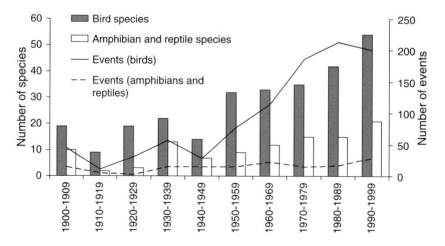

Fig. 8.1 Temporal trends in the number of introduced bird, amphibian and reptile species (bars) and the number of introduction events (lines) in Europe in the 20th century per decade

8.2.2 Biogeographic Patterns

Patterns of spatial spread of the 140 alien birds present in Europe shows that 60% of the species are relatively localised at a continental scale with range limits in two countries at most. Although countries differ in size, we included country in some analyses to enable comparison and enhance political action. Thirteen species (9% of the 140) are omnipresent and have invaded very large areas (more than 16 countries each). The three most widespread species in Europe in terms of their spatial range are mute swan *Cygnus olor*, ring-necked pheasant *Phasianus colchicus* and Canada goose *Branta canadensis*.

As shown in Fig. 8.2, the numbers of established alien birds vary largely between different areas in Europe (Table 8.1). Globally, when analyzing patterns at

Fig. 8.2 Richness of alien bird species in the study area: European countries and their islands (and Israel). We show here established alien species that are breeding (n = 77). The richness on the mainland and the islands of a single country (e.g., Italy and its islands) are shown separately

Table 8.1 Number of introduction attempts and success of introductions of birds, amphibians and reptiles in different areas in Europe (alphabetic order of areas). Data for areas with less than five introduction events is pooled in the given row

	Birds	
	Species	Introduction events
Area	All (successful)	All (successful)
Austria	27 (8)	44 (19)
Azores	9 (5)	14 (7)
Balearic Islands	27 (4)	32 (8)
Belgium	24 (16)	64 (49)
Canary Islands	32 (9)	51 (26)
Cyprus	8 (5)	17 (2)
Czech Republic	20 (6)	25 (8)
Denmark	13 (9)	15 (10)
Russia	9 (6)	16 (10)
Finland	13 (9)	25 (17)
France	43 (20)	115 (75)
Germany	60 (15)	120 (49)
Gibraltar	13 (2)	14 (6)
Greece	6 (6)	13 (13)
Ireland	10 (7)	56 (29)
Israel	29 (13)	43 (25)
Italy	72 (19)	120 (75)
Latvia	12 (6)	13 (6)
Norway	12 (8)	29 (19)
Poland	9 (6)	11 (7)
Portugal	27 (6)	56 (17)
Sardinia	9 (5)	11 (5)
Sicily	15 (5)	18 (6)
Spain	52 (18)	94 (47)
Sweden	18 (13)	29 (22)
Switzerland	13 (7)	20 (11)
The Netherlands	25 (13)	89 (60)
Ukraine	15 (6)	19 (12)
United Kingdom	67 (23)	600 (423)
Other islands	27 (17)	46 (24)
Other countries	69 (52)	75 (57)

	Amphibians and reptiles	
	Species	Introduction events
Area	All (successful)	All (successful)
Austria	13 (2)	16 (4)
Azores	4 (2)	5 (2)
Balearic Islands	15 (13)	18 (15)
Belgium	8 (3)	9 (3)
Canary Islands	10 (5)	15 (10)
Croatia	2 (1)	6 (5)
France	13 (11)	29 (16)
Germany	4 (2)	9 (4)
Gibraltar	7 (1)	11 (3)
Israel	3 (2)	11 (9)

Continued

Table 8.1 (continued)

Area	Amphibians and reptiles	
	Species	Introduction events
	All (successful)	All (successfull)
Italy	16 (10)	37 (26)
Malta	6 (2)	6 (2)
Portugal	3 (2)	7 (2)
Spain	26 (17)	49 (20)
Switzerland	7 (2)	9 (2)
The Netherlands	7 (5)	8 (6)
United Kingdom	20 (14)	88 (23)
Other countries and islands	14 (7)	22 (7)

the same spatial resolution (i.e., 50 × 50 km grid squares), richness of alien birds significantly decreases with longitude (Fig. 8.2). Such patterns in alien bird richness can arise as a result of various factors, including variations in human population density (GHFD 2001), which is often associated with introduction effort, increased habitat modification and high disturbance. In addition, the mean number of alien species on island grids (3.4 ± 0.9) is higher than on mainland (2.2 ± 0.04) grid squares (t-value = 11.9, $p < 0.001$) but this is mainly due to the effect of the UK having very high numbers of alien birds. Efforts of alien introductions to UK explains this difference (no more difference when UK is taken out).

8.2.3 Main Pathways to Europe

Not all stages of the invasion process have been equally studied and knowledge on main pathways of introduction is often sparse (Jenkins 1999). The most important vector for the arrival of alien birds into Europe is deliberate importation (88 species until 2007). Subsequently, the majority of aliens were deliberately released into the wild (74 species). In cases of non-deliberate introductions, 77 species escaped accidentally from captivity (a species can be introduced multiple times by similar or different pathways). This is probably underestimated because the latter cases are less often recorded than deliberate releases (Kark and Sol 2005).

For the 88 bird species deliberately imported, the main purposes were hunting and what was once termed "fauna improvement", aimed at changing the native fauna, which was in the 19th and early 20th centuries considered as environmental "improvement". Most species non-deliberately introduced originated from zoos or bird parks and were held as pets (Table 8.2). The majority of alien species established in Europe are still localised to the country to which they were introduced (83%) and 17% have naturally spread from the original country of introduction to neighbouring countries.

Table 8.2 Primary modes and sources of introduction of alien birds, amphibians and reptiles in Europe. Values are the number of species introduced. A single species may be represented by more than one category. In parentheses are the number of additional species for which it is unknown whether released and/or escaped is the mode of introduction

Mode of introduction	Source or reason of introduction	Number of bird species	Number of amphibians and reptiles
Released	Hunting/Harvesting	24	8
	Fauna "improvement"	11	28
	Biological control	1	2
	Agriculture	0	1
	Others	17	2
	Unknown	29	19
Escaped	From zoos	27 (12)	1
	As pets	15 (4)	9
	For food	0	10
	Others	10 (3)	12
	Unknown	26 (6)	2
Arrived by dispersal from other alien range	Natural spread	14	3

8.2.4 Most Invaded Habitats

Habitat type is an important factor influencing the establishment of alien birds (Duncan et al. 2003). To determine which European habitats host the highest number of alien birds, species were assigned to habitats using the available literature. Habitats were classified according to the European Nature Information System, level 1 (EUNIS according to Davies and Moss 2003). The number of established alien birds per each EUNIS habitat in 51 (of the 55) regions was calculated, giving a total of 165 species-habitat records. On average, established species occurred in two different EUNIS habitats. The largest proportion of the 165 bird introduction records are into arable land, gardens and parks (Table 8.3). These findings suggest that many introduced birds prefer human-modified habitats in their alien range. This is not surprising since alien birds are often generalist species that can occupy multiple habitats, including human-related habitats also in their native ranges. For example, the monk parakeet *Myiopsitta monachus*, rock dove *Columbia livia* and Canada goose **Branta canadensis** that established in Europe are present in their alien distribution range in five or more different habitat types.

8.2.5 Ecological and Economic Impacts

Actual data for impacts of alien birds is often scarce, particularly at regional scales, and part of the information on impacts comes from their native ranges. Most of the

Table 8.3 Total number of established alien birds, amphibians and reptiles found in habitats of 55 European countries with successful introductions (a species may have been introduced in multiple habitats). Habitats are classified according to the EUNIS classification of Davies and Moss (2003), level 1

EUNIS habitats	Number of bird species	Number of amphibian and reptile species
I: Arable land, gardens and parks	38	34
C: Inland surface water habitats	27	32
G: Woodland and forest	25	34
D: Mires, bogs and fens	15	17
E: Grasslands	15	18
B: Coastal habitats	15	9
J: Constructed and industrial	14	18
F: Heathland, hedgerows and scrub	11	17
H: Inland unvegetated or sparsely vegetated areas	5	18
Unknown	28	0

alien bird species with impacts have their native ranges in the Afrotropical, Oriental, and Palearctic biogeographic regions. Over 60% of established alien species originating from Australian, Nearctic and Oriental regions have known negative impacts. It is as yet unclear whether species impacts known from other locations will also be detrimental in Europe. However, it would be useful, for effective control, to consider information from other regions as indication of potential impact rather than wait until a species has spread enough when the evidence of impacts becomes obvious but too costly to deal with. Birds have often been considered less problematic than other aliens (e.g., mammals). Therefore, limited work has been done on their impacts. However, both scientists and policy-makers are becoming increasingly aware that introductions of alien birds can impose serious impacts on native species biodiversity, economic resources and human health (Williamson 1999; Pimentel et al. 2000). Of the 140 alien bird species now breeding in Europe and Israel, 66 species from 18 different bird families have been shown to have negative impacts, with many others unknown. Economic impacts, mainly feeding damage to grain and fruit crops, are the most common. Sixty bird species widely distributed taxonomically across 14 families have known economic impacts. Biodiversity impacts on native bird communities, including competition for resources, predation, and hybridisation with native species are described for 28 species. Biodiversity impacts are more commonly found among species in the Anatidae (ducks and geese), Corvidae (crows), and Psittacidae (parrot) families. Impacts on human health and well-being are known only in a few families, but this area requires much more research.

There are several alien bird species of special concern in Europe. For example, the ruddy duck ***Oxyura jamaicensis*** and the sacred ibis ***Threskiornis aethiopicus***, although still restricted in their distributions, can have impacts on native bird populations (Hughes et al. 2006; Clergeau and Yesou 2006). The ruddy duck, now being actively controlled in several areas, can hybridise with the native white-headed duck *Oxyura leucocephala* and consequently can pose a serious threat to the persistence of this endangered native duck (Hughes et al. 2006). Two other species, the rose-ringed parakeet **Psittacula krameri** and the Canada goose **Branta canadensis**, are known to have negative economic and biodiversity impacts. These species are distributed widely across Europe and are spreading to other areas. Both species are known to inflict heavy damage on agricultural crops and to compete with native species for resources (Blair et al. 2000; Strubbe and Matthysen 2007). The common myna *Acridotheres tristis*, although only recently arrived in Europe and in Israel, is on the IUCN list of the 100 of the World's Worst Species and has the potential to cause large impacts on native biodiversity (Pell and Tidemann 1997; Holzapfel et al. 2006).

8.3 Alien Amphibians and Reptiles

Data on introductions of alien amphibians and reptiles into Europe was gathered mainly from books by Gasc et al. (1997), Lever (2003) and Cox et al. (2006). This was supplemented by a variety of other sources. The compiled dataset included information on 172 introduction events of 29 amphibian species belonging to eight families, and 183 introduction events of 48 reptile species belonging to 14 families. These two datasets were combined and analysed together. The amount of available information differed significantly among different species. The highest number of introduction events for a single species was 38 and the top 10% of species held nearly half of all introduction events. However, about half of all species were represented only once in our database. Although this result partly depends on variation in data availability, it also reflects differences in introduction effort between species. The total numbers of species and introduction events are strongly underestimated, as many remain undocumented. This refers not only to old but also to very recent introductions.

More data was available on successful introduction attempts that led to established populations (159 introductions of 55 species) than on failed ones (54 introductions of 15 species). This bias is a consequence of successful introductions being easier to detect (and document) than unsuccessful ones. The output of 46 events (accounting for 23 species) was unclear, and in as many as 96 cases (44 species) no information on the fate of the introduction was available.

Among 11 species that had five or more introduction events with known outcomes, introductions of five species (*Hyla meridionalis*, *Podarcis muralis*, *Podarcis sicula*, *Testudo graeca*, *Chamaeleo chamaeleon*) were in all cases successful. This can be largely attributed to the fact that the attempts to introduce these species were often undertaken near their native distribution range. For the remaining six species introduction success of the attempts with known result ranged from 8% to 93%. It is more difficult to draw any conclusions on the species that have never been successfully introduced to Europe, as the data is scarcer.

8.3.1 Temporal Trends Since 1900

Based on the available data, similar to the pattern seen for birds, both the number of introduction events and the number of species introduced throughout the 20th century have increased towards later years (Fig. 8.1). The trend is more evident when the century is divided into two halves, with the years 1950–2000 accounting for 66% of all introduction events and 67% of all species introduced. The apparent decline in introduction effort after the first decade of the 20th century can be partly attributed to data reporting issues. However, it is possible that the World Wars I and II led to a decline in introduction effort.

8.3.2 Biogeographic Patterns

Nearly half (46%) of the amphibians and reptiles successfully introduced to Europe have at least part of their native range in other parts of Europe. Asia and Africa account for one fourth of the reptiles and amphibians introduced to Europe (many of these species also have native ranges in Europe). North America holds only 4% of amphibians and reptiles successfully introduced to Europe. Similar analysis done for the locations from which the individuals were actually imported (referred to as donor areas) shows that for most introduction events the animals were not brought to the non-native areas in Europe directly from their native range but rather from other European areas to which they were introduced in the past, this being a secondary introduction.

There was considerable variation in both the numbers of introduction attempts and their success. In general, south European countries received more attempts to introduce a wider variety of alien amphibians and reptiles than areas further north (Table 8.1). This pattern can probably be attributed to climatic conditions, more adverse for these two groups at higher latitudes. There are, however, exceptions in some northern areas, such as the UK, where introduction effort was high, and in some more southern areas, such as Greece, that had few known introduction attempts.

8.3.3 Main Pathways to Europe

Information on pathways of introductions of amphibians and reptiles to Europe is limited (Hulme et al. 2008). Available data indicate, however, that a minimum of 45 species were at least once intentionally transported into new areas as commodities, such as pets, sources of food or biological control agents. Subsequently, 39 species were intentionally released into the wild and nine species escaped from captivity. Six arrived as contaminants of commodities, such as eggs brought with soil, or tadpoles brought with fish for stocking. Another six species arrived as unwanted stowaways of transport vectors such as ships. Some were introduced by more than one of the above pathways. When intentionally introduced, the majority of species was released for what was termed "improvement" of the local fauna (Table 8.2). These species were often originally brought as pets and then set free by their owners. Species that escaped from captivity were originally brought as food for humans or pets.

8.3.4 Most Invaded Habitats

Analysis of the ecosystems most prone to amphibian and reptile invasion was done in the same manner as for birds. The largest proportion of records are in arable land, gardens and parks (EUNIS category I, Table 8.3). This category is closely followed by woodland, forests, inland surface waters and riparian habitats. One should remember, however, that these results are based on data on occurrence of established species and their habitat requirements. They do not take into account the actual impact of species in each habitat. Thus the actual threat from alien amphibians and reptiles to some habitats may be underestimated. This refers particularly to the small areas (e.g., coastal habitats).

8.3.5 Ecological and Economic Impacts

Of the 55 species successfully introduced to Europe, 12 (seven amphibians and five reptiles) were reported to have some adverse impact on native biodiversity. Altogether, there are 35 records of these species having impacts in different parts of Europe. Among the determined mechanisms of impact, predation is the prevailing mechanism, followed by competition with native species. There is one report of an alien amphibian hybridising with its native relative in Spain, and another, the American bullfrog, **Lithobates catesbeianus**, transmitting pathogens. Interestingly, amphibians had almost twice as many reports for impacts (22 cases) than reptiles (13 cases). Amphibians with impacts belong to three of the seven families successfully established in Europe: Pipidae (the one established species is known to have impacts), Ranidae and Salamandridae (71% and 14% of the established species, respectively, are known to have impacts). Pipidae is probably the most harmful amphibian family also at a global scale, even though it is represented by only two established alien species. Ranidae is the third highest impact family worldwide, with 55% of the established species having adverse effects in different parts of the world.

Undoubtedly, one of the most well known invasive amphibians around the world is the American bullfrog **Lithobates catesbeianus**. Adults severely predate on native anurans and other aquatic herpetofauna, such as snakes and turtles, and larvae can have significant impacts on benthic algae, thus perturbing aquatic community structure (Crayon 2005; Detaint and Coïc 2006). In France and the UK this frog is also the vector of the fungus *Batrachochytrium dendrobatidis* (Garner et al. 2005, 2006). This fungus is an agent of chytridiomycosis, a disease severely affecting native amphibians worldwide (Daszak et al. 2004; Hanselmann et al. 2004). There is an urgent need to carry out systematic research on the ecological impact of this species in Europe because available reports are scarce and limited work has been done.

Of nine reptile families successfully established in Europe, Emydidae, Lacertidae and Colubridae, hold 33%, 30% and 14% of the established species, respectively. These were reported as having negative impacts. In contrast to amphibians, the global pattern is quite different than in Europe. Globally, Emydidae and Lacertidae hold the 12th and 13th position among the most invasive families among reptiles, and Colubridae are ninth. Globally, reptile families with highest proportion of species that have negative impacts include Boidae, Alligatoridae, Scincidae, Polychrotidae and Agamidae. These families do not have established representatives in Europe, but the global pattern should serve as an early warning against attempts to introduce them into Europe.

8.4 Management Options and Their Feasibility

Management of alien vertebrates is not a trivial task due to cultural, social, practical and other aspects. The control or eradication of alien populations of vertebrates, and especially of birds is often controversial due to their appreciation by the public and the incomplete knowledge on their impacts (Clergeau and Yesou 2006). For effective management, actions need to be coordinated across countries. In Europe, coordinated alien bird management involves the ruddy duck, aimed at preventing its spread from the UK to western Europe (Hughes et al. 2006; Smith et al. 2005). This bird is controlled annually in the UK, France, Spain and Portugal. The sacred ibis has recently escaped from numerous zoos and bird parks throughout the world, generating concern, and actions are beginning to be organised in Europe (Clergeau and Yesou 2006). Attention to the need to control alien birds has also emerged locally in some cases, for example in the rose-ringed parakeet in Italy, the mute swan in France, and the common myna in Israel.

Prevention is widely considered to be the most cost-effective way to combat invasions (CBD 2002) and needs to be promptly organised at the national, continental and international scales (Clergeau et al. 2004; Simberloff 2005). Intentional introductions should not be allowed before a risk assessment has been carried out at a species-by-species basis. For birds, amphibian and reptiles, the precautionary principle of the "white list" should be fully applied. For highly dangerous species, ban on import, trade and possession should be considered (Miller et al. 2006). For unauthorised and unintentional introductions, it is probably more effective to undertake more comprehensive control of whole pathways that are responsible for introductions of a wide range of different taxa (pathway-based mechanism, Ruiz and Carlton 2003). Over the past few decades one of the major pathways for the introduction of alien amphibians and reptiles to Europe was via the pet trade, which resulted in many unauthorised deliberate releases and unintentional escapes. Keeping alien amphibians and reptiles is becoming more and more popular and the role of this pathway is likely to increase in future. It is vital therefore to restrict pet trade and to build awareness among amateur children and adults, as well as professional amphibian and reptile keepers about the environmental and legal consequences of unauthorised introductions of their pets into the wild, both intentional and resulting from negligence.

If prevention efforts fail, it is vital to early detect the aliens and if feasible, eradicate invasive aliens (CBD 2002). Among very few examples from Europe, there are successful eradications of the American bullfrog (one in the UK and two in Germany, Ficetola et al. 2007a). Success of these actions can be attributed to the fact that they were initiated early, when the populations were still small and limited in terms of numbers and their alien range.

Undoubtedly, the easiest and cheapest way to avoid damages in the long run is to prevent the spread of alien birds by controlling trade and by acting while populations are still local and small. The action has to be coordinated very quickly after the first evidence of reproduction. Therefore, monitoring and reporting systems need to be established. However, for them to be successful, actions must involve clear and direct communication with the public, and should be combined with longer-term educational programs.

8.5 Future Expected Trends

With the rapid increase in globalisation and the resulting homogenisation of biotas (Lockwood and McKinney 2002), quite a few alien species are establishing in large areas around the world. The current trend of increase in Europe in both the number of alien birds and introduction events is likely to continue. While most of the successful introductions up until now were limited to specific countries and to human-modified habitats, we suspect that with the opening of the borders in Europe, the increase in trade and the expansion of urban landscapes, alien species may become even more abundant and successful in Europe. The increase in trade and free movement of goods and people within most of Europe is likely to be followed by the increase in introduction effort of alien amphibians and reptiles, both in terms of the number of species and the number of introduction attempts. The pet trade is likely to remain the main source of intentional and unintentional introductions. Climate change may lead to the decline of some native amphibians and reptiles in Europe (Araújo et al. 2006) and at the same time may cause an expansion of areas suitable for establishment of alien species (Ficetola et al. 2007b). More research into the effects of global changes on aliens should be conducted, combined with applied actions to reduce their negative impacts and establishment.

Acknowledgements The support of this study by the European Commission's Sixth Framework Programme project DAISIE (Delivering alien invasive species inventories for Europe, contract SSPI-CT-2003-511202) is gratefully acknowledged. We thank Phil Hulme for initiating and coordinating the DAISIE project and all the members of DAISIE and the EU FP6 ALARM project teams for their collaboration. We thank Philip Hulme, Wolfgang Nentwig, Petr Pyšek and an anonymous reviewer for comments that improved the manuscript. We would like to thank Anita Gamauf, Anton Kristin, Eran Banker, Piero Genovesi, Ohad Hatzofe, Jelena, Teemu Lehtiniemi, Michael Miltiadous, Yotam Orchan, Milan Paunovic and Assaf Shwartz for information and input on alien birds, and Wieslaw Król, Olivier Lorvelec, Piotr Płonka, Katja Pobolsaj, Zofia Prokop, Riccardo Scalera, Agnieszka Staszczyk and Małgorzata Strzałka for amphibians and reptiles. We thank Nicola Bacetti, Eran Banker, Daniel Bergmann, Michael Braun, Jordi Clavell, Helder Costa, Gert Ottens, Milan Paunovic, Riccardo Scalera, Ondřej Sedláček, Cagan Sekercioglu, Riccardo Scalera, Diederik Strubbe, Wim Van den Bossche, and Birdlife Belgium, Alexandre Vintchevski and Georg Willi and many other experts for verifying local information for birds.

References

Araújo MB, Thuiller W, Pearson RG (2006) Climate warming and the decline of amphibians and reptiles in Europe. J Biogeogr 33:1712–1728

Blair MJ, McKay H, Musgrove AJ, Rehfish MM (2000) Review of the status of introduced non-native waterbird species in the agreement area of the African-Eurasian waterbird agreement research contract CR0219. British Trust for Ornithology, Thetford, Norfolk

Cassey P, Blackburn TM, Sol D, Duncan RP, Lockwood JL (2004) Global patterns of introduction effort and establishment success in birds. Proc R Soc Lond Biol Lett 271:S405–S408

CBD Convention on Biological Diversity (2002) Decision VI/23 Alien species that threaten ecosystems, habitats or species. The Conference of the Parties. Rio de Janeiro

Clergeau P, Yesou P (2006) Behavioural flexibility and numerous potential sources of introduction for the sacred ibis: causes of concern in western Europe? Biol Invasions 8:1381–1388

Clergeau P, Levesque A, Lorvelec O (2004) The precautionary principle and biological invasion: the case of the house sparrow on the Lesser Antilles. Int J Pest Manage 50:83–89

Cox N, Chanson J, Stuart S (2006) The status and distribution of reptiles and amphibians of the Mediterranean Basin. IUCN, Gland

Crayon JJ (2005) *Rana catesbeiana*. In: Global invasive species database. www.issg.org/database/species/ecology.asp?si = 80&fr = 1&sts = sss. Cited Sept 2007

Daszak P, Strieby A, Cunningham AA, Longcore JE, Brown CC, Porter D (2004) Experimental evidence that the bullfrog (*Rana catesbeiana*) is a potential carrier of chytridomycosis, an emerging fungal disease of amphibians. Herp J 14:201–207

Davies CE, Moss D (2003) EUNIS habitat classification, August 2003. European Topic Centre on Nature Protection and Biodiversity, Paris

Detaint M, Coïc C (2006) La grenouille taureau *Rana catesbeiana* dans le sud-ouest de la France. Premiers résultats du programme de lutte. Bull Soc Herp France 117:41–56

Duncan RP, Blackburn TM, Sol D (2003) The ecology of bird introductions. Annu Rev Eco Evol Syst 34:71–98

Ficetola GF, Coïc C, Detaint M, Berroneau M, Lorvelec O, Miaud C (2007a) Pattern of distribution of the American bullfrog *Rana catesbeiana* in Europe. Biol Invasions 9:767–772

Ficetola GF, Thuiller W, Miaud C (2007b) Prediction and validation of the potential global distribution of a problematic alien invasive species – the American bullfrog. Diversity Distrib 13:476–485

Garner TWJ, Walker S, Bosch J, Hyatt AD, Cunningham AA, Fisher MJ (2005) Chytrid fungus in Europe. Emerg Infect Dis 11:1639–1641

Garner TWJ, Perkins MW, Govindarajulu P, Seglie D, Walker S, Cunningham AA, Fisher MC (2006) The emerging amphibian pathogen *Batrachochytrium dendrobatidis* globally infects introduced populations of the North American bullfrog, *Rana catesbeiana*. Biol Lett 2:455–459

Gasc JP, Cabela A, Crnobrnja-Isailovic J (1997) Atlas of amphibians and reptiles in Europe. Collection Patrimoines Naturels, 29, Soc Europ Herp, Mus Natl Hist Nat & Service du Patrimoine Naturel, Paris.

GHFD Global Human Footprint Dataset (Geographic) (2001) Wildlife Conservation (WCS) and Center for International Earth Science Information Network (CIESIN). www.ciesin.columbia. edu. Cited July 2007

Hagemeijer EJM, Blair MJ (ed) (1997) The EBCC atlas of European breeding birds: their distribution and abundance. Poyser, London

Hanselmann R, Rodriguez A, Lampo M, Fajardo-Ramos L, Aguirre AA, Kilpatrick AM, Rodríguez JP, Daszak P (2004) Presence of an emerging pathogen of amphibians in introduced bullfrogs *Rana catesbeiana* in Venezuela. Biol Conserv 120:115–119

Holzapfel C, Levin N, Hatzofe O, Kark S (2006) Colonization of the Middle East by the invasive common myna *Acridotheres tristis*, L., with special reference to Israel. Sandgrouse 28:1–11

Hughes B, Robinson J, Green AJ, Li D, Mundkur T (2006) International single species action plan for the conservation of the white-headed duck *Oxyura leucocephala*. www.unep-aewa.org/publications/technical_series/ts8_ssap_white-headed-duck_complete.pdf. Cited Oct 2006

Hulme PE, Bacher S, Kenis M, Klotz S, Kühn I, Minchin D, Nentwig W, Olenin S, Panov V, Pergl J, Pyšek P, Roque A, Sol D, Solarz W, Vilà M (2008) Grasping at the routes of biological invasions: a framework for integrating pathways into policy. J Appl Ecol 45:403–414

Jenkins P (1999) Trade and exotic species introductions. In: Sandlund OT, Schei PO, Viken A (eds) Invasive species and biodiversity management. Kluwer, Dordrecht

Jeschke J, Strayer DL (2006) Invasion success of vertebrates in Europe and North America. Proc Natl Acad Sci USA 102:7198–7202

Kark S, Sol D (2005) Establishment success across convergent Mediterranean ecosystems: an analysis of bird introductions. Conserv Biol 19:1519–1527

Lever C (1987) Naturalized birds of the World. Longman, London

Lever C (2003) Naturalized reptiles and amphibians of the World. Oxford University Press, Oxford

Lockwood JL, McKinney ML (2002) Biotic homogenization. Kluwer, New York

Long JL (1981) Introduced birds of the world. Universe Books, New York

Miller C, Kettunen M, Shine C (2006) Scope options for EU action on invasive alien species (IAS). Final report for the European Commission. Institute for European Environmental Policy, Brussels

Pell AS, Tidemann CR (1997) The impact of two exotic hollow-nesting birds on two native parrots in savannah and woodland in eastern Australia. Biol Conserv 79:145–153

Pimentel D, Lach L, Zuniga R, Morrison D (2000) Environmental and economic costs of non-indigenous species in the United States. Bioscience 50:53–65

Ruiz GM, Carlton JT (2003) Invasive species: vectors and management strategies. Island Press, Washington, DC

Shirley S, Kark S (2006) Amassing efforts against alien invasive species in Europe. PLoS Biol 4:1311–1313

Simberloff D (2005) The politics of assessing risk for biological invasions: the USA as a case study. Trends Ecol Evol 20:216–222

Smith GC, Henderson IS, Robertson PA (2005) A model of ruddy duck *Oxyura jamaicensis* eradication for the UK. J Appl Ecol 42:546–555

Strubbe D, Matthysen E (2007) Invasive ring-necked parakeets *Psittacula krameri* in Belgium: habitat selection and impact on native birds. Ecography 30:578–588

Williamson M (1999) Invasions. Ecography 22:5–12

Chapter 9
Alien Mammals of Europe

Piero Genovesi, Sven Bacher, Manuel Kobelt, Michel Pascal, and Riccardo Scalera

9.1 Introduction

Mammals are large, charismatic animals that have a mineralised skeleton that may form long lasting fossils. For these reasons, the level of knowledge on this class, together with other vertebrates, is much higher than for any other animal group. Therefore, the available information on introduction patterns, trends of invasions, and detrimental impacts caused to the environment and to human well-being are more detailed than for other groups covered in the DAISIE project.

History of mammal invasions is very long, as anthropogenic introductions of mammals started at least since the beginning of the Neolithic period. Ancient introductions involved wild species commensal of humans (i.e., black rat *Rattus rattus* and house mouse *Mus musculus*), anthropophilous (i.e., lesser white-toothed shrew *Crocidura suaveolens* and wood mouse *Apodemus sylvaticus*) and domestic species (i.e. species domesticated in the Middle East and gone feral, like the Corsican mouflon *Ovis aries*). Data on alien mammals have been collected from available global reviews (Long 2003; Mitchell-Jones et al. 1999; Lever 1985), national inventories (Austria: Englisch 2002; Denmark: Baagøe and Jensen 2007; France: Pascal et al. 2006; Germany: Geiter et al. 2002; Ireland: Stokes et al. 2006; Italy: Andreotti et al. 2001; Scalera 2001; Liechtenstein: Broggi 2006; Scandinavian countries: Weidema 2000; Spain: Nogales et al. 2006; Palomo and Gisbert 2002; Switzerland: Wittenberg 2006; the UK: Battersby and Tracking Mammals Partnership 2005; Weijden et al. 2005). Databases available on the internet were also used as a source of information (i.e. for Belgium, the Nordic countries, etc.). Other data have been collected through inputs of the experts of the DAISIE consortium, but also with the valuable support of many experts of the IUCN Invasive Species Specialist Group and of the Group of Experts on Invasive Alien Species of the Council of Europe. Independent experts have verified each record, which included information on taxonomy, native range, vector and pathway of introduction, date of introduction, status of the species, basic information on population size, distribution and impacts.

Based on the DAISIE database, in the present chapter we present an overview of the main patterns of mammal invasions in Europe, and analyse the main environmental, social and economic correlates to the arrival and successful establishment.

9.2 Description of the Mammal Inventory

For the DAISIE mammal inventory we only took into consideration invasions that occurred since 1500. The DAISIE dataset includes records of 88 mammal species introduced since 1500 in one or several of the 52 European geographic entities considered for the data collection (43 countries, including Russia, and nine islands or island systems, hereafter called regions, i.e. south Aegean islands).

Of the 88 species listed in the database four have an uncertain status, like the European mink *Mustela lutreola*, considered of uncertain origin in France by Pascal and co-authors (2006), or the Russian desman *Desmana moschata* in Moldova and Lithuania, where it is not known if the species was introduced or arrived naturally. Of the remaining 84 species, 64 (75%) are alien to the European continent, while 20 (25%) are native to at least one European region, but introduced in other country or island of the region, outside their native European range. Of the 20 species native to some countries of Europe, 14 (70%) have established self-sustaining populations in the areas of introduction, while four have become extinct. Moreover, the hamster *Cricetus cricetus* in Denmark has an "unknown" status (NOBANIS 2007), and the beluga *Delphinapterus leucas*, the only known case of an alien cetacean introduced into Europe, is considered "not established" (Reeves and Notarbartolo di Sciara 2006).

Of the 64 non-European aliens, 33 (52%) are considered established, that is, they form self-sustaining populations, in one or more regions, and 20 (31%) are reported as extinct. Among the extinct species, the case of the Himalayan porcupine *Hystrix indica*, escaped from a zoological garden and established in the wild in Devon in the late 1970s, is particularly interesting, because this is the only species which has been intentionally eradicated (Genovesi 2005) and as a result it is now no longer present anywhere in Europe. Seven out of these 64 species are listed as present but not established. However it should be noted that some species quoted as extinct according to the DAISIE database (porcupine *Hystrix cristata*, barbary macaque *Macaca sylvanus*, house mouse and black rat) are actually established if we also consider populations introduced before 1500. In conclusion, we consider that the number of extant alien mammals species in Europe is 44, while additional 15 species have been intentionally or accidentally moved from one European country/region to another.

To estimate the proportion of alien species in the European mammal fauna, we integrated and compared the DAISIE dataset on alien mammals introduced after the year 1500, with the results of the European Mammal Assessment (Temple and Terry 2007), that has the same geographic coverage as the DAISIE project. The European Mammal Assessment comprises 312 species, but excluding introduced species, extinct and domestic forms, the number of native mammal species in Europe drops to 245 including the above mentioned species with uncertain origin. We did not exclude species classified as having marginal occurrence by the European Mammal Assessment, because the DAISIE dataset also includes alien species which have a very limited range in Europe (e.g., Pallas squirrel *Callosciurus erythraeus* in Antibes, France). Alien species account for 15% of species present in

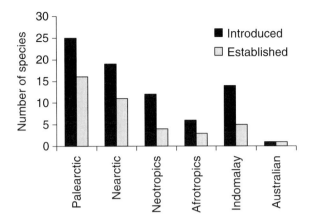

Fig. 9.1 Origin of introduced mammals in Europe

Europe. However, if we consider only terrestrial mammals, the percentage of alien vs. native species raises to 21%.

Of the 64 (75%) non-European aliens, the most frequent area of origin is the non-European Palearctic (29%), followed by Nearctic (22%), Indomalay (16%), Neotropics (14%), Afrotropics (7%), and Australian (1%) (Fig. 9.1). (Several species occur in more than one biome.) With regard to taxonomy, of the 84 species introduced into Europe, 30 belong to the order Rodentia, 24 are Artiodactyla, 12 Carnivora, six Lagomorpha, and two Erinaceomorpha. As previously mentioned, we also recorded the introduction of a cetacean, the beluga *Delphinapterus leucas* escaped from captivity in the Caspian sea, but there are also cases of introductions of two bat species, one of which succeeded, the large Egyptian rousette *Rousettus aegyptiacus* in the Canary islands (Nogales et al. 2006) and one marsupial, the red-necked wallaby *Macropus rufogriseus*, currently established in France and the UK (Long 2003; Pascal et al. 2006) (Fig. 9.2).

9.3 Temporal Trends of Invasion

Considering all introductions reported at the national level for which a date of introduction is known (n = 386 introduction events recorded in Europe), there is an evident exponential increase in the number of introductions since the end of the 19th century (Fig. 9.3). Moreover, considering only the 64 species that are not native of Europe, the number of alien species that has been introduced in this continent has constantly grown in the last centuries (Fig. 9.3).

The rapid rate of invasion is particularly evident if we consider the introductions of non-European aliens after 1850, where there is a clear linear correlation (1850–2000: y = 0.3807x - 706.36, R^2 = 0.99). Moreover, it should be noted that the trend

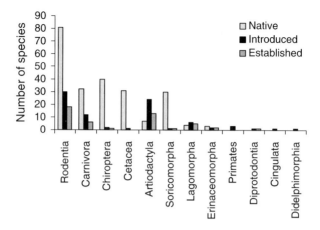

Fig. 9.2 Taxonomy of native, introduced and established mammals in Europe. Number of native species excerpted from the European Mammal Assessment carried on by the IUCN (Temple and Terry 2007). Species introduced before 1500 and domesticated species are not considered

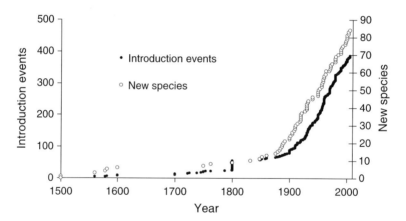

Fig. 9.3 Trends of mammal invasions in Europe: number of introduction events occurred since 1500 (n = 386 with exponential growth y = 1E-07e0.011x) and cumulative number of alien species invading Europe (66 non native alien species established in Europe (y = 4E-0.6e$^{0.0081}$x, R^2 = 0.889)

is constant and does not show any saturation effect and is similar to that found for example in France by Pascal et al. (2006). This pattern appears to confirm that ecological systems rarely show evidence of being saturated with species (Sax et al. 2007). Similarly, the number of alien mammal species successfully established per year confirms the rapid increase of biological invasions in Europe. In fact, this rate has grown from 0.03 species successfully introduced per year before 1800, to one species per year recorded in the last 5 years (Fig. 9.4).

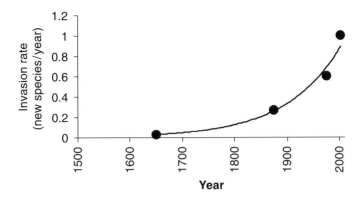

Fig. 9.4 Rate of arrival of new species in Europe (four periods: 1500–1800, 1800–1950, 1950–2000, 2000–2005. Exponential growth: $y = 4E-09e0.0096$, $R^2 = 0.99$)

9.4 Main Pathways to Europe

The most frequent pathway of alien mammals introduction in Europe is intentional release from captivity, which accounts for 35% of all known cases of introduction (n = 411 cases at the national level, 321 with known vector, 90 unknown). If we consider the activities related to invasions by alien species, fur farming has been at the origin of 15% of all recorded cases, hunting 21%, release or escape of pets 10%, and escapes from zoos 6% (n = 266). The so called "fauna improvement" is the wrong concept of "improving" the number of species in an area by intentionally releasing other species. Also "mistaken" reintroduction programmes carried on with species similar to the native ones (i.e., introduction of the Canadian beaver *Castor canadensis* in Finland, due to a confusion between this species and the native European beaver *Castor fiber*) were performed. Both account for 7% of the known introductions. In total 31% of introductions occurred unintentionally, through the inadvertent movement of species due to global trade.

The pathways of introduction have changed very much in the last centuries, showing a decrease in the role of unintentional transport, and an increase of the escapes and of the spread of established populations (Fig. 9.5). The number of intentional introductions is decreasing, probably due to an increasing awareness on the problems related to biological invasions. It is interesting to note that, considering only those introduction events that occurred since 1960 (n = 100), a large proportion of introductions were from fur farms (23%), hunting (17%) and pet trade (15%). Moreover, if we only consider the last 10 years (1997–2007), eight intentional and 11 unintentional introductions have occurred. Cases of intentional introductions include the release of captive animals for ornamental purposes, for "fauna improvement", or for hunting (i.e. aoudad *Ammotragus lervia* and eastern cottontail *Sylvilagus floridanus*).

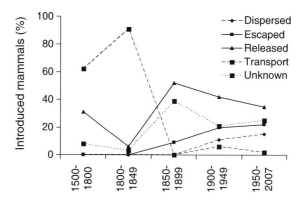

Fig. 9.5 Changes over time of the role of major vectors responsible of the introduction of mammals in Europe

9.5 Biogeographic Patterns

The number of introduced species per country (excluding those of uncertain origin) varies from 1 to 31 (average 7.9), while the number of established species per country ranges from 1 to 12 (average 4.9). Both numbers of introduced and established mammal species per country correlate to country area ($R^2 = 0.33$ and $R^2 = 0.5$ respectively).

The average success rate (calculated as the ratio between number of established and introduced species per country) is 0.72 (±0.22), much higher than what would be expected following the Ten's rule (Williamson 1996). This high invasion success rate can be explained partly with a possible bias in the database (successful introductions are more likely to be reported than unsuccessful ones), but it is also likely due to a high success rate of mammals invasions in suitable environments, confirming the conclusions of previous studies (e.g., Jeschke and Strayer 2005). The most widespread species (established in over 10 European countries) are brown rat *Rattus norvegicus*, muskrat *Ondatra zibethicus*, American mink *Mustela vison*, raccoon dog *Nyctereutes procyonoides*, coypu *Myocastor coypus*, fallow deer *Dama dama*, sika deer *Cervus nippon* and raccoon *Procyon lotor*. Interestingly, all these widespread species are not native to Europe, and all cause rather significant impacts in the invaded areas.

The most invaded countries (Fig. 9.6) are Germany, with 31 species, and the UK with 30, followed by Denmark (18), mainland France (16), the Czech Republic (16), Russia (15) and Italy (14). However the order changes when considering established species only: in this case Russia with 12 species is the most invaded country, followed by mainland France with 10 species, Germany and the UK (9 each), and the Czech Republic, Ukraine, Sweden, Austria and Belgium (8 each). The difference is due to the high number of extinct species recorded in countries like the UK and Germany where 18 and 15 species disappeared after introduction, respectively. It should be noted that the information stored in the DAISIE database

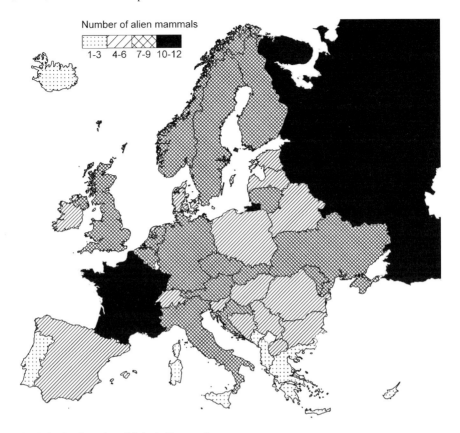

Fig. 9.6 Number of established alien species per country

may be geographically biased, with cases of undetected or unreported invasions in areas with less monitoring activities, for example in some eastern European countries.

If we analyse the number of established alien mammals in European countries in respect to the geographic location of the country, there is a hump-shaped latitudinal pattern, with an increase in the number of species from subtropic to central Europe, and a subsequent decrease in the extreme northern areas (Fig. 9.7); while it does not seem to be any significant relationship with longitude ($R^2 = 0.047$).

9.6 Impacts on Ecology, Ecosystems and Economy

If we consider the whole set of species in the databases, which includes also those not established or with unknown status (some of which could get established in the future), and excluding species extinct in all countries (24 species), the number of

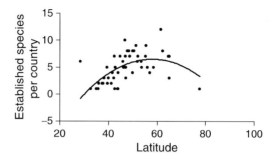

Fig. 9.7 Quadratic correlation between number of established species and latitude (n = 52 regions, $R^2 = 0.33$)

alien mammals causing known ecological impact (in the countries where they have been introduced or elsewhere) is 37 (58%). The number of those with a known impact on human activities (e.g., causing damage to crops, riverbanks, livestock, etc.) or health is 40 species (63%). However, the percentages increase sharply if we limit our analysis to the subset of the 50 established species. In this case at least 34 species (68%) have a known ecological impact. The number of species with a known impact on uses/resources is even higher: 36 (72%). The impact of the remaining species is unknown meaning that it has not been evaluated.

Also in this case the available data on alien mammals in Europe seem to contradict Williamson's Tens rule, confirming the much higher impact caused by this group compared to others. The mechanisms of the impacts caused by alien mammals to European biodiversity are very variable, from direct predation (the American mink threatening native water vole *Arvicola terrestris* in the UK), to competition (American grey squirrel **Sciurus carolinensis** outcompeting the native red squirrel *S. vulgaris*, causing its extinction in all areas of overlap), to hybridisation (sika deer threatening the genetic integrity of red deer in Scotland through interbreeding). Alien mammals also cause huge impacts to European economy and well-being, like coypu damaging crops and weakening the stability of river banks through digging, causing in Italy alone losses over 4 million Euros/year (Panzacchi et al. 2007), not to mention the epidemiological role played by several rodent species, including rats and the Siberian chipmunk **Tamias sibiricus** (Vourc'h et al. 2007).

9.7 Management Options and Their Feasibility

Matching the particularly severe impacts that alien mammals cause, many species are subject to some kind of control management in Europe. In particular, all the most widespread species (coypu, muskrat, rats, American grey squirrel, raccoon, raccoon dog, sika deer, American mink, fallow deer) are controlled in several countries and some have also been subject to eradication programs. Moreover, removal methods by trapping, shooting, etc. are more developed for mammals than for any other vertebrate, and in fact most eradications carried out in Europe have targeted mammal species, such as rats, American minks, coypu, Canadian beaver or Himalayan porcupine (see Genovesi 2005 for a review).

On the basis of the DAISIE dataset, it is evident that there are several species of mammals posing severe threats to regional biodiversity with still very local distribution, making these species very good candidates for eradication programs. Seventeen alien mammals are established in one country only; this is the case of the Pallas squirrel which is only present in Antibes (France); Finlayson's squirrel *Callosciurus fynlaisonii*, only present in Italy with two distinct populations; eastern cottontail, established in Italy; the coati *Nasua nasua* only present in Majorca (Balearic islands, Spain) or three species (Egyptian rousette *Rousettus aegyptiacus*, Algerian hedgehog *Atelerix algirus* and Barbary ground squirrel *Atlantoxerus getulus*) only recorded in the Canary islands (Spain). But the feasibility of an eradication programme should also be considered for several species recorded in more than one country, although still very localised in all or some of the countries; this is the case of the aoudad in Croatia (only present on the Mosor mountain in Dalmatia) and Italy (small population close to Varese), as well as the coypu in northern Spain (present with two very local populations).

Acknowledgements The support of this study by the European Commission's Sixth Framework Programme project DAISIE (Delivering alien invasive species inventories for Europe, contract SSPI-CT-2003-511202) is gratefully acknowledged. We also would like to thank the following experts for having provided us with information, including unpublished data, critical for establishing the DAISIE mammal databases: Giovanni Amori, Sandro Bertolino, Laura Bonesi, Jean-Louis Chapuis, Kristijan Civic, Morris Gosling, Duncan Halley, Ohad Hatzofe, Miklós Heltai, Mark Hill, Djuro Huber, Jasna Jeremic, Melanie Josefsson, Kaarina Kauhala, Colin Lawton, Olivier Lorvelec, Petros Lymberakis, Marco Masseti, Joao Mayol, Robbie A. McDonald, Dan Minchin, Tony Mitchell-Jones, Petri Nummi, Jorge Orueta, Milan Paunovic, Rory Putnam, Ana Isabel Queiroz, David Roy, Patrick Schembri, Vadim E. Sidorovich, Wojciech Solarz, Ondrej Sedlacek, Andrea Stefan, Hennekens Stephan, Jan Stuyck, Valter Trocchi, Goedele Verbeylen and Marten Winter.

References

Andreotti A, Baccetti N, Perfetti A, Besa M, Genovesi P, Guberti V (2001) Mammiferi ed uccelli esotici in Italia: analisi del fenomeno, impatto sulla biodiversità e linee guida gestionali. Quad Cons Natura, 2, Min. Ambiente, Ist Naz Fauna Selvatica

Baagøe HJ, Jensen TS (2007) Dansk pattedyratlas. Gyldendal, Copenhagen

Battersby J, Tracking Mammals Partnership (2005) UK mammals: species status and population trends. First report by the tracking mammals partnership. JNCC/Tracking Mammals Partnership, Peterborough

Broggi MF (2006) Säugetiereneozoen im Fürstentum Liechtenstein. Neobiota im Fürstentum Liechtenstein. Amt für Wald, Natur, Landschaft, 113–117

Englisch H (2002) Säugetiere (Mammalia) In: Essl F, Rabitsch W (eds) Neobiota in Österreich. Umweltbundesamt, Wien, 214–221

Geiter O, Homma S, Kinzelbach R (2002) Bestandsaufnahme und Bewertung von Neozoen in Deutschland. Forschungsbericht 296, Umweltbundesamt, Berlin

Genovesi P (2005) Eradications of invasive alien species in Europe: a review. Biol Invasions 7:127–133

Jeschke JM, Strayer DL (2005) Invasion success of vertebrates in Europe and North America. Proc Natl Acad Sci USA 102:7198–7202

Lever C (1985) Naturalized mammals of the world. Longman, New York

Long JL (2003) Introduced mammals of the world. CABI/CSIRO Publishing, Melbourne

Mitchell-Jones AJ, Amori G, Bogdanowicz W, Krystufek B, Reijnders PJH, Spitzenberger F, Stubbe M, Thissen JBM, Vohralik V, Zima J (1999) The atlas of European mammals. Poyser Natural History, Academic, London

NOBANIS (2007) North European and Baltic Network on Invasive Alien Species. www.nobanis.org. Cited Dec 2007

Nogales M, Rodriguez-Luengo JL, Marrero P (2006) Ecological effects and distribution of invasive non-native mammals on the Canary Islands. Mammal Rev 36:1–49

Palomo LJ, Gisbert J (eds) (2002) Atlas de los mamíferos terrestres de España. Dirección General de Conservación de la Naturaleza. SECEM-SECEMU, Madrid

Panzacchi M, Bertolino S, Cocchi R, Genovesi P (2007) Population control of coypu in Italy vs. eradication in UK: a cost/benefit analysis. Wildlife Biol 13:159–171

Pascal M, Lorvelec O, Vigne JD (2006) Invasions biologiques et extinctions. 11,000 ans d'histoire des vertébrés en France. Quae-Belin Editions, Paris

Reeves R, Notarbartolo di Sciara G (eds) (2006) The status and distribution of cetaceans in the Black Sea and Mediterranean Sea. IUCN Centre for Mediterranean Cooperation, Malaga

Sax DF, Stachowicz JJ, Brown JH, Bruno JF, Dawson MN, Gaines SD, Grosberg RK, Hastings A, Holt RD, Mayfield MM, O'Connor MI, Rice WR (2007) Ecological and evolutionary insights from species invasions. Trends Ecol Evol 22:465–471

Scalera R (2001) Invasioni biologiche. Le introduzioni di vertebrati in Italia: un problema tra conservazione e globalizzazione. Collana Verde, 103. Corpo Forestale dello Stato, Ministero per le politiche agricole e forestali, Rome

Stokes K, O'Neill K, McDonald RA (2006) Invasive species in Ireland. Report to environment and heritage service and national parks and wildlife service by Quercus, Queens University, Environment and Heritage Service, Belfast and National Parks and Wildlife Service, Dublin

Temple HJ, Terry A (eds) (2007) The status and distribution of European mammals. Office for Official Publications of the European Communities. Luxembourg

Vourc'h G, Marmet J, Chassagne M, Bord S, Chapuis JL (2007) *Borrelia burgdorferi* sensu lato in Siberian chipmunks (*Tamias sibiricus*) introduced in suburban forests in France. Vector Borne Zoonotic Dis 7:637–642

Weidema IR (ed) (2000) Introduced species in the Nordic Countries. Nord Environ 13:1–24

Weijden WJ van der, Leeuwis R, Bol P (2005) Biologische globalisering. Omvang, oorzaken, gevolgen, handelingsperspectieven. Achtergronddocument voor de Beleidsnota Invasieve Soorten van het ministerie van Landbouw, Natuur en Voedselkwaliteit. CLM, Milieu en Natuurplanbureau en TU Delft, Culemborg

Williamson M (1996) Biological invasions. Chapman & Hall, London

Wittenberg R (ed) (2006) Invasive alien species in Switzerland. An inventory of alien species and their threat to biodiversity and economy in Switzerland. Environmental Studies 29, Federal Office for the Environment, Bern

Chapter 10
Introduction to the List of Alien Taxa

Sergej Olenin and Viktoras Didžiulis

10.1 Content of the List of Alien Taxa

The list of alien taxa presented in this book is an extraction from the European Alien Species Database developed within the course of the DAISIE project. The overall number of alien species registered in the DAISIE database by February 2007 is 10,771. The alien species list represents a broad taxonomic spectrum of terrestrial and aquatic free living and parasitic organisms. In total, there are 46 phyla, 101 classes, 334 orders, 1,267 families and 4,492 genera recorded in the database. The phyla of Magnoliophytina, Arthropoda and Chordata contain the highest numbers of the alien species: 61%, 23% and 6%, accordingly (Fig. 10.1). Most of the phyla (28) contain less than 10 recorded species; and there are 10 phyla represented by only one recorded alien species.

The list as it is printed in this book includes only taxonomic information on recorded alien species, subspecies and hybrids. The DAISIE database (www.europe-aliens.org), however, contains documented introduction records of the alien taxa for 71 terrestrial and nine marine regions of Europe, the Levantine Basin and the North African coast of the Mediterranean Sea. These regions mostly match political borders of countries as they are known today. However it needs to be mentioned that where data were available at a finer political or biogeographic level these are also presented e.g., the administrative regions of the United Kingdom (Wales, Scotland, England, northern Ireland) or islands like Corsica, Sicily, Crete, Greenland or Svalbard. The total number of species-region records is 45,211, including 41,347 records for terrestrial regions and 3,864 for aquatic regions (Table 10.1).

The classification scheme of the terrestrial and aquatic regions used in the DAISIE database was based on publicly available World Geographical Scheme for Recording Plant Distributions (TDWG 1992) developed by the TDWG Biological Information Standards group formerly known as the Taxonomic Database Working Group (for terrestrial regions); for the marine part we adopted the classification known as Limits of Oceans and Seas (S-23) created by the International Hydrographic Organization (IHO 1986). Additionally the largest inland water bodies were added to allow for the more precise specification of geographic information on occurrence

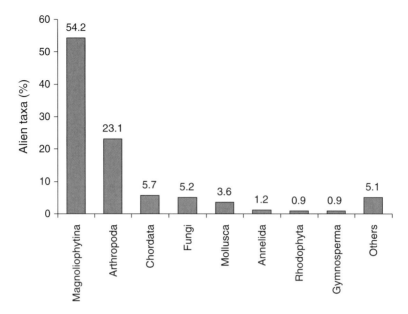

Fig. 10.1 Percentage of the total number of alien species recorded in the DAISIE database in the largest phyla (total N = 10,771 as of February 2008)

Table 10.1 Numbers of species-region records representing different taxonomic/biom groups in the DAISIE database (as of February 2008)

Taxonomic/biom group	Number of introduction events
Aquatic marine	2,777
Aquatic inland	1,087
Terrestrial invertebrates	11,776
Terrestrial vertebrates	1,478
Terrestrial plants	28,093

of aquatic species. This hierarchical system of regions has four levels and there are in total 1,039 regions described. Such approach allows flexibility in dealing with uncertainties in the knowledge of experts and publications. The same system was applied to describe donor areas, i.e. the region(s) from which alien species were introduced as well as their native ranges (biogeographic origin of the species).

Besides the taxonomy and data on regional introduction records for each alien species the DAISIE database contains information on synonyms, ecofunctional grouping (such as phytoplankton, zooplankton, macroalgae, terrestrial plant, herbivorous fish, etc.), species status (native in some parts of Europe but alien in other parts of Europe, alien to Europe, or cryptogenic (sensu Carlton 1996), i.e. origin is unknown and it cannot be distinguished whether it is native or alien), introduction time, data on ecological and economic impacts, invasion history, population status as well as comments as free text. Sources of the data can be backtracked to original references and contributors.

10.2 Data Collation Process

The collation of the alien species list and development of the database was the result of the dedicated efforts of the DAISIE project partners and more than 300 collaborators from Europe and neighbouring countries. Data collection, quality assessment of the datasets and their integration in the database was a challenging task for DAISIE. Specialists involved in the project represented different fields of expertise and academic schools; therefore, at the initial phase of the project, there was a need to harmonise the alien species terminology and vocabulary used in the database (Pyšek et al. 2008).

The DAISIE partners represented 15 countries, thus the DAISIE project goals and principles were presented at several scientific meetings and e-conferences, that yielded new contacts with collaborators outside the DAISIE consortium agreeing to gratuitously share their data with the project. As a result an essential number of checklists for countries were compiled, and the project is deeply grateful to the collaborators, whose expertise and dedication helped to essentially populate the European Alien Species Database. Both project partners and collaborators have invested considerable time in libraries, natural history museums, environmental data centers and research laboratories in order to mine data, check information quality and prepare the datasets for integration into the Database.

The DAISIE partners and collaborators created a hierarchical contribution network to collect the data in a most efficient way. The experts were asked to submit the alien species checklists for their countries/regions according to their taxonomic competence to five focal points responsible for compilation of terrestrial plants, terrestrial invertebrates, terrestrial vertebrates, aquatic inland and marine datasets. The taxonomic checklists were further analyzed and checked for technical errors, such as typographical errors, incorrect assignments of attributes, duplicate entries or data format mismatches. The corrected, re-classed, normalised data were fed into a relational database. Considerable effort was spent on the problems of synonymies and the most recent taxonomy as well as on finding and defining native ranges and donor areas for each species.

10.3 Concluding Remarks

It should be noted that not each and every species in this list is alien in every region of Europe. This means that quite many species in the list are native and even may be endangered or endemic in some areas of Europe but are alien in others. The list of alien taxa as presented in this book is different from the DAISIE database insofar as it contains only species, subspecies and hybrid names. We marked alien taxa from outside Europe with an A, European species which became alien outside their native range with an E, and all species of unknown origin (i.e. either type A or E) with a C for cryptogenic. All further information was excluded. This additional information which represents the main richness of the DAISIE project can be found in the internet via the DAISIE project's portal at www.europe-aliens.org. Also, as the database is being updated, the species numbers presented in this book will deviate from the future state of the database.

The current total number of records in the DAISIE database is around 100,000. This number is constantly growing due to increasing knowledge, arrival of new alien species, their spread into new regions (i.e. addition of new introduction records), changes in species taxonomy, etc. Therefore it is advisable to visit the DAISIE portal at www.europe-aliens.org for the most up-to-date information and use the database exploration tools presented therein.

The DAISIE project partners and collaborators tried as far as possible to avoid mistakes and uncertainties in the database to make it robust and reliable. However due to the dynamic nature and enormous amount of the collected information further specifications, corrections and additions will be definitely needed. The DAISIE project encourages all users of the European alien species database to contribute to its improvement by submitting of new and correction of existing data in order to make it a trustworthy collective tool for practical and academic purposes related to biological invasions in Europe and other regions of the world.

References

Carlton JT (1996) Biological invasions and cryptogenic species. Ecology 77:1653–1655
IHO (1986) Limits of oceans and seas. Special publication S-23. International Hydrographic Organization, Monaco. www.iho.shom.fr/welcome.htm. Cited Dec 2007
Pyšek P, Hulme PE, Nentwig W (2008) Glossary of the main technical terms used in the Handbook. In: DAISIE Handbook of alien species in Europe. Springer, Dordrecht, 375–379
TDWG (1992) World geographic scheme for recording plant distributions. Biodiversity Information Standards TDWG. www.tdwg.org. Cited Dec 2007

Chapter 11
List of Species Alien in Europe and to Europe

Species printed in **bold** belong to the 100 most invasive alien species in Europe (chapter 13).

A   Alien taxon from outside Europe
E   European species which became alien outside its native range
C   Cryptogenic species

Bacteria
Flavobacteriaceae
    A   *Aeromonas salmonicida* (Lehmann & Neumann, 1896) Griffin et al., 1953
Aerococcaceae
    A   *Aerococcus viridans* Williams et al., 1953
Vibrionaceae
    A   *Vibrio cholerae* Pacini, 1854
    A   *Vibrio vulnificus* (Reichelt et al., 1979) Farmer, 1980

Protista
Ceratiaceae
    A   *Ceratium candelabrum* (Ehrenberg) Stein
Chatonellaceae
    A   *Olisthodiscus luteus* N. Carter
Dinophysiaceae
    C   *Dinophysis acuta* Ehrenberg
Haplosporidiidae
    A   *Bonamia ostreae* Pichot, Comps, Tige, Grizel & Rabouin, 1980
    A   *Haplosporidium armoricanum* (Van Banning, 1977)
    A   *Haplosporidium nelsoni* Haskin, Stauber, & Mackin
Oodiniaceae
    C   *Pfiesteria piscicida* K.A. Steidinger & J.M. Burkholder
    C   *Pfiesteria shumwayae* H.B. Glasgow & J.M. Burkholder
Peridiniaceae
    A   *Pentapharsodinium tyrrhenicum* (Balech, 1990) Montresor et al.
    A   *Scrippsiella hangoei* (J. Schiller) J. Larsen
Perkinsidae
    C   *Marteilia refringens* (Grizel et al., 1974)
    A   *Perkinsus atlanticus* Azevedo, 1989
    A   *Perkinsus olseni* Lester & Davis, 1981
Scyphidiidae
    A   *Ambiphrya ameiuri* (Thompson, Kirkegaard & Jahn, 1947)

Chromista
Albuginales
    C   *Albugo candida* (Pers.) Roussel (1806)
Peronosporales
    A   *Basidiophora entospora* Roze & Cornu (1869)
    C   *Bremia lactucae* Regel (1843)
    C   *Hyaloperonospora parasitica* (Pers.) Constant. (2002)
    C   *Paraperonospora leptosperma* (de Bary) Constant. (1989)
    E   *Peronospora aestivalis* Syd. (1923)
    C   *Peronospora affinis* Rossmann (1863)
    C   *Peronospora arborescens* (Berk.) de Bary (1855)
    A   *Peronospora arthurii* Farlow
    C   *Peronospora barbareae* Gäum. (1918)
    C   *Peronospora berteroae* Gäum.
    C   *Peronospora buniadis* Gäum. (1918)
    E   *Peronospora camelinae* Gäum. (1918)
    C   *Peronospora chlorae* de Bary (1872)
    E   *Peronospora cochleariae* Gäum (1923)
    C   *Peronospora conglomerata* Fuckel (1863)

C *Peronospora corydalis-intermediae* Gäum. (1923)
A *Peronospora destructor* (Berk.) Casp. ex Berk. (1860)
C *Peronospora digitalis* Gäum. (1923)
C *Peronospora ducometii* Siemaszko & Jankowska
E *Peronospora erodii* Fuckel, 1867
C *Peronospora erysimi* Gäum. (1918)
A *Peronospora fagopyri* Tanaka (1934)
C *Peronospora farinosa* (Fr.) Fr., 1849
C *Peronospora grisea* (Unger) de Bary (1863)
E *Peronospora hesperidis* Gäum. (1918)
E *Peronospora iberidis* Gäum. (1927)
C *Peronospora jaapiana* Magnus (1910)
A *Peronospora jacksonii* C.G. Shaw (1951)
E *Peronospora kochiae* Gäum. (1919)
C *Peronospora lagerheimii* Gäum. (1923)
A *Peronospora lamii* A. Braun, 1857
A *Peronospora manshurica* (Naumov) Syd. in Gäumann (1923)
E *Peronospora matthiolae* Gäum. (1918)
C *Peronospora mayorii* Gäum. (1923)
C *Peronospora media* Gäum. (1920)
C *Peronospora myosotidis* de Bary (1863)
C *Peronospora obovata* Bonord. (1890)
E *Peronospora ornithopi* Gäum. (1923)
C *Peronospora phyteumatis* Fuckel (1865)
C *Peronospora scleranthi* Rabenh. (1850)
C *Peronospora sisymbrii-officinalis* Gäum. (1918)
A *Peronospora sparsa* Berk. (1862)
A *Peronospora spinaciae* Laubert
C *Peronospora statices* Lobik
A *Peronospora tomentosa* Fuckel (1863)
C *Peronospora trifoliorum* de Bary (1863)
E *Peronospora viciae* (Berk.) Casp., 1885
C *Peronospora violae* de Bary (1863)
A *Plasmopara angustiterminalis* Novot. (1962)
A *Plasmopara halstedii* (Farl.) Berl. & De Toni (1888)
C *Plasmopara nivea* (Unger) J. Schröt. (1886)
A *Plasmopara pastinacae* Săvul. & O. Săvul. (1951)
A *Plasmopara viticola* (Berk. & M.A. Curtis) Berl. & De Toni (1888)
A *Pseudoperonospora cannabina* (G.H. Otth) Curzi (1926)
A *Pseudoperonospora cubensis* (Berk. & Curtis) Rostovzev (1903)
A *Pseudoperonospora humuli* (Miyabe & Takah.) G.W. Wilson (1914)

Pythiales

C *Phytophthora alni* Brasier & S.A. Kirk (2004)
E *Phytophthora cactorum* (Lebert & Cohn) J. Schröt. (1886)
C *Phytophthora cambivora* (Petri) Buisman (1927)
C *Phytophthora capsici* Leonian (1922)
A **Phytophthora cinnamomi** Rands (1922)
C *Phytophthora citricola* Sawada (1927)
C *Phytophthora citrophthora* Sawada (1927)
C *Phytophthora cryptogea* Pethybr. & Laff., 1919
C *Phytophthora drechsleri* Tucker (1931)
C *Phytophthora erythroseptica* Pethybr. (1913)
C *Phytophthora hedraiandra* De Cock & Man in 't Veld (2004)
C *Phytophthora hibernalis* Carne, 1925
C *Phytophthora ilicis* Buddenh. & Roy A. Young (1957)
A *Phytophthora infestans* (Mont.) de Bary (1876)
C *Phytophthora inundata* Brasier, Sánch. Hern. & S.A. Kirk (2003)
C *Phytophthora kernoviae* Brasier, Beales & S.A. Kirk (2005)
A *Phytophthora lateralis* Tucker & Milbrath (1942)
C *Phytophthora megasperma* Drechsler (1931)
A *Phytophthora nicotianae* Breda de Haan (1896)
A *Phytophthora palmivora* (E.J. Butler) E.J. Butler (1919)
C *Phytophthora quercina* T. Jung (1999)
C *Phytophthora ramorum* Werres De Cock & Man in 't Veld (2001)
C *Phytophthora sojae* Kaufm. & Gerd. (1958)
C *Phytophthora syringae* (Kleb.) Kleb.
C *Pythium dimorphum* F.F. Hendrix & W.A. Campb. (1971)
C *Pythium mamillatum* Meurs (1928)
A *Pythium prolatum* W.A. Campb. & F.F. Hendrix (1969)
A *Pythium splendens* Hans Braun (1925)
C *Pythium sylvaticum* W.A. Campb. & F.F. Hendrix (1967)

Saprolegniales

A **Aphanomyces astaci** Schikora (1906)
C *Aphanomyces euteiches* Drechsler (1925)
A *Aphanomyces raphani* J.B. Kendr. (1927)

A *Ostracoblabe implexa* Bornet & Flahault, 1889

Sclerosporales
C *Sclerophthora macrospora* (Sacc.) Thirum., C.G. Shaw & Naras. (1953)

Fungi

Agaricales
C *Agaricus bernardii* Quél. (1887)
E *Agaricus bisporus* (J.E. Lange) Pilát, 1951
C *Agaricus placomyces* Peck (1878)
A *Agrocybe putaminum* (Maire) Singer, 1936
C *Amanita inopinata* D.A. Reid & Bas (1987)
C *Clavulinopsis daigremontiana* (Boud.) Corner (1950)
A *Collybia biformis* (Peck) Singer (1962)
A *Conocybe crispella* (Murrill) Singer, 1950
C *Conocybe intrusa* (Peck) Singer (1950)
A *Crepidotus crocophyllus* (Berk. & M.A. Curtis) Sacc. (1887)
C *Cyathus stercoreus* (Schwein.) De Toni (1888)
A *Descolea antarctica* Singer (1950)
A *Descolea maculata* Bougher (1986)
C *Gerhardtia piperata* (A.H. Smith) Bon (1994)
C *Hydnangium carneum* Wallroth (1839)
A *Laccaria fraterna* (Cooke & Massee) Pegler (1965)
A *Lentinula edodes* (Berk.) Pegler (1976)
A *Lepiota elaiophylla* Vellinga & Huijser (1998)
C *Lepiota × anthophylla* P.D. Orton (1960)
A *Leucoagaricus americanus* (Peck) Vellinga (2000)
A *Leucoagaricus marginatus* (Burl.) Boisselet (2002)
C *Leucoagaricus melanotrichus* (Malençon & Bertault) Trimbach (1975)
A *Leucocoprinus birnbaumii* (Corda) Singer (1962)
C *Leucocoprinus bresadolae* (Schulzer) M.M. Moser (1967)
A *Leucocoprinus cepistipes* (Sowerby) Pat., 1889
A *Leucocoprinus cretaceus* (Bull.) Locq., 1945
A *Leucocoprinus ianthinus* (Cooke) Locq. (1945)

C *Leucocoprinus lilacinogranulosus* (Henn.) Locq. (1943)
A *Leucocoprinus straminellus* (Bagl.) Narducci & Caroti (1995)
C *Leucocoprinus tenellus* (Boud.) Locq., 1943
C *Leucocoprinus zeylanicus* (Berk.) Boedijn (1940)
C *Marasmius buxi* Fr., 1872
A *Melanotus flavolivens* (Berk. & M.A. Curtis) Singer (1946)
A *Mycena alphitophora* (Berk.) Sacc. (1875)
C *Omphalotus olearius* (DC.) Singer, 1946
A *Pluteus variabilicolor* Babos (1978)
A *Psilocybe cyanescens* Wakef. (1946)
C *Queletia mirabilis* Fr. (1871)
A *Sclerotium hydrophilum* Sacc. (1899)
A *Setchelliogaster rheophyllus* (Bertault & Malençon) G. Moreno & Kreisel (1997)
A *Setchelliogaster tenuipes* (Setch.) Pouzar, 1958
A *Squamanita odorata* (Cool) Imbach, 1946
C *Strobilurus esculentus* (Wulfen) Singer, 1962
A *Stropharia aurantiaca* (Cooke) M. Imai (1938)
E *Stropharia percevalii* (Berk. & Broome) Sacc., 1887
A *Stropharia rugosoannulata* Farl. ex Murrill (1922)
A *Tricholoma caligatum* (Viv.) Ricken (1914)
C *Tricholoma psammopus* (Kalchbr.) Quél. (1875)
C *Typhula trifolii* Rostr. (1890)

Amanitales
A *Amanita asteropus* Sabo (1963)
C *Amanita singeri* Bas (1969)

Atheliales
A *Athelia rolfsii* (Curzi) C.C. Tu & Kimbr. (1978)

Blastocladiales
C *Physoderma alfalfae* (Pat. & Lagerh.) Karling (1950)

Boletales
A *Boletinus asiaticus* Singer (1938)
C *Boletinus paluster* (Peck) Peck (1889)
A *Boletinus pictus* (Peck) Lj.N. Vassiljeva, 1978
C *Chroogomphus helveticus* (Singer) M.M. Moser, 1967

- A *Descomyces albus* (Klotzsch) Bougher & Castellano, 1993
- E *Gastrosporium simplex* Mattir. (1903)
- C *Gomphidius glutinosus* (Schaeff.) Fr. (1838)
- C *Gomphidius maculatus* (Scop. : Fr.) Fr.(1838)
- C *Pisolithus arrhizus* (Scop.) Rauschert (1959)
- A *Rhizopogon villosulus* Zeller (1941)
- C *Serpula lacrymans* (Wulfen) J. Schröt. (1885)
- A *Suillus amabilis* (Peck) Singer (1966)
- C *Suillus bovinus* (Pers.) Roussel, 1898
- C *Suillus cavipes* (Opat.) A.H. Sm. & Thiers (1964)
- C *Suillus clintonianus* (Peck) Kuntze (1898)
- C *Suillus flavidus* (Fr.) J. Presl, 1846
- C *Suillus grevillei* (Klotzsch) Singer, 1951
- C *Suillus luteus* (L.) Roussel, 1796
- C *Suillus placidus* (Bonord.) Singer, 1945
- A *Suillus spectabilis* (Peck) Kuntze (1898)
- C *Suillus variegatus* (Sw.) Kuntze, 1898
- C *Suillus viscidus* (L.) Fr., 1796

Botryosphaeriales
- C *Diplodia lilacis* Westend. (1852)
- C *Dothidotthia celtidis* (Ellis & Everh.) M.E. Barr (1989)
- C *Guignardia philoprina* (Berk. & M.A. Curtis) Aa, 1973
- A *Macrophomina phaseolina* (Tassi) Goid. (1947)
- C *Phyllosticta mahoniicola* Pass.
- C *Phyllosticta violae* Desm. (1847)

Capnodiales
- C *Cercospora apii* Fresen. (1863)
- C *Cercospora thalictri* Thüm.
- C *Cladosporium macrocarpum* Preuss (1848)
- C *Cladosporium phlei* (C.T. Greg.) G.A. de Vries (1952)
- C *Cymadothea trifolii* (Pers.) F.A. Wolf (1935)
- C *Davidiella carinthiaca* (Jaap) Aptroot (2006)
- C *Mycosphaerella cydoniae* Grove (1918)
- A *Mycosphaerella linicola* Naumov (1926)
- C *Mycosphaerella pinodes* (Berk. & A. Bloxam) Vestergr. (1912)
- C *Mycosphaerella podagrariae* (Fr.) Petr., 1921
- C *Mycosphaerella rhododendri* Lindau, 1903
- C *Mycosphaerella ribis* (Fuckel) Lindau (1903)
- C *Mycovellosiella concors* (Casp.) Deighton (1974)
- A *Mycovellosiella fulva* (Cooke) Arx (1983)
- C *Ramularia collo-cygni* B. Sutton & J.M. Waller (1988)
- C *Ramularia matronalis* Sacc. (1886)
- C *Ramularia rhei* Allesch. (1896)
- C *Ramularia sambucina* Sacc. (1882)
- C *Ramularia sonchi-oleracei* Fautrey (1891)
- C *Septoria betulae* Pass. (1867)
- C *Septoria erysimi* Niessl (1864)
- C *Septoria lamiicola* Sacc. (1884)
- C *Septoria petroselini* Desm. (1843)
- C *Septoria phleina* Baudyš & Picb. (1926)
- C *Septoria phlogis* Sacc. & Speg. (1878)
- C *Septoria scleranthi* Desm. (1857)
- C *Septoria secalis* Prill. & Delacr. (1889)
- C *Septoria stellariae* Roberge ex Desm. (1847)

Chytridiales
- A *Synchytrium endobioticum* (Schilb.) Percival (1909)
- C *Synchytrium laetum* J. Schröt. (1870)

Cortinariales
- C *Bolbitius coprophilus* (Peck) Hongo, 1959
- A *Bolbitius incarnatus* Hongo (1958)

Dacrymycetales
- A *Calocera pallidospathulata* D.A. Reid (1974)

Diaporthales
- C *Apiognomonia errabunda* (Roberge ex Desm.) Höhn. (1918)
- C *Apiognomonia veneta* (Sacc. & Speg.) Höhn. (1920)
- A *Cryphonectria parasitica* (Murrill) M.E. Barr (1978)
- C *Cryptodiaporthe aesculi* (Fuckel) Petr. (1921)
- C *Cryptodiaporthe castanea* (Tul. & C. Tul.) Wehm., 1933
- C *Cryptodiaporthe robergeana* (Desm.) Wehm., 1933
- C *Cryptosporella platanigera* (Berk. & Broome) Sacc. (1882)
- C *Diaporthe aucubae* Sacc. (1878)
- C *Diaporthe citri* F.A. Wolf (1926)

A *Diaporthe helianthi* Munt.-Cvetk. Mihaljč. & M. Petrov (1981)
C *Diaporthe ilicis* (Ellis & Everh.) Wehm. (1933)
C *Diaporthe nobilis* Sacc. & Speg. (1878)
C *Diaporthe oncostoma* (Duby) Fuckel (1870)
C *Diaporthe perniciosa* Marchal & É.J. Marchal (1921)
C *Diaporthe skimmiae* Grove, 1933
C *Diaporthe strumella* (Fr.) Fuckel (1870)
A *Diaporthe vaccinii* Shear, 1931
E *Diplodina aesculi* (Sacc.) B. Sutton (1980)
A *Diplodina rhoina* Hollós
A *Discula destructiva* Redlin (1991)
C *Gnomonia leptostyla* (Fr.) Ces. & De Not. (1863)
C *Gnomonia tetraspora* G. Winter, 1872
C *Hapalocystis berkeleyi* Auersw. ex Fuckel, 1863
C *Leucostoma curreyi* (Nitschke) Défago (1942)
C *Leucostoma kunzei* (Fr.) Munk, 1953
C *Melanconis modonia* Fuckel (1863)
A *Melanconium oblongum* Berk. (1874)
A *Phomopsis juniperivora* G. Hahn (1920)
C *Phomopsis limonii* Vegh (1994)
C *Valsa laurocerasi* Tul. & C. Tul., 1863

Dothideales
A *Guignardia aesculi* (Peck) V.B. Stewart (1916)
A *Guignardia bidwellii* (Ellis) Viala & Ravaz (1892)
A *Kabatiella caulivora* (Kirchn.) Karak. (1923)
E *Kabatiella lini* (Laff.) Karak., 1957
A *Kabatiella zeae* Narita & Y. Hirats. (1959)
A *Mycosphaerella mori* (Fuckel) F.A. Wolf, 1935
C *Otthia spiraeae* (Fuckel) Fuckel (1870)
A *Stigmina thujina* (Dearn.) B. Sutton (1972)

Entylomatales
E *Entyloma borraginis* (Cif, 1924)
E *Entyloma calendulae* (Oudem.) de Bary
C *Entyloma compositarum* Farl., 1883
A *Entyloma dahliae* Syd. & P. Syd.
C *Entyloma fergussonii* (Berk. & Broome) Plowr. (1889)
C *Entyloma fuscum* J. Schröt. (1877)
A *Entyloma gaillardianum* Vánky
A *Entyloma ludwigianum* Syd. (1932)

Erysiphales
E *Arthrocladiella mougeotii* (Lév.) Vassilkov
C *Blumeria graminis* (D+D241C.) Speer (1975)
C *Erysiphe alphitoides* (Griffon & Maubl.) U. Braun & S. Takam. (2000)
A *Erysiphe arcuata* U. Braun V.P. Heluta & S. Takamatsu (2006)
A *Erysiphe australiana* (McAlpine) U. Braun & S. Takam. (2000)
C *Erysiphe azaleae* (U. Braun) U. Braun & S. Takamatsu
A *Erysiphe begoniicola* U. Braun & S. Takam. (2000)
E *Erysiphe betae* (Vanha) Weltzier
A *Erysiphe catalpae* Simonjan
C *Erysiphe cruciferarum* Opiz ex L. Junell (1967)
A *Erysiphe deutziae* (Bunkina) U. Braun & S. Takam. (2000)
C *Erysiphe elevata* (Burrill) U. Braun & S. Takam.
A *Erysiphe euonymi-japonici* (Vienn.-Bourg.) U. Braun & S. Takamatsu
A *Erysiphe flexuosa* (Peck) U. Braun & S. Takam. (2000)
C *Erysiphe heraclei* DC. (1815)
A *Erysiphe howeana* U. Braun
C *Erysiphe hypophylla* (Nevod.) U. Braun & Cunningt.
C *Erysiphe lonicerae* DC.
A *Erysiphe magnicellulata* U. Braun
A *Erysiphe necator* Schwein.
E *Erysiphe paeoniae* R.Y. Zheng & G.Q. Chen (1981)
A *Erysiphe palczewskii* (Jacz.) U. Braun & S. Takamatsu
C *Erysiphe platani* (Howe) U. Braun & S. Takam. (2000)
C *Erysiphe rayssiae* (Mayor) U. Braun & S. Takam., 2000
A *Erysiphe russellii* (Clinton) U. Braun & S. Takamatsu
A *Erysiphe sedi* U. Braun (1981)
C *Erysiphe symphoricarpi* (Howe) U. Braun & S. Takamatsu
C *Erysiphe syringae* Schwein.
C *Erysiphe syringae-japonicae* (U. Braun) U. Braun & S. Takamatsu
A *Erysiphe thermopsidis* (U. Braun) U. Braun & S. Takam., 2000
C *Erysiphe verbenicola* U. Braun & S. Takam. (2000)

A *Golovinomyces orontii* (Castagne) V.P. Heluta (1988)
C *Microsphaera berberidis* (DC.) Lév. (1851)
C *Microsphaera grossulariae* (Wallr.) Lév. (1851)
C *Microsphaera jaczewskii* U. Braun (1981)
C *Microsphaera pseudacaciae* (P.D. Marchenko) U. Braun [as 'pseudoacaciae'] (1981)
C *Neoerysiphe galeopsidis* (DC.) U. Braun (1999)
A *Oidium chrysanthemi* Rabenh.
E *Oidium dianthi* Jacz. (1926)
A *Oidium hortensiae* Jørst. (1925)
E *Oidium hyssopi* Erikss.
A *Oidium lycopersici* Cooke & Massee 1887
C *Oidium neolycopersici* L. Kiss (2001)
E *Oidium verbenae* Thüm. & P.C. Bolle (1927)
C *Podosphaera clandestina* (Wallr.) Lév. (1851)
C *Podosphaera leucotricha* (Ellis & Everh.) E.S. Salmon (1900)
A *Podosphaera mors-uvae* (Schwein.) U. Braun & S. Takam (2000)
C *Sawadaea tulasnei* (Fuckel) Homma (1937)
C *Sphaerotheca fuliginea* (Schltdl.) Pollacci(1913)
C *Sphaerotheca verbenae* S vul. & Negru (1955)
A *Uncinula kusanoi* Syd. & P. Syd.

Exobasidiales
C *Exobasidium rhododendri* (Fuckel) Cram. ap. Geyl.
A *Graphiola phoenicis* (Moug.) Poit. (1824)

Geastrales
C *Geastrum campestre* Morgan (1887)
E *Geastrum floriforme* Vittad. (1842)
A *Geastrum morganii* Lloyd
A *Geastrum smardae* V.J. Stan k (1956)

Helotiales
C *Arachnopeziza aranea* (De Not.) Boud. (1907)
C *Blumeriella jaapii* (Rehm) Arx (1961)
C *Botryotinia draytonii* (Buddin & Wakef.) Seaver, 1951
C *Botryotinia sphaerosperma* (P.H. Greg.) N.F. Buchw., 1949
C *Botryotinia squamosa* Vienn.-Bourg. (1953)

C *Botrytis galanthina* (Berk. & Broome) Sacc. (1886)
C *Ceuthospora foliicola* (Lib.) Jaap (1912)
A *Ciborinia camelliae* L.M. Kohn (1979)
C *Cryptocline taxicola* (Allesch.) Petr. (1925)
C *Dasyscyphella castaneicola* (Graddon) Raitv., 2002
A *Didymascella thujina* (E.J. Durand) Maire (1927)
C *Discohainesia oenotherae* (Cooke & Ellis) Nannf., 1932
C *Drepanopeziza populi-albae* (Kleb.) Nannf., 1932
A *Drepanopeziza punctiformis* Gremmen (1965)
A *Eupropolella arundinariae* (E.K. Cash) Dennis, 1975
C *Eupropolella britannica* Greenh. & Morgan-Jones, 1972
E *Gloeosporium aroniae* Kill. (1928)
C *Hyalopeziza spinicola* Graddon, 1977
C *Hyaloscypha mirabilis* Velen., 1934
C *Hysterostegiella lauri* (Caldesi) B. Hein, 1983
C *Lachnellula occidentalis* (G.G. Hahn & Ayers) Dharne, 1965
C *Lachnellula resinaria* (Cooke & W. Phillips) Rehm, 1893
C *Lachnellula willkommii* (Hartig) Dennis, 1962
C *Lanzia coracina* (Durieu & Lév.) Spooner (1981)
C *Lanzia echinophila* (Bull.) Korf, 1982
A *Marssonina clematidis* (Allesch.) Magnus (1906)
C *Meria laricis* Vuill. (1896)
C *Moellerodiscus advenulus* (W. Phillips) Dumont, 1976
A *Monilinia fructicola* (G. Winter) Honey (1928)
C *Monilinia mespili* (Schellenb.) Whetzel, 1945
C *Neobulgaria undata* (W.G. Sm.) Spooner & Y.J. Yao (1995)
C *Neofabraea perennans* Kienholz (1939)
C *Niptera subbiatorina* Rehm, 1891
C *Pezicula houghtonii* (W. Phillips) J.W. Groves, 1946
C *Peziza ostracoderma* Korf (1961)
C *Psilachnum auranticolor* Graddon, 1977
A *Rhabdocline pseudotsugae* Syd. (1922)
C *Sarcotrochila alpina* (Fuckel) Höhn., 1917

A *Septotis podophyllina* (Ellis & Everh.) B. Sutton (1970)
C *Stromatinia cepivora* (Berk.) Whetzel (1945)
C *Stromatinia gladioli* (Drayton) Whetzel, 1945
C *Trochila laurocerasi* (Desm.) Fr. (1849)
C *Tympanis laricina* (Fuckel) Sacc. (1889)

Hymenochaetales
A *Inonotus rickii* (Pat.) D.A. Reid, 1957

Hypocreales
C *Calonectria kyotensis* Terash. (1968)
C *Cordyceps tuberculata* (Lebert) Maire (1917)
A *Cylindrocladium buxicola* Henricot (2002)
C *Cylindrocladium pauciramosum* C.L. Schoch & Crous (1999)
C *Epichloë typhina* (Pers.) Tul. & C. Tul. (1865)
C *Fusarium foetens* Schroers, O'Donnell, Baayen & Hooftman (2004)
A *Gibberella circinata* Nirenberg & O'Donnell (1998)
A *Myrothecium roridum* Tode (1790)
C *Nectria auriger* Berk. & Ravenel
C *Nectriella consolationis* (Sacc.) E. Müll., 1962
C *Neonectria galligena* (Bres.) Rossman & Samuels (1999)
A *Pseudonectria pachysandricola* B.O. Dodge, 1944
A *Sporophagomyces chrysostomus* (Berk. & Broome) K. Põldmaa & Samuels (1999)
C *Verticillium dahliae* Kleb. (1913)
C *Volutella buxi* (Corda) Berk. (1850)

Hysterangiales
A *Hysterangium inflatum* Rodway (1918)

Incertae sedis
C *Ascochyta aquilegiae* (Roum. & Pat.) Sacc. (1884)
C *Ascochyta dianthi* (Alb. & Schwein.) Berk. (1860)
C *Ascochyta doronici* Allesch. (1897)
C *Ascochyta grossulariae* Oudem. (1898)
C *Ascochyta pisi* Lib. (1830)
C *Ascochyta sorghi* Sacc., 1875
C *Ascochyta syringae* Bres. (1921)
C *Ascochyta trifolii* Siemaszko (1914)
A *Aulographina eucalypti* (Cooke & Massee) Arx & E. Müll., 1960
C *Cerebella andropogonis* Ces. (1851)
C *Circinotrichum britannicum* P.M. Kirk, 1981

A *Cryptostroma corticale* (Ellis & Everh.) P.H. Greg. & S. Waller (1951)
C *Epibelonium gaeumannii* E. Müll., 1963
C *Hymenula cerealis* Ellis & Everh., 1894
C *Kabatina thujae* R. Schneid. & Arx (1966)
C *Lembosina aulographoides* (E. Bommer, M. Rousseau & Sacc.) Theiss. (1913)
A *Magnaporthe grisea* (T.T. Hebert) M.E. Barr (1977)
A *Magnaporthe salvinii* (Catt.) R.A. Krause & R.K. Webster (1972)
C *Mastigosporium kitzebergense* U. Schlöss., 1970
C *Morenoina chamaecyparidis* J.P. Ellis, 1980
C *Mycocentrospora acerina* (R. Hartig) Deighton (1972)
A *Nattrassia mangiferae* (Syd. & P. Syd.) B. Sutton & Dyko (1989)
A *Nematospora coryli* Peglion (1901)
C *Nematostoma parasiticum* (R. Hartig) M.E. Barr (1997)
C *Pseudoseptoria donacis* (Pass.) B. Sutton (1977)
A *Pycnostysanus azaleae* (Peck) E.W. Mason (1941)
C *Rhynchosporium orthosporum* Caldwell (1937)
A *Spermospora subulata* (R. Sprague) R. Sprague, 1948
C *Sphaeropsis sapinea* (Fr.) Dyko & B. Sutton (1980)
A *Stenocarpella macrospora* (Earle) B. Sutton, 1977
A *Stenocarpella maydis* (Berk.) B. Sutton, 1980
C *Stigmatea aegopodii* (Fr.) Oudem., 1876
C *Stomiopeltis cupressicola* J.P. Ellis, 1977
C *Strasseria geniculata* (Berk. & Broome) Höhn. (1919)
C *Vialaea insculpta* (Oudem.) Sacc. (1896)

Labyrinthulales
A *Labyrinthula zosterae* Porter & Muehlstein, 1989

Leotiales
A *Ciboria americana* E.J. Durand (1902)

Microascales
A *Ceratocystis paradoxa* (Dade) C. Moreau (1952)

- A *Ceratocystis platani* (J.M. Walter) Engelbr. & T.C. Harr. (2005)
- C *Chalara populi* Veldeman ex Kiffer & Delon (1983)

Microbotryales
- E *Microbotryum violaceum* (Pers.) G. Deml & Oberw. (1982)
- C *Sphacelotheca destruens* (Schltdl.) J.A. Stev. & Aar.G. Johnson, 1944
- A *Sphacelotheca reiliana* (J.G. Kühn) G.P. Clinton (1902)

Microstromatales
- E *Microstroma juglandis* (Berenger) Sacc. (1886)

Microthyriales
- C *Lichenopeltella fimbriata* (J.P. Ellis) P.M. Kirk ined.
- C *Microthyrium lauri* Höhn., 1919

Mycosphaerellales
- A *Mycosphaerella dearnessii* M.E. Barr (1972)
- C *Mycosphaerella isariophora* (Desm.) Johanson (1884)
- C *Mycosphaerella pini* Rostr. (1957)
- A *Phloeospora robiniae* (Lib.) Höhn. (1905)

Myriangiales
- E *Elsinoë ampelina* Shear (1929)
- C *Sphaceloma murrayae* Jenkins & Grodz. (1943)

Ophiostomatales
- A **Ophiostoma novo-ulmi** Brasier, 1991
- A *Ophiostoma ulmi* (Buisman) Nannf., 1934

Orbiliales
- C *Orbilia retrusa* (W. Phillips & Plowr.) Sacc., 1889

Pezizales
- A *Geopora sumneriana* (Cooke) M. Torre (1976)
- A *Hydnotrya cubispora* (E.A. Bessey & B.E. Thomps.) Gilkey, 1939
- A *Paurocotylis pila* Berk., 1858
- C *Pithya cupressina* Fuckel (1870)
- A *Tuber gibbosum* Harkn. (1899)

Phallales
- A *Aseroë rubra* Labill. (1800)
- A *Clathrus archeri* (Berk.) Dring (1980)
- E *Clathrus ruber* P. Micheli ex Pers., 1801
- A *Ileodictyon cibarium* Tul., 1844
- C *Lysurus cruciatus* (Lepr. & Mont.) Henn., 1902
- C *Lysurus mokusin* Cibot (1775)
- A *Mutinus elegans* (Mont.) E. Fisch. (1888)
- A *Mutinus ravenelii* (Berk. & M.A. Curtis) E. Fisch. (1888)
- C *Phallus impudicus* L., 1753

Phyllachorales
- C *Colletotrichum dematium* (Pers.) Grove (1918)
- A *Colletotrichum lindemuthianum* (Sacc. & Magnus) Briosi & Cavara (1889)
- E *Colletotrichum linicola* Pethybr. & Laff. (1918)
- C *Colletotrichum orbiculare* (Berk. & Mont.) Arx, 1957
- E *Colletotrichum villosum* Weimer (1945)
- A *Glomerella acutata* Guerber & J.C. Correll (2001)
- A *Glomerella cingulata* (Stoneman) Spauld. & H. Schrenk (1903)
- A *Phyllachora ambrosiae* (Berk. & M.A. Curtis) Sacc. (1883)
- A *Phyllachora pomigena* (Schwein.) Sacc. (1883)

Pleosporales
- A *Alternaria gaisen* Nagano (1928)
- C *Alternaria radicina* Meier, Drechsler & E.D. Eddy (1922)
- C *Alternaria saponariae* (Peck) Neerg. (1938)
- C *Astrosphaeriella trochus* (Penzig & Saccardo) D. Hawksworth (1981)
- A *Cochliobolus heterostrophus* (Drechsler) Drechsler, 1934
- C *Coniothyrium laburnophilum* Oudem.
- C *Coniothyrium ribis* Brunaud (1889)
- C *Cucurbitaria laburni* (Pers.) De Not. (1862)
- C *Cucurbitaria piceae* Borthw., 1909
- A *Didymella ligulicola* (K.F. Baker, Dimock & L.H. Davis) Arx (1962)
- C *Didymella lycopersici* Kleb. (1921)
- C *Drechslera phlei* (J.H. Graham) Shoemaker (1959)
- C *Drechslera poae* (Baudyš) Shoemaker (1962)
- C *Leptosphaeria lindquistii* Fressi (1968)
- C *Leptosphaeria lunariae* (Berk. & Broome) Sacc., 1883
- A *Leptosphaerulina trifolii* (Rostovzev) Petr. (1959)
- C *Melanomma rhododendri* Rehm, 1881
- A *Phaeocryptopus gaeumannii* (T. Rohde) Petr. (1938)
- C *Phaeocryptopus nudus* (Peck) Petr. (1938)
- C *Phaeosphaeria eustoma* (Fuckel) L. Holm (1957)

C *Phaeosphaeria salebricola* (E. Bommer, M. Rousseau & Sacc.) Leuchtm. (1984)
C *Phoma phlogis* Roum. (1884)
A *Phoma tracheiphila* (Petri) L.A. Kantsch. & Gikaschvili (1948)
C *Pleospora herbarum* (Pers.) Rabenh. (1854)
C *Protoventuria arxii* (E. Müll.) M.E. Barr (1971)
C *Pyrenophora bromi* (Died.) Drechsler (1923)
C *Pyrenophora chaetomioides* Speg. (1899)
C *Pyrenophora lolii* Dovaston (1948)
C *Rhizosphaera kalkhoffii* Bubák (1914)
C *Stemphylium sarciniforme* (Cavara) Wiltshire (1938)
C *Venturia chlorospora* (Ces.) P. Karst. (1873)
C *Venturia inaequalis* (Cooke) G. Winter (1875)
C *Venturia pirina* Aderh. (1896)
E *Venturia populina* (Vuill.) Fabric. (1902)
C *Venturia saliciperda* Nüesch, 1960
C *Xenomeris nicholsonii* (Cooke) Petr. (1928)
Polyporales
  A *Flaviporus brownei* (Humb.) Donk (1960)
  C *Ganoderma pfeifferi* Bres., 1889
  C *Ganoderma resinaceum* Boud., 1890
  A *Laccocephalum hartmannii* (Cooke) Núñez & Ryvarden (1995)
  A *Perenniporia ochroleuca* (Berk.) Ryvarden
  A *Phanerochaete salmonicolor* (Berk. & Broome) Jülich (1975)
  A *Pycnoporellus fulgens* (Fr.) Donk (1971)
  A *Rigidoporus lineatus* (Pers.) Ryvarden, 1972
Pyrenulales
  C *Aglaospora profusa* (Fr.) De Not. (1844)
  C *Anisomeridium polypori* (Ellis & Everh.) M.E. Barr, 1996
Rhizophydiales
  C *Batrachochytrium dendrobatidis* Longcore, Pessier & D.K. Nichols, 1999
Rhytismatales
  C *Lophodermium piceae* (Fuckel) Höhn., 1917
  C *Lophodermium pini-excelsae* S. Ahmad, 1954
  C *Lophodermium vagulum* M. Wilson & N.F. Robertson, 1947
  C *Lophomerum ponticum* Minter, 1980

C *Pseudophacidium piceae* E. Müll., 1963
C *Rhytisma acerinum* (Pers.) Fr. (1819)
Russulales
  C *Lactarius deterrimus* Gröger (1968)
  C *Russula melliolens* Quél. (1901)
Taphrinales
  C *Taphrina bullata* (Berk. & Broome) Tul. (1866)
  C *Taphrina cerasi* (Fuckel) Sadeb. (1890)
  C *Taphrina crataegi* Sadeb. (1890)
  C *Taphrina deformans* (Berk.) Tul., 1866
  C *Taphrina pruni* Tul. (1866)
Techisporales
  C *Trechispora nivea* (Pers.) K.H. Larss.(1995)
Tilletiales
  C *Tilletia bromi* (Brockmann) Brockmann (1959)
  A *Tilletia caries* (DC.) Tul. & C. Tul. (1847)
  A *Tilletia controversa* J.G. Kühn (1874)
Tremellales
  C *Guepinia helvelloides* (DC.) Fr. (1828)
Tricholomatales
  A *Collybia luxurians* Peck (1897)
  C *Marasmiellus virgatocutis* Robich Esteve-Rav. & G. Moreno (1994)
Trichosphaeriales
  C *Khuskia oryzae* H.J. Huds., 1963
Tulostomatales
  E *Tulostoma giovanellae* Bres.
Uredinales
  E *Caeoma laricis* (Westend.) R. Hartig (1874)
  C *Chrysomyxa abietis* (Wallr.) Unger, 1840
  A *Coleosporium clematidis* Barclay (1890)
  E *Coleosporium tussilaginis* (Pers.) Lév. (1849)
  E *Cronartium flaccidum* (Alb. & Schwein.) G. Winter (1880)
  A *Cronartium ribicola* A. Dietr. (1856)
  A *Cumminsiella mirabilissima* (Peck) Nannf. (1947)
  C *Endophyllum sempervivi* (Alb. & Schwein.) de Bary (1863)
  C *Gymnoconia nitens* (Schwein.) F. Kern & Thurst. (1929)
  A *Gymnosporangium asiaticum* Miyabe ex G. Yamada, 1904
  C *Gymnosporangium clavariiforme* (Jacq.) DC. (1805)
  C *Gymnosporangium confusum* Dietel (1889)
  C *Gymnosporangium cornutum* Arthur ex F. Kern (1911)

- A *Gymnosporangium juniperi-virginianae* Schwein. (1822)
- A *Gymnosporangium sabinae* (Dicks.) G. Winter (1884)
- C *Gymnosporangium tremelloides* R. Hartig (1882)
- C *Kuehneola uredinis* (Link) Arthur (1906)
- E *Leucotelium cerasi* (Berenger) Tranzschel, 1935
- E *Melampsora allii-populina* Kleb. (1902)
- C *Melampsora amygdalinae* Kleb., 1900
- E *Melampsora caprearum* Thüm., 1879
- C *Melampsora euphorbiae* (C. Schub.) Castagne (1843)
- E *Melampsora hypericorum* (DC.) J. Schröt. (1871)
- E *Melampsora laricis-pentandrae* Kleb. (1897)
- E *Melampsora laricis-populina* Kleb., 1902
- A *Melampsora medusae* Thüm. (1878)
- E *Melampsora populnea* (Pers.) P. Karst. (1879)
- C *Melampsora ribesii-viminalis* Kleb. (1900)
- C *Melampsora salicis-albae* Kleb. (1901)
- E *Melampsorella caryophyllacearum* (DC.) J. Schröt. (1874)
- C *Melampsorella symphyti* (DC.) Bubák (1903)
- C *Melampsoridium hiratsukanum* S. Ito ex Hirats. f. (1927)
- C *Mikronegeria fagi* Dietel & Neger, 1899
- C *Nyssopsora echinata* (Lév.) Arthur (1906)
- C *Phragmidium fusiforme* J. Schröt., 1870
- C *Phragmidium potentillae* (Pers.) P. Karst. (1879)
- C *Phragmidium violaceum* (Schultz) G. Winter (1880)
- C *Puccinia aegopodii* (Schumach.) Link (1817)
- C *Puccinia allii* DC.) F. Rudolphi (1829)
- A *Puccinia antirrhini* Dietel & Holw. (1897)
- C *Puccinia arenariae* (Schumach.) J. Schröt. (1880)
- C *Puccinia asarina* Kunze (1817)
- E *Puccinia asparagi* DC. (1805)
- E *Puccinia buxi* Sowerby (1809)
- C *Puccinia calcitrapae* DC. (1805)
- A *Puccinia callistephi* S vul. (1939)
- C *Puccinia chaerophylli* Purton (1821)
- A *Puccinia chrysanthemi* Roze (1900)
- C *Puccinia cnici-oleracei* Pers. (1823)
- C *Puccinia conii* (F. Strauss) Fuckel (1870)
- C *Puccinia coronata* Corda (1837)
- C *Puccinia crepidis* J. Schröt. (1889)
- E *Puccinia cribrata* Arthur & Cummins (1933)
- C *Puccinia cyani* Pass., 1874
- C *Puccinia difformis* Kunze (1817)
- A *Puccinia distincta* McAlpine (1896)
- C *Puccinia festucae* Plowr. (1893)
- C *Puccinia gentianae* (F. Strauss) Link (1824)
- C *Puccinia gladioli* Castagne (1842)
- C *Puccinia graminis* Pers. (1794)
- A *Puccinia helianthi* Schwein. (1822)
- A *Puccinia hemerocallidis* Thüm. (1880)
- C *Puccinia heraclei* Grev. (1823)
- A *Puccinia hordei* G.H. Otth, 1871
- A *Puccinia horiana* Henn. (1901)
- C *Puccinia iridis* Wallr. (1844)
- A *Puccinia komarovii* Tranzschel
- A *Puccinia kusanoi* Dietel, 1899
- A *Puccinia lagenophorae* Cooke (1884)
- C *Puccinia ljulinica* Hinkova & Koeva, 1966
- A *Puccinia longicornis* Pat. & Har., 1891
- A *Puccinia malvacearum* Bertero ex Mont. (1852)
- C *Puccinia mariana* Sacc. (1915)
- C *Puccinia menthae* Pers. (1801)
- E *Puccinia minussensis* Thüm.
- C *Puccinia nitida* (F. Strauss) Röhl. (1813)
- A *Puccinia oxalidis* Dietel & Ellis (1895)
- A *Puccinia pelargonii-zonalis* Doidge (1926)
- C *Puccinia phragmitis* (Schumach.) Körn. (1876)
- C *Puccinia pimpinellae* (F. Strauss) Link (1824)
- C *Puccinia poarum* E. Nielsen (1877)
- E *Puccinia prostii* Duby, 1830
- C *Puccinia recondita* Dietel & Holw. (1857)
- C *Puccinia ribis* DC. (1805)
- C *Puccinia smyrnii* Biv., 1894
- A *Puccinia sorghi* Schwein. (1832)
- C *Puccinia tanaceti* Fuckel
- A *Puccinia tasmanica* Dietel (1903)
- C *Puccinia thymi* (Fuckel) P. Karst. (1884)
- E *Puccinia tulipae* J. Schröt. (1875)
- C *Puccinia vincae* Berk. (1836)
- C *Puccinia violae* (Schumach.) DC. (1815)

11 List of Species Alien in Europe and to Europe

C *Pucciniastrum areolatum* (Fr.) G.H. Otth (1863)
C *Pucciniastrum epilobii* G.H. Otth (1861)
C *Trachyspora intrusa* (Grev.) Arthur (1934)
C *Tranzschelia anemones* (Pers.) Nannf. (1939)
C *Uromyces aecidiiformis* (F. Strauss) C.C. Rees (1917)
C *Uromyces anthyllidis* (Grev.) J. Schröt. (1875)
A *Uromyces appendiculatus* F. Strauss (1833)
E *Uromyces beticola* (Bellynck) Boerema, Loer. & Hamers (1987)
A *Uromyces caraganicola* Henn. (1901)
C *Uromyces colchici* Massee, 1892
C *Uromyces dactylidis* G.H. Otth (1861)
C *Uromyces dianthi* (Pers.) Niessl, 1872
C *Uromyces erythronii* (DC.) Pass. (1867)
C *Uromyces limonii* (DC.) Berk.
E *Uromyces lupinicola* Bubák (1902)
E *Uromyces pisi-sativi* (Pers.) Liro, 1908
C *Uromyces polygoni-avicularis* (Pers.) P. Karst. (1879)
C *Uromyces rumicis* (Schumach.) G. Winter (1884)
A *Uromyces silphii* Arthur
A *Uromyces transversalis* (Thüm.) G. Winter (1884)
C *Uromyces trifolii-repentis* (Castagne) Liro (1906)
C *Uromyces viciae-craccae* Const. (1904)
C *Zaghouania phillyreae* Pat., 1901
Urocystidales
  C *Melanotaenium hypogaeum* (Tul. & C. Tul.) Schellenb., 1911
  C *Melanotaenium jaapii* Magnus, 1911
  C *Urocystis agropyri* (Preuss) A.A. Fisch. Waldh. (1867)
  A *Urocystis cepulae* Frost (1877)
  E *Urocystis eranthidis* (Pass.) Ainsw. & Sampson, 1950
  C *Urocystis gladiolicola* Ainsw. (1950)
  A *Urocystis occulta* (Wallr.) Rabenh. (1870)
  C *Urocystis syncocca* (L.A. Kirchn.) B. Lindeb. (1959)
  C *Vankya ornithogali* (J.C. Schmidt & Kunze) (2000)
Ustilaginales
  A *Kochmania oxalidis* (Ellis & Tracy) Pi&aogonek;tek (2005)
  E *Macalpinomyces spermophorus* (Berk. & M.A. Curtis ex de Toni) Vánky, (2003)

A *Melanopsichium pennsylvanicum* Hirschh.
A *Sporisorium sorghi* Ehrenb. ex Link, 1825
E *Ustilago avenae* (Pers.) Rostr., 1890
C *Ustilago cynodontis* (Pass.) Henn., 1893
A *Ustilago hordei* (Pers.) Lagerh., 1889
A *Ustilago maydis* (DC.) Corda (1842)
C *Ustilago trichophora* (Link) Körn.
A *Ustilago tritici* (Pers.) C.N. Jensen, Kellerm. & Swingle
Xylariales
  C *Anthostomella trachycarpi* S.M. Francis (1980)
  C *Discostroma corticola* (Fuckel) Brockmann (1976)
  A *Entoleuca mammata* (Wahlenb.) J.D. Rogers & Y.M. Ju (1996)
  A *Eutypella parasitica* R.W. Davidson & R.C. Lorenz (1938)
  C *Pestalotiopsis guepinii* (Desm.) Steyaert (1949)
  C *Pseudomassaria thistletonia* (Cooke) Arx, 1952
  A *Seimatosporium parasiticum* (Dearn. & House) Shoemaker, 1964
  C **Seiridium cardinale** (W.W. Wagener) B. Sutton & I.A.S. Gibson (1972)
  C *Xylaria arbuscula* Sacc. (1878)

Cyanobacteria
Nostocaceae
  C *Nodularia spumigena* Mertens *ex* Bornet & Flahault, 1886

Bacillariophyta
Bacillariaceae
  A *Pseudonitzschia seriata obtusa* (Hasle) Hasle, 1993
Corethraceae
  C *Corethron criophilum* Castracane, 1886
Pleurosigmataceae
  A *Pleurosigma simonsenii* G.R. Hasle
Triceratiaceae
  C *Pleurosira laevis* (Ehrenberg) Compère

Haptophyta
Gephyrocapsaceae
  A *Isochrysis galbana* Parke
Phaeocystaceae
  A *Phaeocystis pouchetii* (Hariot) Lagerheim

Ochrophyta
Alariaceae
- A *Alaria esculenta* (Linnaeus) Greville
- E **Undaria pinnatifida** (Harvey) Suringar, 1873

Asterolampraceae
- A *Asteromphalus sarcophagus* Wallich

Chaetocerotaceae
- A *Chaetoceros calcitrans* (Paulsen) Takano, 1968
- A *Chaetoceros peruvianus* Brightwell
- A *Chaetoceros rostratus* Lauder

Chordaceae
- E *Chorda filum* (Linnaeus) Stackhouse, 1797

Chordariaceae
- E *Acrothrixgracilis* Kylin, 1907
- E *Asperococcus scaber* Kuckuck
- A *Botrytella parva* (Takamatsu) Kim, 1996
- A *Cladosiphon zosterae* (J. Agardh) Kylin
- A *Halothrix lumbricalis* (Kützing) Reinke, 1888
- E *Leathesia difformis* (Linnaeus) Areschoug, 1847
- A *Leathesia verruculiformis* Y. Lee & I.K. Lee
- A *Sphaerotrichia divaricata* (C. Agardh) Kylin
- A *Sphaerotrichia firma* (Gepp) Zinova

Coscinodiscaceae
- A **Coscinodiscus wailesii** Gran & Angst

Cutleriaceae
- C *Zanardinia prototypus* (Nardo) G. Furnari

Desmarestiaceae
- A *Desmarestia viridis* (O.F. Müller) J.V. Lamouroux, 1813

Dictyotaceae
- A *Dictyota ciliolata* Kützing, 1859
- A *Dictyota okamura* (E.Y. Dawson) I. Hörnig, R. Schnetter & W.F. Prud'homme van Reine, 1992
- A *Padina boergesenii* Allender & Kraft, 1983
- A *Padina boryana* Thivy, 1966
- E *Stypopodium schimperi* (Buchinger ex Kützing) Verlaque & Boudouresque, 1991

Ectocarpaceae
- C *Litosiphon laminariae* (Lyngbye) Harvey

Fucaceae
- E *Ascophyllum nodosum* (Linnaeus) Le Jolis
- C *Fucus disticus* Linnaeus, 1767
- C *Fucus evanescens* C. Agardh
- E *Fucus spiralis* Linnaeus, 1753

Hemiaulaceae
- A *Eucampia cornuta* (Cleve) Grunow, 1883

Hemidiscaceae
- E *Actinocyclus normanii* (W. Greg.) Hust.

Heterosigmataceae
- C *Heterosigma akashiwo* (Y. Hada) Y. Hada ex Y. Hara & M. Chihara, 1967
- C *Heterosigma carterae* (Hulbert) Taylor

Laminariaceae
- A *Laminaria japonica* Areschoug, 1851
- A *Laminaria ochotensis* Miyabe
- A *Laurencia brongniartii* J. Agardh

Leathesiaceae
- C *Corynophlaea umbellata* (C. Agardh) Kützing

Lessoniaceae
- A *Macrocystis pyrifera* (Linnaeus) C. Agardh

Oxytoxaceae
- C *Oxytoxum criophilum* Balech & El-Sayed, 1965

Pilayellaceae
- A *Pilayella littoralis* (Linnaeus) Kjellman

Ralfsiaceae
- C *Pseudolithoderma roscoffense* Loiseaux

Rhizosoleniaceae
- A *Pseudosolenia calcar-avis* (Schultze) Sundström, 1986
- C *Rhizosolenia indica* H. Peragallo

Sargassaceae
- A *Sargassum muticum* (Yendo) Fensholt, 1955

Scytosiphonaceae
- A *Colpomenia peregrina* Sauvageau, 1927
- A *Endarachne binghamiae* J. Agardh
- C *Falkenbergia rufolanosa* (Harvey) Schmitz
- A *Scytosiphon dotyi* M.J. Wynne, 1969

Skeletonemaceae
- E *Skeletonema subsalsum* (A. Cleve) Bethge

Stephanopyxidaceae
- A *Stephanopyxis palmeriana* (Greville) Grunow

Thalassiosiraceae
- A *Thalassiosira incerta* Makarova
- C *Thalassiosira hendeyi* Hasle & G. Fryxell
- A *Thalassiosira punctigera* (Castracane) Hasle
- A *Thalassiosira tealata* Takano

Triceratiaceae
 C *Odontella sinensis* (Greville) Grunow
Vacuolariaceae
 A *Chattonella antiqua* (Hada) Ono
 A *Chattonella marina* (Subrahmanyan) Hara & Chihara
 A *Fibrocapsa japonica* S. Toriumi & H. Takano

Dinophyta
Actiniscaceae
 C *Dicroerisma psilonereiella* Taylor & Cattell
Goniodomataceae
 C *Alexandrium angustitabulatum* F.J.R. Taylor, nom. illeg.
 A **Alexandrium catenella** (Whedon & Kofoid) E. Balech
 A *Alexandrium leei* E. Balech
 C *Alexandrium minutum* Halim
 C *Alexandrium tamarense* (Lebour) E. Balech
 E *Alexandrium taylori* Balech
Gonyaulacaceae
 C *Pyrodinium bahamense* Plate
Gymnodiniaceae
 C *Gymnodinium aureolum* (E.M. Hulburt) G. Hansen
 A *Gymnodinium catenatum* L.W. Graham
 C *Gyrodinium corallinum* Kofoid & Swezy, 1921
 A *Karlodinium micrum* (B. Leadbeater & J.D. Dodge) J. Larsen
 C *Karenia mikimotoi* (Miyake & Kominami ex Oda) G. Hansen & Ø. Moestrup
Noctilucaceae
 C *Noctiluca scintillans* (Macartney) Ehrenberg
Prorocentraceae
 C *Prorocentrum minimum* (Pavillard) J. Schiller
 C *Prorocentrum triestinum* J. Schiller
Pyrocystaceae
 E *Dissodinium pseudocalani* (Gonnert) Drebes ex Elbrachter & Drebes
Thecadiniaceae
 A *Thecadinium yashimaense* Yoshimatsu, Toriumi & Dodge

Chlorophyta
Caulerpaceae
 A *Caulerpa mexicana* Sonder ex Kützing, 1849
 A **Caulerpa racemosa cylindracea** (Sonder) Verlaque, Huisman & Boudouresque, 2003
 A *Caulerpa racemosa* [*lamourouxii*] f. *requienii* (Montagne) Weber-van Bosse, 1898
 A *Caulerpa racemosa turbinata uvifera*
 E *Caulerpa scalpelliformis* (Brown ex Turner) C. Agardh, 1817
 A **Caulerpa taxifolia** (M. Vahl) C. Agardh, 1817
 A *Caulerpa webbiana* Montagne
Characeae
 E *Chara connivens* Salzmann ex A. Braun
Cladophoraceae
 C *Cladophora hutchinsioides* Hoek & Womersley, 1984
Codiaceae
 C *Codium adherens* C. Agardh
 A *Codium fragile atlanticum* (Cotton) P. Silva
 A *Codium fragile scandinavicum* P.C. Silva
 A **Codium fragile tomentosoides** (van Goor) P. Silva
 E *Codium taylorii* P.C. Silva, 1960
 E *Codium vermilara* (Olivi) Delle Chiaje
Dasycladaceae
 C *Neomeris annulata* Dickie, 1874
Derbesiaceae
 A *Derbesia boergesenii* (Iyengar & Ramanathan) Mayhoub, 1976
 A *Derbesia rhizophora* Yamada, 1961
Mamiellaceae
 C *Mantoniella squamata* (I. Manton & M. Parke) T.V. Desikachary
Monostromataceae
 A *Monostroma obscurum* (Kützing) J. Agardh, 1883
 A *Protomonostroma undulatum* (Wittrock) K.L. Vinogradova
Polyphysaceae
 A *Acetabularia calyculus* J.V. Lamouroux in Quoy & Gaimard, 1824
Siphonocladaceae
 A *Cladophoropsis javanica* (Kützing) P.C. Silva, 1996
 C *Cladophoropsis membranacea* (Hofman Bang ex C. Agardh) Børgesen
Ulvaceae
 A *Ulva fasciata* Delile
 A *Ulva ohnoi* Hiraoka & Shimada, 2004
 A *Ulva pertusa* Kjellman, 1897
 E *Ulva scandinavica* Bliding

Rhodophyta
Acrochaetiaceae
- C *Acrochaetium balticum* (Rosenvinge), Aleem & Schulz
- A *Acrochaetium catenulatum* M.A. Howe
- A *Audouinella robusta* (Børgesen) Garbary, 1979
- A *Audouinella spathoglossi* (Børgesen) Garbary, 1987
- A *Audouinella subseriata* (Børgesen) Garbary, 1987
- A *Colaconema dasyae* (Collins) Stegenga, I. Mol, Prud'homme van Reine & Lokhorst

Areschougiaceae
- A *Agardhiella subulata* (C. Agardh) Kraft & M.J. Wynne, 1979
- A *Sarconema filiforme* (Sonder) Kylin, 1932
- A *Sarconema scinaioides* Børgesen, 1934
- A *Solieria chordalis* (C. Agardh) J. Agardh
- A *Solieria dura* (Zanardini) F. Schmitz, 1895
- A *Solieria filiformis* (Kützing) P.W. Gabrielson, 1985

Bangiaceae
- C *Porphyra tenera* Kjellman
- C *Porphyra umbilicalis* (Linnaeus) Kützing
- A *Porphyra yezo*ensis Ueda, 1932

Bonnemaisoniaceae
- A *Asparagopsis armata* Harvey, 1855
- A *Asparagopsis taxiformis* (Delile) Trevisan de Saint-Léon, 1845
- A **Bonnemaisonia hamifera** Hariot, 1891
- C *Falkenbergia rufolanosa* (Harvey) F. Schmitz

Caulacanthaceae
- A *Caulacanthus us*tulatus (Mertens ex Turner) Kützing

Ceramiaceae
- A *Acrothamnion preissii* (Sonder) E.M. Wollaston, 1968
- E *Aglaothamnion feldmanniae* Halos, 1965
- A *Aglaothamnion halliae* (F.S. Collins), Aponte, Ballantine & J.N. Norris
- E *Anotrichium furcellatum* (J. Agardh) Baldock, 1976
- A *Antithamnion amphigeneum* A. Millar, 1990
- C *Antithamnion densum* (Suhr) M.A. Howe
- A *Antithamnion diminuatum* Wollaston
- A *Antithamnion nipponicum* Yamada & Inagaki, 1935
- E *Antithamnion pectinatum* (Montagne) Brauner ex Athanasiadis & Tittley, 1994
- C *Antithamnionella elegans* (Berthold) J.H. Price & D.M. John, 1986
- A *Antithamnionella spirographidis* (Schiffner) Wollaston, 1968
- A *Antithamnionella ternifolia* (J.D. Hooker & Harvey) Lyle, 1922
- A *Ceramium strobiliforme* G.W. Lawson & D.M. John, 1982
- C *Grallatoria reptans* M.A. Howe
- A *Griffithsia corallinoides* (Linnaeus) Trevisan, 1845
- A *Gymnophycus hapsiphorus* Huisman & Kraft
- A *Pleonosporium caribaeum* (Børgesen) R.E. Norris, 1985
- A *Scageliopsis patens* Wolaston

Corallinaceae
- A *Lithophyllum yessoense* Foslie, 1909

Dasyaceae
- E *Dasya baillouviana* (S.G. Gmelin) Montagne
- A *Grateloupia asiatica* S. Kawaguchi & H.W. Wang, 2001
- C *Grateloupia doryphora* (Montagne) M.A. Howe
- A *Grateloupia filicina* A. Gepp & E.S. Gepp, 1906
- A *Grateloupia lanceolata* (Okamura) Kawaguichi, 1997
- A *Grateloupia luxurians* A. Gepp & E.S. Gepp, 1906
- A *Grateloupia patens* (Okamura) Kawaguichi & H.W. Wang
- A *Grateloupia turururu* Yamada, 1941
- A *Heterosiphonia japonica* Yendo, 1920

Delesseriaceae
- C *Nitophyllum stellato-corticatum* Okamura, 1932

Dumontiaceae
- A *Pikea californica* Harvey

Galaxauraceae
- A *Galaxaura rugosa* (Ellis & Solander) Lamouroux, 1816

Gigartinaceae
- C *Chondrus giganteus* Mikami

Gracilariaceae
- A *Gracilaria armata* (C. Agardh) Greville, 1830
- A *Gracilaria disticha* (J. Agardh) J. Agardh

- C *Gracilaria gracilis* (Stackhouse) Steentoft, L.M. Irvine & Farnham
- C *Gracilaria multipartita* (clemente y Rubio) Harvey
- A *Gracilaria vermiculophylla* (Ohmi) Papenfuss

Halymeniaceae
- A *Cryptonemia hibernica* Guiry & L. Irvine, 1974
- C *Cryptonemia lomation* (Bertoloni) J. Agardh
- A *Prionitis patens* Okamura, 1899

Hypneaceae
- A *Hypnea cornuta* (Kützing) J. Agardh, 1851
- A *Hypnea esperi* Bory de Saint-Vincent, 1828
- C *Hypnea muciformis* (Wulfen) J.V. Lamouroux
- A *Hypnea spicifera* (Suhr) Harvey, 1847
- A *Hypnea spinella* (C. Agardh) Kützing, 1847
- A *Hypnea valentiae* (Turner) Montagne, 1841

Lomentariaceae
- C *Lomentaria flaccida* Tanaka, 1944
- A *Lomentaria hakodatensis* Yendo, 1920

Nemastomataceae
- A *Predaea huismanii* Kraft

Palmariaceae
- E *Devaleraea ramentacea* (Linnaeus) Guiry

Phragmonemataceae
- A *Goniotrichiopsis sublittoralis* G.M. Smith, 1943

Phyllophoraceae
- A *Ahnfeltiopsis flabelliformis* (Harvey) Masuda, 1993

Plocamiaceae
- A *Plocamium secundatum* (Kützing) Kützing, 1866

Rhodomelaceae
- E *Chondria coerulescens* (J. Agardh) Falkenberg
- A *Chondria collinsiana* M.A. Howe, 1920
- A *Chondria curvilineata* Collins & Hervey, 1917
- A *Chondria polyrhiza* Collins & Hervey, 1917
- A *Chondria pygmaea* Garbary & Vandermeulen, 1990
- A *Dipterosiphonia dendritica* (C. Agardh) F. Schmitz
- A *Herposiphonia parca* Setchell, 1926
- A *Laurencia caduciramulosa* Masuda & Kawaguchi, 1997
- A *Laurencia okamurae* Yamada, 1931
- A *Lophocladia lallemandii* (Montagne) F. Schmitz, 1893
- C *Neosiphonia harveyi* (J.W. Bailey) M.S. Kim, H.G. Choi, Guiry & G.W. Saunders, 2001
- A *Neosiphonia sphaerocarpa* (Børgesen) M.S. Kim & I.K. Lee, 1999
- A *Polysiphonia atlantica* Kapraun & J.N. Norris
- A *Polysiphonia fucoides* (Hudson) Greville
- A *Polysiphonia japonica* Harvey
- A *Polysiphonia morrowii* Harvey, 1856
- A *Polysiphonia paniculata* Montagne, 1842
- A *Polysiphonia senticulosa* Harvey
- A *Pterosiphonia tanakae* Uwai & Masuda, 1999
- E *Symphyocladia marchantioides* (Harvey) Falkenberg, 1897
- A *Womersleyella setacea* (Hollenberg) R.E. Norris, 1992

Rhodymeniaceae
- A *Botryocladia madagascariensis* Feldmann-Mazoyer, 1945
- A *Chrysymenia wrightii* (Harvey) Yamada, 1932
- A *Rhodymenia ery*thraea Zanardini, 1858

Sarcomeniaceae
- A *Platysiphonia caribaea* D.L. Ballantine & M.J. Wynne

Sphaerococcaceae
- C *Sphaerococcus coronopifolius* Stackhouse

Squamariaceae
- E *Rhodophysema georgii* Batters, 1900

Pteridophyta

Adiantaceae
- E *Adiantum capillus-veneris* L.
- A *Adiantum hispidulum* Sw.
- A *Adiantum raddianum* C. Presl
- A *Ceratopteris thalictroides* (L.) Brongn.
- A *Onychium japonicum* (Thunb.) G. Kunze

Aspleniaceae
- E *Asplenium foreziense* Legrand [in Sched.]
- E *Ceterach officinarum* Willd.
- E *Phyllitis scolopendrium* (L.) Newman

Athyriaceae
   A *Deparia petersenii* (Kunze) M. Kato
Azollaceae
   A *Azolla caroliniana* Willd.
   A *Azolla filiculoides* Lam.
   A *Azolla mexicana* Schlecht. & Cham. ex K. Presl
Blechnaceae
   C *Blechnum capense* (L.) Schlecht.
   A *Blechnum cordatum* (Desv.) Hieron
   C *Blechnum occidentale* L.
   E *Blechnum spicant* (L.) Roth
   A *Doodia caudata* (Cav.) R. Br.
Cytheaceae
   A *Sphaeropteris cooperi* (F. Müll.) R.M. Tryon
Davalliaceae
   A *Nephrolepis cordifolia* (L.) C. Presl
   A *Nephrolepis exaltata* (L.) Schott
Dennstaedtiaceae
   A *Dennstaedtia punctilobula* (Michx.) Moore
   A *Microlepia platyphylla* (D. Don) J.E. Sm.
Dicksoniaceae
   A *Dicksonia antarctica* Labill.
Dryopteridaceae
   A *Cyrtomium falcatum* (L. f.) C. Presl
   A *Cyrtomium fortunei* J. Smith
   A *Cystopteris bulbifera* (L.) Bernh.
   E *Dryopteris oreades* Fomin
   E *Polystichum aculeatum* (L.) Roth
   E *Polystichum braunii* (Spenn.) Feé
   E *Polystichum lonchitis* (L.) Roth
   A *Polystichum munitum* (Kaulf.) C. Presl
   A *Rumohra adiantiformis* (Forster) Ching
Equisetaceae
   E *Equisetum arvense* L.
   E *Equisetum palustre* L.
   E *Equisetum ramosissimum* Desf.
   E *Equisetum variegatum* Schleich.
Hypolepidaceae
   E *Pteridium aquilinum* (L.) Kuhn ex Kersten
Lycopodiaceae
   C *Lycopodiella cernua* (L.) Pichi Serm.
Marsileaceae
   E *Marsilea quadrifolia* L.
Ophioglossaceae
   E *Botrychium simplex* Hitchc.
Osmundaceae
   E *Osmunda regalis* L.
Polypodiaceae
   A *Phlebodium aureum* (L.) J. Smith
   A *Phymatosorus diversifolius* (Willd.) Pichi Serm.
   A *Platycerium bifurcatum* (Cav.) C. Chr.
   A *Polypodium hesperium* Maxon
Pteridaceae
   A *Pellaea viridis* (Forssk.) Prantl
   A *Pityrogramma calomelanos* (L.) Link
   A *Pityrogramma ebenea* (L.) Proctor
   E *Pteris cretica* L.
   A *Pteris multifida* Poir.
   A *Pteris tremula* R.Br.
   E *Pteris vittata* L.
Salviniaceae
   A *Salvinia auriculata* Aubl.
   A *Salvinia* × *molesta* D.S. Mitch.
   E *Salvinia natans* (L.) All.
Selaginellaceae
   A *Selaginella apoda* (L.) Spring
   A *Selaginella kraussiana* (G. Kunze) A. Braun
Thelypteridaceae
   A *Christella dentata* (Forssk.) Brownsey & Jermy
Woodsiaceae
   A *Diplazium esculentum* (Retz.) Sw.
   E *Matteuccia struthiopteris* (L.) Tod.
   A *Onoclea sensibilis* L.

Bryophytoa
Brachytheciaceae
   A *Myuroclada maximowiczii* (G.G. Borshch.) Steere & W.B. Schofield
Bryaceae
   C *Bryum apiculatum* Schwägr.
   A *Bryum valparaisense* Thér.
Calomniaceae
   A *Calomnion complanatum* (Hook. f. & Wilson) Lindb.
Daltoniaceae
   A *Acrophyllum dentatum* (Hook. f. & W.) Vitt & Crosby
   A *Calyptrochaeta apiculata* Hook. f. & Wilson
Dicranaceae
   C *Campylopus incrassatus* Müll. Hall.
   A **Campylopus introflexus** (Hedw.) Brid.
Ditrichaceae
   E *Ceratodon purpureus* (Hedw.) Brid.
   C *Ditrichum punctulatum* Mitten
Fossombroniaceae
   A *Fossombronia crispa* Nees
Geocalycaceae
   A *Heteroscyphus fissistipus* (Hook. f. & Taylor) Schiffn.

    A *Lophocolea bispinosa* (Hook. f.
       & Taylor) Gottsche, Lindenb. & Nees
    A *Lophocolea brookwoodiana* Paton
       & Sheahan
    A *Lophocolea semiteres* (Lehm.) Mitt.
Gigaspermaceae
    E *Oedopodiella australis* (Wager
       & Dixon) Dixon
Hypopterygiaceae
    A *Hypopterygium tamarisci* (Sw.) Brid.
       ex Müll. Hal.
Lepidoziaceae
    A *Lophozia herzogiana* E.A. Hodg.
       & Grolle
    A *Telaranea murphyae* Paton
    A *Telaranea tetradactyla* (Hook.
       f. & Taylor) Hodgs.
Lunulariaceae
    E *Lunularia cruciata* (L.) Dumort. ex
       Lindb.
Orthodontiaceae
    A *Orthodontium lineare* Schwägr.
    C *Orthodontium pelucens* (Hook.) Bruch.
       & Schimp.
Pelliaceae
    E *Pellia epiphylla* (L.) Corda
    A *Plagiochila retrorsa* Gottsche
Polytrichaceae
    A *Atrichum crispum* P. Beauv.
Pottiaceae
    C *Bryoerythrophyllum inaeqalifolium*
       (Taylor) R.H. Zander
    C *Didymodon australasiae* (Hook.
       & Grev.) R.H. Zander
    C *Hennediella macrophylla* (R. Br. ter.)
       Par.
    A *Hennediella stanfordensis* (Steere)
       Blockeel
    A *Leptophascum leptophyllum* (Müll.
       Hal.) J. Guerra & M.J. Cano
    A *Pleurochaete malacophylla* (Müll.
       Hal.) Broth.
    C *Scopelophila cataractae* (Mitt.) Broth.
    A *Syntrichia bogotensis* (Hampe) R.H.
       Zander
    A *Tortula amplexa* (Lesq.) Steere
    A *Tortula bogosica* (Müll. Hal.) R.H. Zander
    A *Tortula bolanderi* (Lesq. & James)
       M. Howe
Rhabdoweisiaceae
    C *Dicranoweisia cirrata* (Hedw.) Lindb.
Rhizogoniaceae
    A *Leptotheca gaudichaudii* Schwägr.
Ricciaceae
    E *Riccia rhenana* Lorb. ex Müll. Frib.
    C *Ricciocarpos natans* (L.) Corda
Sematophyllaceae
    A *Sematophyllum adnatum* (Michx.)
       E. Britton
    E *Splachnobryum obtusum* (Brid.) Müll.
       Hal.
Sphaerocarpaceae
    C *Sphaerocarpos stipitatus* Bisch. ex
       Lindenb.
Thuidiaceae
    A *Thuidiopsis sparsa* (Hook. f. & Wilson)
       Broth.

Lichenes
Monoblastiaceae
    C *Anisomeridium polypori* (Ellis &
       Everhart) Barr
Lecanoraceae
    C *Lecanora conizaeoides* Nyl. ex Crombie
Physciaceae
    C *Phaeophyscia rubropulchra* (Degel.)
       Essl.
Parmeliaceae
    C *Parmelia elegantula* (Zahlbr.) Szat.
    C *Parmelia exasperatula* Nyl.
    C *Parmelia laciniatula* (Flagey ex
       Olivier) Zahlbr.
    C *Parmelia submontana* Nadv. ex. Hale

Gymnospermae
Araucariaceae
    A *Agathis brownii* (Lem.) L.H. Bailey
    A *Araucaria araucana* (Molina) K. Koch
    A *Araucaria bidwillii* Hook.
    A *Araucaria heterophylla* (Salisb.) Franco
Cephalotaxaceae
    A *Cephalotaxus fortunei* Hook.
Cupressaceae
    A *Calocedrus decurrens* (Torr.) Florin
    E *Chamaecyparis la*wsoniana
       (A. Murray) Parl.
    E *Chamaecyparis nootkatensis* (D. Don)
       Spach
    A *Chamaecyparis pisifera* (Siebold &
       Zucc.) Endl.
    A *Chamaecyparis thyoides* (L.) Britton,
       Sterns & Poggenb.
    A *Cryptomeria japonica* (L.f.) D. Don
    A *Cupressus arizonica* Greene
    A *Cupressus goveniana* Gordon
    A *Cupressus lusitanica* Mill.
    A *Cupressus macrocarpa* Hartw.
    E *Cupressus sempervirens* L.

- A × *Cuprocyparis leylandii* (A.B. Jacks. & Dallim.) Farjon & Hiep
- A *Juniperus chinensis* L.
- E *Juniperus sabina* L.
- A *Juniperus virginiana* L.
- A *Platycladus orientalis* (L.) Franco
- A *Tetraclinis articulata* (Vahl) Mast.
- E *Thuja occidentalis* L.
- E *Thuja plicata* D.Don ex Lamb.

Cycadaceae
- A *Cycas circinalis* L.
- A *Cycas revoluta* Thunb.

Ephedraceae
- A *Ephedra altissima* Desf.
- E *Ephedra distachya* L.

Ginkgoaceae
- A *Ginkgo biloba* L.

Pinaceae
- E *Abies alba* Mill.
- A *Abies balsamea* (L.) Mill.
- E *Abies cephalonica* J.W. Loudon
- A *Abies concolor* (Gordon & Glend.) Lindl. E. Hildebr.
- A *Abies fraseri* (Pursh) Poir.
- A *Abies grandis* (Douglas ex D. Don) Lindl.
- A *Abies* × *insignis* Carrire ex Bailly
- A *Abies lasiocarpa* (Hook.) Nutt.
- E *Abies nordmanniana* (Steven) Spach
- E *Abies pinsapo* Boiss.
- A *Abies procera* Rehder
- E *Abies sibirica* Ledeb.
- A *Abies veitchii* Lindl.
- A *Cedrus atlantica* (Manetti ex Endl.) Carrire
- E *Cedrus deodara* (Roxb. ex D. Don) G. Don
- A *Cedrus libani* A. Rich.
- E *Larix decidua* Mill.
- A *Larix gmelinii* (Rupr.) Kuzen.
- A *Larix kaempferi* (Lindl.) Carrire
- A *Larix laricina* (Du Roi) K.Koch
- A *Larix* × *marschlinsii* Coaz
- E *Larix sibirica* Ledeb.
- E *Picea abies* (L.) H.Karst.
- A *Picea alcockiana* Carrire
- A *Picea engelmannii* Parry ex Engelm.
- E *Picea glauca* (Moench) Voss
- A *Picea mariana* Britton., Stern & Poggenb.
- E *Picea omorika* (Pancic) Purkyne
- A *Picea orientalis* (L.) Link
- A *Picea pungens* Engelm.
- E *Picea sitchensis* (Bong.) Carrire
- A *Pinus banksiana* Lamb.
- E *Pinus canariensis* C. Sm.
- A *Pinus cembra sibirica* (Du Tour) E.Murray
- E *Pinus contorta* Douglas ex J.W. Loudon
- A *Pinus coulteri* D. Don
- E *Pinus halepensis* Mill.
- A *Pinus leucodermis* Antoine
- E *Pinus mugo* Turra
- E *Pinus nigra* Arnold
- E *Pinus peuce* Griseb.
- E *Pinus pinaster* Aiton
- E *Pinus pinea* L.
- A *Pinus ponderosa* Douglas ex Lawson & P. Lawson
- A *Pinus radiata* D. Don
- A *Pinus rigida* Mill.
- A *Pinus* × *rotundata* Link
- A *Pinus strobus* L.
- E *Pinus sylvestris* L.
- E *Pinus uncinata* Ramond ex DC.
- A *Pinus wallichiana* A.B. Jacks.
- A *Pseudotsuga menziesii* (Mirb.) Franco
- A *Tsuga canadensis* (L.) Carrire
- E *Tsuga heterophylla* (Raf.) Sarg.
- A *Tsuga mertensiana* (Bong.) Carrire

Taxaceae
- A *Sequoia sempervirens* (D. Don) Endl.
- E *Taxus baccata* L.
- A *Taxus* × *media* Rehder

Taxodiaceae
- A *Metasequoia glyptostroboides* Hu & W.C. Cheng
- A *Sciadopitys verticillata* (Thunb.) Siebold & Zucc.
- A *Sequoiadendron giganteum* (Lindl.) J. Buchholz
- A *Taxodium distichum* (L.) L.C.M. Rich.

Zamiaceae
- A *Encephalartos lehmannii* Lehm.
- A *Encephalartos transvenosus* Stapf & Burtt Davy

Magnoliophytina, Dicotyledonae, Apiales

Apiaceae
- E *Aegopodium podagraria* L.
- E *Aethusa cynapium* L.
- E *Ammi majus* (L.) Lam.
- E *Ammi visnaga* (L.) Lam.
- E *Anethum graveolens* L.
- E *Angelica archangelica* L.
- A *Angelica pachycarpa* Lange
- E *Anthriscus caucalis* Bieb.
- E *Anthriscus cerefolium* (L.) Hoffm.

11 List of Species Alien in Europe and to Europe

E *Anthriscus cerefolium longirostris* (Bertol.) Cannon
E *Anthriscus fumarioides* (Waldst. & Kit.) Sprengel
E *Anthriscus sylvestris* (L.) Hoffm
E *Apium graveolens* L.
E *Apium leptophyllum* (Pers.) F. Müller ex Bentham
E *Apium nodiflorum* (L.) Lag.
E *Astrantia major* L.
E *Astrodaucus orientalis* (L.) Drude
E *Athamanta sicula* L.
E *Bifora radians* Bieb.
E *Bifora testiculata* (L.) Roth
E *Bowlesia incana* Ruiz & Pavn
E *Bunium bulbocastanum* L.
E *Bupleurum baldense* Turra
E *Bupleurum croceum* Fenzl
E *Bupleurum falcatum* L.
E *Bupleurum fontanesii* Guss. ex Caruel
E *Bupleurum fruticosum* L.
E *Bupleurum gerardi* All.
E *Bupleurum lancifolium* Hornem.
E *Bupleurum longifolium* L.
A *Bupleurum odontites* L.
E *Bupleurum praealtum* L.
E *Bupleurum rotundifolium* L.
E *Carum carvi* L.
E *Caucalis bischofii* Kozo-Poljanskij
E *Chaerophyllum aromaticum* L.
E *Chaerophyllum aureum* L.
E *Chaerophyllum bulbosum* L.
E *Chaerophyllum byzantinum* Boiss.
E *Chaerophyllum hirsutum* L.
E *Chaerophyllum temulentum* L.
E *Cnidium silaifolium* (Jacq.) Simonk.
E *Conium maculatum* L.
E *Conopodium majus* (Gouan) Loret
E *Coriandrum sativum* L.
A *Cryptotaenia canadensis* L.
E *Cuminum cyminum* L.
E *Cyclospermum leptophyllum* (Pers.) Sprague ex Britt. & Wilson
E *Daucosma laciniata* Engelm. & Gray
E *Daucus aureus* Desf.
E *Daucus broteri* Ten.
E *Daucus carota* L.
E *Daucus durieua* Lange
A *Daucus glochidiatus* (Labill.) Fischer, C. Meyer & Ave-Lall.
E *Daucus muricatus* (L.) L.
C *Distichoselinum tenuifolium* (Lag.) García-Martín & Silvestre
E *Eryngium amethystinum* L.
E *Eryngium campestre* L.
E *Eryngium giganteum* M. Bieb.
A *Eryngium pandanifolium* Cham. & Schlecht.
E *Eryngium planum* L.
A *Eryngium tricuspidatum* L.
E *Falcaria vulgaris* Bernh.
E *Fedia cornucopiae* (L.) Gaertn.
E *Ferula communis* L.
E *Ferulago campestris* (Besser) Grec
A *Ferulago sylvatica* (Besser) Reichb.
E *Foeniculum vulgare* (L.) Mill.
E *Hacquetia epipactis* (Scop.) DC.
E *Heracleum laciniatum* Sommier & Lévier
E **Heracleum mantegazzianum** Sommier & Lévier
A *Heracleum persicum* Desf.
A *Heracleum pubescens* (Hoffm.) M. Bieb.
A *Heracleum sosnowskyi* Manden.
A *Hydrocotyle bonariensis* Commerson ex Lam K.
A *Hydrocotyle moschata* G. Forster
A *Hydrocotyle novae-zeelandiae* DC.
A *Hydrocotyle ramiflora* Maximow (1887)
A *Hydrocotyle ranunculoides* L. f.
A *Hydrocotyle sibthorpioides* Lam.
A *Hydrocotyle verticillata* Thunb.
E *Imperatoria ostruthium* L.
C *Kadenia dubia* (Schkuhr) Lavrova & V.N. Tikhom.
E *Kundmannia sicula* (L.) DC.
E *Lagoecia cuminoides* L.
E *Laserpitium siler* L.
E *Levisticum officinale* W.D.J. Koch
E *Ligusticum scoticum* L.
E *Lilaeopsis attenuata* (Hooker & Arnott) Fernald
A *Magydaris pastinacea* (Lam.) Paol.
E *Meum athamaticum* Jacq.
E *Molopospermum peloponnesiacum* (L.) Koch
E *Myrrhis odorata* (L.) Scop.
A *Myrrhoides nodosa* (L.) Cannon
E *Naufraga balearica* Constance & Cannon
A *Oenanthe javanica* DC.
E *Oenanthe pimpinelloides* L.
E *Oenanthe silaifolia* M. Bieb.
E *Orlaya daucoides* (L.) Greuter
E *Orlaya daucorlaya* Murb.
E *Pastinaca sativa* L.

- E *Petroselinum crispum* (Mill.) A.W. Hill
- E *Petroselinum segetum* (L.) Koch
- A *Peucedanum austriacum* (Jacq.) Koch
- A *Peucedanum chryseum* (Boiss. & Heldr.) D.F. Chamb.
- A *Peucedanum verticillare* (L.) Koch ex DC.
- E *Pimpinella anisum* L.
- E *Pimpinella cretica* Poir. Ex Lam.
- E *Pimpinella lutea* Desf.
- E *Pimpinella major* (L.) Hudson
- E *Pimpinella peregrina* L.
- E *Pimpinella saxifraga* L.
- A *Pimpinella villosa* Schousb.
- A *Ptychotis saxifraga* (L.) Lor. & Barr.
- E *Ridolfia segetum* (Guss.) Moris
- E *Scandix balansae* Reut. ex Boiss.
- E *Scandix iberica* M. Bieb.
- E *Scandix pecten-veneris* L.
- E *Scandix stellata* Banks & Sol.
- E *Seseli libanotis* (L.) W.D.J. Koch
- E *Seseli montanum* L.
- E *Seseli tortuosum* L.
- E *Seseli varium* Trevir.
- E *Sison amomum* L.
- A *Sium sisarum* L.
- E *Smyrnium olusatrum* L.
- E *Smyrnium perfoliatum* L.
- E *Thapsia villosa* L.
- E *Tordylium apulum* L.
- E *Tordylium maximum* L.
- E *Torilis arvensis* (Huds.) Link
- E *Torilis japonica* (Houtt.) DC.
- E *Torilis leptophylla* (L.) Reichb. f.
- E *Torilis nodosa* (L.) Gaertn.
- A *Torilis tenella* (Delile) Reichb. f.
- E *Torilis ucranica* Spreng.
- A *Trachyspermum ammi* (L.) Sprague ex Turrill
- E *Turgenia latifolia* (L.) Hoffm.
- A *Tyrimnus leucographus* (L.) Cass.

Araliaceae
- A *Aralia chinensis* L.
- A *Aralia elata* (Miq.) Deeman
- A *Aralia racemosa* L.
- A *Fatsia japonica* (Thunb.) Decne. & Planchon
- E *Hedera algeriensis* Hibberd
- E *Hedera colchica* (K. Koch) K. Koch
- A *Schefflera actinophylla* (Endl.) Harms
- A *Tetrapanax papyriferus* (Hook.) K. Koch

Pittosporaceae
- A *Hymenosporum flavum* (Hook.) F. Müll.
- A *Pittosporum crassifolium* Banks & Sol. ex A. Cunn.
- A *Pittosporum tobira* (Thunb.) W.T. Aiton
- A *Pittosporum undulatum* Vent.
- A *Sollya fusiformis* (Labill.) Briq.

Magnoliophytina, Dicotyledonae, Aquifoliales
Aquifoliaceae
- A *Ilex × altaclerensis* (Loudon) Dallimore
- E *Ilex aquifolium* L.

Magnoliophytina, Dicotyledonae, Asterales
Asteraceae
- A *Acanthospermum hispidum* DC.
- A *Acanthospermum humile* (Sw.) DC.
- E *Achillea ageratum* L.
- E *Achillea biebersteinii* Afan.
- E *Achillea cartilaginea* Ledeb. ex Reichb.
- E *Achillea crithmifolia* Waldst. & Kit.
- E *Achillea distans* Waldst. & Kit. ex Willd.
- A *Achillea filipendulina* Lam.
- E *Achillea grandifolia* Friv.
- E *Achillea ligustica* All.
- E *Achillea micrantha* Willd.
- E *Achillea millefolium* L.
- E *Achillea nobilis* L.
- E *Achillea pannonica* Scheele
- E *Achillea ptarmica* L.
- E *Achillea punctata* L.
- E *Achillea roseoalba* Ehrend.
- E *Achillea setacea* Waldst. & Kit.
- A *Achillea stricta* (Koch) Schleich. ex Gremli
- E *Acroptilon repens* (L.) DC.
- E *Adenostyles alliariae* (Gouan) A. Kerner
- E *Aetheorhiza bulbosa* (L.) Cass.
- A *Ageratina riparia* (Regel) King & H.E. Robins.
- A *Ageratum conyzoides* L.
- E *Ageratum houstonianum* Mill.
- A *Amberboa moschata* (L.) DC.
- A **Ambrosia artemisiifolia** L.
- A *Ambrosia confertiflora* DC.
- A *Ambrosia coronopifolia* Torrey & A. Gray
- A *Ambrosia tenuifolia* Sprengel
- A *Ambrosia trifida* L.
- A *Amphiachyris dracunculoides* (DC.) Nutt.
- E *Anacyclus clavatus* (Desf.) Pers.
- E *Anacyclus radiatus* Lois.
- E *Anacyclus valentinus* L.

- E *Anaphalis margaritacea* (L.) Bentham
- E *Andryala integrifolia* L.
- E *Anthemis altissima* L.
- E *Anthemis arvensis* L.
- E *Anthemis austriaca* Jacq.
- E *Anthemis cotula* L.
- E *Anthemis punctata* Vahl
- E *Anthemis ruthenica* M. Bieb.
- E *Anthemis segetalis* L.
- E *Anthemis tinctoria* L.
- A *Anthemis tomentosa* L.
- A *Anthemis triumfettii* (L.) DC.
- E *Aposeris foetida* (L.) Less.
- A *Arctium* × *ambiguum* (Celak.) Beck
- A *Arctium* × *cimbricum* (Krause) Hayek
- E *Arctium lappa* L.
- A *Arctium* × *maassii* (M. Schulye) Rouy
- E *Arctium minus* (Hill.) Bernh.
- A *Arctium* × *mixtum* (Simk.) Nyman
- A *Arctium* × *neumannii* Rouy
- A *Arctium* × *nothum* (Ruhmer) Weiss
- E *Arctium tomentosum* Mill.
- A *Arctotheca calendula* (L.) Levyns
- A *Arctotis stoechadifolia* Berg.
- A *Arctotis venusta* Norl.
- A *Argyranthemum frutescens* (L.) Sch.-Bip. ex Webb. & Berth.
- A *Argyranthemum pinnatifidum* (L. f.) R.T. Lowe
- E *Arnica chamissonis* Less.
- E *Arnica montana* L.
- E *Arnoseris minima* (L.) Schweigger & Koerte
- E *Artemisia abrotanum* L.
- E *Artemisia absinthium* L.
- A *Artemisia afra* Jacq. ex Willd.
- A *Artemisia alba* Turra
- E *Artemisia annua* L.
- E *Artemisia arborescens* L.
- A *Artemisia argyi* H. Lév. & Vaniot
- E *Artemisia austriaca* Jacq.
- E *Artemisia biennis* Willd.
- E *Artemisia campestris campestris* L.
- E *Artemisia campestris maritima* Arcangeli
- E *Artemisia canariensis* Less.
- A *Artemisia capillaris* Thunb.
- A *Artemisia codonocephala* Diels
- E *Artemisia dracunculus* L.
- E *Artemisia dubia* Wall.
- A *Artemisia gnaphalodes* Nutt.
- A *Artemisia integrifolia* L.
- A *Artemisia latifolia* Ledeb.
- A *Artemisia ludoviciana* Nutt.
- E *Artemisia pontica* L.
- A *Artemisia schrenkiana* Ledeb.
- E *Artemisia scoparia* Waldst. & Kit.
- E *Artemisia siversiana* Ehrh. ex Willd.
- E *Artemisia stellerana* Besser
- E *Artemisia tournefortiana* Reichb.
- E *Artemisia verlotiorum* Lamotte
- E *Artemisia vulgaris* L.
- E *Aster amellus* L.
- A *Aster bellidiastrum* (L.) Scop.
- E *Aster chinensis* L.
- A *Aster cordifolius* L.
- A *Aster divaricatus* L.
- A *Aster dumosus* L.
- A *Aster ericoides* L.
- A *Aster laevis* L.
- A *Aster lanceolatus* Willd.
- A *Aster lateriflorus* (L.) Britt.
- A *Aster macrophyllus* L.
- A *Aster novae-angliae* L.
- A *Aster novi-belgii* L.
- A *Aster patulus* Lam.
- A *Aster pilosus* Willd.
- A *Aster riparius* Nees
- E *Aster* × *salignus* Willd.
- A *Aster schreberi* Nees
- E *Aster sedifolius* L.
- E *Aster sibiricus* L.
- E *Aster squamatus* (Spreng.) Hieron.
- A *Aster* × *versicolor* Willd.
- E *Asteriscus aquaticus* (L.) Less.
- A *Athanasia trifurcata* L.
- A *Baccharis halimifolia* L.
- E *Bellis perennis* L.
- A *Bidens aristosa* (Michx.) Britton
- E *Bidens aurea* (Aiton) Sherff
- E *Bidens bipinnata* L.
- A *Bidens biternata* (Lour.) Merr. & Sherff
- A *Bidens connata* Muhl. ex Willd.
- A *Bidens frondosa* L.
- E *Bidens pilosa* L.
- E *Bidens subalternans* DC.
- E *Bidens tripartita* L.
- A *Bidens triplinervia* Humb., Bonpl. & Kunth
- A *Bidens vulgata* E.L. Greene
- A *Boltonia asteroides* (L.) L'Hér.
- A *Bombycilaena erecta* (L.) Smolj.
- A *Borrichia frutescens* (L.) DC.
- A *Brachyactis ciliata* (Ledeb.) Ledeb.
- A *Brachyglottis repanda* Forster & G. Forster
- E *Calendula arvensis* L.
- E *Calendula officinalis* L.

- E *Calendula stellata* Cav.
- E *Calendula suffruticosa* Vahl
- A *Calotis cuneifolia* R. Br.
- A *Calotis hispidula* (F. Müll.) F. Müll.
- A *Calotis lappulacea* Benth.
- E *Carduus acanthoides* L.
- C *Carduus argemone* Pourr. ex Lam.
- A *Carduus bourgeanus* Boiss. & Reuter
- E *Carduus crispus* L.
- E *Carduus hamulosus* Ehrh.
- A *Carduus* × *leptocephalus* Peterm.
- E *Carduus nutans* L.
- A *Carduus* × *orthocephalus* Wallr.
- E *Carduus pycnocephalus* L.
- E *Carduus* × *sepincola* Hausskn.
- A *Carduus* × *stangii* Buek
- A *Carduus tenuiflorus* Curtis
- E *Carduus* × *theriotii* Rouy
- E *Carduus thoermeri* Weinm.
- E *Carduus vivariensis* Jord.
- E *Carlina acanthifolia* All.
- E *Carlina acaulis* L.
- E *Carlina vulgaris* L.
- A *Carpesium cernuum* L.
- E *Carthamus dentatus* (Forsk.) Vahl.
- E *Carthamus lanatus* L.
- E *Carthamus tinctorius* L.
- E *Catananche caerulea* L.
- E *Catananche lutea* L.
- A *Centaurea acaulis* L.
- E *Centaurea adamii* Willd.
- E *Centaurea alba* L.
- A *Centaurea algeriensis* Durieu & Coss.
- E *Centaurea arenaria* Bieb. ex Willd.
- E *Centaurea aspera* L.
- E *Centaurea biebersteinii* DC.
- E *Centaurea calcitrapa* L.
- C *Centaurea cineraria* L.
- E *Centaurea cyanus* L.
- A *Centaurea dealbata* Willd.
- E *Centaurea depressa* M.Bieb.
- E *Centaurea dichroantha* A. Kerner
- E *Centaurea diffusa* Lam.
- E *Centaurea diluta* Aiton
- E *Centaurea elegans* Salisb.
- A *Centaurea eriophora* L.
- A *Centaurea* × *extranea* Beck
- A *Centaurea* × *gerstlaueri* Erdner
- E *Centaurea glaberrima* Tausch
- E *Centaurea hyalolepis* Boiss.
- E *Centaurea iberica* Trev. ex Sprengel
- E *Centaurea jacea* L.
- E *Centaurea macroptilon* Borbs
- E *Centaurea melitensis* L.
- E *Centaurea montana* L.
- A *Centaurea napifolia* L.
- A *Centaurea* × *nemenyiana* J. Wagner
- E *Centaurea nicaeensis* All.
- E *Centaurea nigra* L.
- E *Centaurea nigrescens* Willd.
- E *Centaurea orientalis* L.
- E *Centaurea paniculata* L.
- E *Centaurea pannonica* (Heuff.) Simonk.
- E *Centaurea pectinata* L.
- E *Centaurea phrygia* L.
- A *Centaurea* × *psammogena* (Gyer) Holub
- A *Centaurea pullata* L.
- A *Centaurea ragusina* L.
- E *Centaurea rupestris* L.
- E *Centaurea ruthenica* Lam.
- A *Centaurea salicifolia* M. Bieb. ex Willd.
- E *Centaurea scabiosa* L.
- E *Centaurea solstitialis* L.
- E *Centaurea sonchifolia* L.
- E *Centaurea stoebe* L.
- E *Centaurea triumfetti* All.
- A *Centipeda minima* (L.) A. Braun & Aschers.
- E *Chamaemelum mixtum* (L.) All.
- E *Chamaemelum nobile* (L.) All.
- A *Chartolepis glastifolia* (L.) Cass.
- C *Cheirolophus intybaceus* (Lam.) Dostl
- E *Cheirolophus sempervirens* (L.) Pomel
- E *Chondrilla juncea* L.
- E *Chondrilla pauciflora* Ledeb.
- A *Chrysanthemoides monilifera* (L.) T. Norlindh
- A *Chrysanthemum carinatum* Schousboe
- E *Chrysanthemum coronarium* L.
- A *Chrysanthemum indicum* Thunb.
- A *Chrysanthemum praealtum* Vent.
- E *Chrysanthemum segetum* L.
- A *Chrysocoma coma-aurea* L.
- A *Chrysocoma tenuifolia* P. Bergius
- E *Cicerbita alpina* (L.) Wallr.
- E *Cicerbita bourgaei* (Boiss.) Beauverd
- E *Cicerbita macrophylla* (Willd.) Wallr.
- E *Cicerbita plumieri* (L.) Kirschl.
- A *Cichorium calvum* Ascherson
- E *Cichorium endivia* L.
- E *Cichorium intybus* L.
- E *Cirsium arvense* (L.) Scop.
- A *Cirsium* × *aschersonianum* Celak.
- A *Cirsium* × *bipontinum* F.W. Schultz
- E *Cirsium canum* (L.) All.
- A *Cirsium* × *celakovskyanum* Knaf
- E *Cirsium dissectum* (L.) Hill
- A *Cirsium echinus* (Bieb.) Hand.-Mazz.

E *Cirsium erisithales* (Jacq.) Scop.
A *Cirsium* × *grandiflorum* Kitt.
E *Cirsium helenioides* (L.) Hill
E *Cirsium heterophyllum* (L.) Hill
E *Cirsium oleraceum* (L.) Scop.
E *Cirsium palustre* (L.) Scop.
A *Cirsium* × *preiseri* Uechtr.
A *Cirsium* × *reichenbachianum* Löhr.
E *Cirsium rivulare* (Jacq.) All.
A *Cirsium* × *sabaudum* Löhr.
A *Cirsium scabrum* (Poir.) Bonnet & Barratte
A *Cirsium* × *sextinum* Huter
A *Cirsium* × *soroksarense* Wagner
A *Cirsium* × *subalpinum* Gaudin
A *Cirsium* × *subspinuligerum* Peterm.
A *Cirsium tuberosum* (L.) All.
E *Cirsium vulgare* (Savi) Ten.
E *Cnicus benedictus* L.
E *Coleostephus myconis* (L.) Reichenb. fil.
A *Conyza bilbaoana* J. Rémy
A *Conyza blakei* (Cabrera) Cabrera
A *Conyza bonariensis* (L.) Cronquist
A *Conyza canadensis* (L.) Cronquist
A *Conyza floribunda* Kunth
A *Conyza ivifolia* (L.) Less.
A *Conyza* × *mixta* Foucaud & Neyraut
A *Conyza pampeana* (Parodi) Cabrera
A *Conyza primulaefolia* (Lam.) Cuatrec
A *Conyza sumatrensis* (Retz.) E. Walker
A *Conyza triloba* Decne.
A × *Conyzigeron huelsenii* (Vatke) Rauschert
A *Coreopsis bicolor* Reichenb.
A *Coreopsis grandiflora* Hogg ex Sweet
A *Coreopsis lanceolata* L.
A *Coreopsis verticillata* L.
A *Cosmos bipinnatus* Cav.
A *Cotula anthemoides* L.
A *Cotula australis* (Sieber ex Sprengel) Hook. f.
A *Cotula coronopifolia* L.
A *Cotula dioica* (Hook. f.) Hook. f.
A *Cotula mexicana* (DC.) Cabrera.
A *Cotula squalida* (Hook. f.) Hook. f.
A *Cotula turbinata* L.
E *Cousinia tenella* Fisch. & C.A. Mey.
A *Crassocephalum crepidioides* (Bentham) S. Moore (1912)
A *Crassocephalum subscandens* (Hochst. ex A. Rich.) S. Moore
E *Crepis alpina* L.
E *Crepis aurea* (L.) Cass.
E *Crepis bellidifolia* Loisel.

E *Crepis biennis* L.
E *Crepis bursifolia* L.
E *Crepis capillaris* (L.) Wallr.
E *Crepis dioscoridis* L.
E *Crepis foetida* L.
E *Crepis foetida rhoeadifolia* (M. Bieb.) Celak.
E *Crepis micrantha* Czerep.
E *Crepis neglecta* L.
E *Crepis nicaeensis* Balbis
E *Crepis pulchra* L.
E *Crepis sancta* (L.) Babcock
E *Crepis setosa* Haller fil.
E *Crepis tectorum* L.
E *Crepis vesicaria* L.
E *Crepis zacintha* (L.) Babcock
E *Cynara cardunculus* L.
E *Cynara scolymus* L.
A *Dahlia coccinea* Cav.
A *Dahlia* × *cultorum* Thorsrud & Reisaeter
A *Dahlia pinnata* Cav.
A *Dendranthema* × *grandiflorum* (Ramat.) S. Kitamura
A *Dendranthema indicum* (L.) Desmoulins
A *Dichrocephala integrifolia* (L.) Kuntze
E *Dittrichia graveolens* (L.) Greuter
E *Dittrichia viscosa* (L.) W. Greuter
E *Doronicum columnae* Ten.
A *Doronicum* × *excelsum* (N.E. Br.) Stace
E *Doronicum orientale* Hoffm.
E *Doronicum pardalianches* L.
E *Doronicum plantagineum* L.
E *Doronicum* × *willdenowii* (Rouy) A.W. Hill
E *Echinops banaticus* Rochel ex Schrader
E *Echinops exaltatus* Schrader
E *Echinops ruthenicus* M. Bieb.
E *Echinops sphaerocephalus* L.
E *Eclipta prostrata* (L.) L.
A *Eleutheranthera ruderalis* (Sw.) Sch.-Bip.
E *Erechtites hieraciifolia* (L.) Raf. ex DC.
A *Erigeron annuus* (L.) Pers.
A *Erigeron annuus septentrionalis* (Fern.& Wieg.) Wagenitz
A *Erigeron annuus strigosus* (Mühlenb.) Wagenitz
A *Erigeron ciliatus* Ledeb.
A *Erigeron glaucus* Ker-Gawl.
E *Erigeron karvinskianus* DC.
A *Erigeron philadelphicus* L.
A *Erigeron speciosus* (Lindl.) DC.
A *Eriocephalus africanus* L.
A *Eriophyllum lanatum* (Pursh) Forbes

- A *Eupatorium adenophorum* Spreng.
- A *Eupatorium ageratoides* L. f.
- E *Eupatorium cannabinum* L.
- A *Facelis retusa* (Lam.) Schultz-Bip.
- A *Felicia tenella* (L.) Nees
- E *Filago arv*ensis L.
- E *Filago gall*ica L.
- E *Filago lutescens* Jord.
- E *Filago minima* (Sm.) Pers.
- E *Filago pyramidata* L.
- E *Filago vulgaris* Lam.
- E *Flaveria bidentis* (L.) O. Kuntze
- A *Gaillardia aristata* Pursh
- A *Gaillardia* × *grandiflora* Van Houtte
- A *Gaillardia pulchella* Foug.
- E *Galactites tomentosa* Moench
- E *Galatella sedifolius dracunculoides* (Lam.) Merxm.
- E *Galatella sedifolius sedifolius* L.
- A *Galinsoga* × *mixta* J. Murray
- A *Galinsoga parviflora* Cav.
- A *Galinsoga quadriradiata* Ruiz & Pavón
- A *Galinsoga urticifolia* (Humb., Bonpl. & Kunth) Benth.
- A *Gamochaeta americana* (Mill.) Wedd.
- A *Gamochaeta falcata* (Lam.) Cabrera
- A *Gamochaeta filaginea* (DC.) Cabrera
- A *Gamochaeta pensylvanica* (Willd.) Cabrera
- A *Gamochaeta purpurea* (L.) Cabrera
- A *Gamochaeta subfalcata* (Cabrera) Cabrera
- A *Gazania linearis* Druce
- A *Gazania rigens* (L.) Gaertner
- E *Gnaphalium antillanum* Urban
- E *Gnaphalium calviceps* Fernald
- A *Gnaphalium gaudichaudianum* DC.
- A *Gnaphalium sphaericum* Willd.
- E *Gnaphalium supinum* L.
- E *Gnaphalium sylvaticum* L.
- E *Gnaphalium uliginosum* L.
- A *Gnaphalium undulatum* L.
- E *Grindelia squarrosa* (Pursh) Dunal
- A *Grindelia stricta* DC.
- A *Guizotia abyssinica* (L.f.) Cass.
- A *Gymnocoronis spilanthoides* DC.
- A *Gymnostyles stolonifera* (Brot.) Tutin
- E *Hedypnois cretica* (L.) Dum.-Cours.
- A *Helenium autumnale* L.
- A *Helenium* × *clementii* Verloove & Lambinon
- A *Helenium mexicanum* Humb., Bonpl. & Kunth
- A *Helianthus angustifolius* L.
- A *Helianthus annuus* L.
- A *Helianthus cucumerifolius* Torrey & A. Gray
- A *Helianthus debilis* Nutt.
- A *Helianthus decapetalus* L.
- A *Helianthus* × *laetiflorus* Pers.
- A *Helianthus* × *multiflorus* L.
- A *Helianthus pauciflorus* Nutt.
- A *Helianthus petiolaris* Nutt.
- A *Helianthus rigidus* (Cass.) Desf.
- A *Helianthus salicifolius* A. Dietr.
- A *Helianthus strumosus* L.
- A *Helianthus tuberosus* L.
- E *Helichrysum arenarium* (L.) Moench
- A *Helichrysum bellidioides* (G. Forster) Willd.
- A *Helichrysum bracteatum* (Vent.) Andrews
- A *Helichrysum foetidum* (L.) Cass.
- A *Helichrysum orbiculare* Druce
- E *Helichrysum orientale* (L.) Gaertn.
- A *Helichrysum petiolare* Hilliard & B.L. Burtt
- E *Helichrysum stoechas* (L.) Moench
- A *Heliopsis helianthoides* (L.) Sweet
- A *Heliopsis scabra* Dunal
- A *Helipterum floribundum* DC.
- A *Helipterum strictum* (Lindl.) Benth.
- A *Hemizonia fasciculata* (DC.) Torr. & A. Gray
- A *Hemizonia pungens* (Hook. & Arnott) Torr. & A. Gray
- A *Heterotheca subaxillaris* (Lam.) Britt. & Rusby
- E *Hieracium amplexicaule* L.
- A *Hieracium brachiatum* Bertol. ex DC.
- A *Hieracium* × *duplex* Peter
- E *Hieracium flagellare* Willd. ex Schlecht
- E *Hieracium* × *florentoides* Arv.-Touv.
- E *Hieracium grandidens* Dahlst.
- E *Hieracium lanatum* Vill.
- E *Hieracium murorum* L.
- E *Hieracium oblongum* Jord.
- A *Hieracium pannosum* Boissier
- E *Hieracium piloselloides* Vill.
- E *Hieracium pilosum* Schleich. ex Froel.
- A *Hieracium* × *prussicum* Naeg. & Peter
- E *Hieracium pulmonarioides* Vill.
- E *Hieracium scotostictum* Hyl.
- E *Hieracium speciosum* Willd. ex Hornem.
- E *Hieracium speluncarum* Arv.-Touv.
- E *Hypochaeris achyrophorus* L.
- E *Hypochaeris glabra* L.
- A *Hypochaeris microcephala* (Schultz-Bip.) Cabrera

- E *Hypochaeris radicata* L.
- E *Inula britannica* L.
- E *Inula conyzae* (Griess.) Meikle
- E *Inula ensifolia* L.
- E *Inula germanica* L.
- E *Inula helenium* L.
- E *Inula hirta* L.
- E *Inula spiraeifolia* L.
- E *Iva* × *anthiifolia* Nutt.
- A *Kleinia aizoides* DC.
- E *Kleinia neriifolia* Cav.
- A *Kleinia repens* (L.) Haw.
- E *Lactuca perennis* L.
- E *Lactuca saligna* L.
- A *Lactuca sativa* L.
- E *Lactuca serriola* L.
- E *Lactuca sibirica* (L.) Benth. ex Maxim.
- E *Lactuca tatarica* (L.) C. Meyer
- E *Lactuca virosa* L.
- E *Lapsana communis* L.
- E *Leontodon hispidus* L.
- E *Leontodon saxatilis* Lam.
- E *Leontopodium alpinum* Cass.
- A *Lepidotheca aurea* (L.) Kovalevsk.
- E *Leucanthemella serotina* (L.) Tzvelev
- E *Leucanthemum lacustre* (Brot.) Samp.
- E *Leucanthemum maximum* (Ramond) DC.
- E *Leucanthemum paludosum* (Poiret) Bonnet & Barratte
- E *Leucanthemum praecox* (Horvatic) Horvatíc
- E *Leucanthemum* × *superbum* (J.W. Ingram) Berg. ex Kent.
- E *Leucanthemum vulgare* Lam.
- A *Leysera tenella* DC.
- E *Ligularia dentata* (A.Gray) H.Hara
- A *Ligularia przewalskii* (Maxim.) Diels
- E *Ligularia sibirica* (L.) Cass.
- A *Ligularia tussilaginea* (Burm. f.) Makino
- E *Madia sativa* Molina
- E *Mantisalca salmantica* (L.) Briq.
- E *Matricaria discoidea* DC.
- E *Matricaria recutita* L.
- A *Matricaria suffruticosa* (L.) Druce
- A *Matricaria trichophylla* (Boiss.) Boiss.
- E *Melampodium montanum* Benth.
- E *Microlonchus salmanticus* (L.) DC.
- E *Micropus supinus* L.
- A *Montanoa bipinnatifida* (Kunth) C. Koch
- E *Mycelis muralis* (L.) Dumort.
- E *Notobasis syriaca* (L.) Cass.
- A *Olearia avicenniifolia* (Raoul) Hook. f.
- A *Olearia* × *haastii* Hook. f.
- A *Olearia macrodonta* Baker
- A *Olearia paniculata* (Forster & G. Forster) Druce
- A *Olearia traversii* (F. Muell.) Hook. f.
- E *Onopordum acanthium* L.
- E *Onopordum* × *beckianum* John
- E *Onopordum bracteatum* Boiss. & Heldr.
- E *Onopordum illyricum* L.
- E *Onopordum macracanthum* Schousb.
- E *Onopordum nervosum* Boiss.
- A *Osteospermum barberae* (Harv.) Norl.
- A *Osteospermum ecklonis* (DC.) Norl.
- A *Osteospermum jucundum* (Phillips) Norl.
- E *Otanthus maritimus* (L.) Hoffm. & Link
- E *Pallenis spinosa* (L.) Cass.
- A *Parthenium hysterophorus* L.
- E *Petasites albus* (L.) Gaertner
- E *Petasites fragans* (Vill.) C. Presl
- E *Petasites hybridus* (L.) P. Gaertn., B. Mey. & Scherb
- E *Petasites japonicus* (Sieb. & Zucc.) Maxim.
- E *Petasites spurius* (Retz.) Reichb.
- E *Picnomon acarna* (L.) Cass.
- E *Picris altissima* Delile
- E *Picris echioides* L.
- E *Picris hieracioides* L.
- E *Picris pauciflora* Willd.
- A *Picris sprengerana* (L.) Poir.
- E *Pilosella aurantiaca* (L.) F. Schultz & Schultz-Bip.
- E *Pilosella bauhinii* (Schultes) Arvet-Touvet
- E *Pilosella caespitosa* (Dumort.) Sell & C. West
- E *Pilosella carpathicola* (Nägeli & Peter) Czerep.
- A *Pilosella* × *floribunda* (Wimmer & Grab.) Arv.-Touv.
- E *Pilosella praealta* (Villars ex Gochnat) F. Schultz & Schultz-Bip.
- A *Pilosella* × *stoloniflora* (Waldst. & Kit.) F.W. Schultz & Sch.-Bip.
- A *Plecostachys serpyllifolia* (Berg.) Hilliard & B.L. Burtt
- A *Podolepis auriculata* DC.
- E *Prenanthes purpurea* L.
- E *Ptilostemon afer* (Jacq.) Greuter
- E *Ptilostemon casabonae* (L.) W. Greuter
- A *Pulicaria arabica* (L.) Cass.
- E *Pulicaria dysenterica* (L.) Bernh.
- E *Pulicaria paludosa* Link

E *Pulicaria vulgaris* Gaertn.
A *Pyrethrum coccineum* (Willd.) Vorosch.
E *Reichardia picroides* (L.) Roth
E *Rhagadiolus stellatus* (L.) Gaertn.
A *Rhodanthe manglesii* (O.F. Müller) Lindl.
A *Rudbeckia drummondii* Paxton
A *Rudbeckia fulgida* Aiton
A *Rudbeckia hirta* L.
A *Rudbeckia laciniata* L.
E *Rudbeckia triloba* L.
E *Santolina chamaecyparissus* L.
E *Santolina rosmarinifolia* L.
A *Sanvitalia procumbens* Lam.
E *Saussurea amara* (L.) DC.
E *Schkuhria pinnata* (Lam.) O. Kuntze
E *Scolymus grandiflorus* Desf.
E *Scolymus hispanicus* L.
E *Scolymus maculatus* L.
E *Scorzonera hispanica* L.
E *Scorzonera laciniata* L.
A *Senecio × albescens* Burbidge & Colgan
E *Senecio angulatus* L. f.
A *Senecio × baxteri* Druce
E *Senecio cannabifolius* L.
E *Senecio cineraria* DC.
A *Senecio deltoideus* Less.
E *Senecio doria* L.
A *Senecio doronicum* (L.) L.
A *Senecio dubitabilis* C. Jefferey & Y.L. Chen
E *Senecio eboracensis* R.J. Abbott & A.J. Lowe
A *Senecio elegans* L.
A *Senecio erraticus* Bertol.
E *Senecio erucifolius* L.
E *Senecio fluviatilis* Wallr.
E *Senecio gallicus* Chaix
A *Senecio glastifolius* L. f.
A *Senecio glossanthus* (Sonder) Belcher
A *Senecio grandiflorus* P. Bergius
A *Senecio × helwingii* Beger ex Hegi
A *Senecio hybridus* (Willd.) Regel
E *Senecio inaequidens* DC.
E *Senecio jacobaea* L.
A *Senecio lineatus* DC.
A *Senecio macroglossus* DC.
A *Senecio mikanioides* Otto ex Walp.
A *Senecio monroi* Hook. f.
E *Senecio parviflora* All.
A *Senecio petasitis* (Sims) DC.
A *Senecio platensis* (Hieron.) Arechav.
A *Senecio pterophorus* DC.

A *Senecio quadridentatus* Labill.
E *Senecio rupestris* Waldst. & Kit.
A *Senecio smithii* DC.
A *Senecio soldanella* A. Gray
E *Senecio squalidus* L.
A *Senecio × subnebrodensis* Simonkai
E *Senecio sylvaticus* L.
A *Senecio tamoides* DC.
A *Senecio × thuretii* Briq. & Cavill.
E *Senecio vernalis* Waldst. & Kit.
A *Senecio × viscidulus* Scheele
E *Senecio viscosus* L.
E *Senecio vulgaris* L.
A *Serratula quinquefolia* M. Bieb. ex Willd.
E *Serratula tinctoria* L.
A *Shinnersia rivularis* (A. Gray) R.M. King & H.E. Robinson
A *Sigesbeckia orientalis* L.
A *Sigesbeckia serrata* DC.
A *Silphium perfoliatum* L.
E *Silybum marianum* (L.) Gaertn.
E *Sinacalia tangutica* (Maxim.) R. Nordenstam
A *Solidago canadensis* L.
A *Solidago gigantea* Aiton
A *Solidago graminifolia* (L.) Salisb.
A *Solidago × niederederi* Khek
A *Solidago rugosa* Mill.
E *Solidago sempervirens* L.
A *Solidago shortii* Torrey & A. Gray
A *Soliva anthemifolia* (Juss.) R. Br. ex Less.
E *Soliva pterosperma* (Juss.) Less.
E *Sonchus arvensis* L.
E *Sonchus asper* (L.) Hill.
E *Sonchus oleraceus* L.
E *Sonchus tenerrimus* L.
A *Stuartina muelleri* Sond.
E *Tagetes erecta* L.
A *Tagetes minuta* L.
E *Tagetes patula* L.
E *Tanacetum balsamita* L.
E *Tanacetum cineariifolium* (Trev.) Schultz-Bip.
E *Tanacetum corymbosum* (L.) Schultz-Bip.
E *Tanacetum macrophyllum* (Waldst. & Kit.) Schultz-Bip.
A *Tanacetum millefolium* (L.) Tzvelev
A *Tanacetum partheniifolium* (Willd.) Sch.-Bip.
E *Tanacetum parthenium* (L.) Schultz-Bip.

- E *Tanacetum vulgare* L.
- A *Taraxacum ekmanii* Dahlst.
- A *Taraxacum latisectum* Lindb.
- A *Taraxacum maderense* Sahlin & Soest
- E *Taraxacum officinale* Weber
- A *Taraxacum perssonii* Planglund ex Sahlin & Soest
- A *Taraxacum simile* Raunk.
- E *Telekia speciosa* (Schreber) Baumg.
- E *Tolpis barbata* (L.) Gaertn.
- A *Tolpis staticifolia* (All.) Schultz-Bip.
- A *Tolpis umbellata* Bertol.
- E *Tragopogon crocifolius* L.
- E *Tragopogon dubius* Scop.
- E *Tragopogon graminifolius* DC.
- E *Tragopogon hybridus* L.
- A *Tragopogon × mirabilis* Rouy
- E *Tragopogon porrifolius* L.
- A *Tripleurospermum disciforme* (C.A. Mey.) Sch.-Bip.
- E *Tripleurospermum inodorum* (L.) Schultz-Bip.
- E *Tussilago farfara* L.
- E *Urospermum picroides* (L.) Scop. ex F.W. Schmidt
- A *Verbesina alternifolia* (L.) Britton
- A *Verbesina encelioides* (Cav.) Bentham & Hooker f. ex A. Gray
- A *Verbesina occidentalis* (L.) Walter
- A *Wedelia glauca* (Ortega) Blake
- E *Xanthium albinum* (Widder) Scholz & Sukkop
- A *Xanthium albinum riparium* (Celak.) Widder & Wagenitz
- A *Xanthium ambrosioides* Hook. & Arn.
- E *Xanthium orientale* L.
- A *Xanthium pungens* Wallr.
- A *Xanthium saccharatum* Wallr.
- A *Xanthium sibiricum* Patrin ex Widder
- E *Xanthium spinosum* L.
- E *Xanthium strumarium* L.
- E *Xeranthemum annuum* L.
- E *Xeranthemum cylindraceum* Sibth. & Sm.
- A *Zinnia angustifolia* Humb., Bonpl. & Kunth
- A *Zinnia elegans* Jacq.
- A *Zoegea crinita baldschuanica* Boiss.

Campanulaceae
- E *Campanula alliariifolia* Willd.
- A *Campanula aparinoides* Pursh
- E *Campanula carpatica* Jacq.
- E *Campanula erinus* L.
- E *Campanula fenestrellata* Feer.
- A *Campanula fragilis* Cirillo
- E *Campanula garganica* Ten.
- E *Campanula glomerata* L.
- E *Campanula × iserana* Kovanda
- E *Campanula lactiflora* M. Bieb.
- E *Campanula latifolia* L.
- E *Campanula lusitanica* Loefl.
- E *Campanula medium* L.
- E *Campanula patula* L.
- E *Campanula persicifolia* L.
- E *Campanula portenschlagiana* Schultes
- E *Campanula poscharskyana* Degen.
- E *Campanula pyramidalis* L.
- E *Campanula rapunculoides* L.
- E *Campanula rapunculus* L.
- E *Campanula rhomboidalis* L.
- A *Downingia elegans* (Douglas ex Lindley) Torrey
- E *Jasione montana* L.
- A *Legousia castellana* (Lange) Samp.
- E *Legousia hybrida* (L.) Delarbre
- E *Legousia pentagonia* (L.) Druce
- E *Legousia speculum-veneris* (L.) Chaix
- A *Lobelia erinus* L.
- A *Lobelia inflata* L.
- A *Lobelia laxiflora* Humb., Bonpl. & Kunth
- A *Lobelia pinifolia* L.
- A *Lobelia siphilitica* L.
- E *Lobelia urens* L.
- E *Phyteuma nigrum* F.W. Schmidt
- E *Phyteuma scheuchzeri* All.
- E *Phyteuma spicatum* L.
- A *Pratia angulata* (G. Forster) Hook. f.
- A *Pratia pedunculata* (R. Br.) Benth.
- E *Symphyandra hofmannii* Pant.
- E *Trachelium caeruleum* L.

Menyanthaceae
- E *Nymphoides peltata* (S.G. Gmel.) Kuntze

Magnoliophytina, Dicotyledonae, Brassicales

Brassicaceae
- E *Alliaria petiolata* (M. Bieb.) Cavara & Grande
- E *Alyssum alyssoides* (L.) L.
- A *Alyssum argenteum* All.
- E *Alyssum corsicum* Duby
- E *Alyssum desertorum* Stapf
- E *Alyssum hirsutum* M. Bieb.
- A *Alyssum libycum* (Viv.) Cosson
- E *Alyssum linifolium* Stephan ex Willd.
- E *Alyssum montanum* L.
- E *Alyssum murale* Waldst. & Kit.

- A *Alyssum rostratum* Steven
- A *Alyssum scutigerum* Durieu ex Cosson & Durieu
- E *Alyssum simplex* Rudolphi
- E *Alyssum tortuosum* M. Bieb.
- A *Alyssum wierzbickii* Heuff.
- A *Arabidopsis pumila* (Stephan ex Willd.) N. Busch
- E *Arabidopsis suecica* (Fr.) Norrl.
- E *Arabidopsis thaliana* (L.) Heynh.
- E *Arabis alpina* L.
- E *Arabis auriculata* Lam.
- E *Arabis caucasica* Willd.
- E *Arabis collina* Ten.
- E *Arabis pendula* L.
- E *Arabis planisiliqua* (Pers.) Reichb.
- E *Arabis procurrens* Waldst. & Kit.
- E *Arabis turrita* L.
- E *Armoracia rusticana* P. Gaertn., B. Mey & Scherb.
- E *Aubrieta columnae* Guss.
- A *Aubrieta × cultorum* Bergm.
- E *Aubrieta deltoidea* (L.) DC.
- E *Aurinia petraea* (Ard.) Schur
- E *Aurinia saxatilis* (L.) Desv.
- A *Aurinia sinuata* (L.) Griseb.
- E *Barbarea intermedia* Boreau
- A *Barbarea × krausei* P. Fourn.
- E *Barbarea stricta* Andrz.
- E *Barbarea verna* (Mill.) Aschers.
- E *Barbarea vulgaris arcuata* (Opiz) Hayek
- E *Barbarea vulgaris vulgaris* R. Br.
- E *Berteroa incana* (L.) DC.
- E *Berteroa mutabilis* (Vent.) DC.
- E *Biscutella auriculata* L.
- E *Biscutella eriocarpa* DC.
- A *Boreava aptera* Boiss. & Heldr.
- A *Boreava orientalis* Jaub. & Spach
- A *Brassica barrelieri* (L.) Janka
- A *Brassica carinata* A. Braun
- E *Brassica elongata* Ehrh.
- A *Brassica × harmsiana* O. E. Schulz
- E *Brassica juncea* (L.) Czern.
- E *Brassica napus napus* L.
- E *Brassica napus rapifera* Metzg.
- E *Brassica nigra* (L.) W.D.J. Koch
- E *Brassica oleracea* L.
- A *Brassica procumbens* (Poir.) O.E. Schulz
- E *Brassica rapa* L.
- E *Brassica rapa campestris* (L.) A.R. Clapham
- E *Brassica rapa oleifera* (DC.) Metzger
- E *Brassica rapa rapa* L.
- E *Brassica tournefortii* Gouan
- E *Bunias erucago* L.
- E *Bunias orientalis* L.
- A *Cakile edentula* (Bigelow) Hook.
- E *Cakile maritima* Scop.
- E *Calepina irregularis* (Asso) Thell.
- E *Camelina alyssum* (Mill.) Thell.
- A *Camelina laxa* C.A. Meyer
- E *Camelina microcarpa* Andrz. ex DC.
- E *Camelina microcarpa microcarpa* Andrz.
- E *Camelina microcarpa sylvestris* (Wallr.) Hiitonen
- E *Camelina rumelica* Velen.
- E *Camelina sativa* (L.) Crantz.
- E *Camelina sativa pilosa* (DC.) N. Zinger
- E *Capsella bursa-pastoris* (L.) Medik.
- A *Capsella × gracilis* Gren.
- E *Capsella grandiflora* (Fauch & Chaub.) Boiss.
- E *Capsella rubella* L. (Reuter)
- E *Cardamine bulbifera* (L.) Crantz
- A *Cardamine chelidonia* L.
- A *Cardamine corymbosa* Hook. f.
- E *Cardamine glanduligera* O. Schwarz
- E *Cardamine heptaphylla* (Villars) O. Schulz
- E *Cardamine hirsuta* L.
- E *Cardamine impatiens* L.
- A *Cardamine parviflora* L.
- E *Cardamine pratensis* L.
- E *Cardamine raphanifolia* Pourret
- E *Cardamine trifolia* L.
- E *Cardaminopsis arenosa* (L.) Hayek
- E *Cardaminopsis halleri* (L.) Hayek
- E *Cardaria chalepensis* (L.) Hand.-Mazz.
- E *Cardaria draba* (L.) Desv.
- A *Cardaria pubescens* (C.A. Mey.) Jarmolenko
- E *Carrichtera annua* (L.) DC.
- E *Cheiranthus canus* Piller & Mitterp.
- E *Chorispora tenella* (Pallas) DC.
- A *Clypeola eriocarpa* Cav.
- A *Clypeola jonthlaspi* L.
- A *Cochlearia glastifolia* L.
- E *Cochlearia megalosperma* (Maire) Vogt
- E *Cochlearia officinalis* L.
- E *Coincya monensis* (L.) Greuter & Burdet
- E *Conringia austriaca* (Jacq.) Sweet
- E *Conringia orientalis* (L.) Dumort.
- A *Conringia planisiliqua* Fischer & C.A. Meyer
- A *Cordylocarpus muricatus* Desf.

- E *Coronopus didymus* (L.) Sm
- E *Coronopus squamatus* (Forssk.) Asch.
- A *Crambe abyssinica* Hochst. ex R.E. Fries
- E *Crambe cordifolia* Steven
- E *Crambe hispanica* L.
- A *Crambe koktebelica* (Junge) N. Busch
- E *Crambe maritima* L.
- A *Descurainia appendiculata* (Gris.) Schulz
- A *Descurainia pinnata* (Walter) Britton
- E *Descurainia sophia* (L.) Webb ex Prantl
- A *Diplotaxis assurgens* (Delile) Thell.
- E *Diplotaxis catholica* (L.) DC.
- E *Diplotaxis erucoides* (L.) DC.
- E *Diplotaxis muralis* (L.) DC.
- E *Diplotaxis tenuifolia* (L.) DC.
- E *Diplotaxis tenuisiliqua* Delile
- E *Diplotaxis viminea* (L.) DC.
- E *Diplotaxis virgata* (Cav.) DC.
- E *Draba aizoides* L.
- E *Draba muralis* L.
- E *Draba nemorosa* L.
- E *Draba sibirica* (Pall.) Thell.
- E *Enarthrocarpus clavatus* Delile ex Godr.
- E *Enarthrocarpus lyratus* (Forsskal) DC.
- E *Erophila verna* (L.) DC.
- E *Eruca sativa* (L.) Mill.
- E *Eruca vesicaria* (L.) Cav.
- A *Erucaria grandiflora* Boiss.
- E *Erucaria hispanica* (L.) Druce
- E *Erucastrum gallicum* (Willd.) O.E. Schulz
- E *Erucastrum nasturtiifolium* (Poiret) O.E. Schulz
- A *Erysimum argillosum* (Greene) Rydberg
- E *Erysimum aureum* M. Bieb.
- E *Erysimum cheiranthoides* L.
- E *Erysimum cheiri* (L.) Crantz
- E *Erysimum cuspidatum* (M. Beib.) DC.
- E *Erysimum diffusum* Ehrh.
- E *Erysimum hungaricum* Zapal.
- E *Erysimum marschallianum* Andrz. ex DC.
- A *Erysimum × marshallii* (Henfry) Bois
- A *Erysimum perofskianum* Fisch. & C.A. Mey.
- E *Erysimum repandum* L.
- E *Erysimum strictum* Gaertn., B. Mey & Scherb.
- E *Euclidium syriacum* (L.) R. Br.
- E *Fibigia clypeata* (L.) Medik.
- A *Heliophila pusilla* L. f.
- E *Hesperis matronalis* L.
- E *Hesperis matronalis nivea* (Baumg.) Perrier
- A *Hesperis oblongifolia* Schur
- E *Hesperis pycnotricha* Borbs & Degen
- E *Hesperis tristis* L.
- E *Hirschfeldia incana* (L.) Lagrze-Fossat
- E *Hornungia petraea* (L.) Reichb.
- E *Hymenolobus procumbens* (L.) Nutt. ex Torr. & A. Gray
- E *Iberis amara* L.
- E *Iberis odorata* L.
- E *Iberis pinnata* L.
- E *Iberis sempervirens* L.
- E *Iberis umbellata* L.
- E *Isatis tinctoria* L.
- E *Isatis tinctoria canescens* (DC.) Nyman
- E *Isatis tinctoria tinctoria* L.
- E *Jonopsidium acaule* (Desf.) Reichb.
- E *Jonopsidium albiflorum* Durieu
- A *Lepidium africanum* (Burm. fil.) DC.
- A *Lepidium bonariense* L.
- A *Lepidium calycinum* Godron
- E *Lepidium campestre* (L.) R. Br.
- A *Lepidium densiflorum* Schrader
- A *Lepidium desertorum* Eckl. & Zeyh.
- A *Lepidium fasciculatum* Thell.
- E *Lepidium graminifolium* L.
- E *Lepidium heterophyllum* Bentham
- A *Lepidium hyssopifolium* Desv.
- E *Lepidium latifolium* L.
- A *Lepidium niloticum* Sieber ex Steud.
- E *Lepidium perfoliatum* L.
- A *Lepidium pinnatifidum* Ledeb.
- E *Lepidium ruderale* L.
- A *Lepidium sativum* L.
- E *Lepidium sativum sativum* L.
- A *Lepidium schinzii* Thell.
- E *Lepidium spinosum* Ard.
- E *Lepidium virginicum* L.
- E *Lobularia maritima* (L.) Desv.
- E *Lunaria annua* L.
- E *Malcolmia africana* (L.) R. Br.
- A *Malcolmia chia* (L.) DC.
- A *Malcolmia crenulata* (DC.) Boiss.
- E *Malcolmia flexuosa* (Sm.) Sm.
- A *Malcolmia laxa* (Lam.) DC.
- E *Malcolmia maritima* (L.) R. Br. ex Aiton
- A *Malcolmia torulosa* (Desf.) Boiss.
- E *Matthiola incana* (L.) R. Br.
- E *Matthiola longipetala* (Vent.) DC.
- A *Matthiola lunata* DC.
- E *Moricandia arvensis* (L.) DC.
- E *Myagrum perfoliatum* L.

E *Neslia apiculata* Fisch. & C.A. Mey.
E *Neslia paniculata* (L.) Desv.
A *Peltaria alliacea* Jacq.
A *Raphanus* × *micranthus* O.E. Schulz
E *Raphanus raphanistrum* L.
E *Raphanus raphanistrum landra* (Moretti ex DC.) Bonnier & Layens
E *Raphanus raphanistrum maritimus* (Sm.) Thell.
E *Raphanus raphanistrum raphanistrum* L.
E *Raphanus sativus* L.
E *Rapistrum perenne* (L.) All.
E *Rapistrum rugosum* (L.) Bergeret
E *Rapistrum rugosum orientale* (L.) Arcang.
E *Rapistrum rugosum rugosum* (L) J.P. Bergeret
E *Rorippa* × *armoracioides* (Tausch) Fuss
E *Rorippa austriaca* (Crantz) Besser
A *Rorippa curvisiliqua* (Hook.) Bessey ex Britton
E *Rorippa* × *hungarica* Borbás
E *Rorippa lippizensis* (Wulfen) Reichb.
E *Rorippa microphylla* (Boenn.) Hyl. ex & D. Löve
E *Rorippa nasturtium-aquaticum* (L.) Hayek
E *Rorippa palustris* (L.) Besser
A *Rorippa* × *prostrata* (Bergeret) Schinz & Thellung
E *Rorippa pyrenaica* (Lam.) Reichb.
E *Rorippa* × *sterilis* Airy Shaw
E *Rorippa sylvestris* (L.) Besser
E *Sinapis alba* L.
E *Sinapis alba dissecta* (Lag.) Simonk.
E *Sinapis arvensis* L.
E *Sisymbrella aspera* (L.) Spach
E *Sisymbrium altissimum* L.
A *Sisymbrium assoanum* Loscos & J. Pardo
E *Sisymbrium austriacum* Jacq.
E *Sisymbrium erysimoides* Desf.
E *Sisymbrium irio* L.
E *Sisymbrium loeselii* L.
A *Sisymbrium luteum* (Maxim.) O.E. Schulz
E *Sisymbrium officinale* (L.) Scop.
E *Sisymbrium orientale* L.
E *Sisymbrium polyceratium* L.
E *Sisymbrium polymorphum* (Murray) Roth
E *Sisymbrium runcinatum* Lag. ex DC.
A *Sisymbrium septulatum* DC.
E *Sisymbrium strictissimum* L.
E *Sisymbrium supinum* L.
E *Sisymbrium volgense* M. Bieb. ex E. Fourn.

E *Teesdalia nudicaulis* (L.) R. Br.
E *Thlaspi alliaceum* L.
E *Thlaspi arvense* L.
E *Thlaspi brachypetalum* Jordan
E *Thlaspi caerulescens* J. & C. Presl.
A *Thlaspi kovatsii* Heuff.
E *Thlaspi macrophyllum* Hoffm.
E *Thlaspi perfoliatum* L.

Capparaceae
E *Capparis spinosa* L.
A *Cleome grav*eolens Raf.
A *Cleome hassleriana* Chodat
A *Cleome speciosa* Raf.
A *Cleome spinosa* Jacq.
E *Cleome violacea* L.

Caricaceae
A *Carica papaya* L.

Resedaceae
E *Reseda alba* L.
A *Reseda inodora* Rchb.
E *Reseda lutea* L.
E *Reseda luteola* L.
E *Reseda media* Lag.
E *Reseda odorata* L.
E *Reseda phyteuma* L.

Tropaeolaceae
A *Tropaeolum majus* L.
A *Tropaeolum pentaphyllum* Lam.
A *Tropaeolum polyphyllum* Cav.
A *Tropaeolum speciosum* Poeppig & Endl.

Magnoliophytina, Dicotyledonae, Buxales
Buxaceae
E *Buxus sempervirens* L.
A *Pachysandra terminalis* Siebold & Zucc.

Magnoliophytina, Dicotyledonae, Canellales
Winteraceae
A *Maireana brevifolia* (R. Br.) P.G. Wilson

Magnoliophytina, Dicotyledonae, Caryophyllales
Aizoaceae
E *Aizoon canariense* L.
E *Aizoon hispanicum* L.
A *Aptenia cordifolia* (L. f.) Schwantes
A *Aptenia lancifolia* Bolus
A *Carpobrotus acinaciformis* (L.) Bolus
A *Carpobrotus chilensis* (Molina) N.E. Br.
A **Carpobrotus edulis** (L.) N.E. Br.
A *Carpobrotus glaucescens* (Haw.) Schwantes
A *Carpobrotus quadrifidus* L. Bolus
A *Delosperma cooperi* (Hook. f.) L. Bolus

A *Disphyma crassifolium* (L.) L. Bolus
A *Drosanthemum floribundum* (Haw.) Schwantes
A *Drosanthemum hispidum* (L.) Schwantes
A *Erepsia heteropetala* (Haw.) Schwantes
A *Galenia officinalis* L.
A *Galenia secunda* (L. fil.) Sond.
A *Lampranthus deltoides* (L.) Glen
A *Lampranthus falciformis* (Haw.) N.E. Br.
A *Lampranthus glaucus* (L.) N.E. Br.
A *Lampranthus multiradiatus* (Jacq.) N.E. Br.
A *Lampranthus roseus* (Willd.) Schwantes
A *Lampranthus spectabilis* (Haw.) N.E. Br.
A *Mesebryanthemum purpuro-crocea* Haw.
A *Mesembryanthemum crystallinum* L.
A *Mesembryanthemum nodiflorum* L.
A *Mesembryanthemum tumidulum* Haw.
A *Ruschia caroli* (L. Bolus) Schwantes
A *Sesuvium portulacastrum* (L.) L.

Amaranthaceae

E *Achyranthes aspera* L.
E *Achyranthes sicula* (L.) All.
A *Alternanthera angustifolia* R. Br.
E *Alternanthera caracasana* Humb., Bonpl. & Kunth
A *Alternanthera denticulata* R. Br.
A *Alternanthera nodiflora* R. Br
A *Alternanthera paronychioides* A. St. Hil.
A *Alternanthera philoxeroides* (Mart.) Griseb.
A *Alternanthera pungens* Humb., Bonpl. & Kunth
A *Alternanthera sessilis* R. Br.
A *Alternanthera tenella* Colla
A *Amaranthus acutilobus* Uline & Bray
E *Amaranthus albus* L.
A *Amaranthus × alleizettei* Aellen
A *Amaranthus blitoides* S. Watson
E *Amaranthus blitum emarginatus* (Moq. ex Uline & Bray) Carretero, Muñoz Garmendia & Pedrol
E *Amaranthus bouchonii* Thell.
A *Amaranthus capensis* Thell.
A *Amaranthus caudatus* L.
A *Amaranthus clementii* Domin
A *Amaranthus crispus* (Lesp. & Thevenau) N. Terracc.
E *Amaranthus cruentus* L.
A *Amaranthus deflexus* L.
A *Amaranthus × dobrogensis* Morariu

A *Amaranthus × galii* Sennen & Gonzalo
A *Amaranthus gracilis* Poir. in Lam.
A *Amaranthus graecizans* L.
A *Amaranthus hybridus* L.
A *Amaranthus hypochondriacus* L.
E *Amaranthus lividus* L.
A *Amaranthus mitchellii* Benth.
A *Amaranthus muricatus* (Gillies ex Moq.) Hieron.
A *Amaranthus × ozanonii* Thell.
A *Amaranthus palmeri* S. Watson
A *Amaranthus × polgarianus* Priszter & Kárpáti
A *Amaranthus polygonioides* L.
A *Amaranthus quitensis* Humb., Bonpl. & Kunth
A *Amaranthus × ralletii* Contré ex d'Alleiz. & Loiseau
A *Amaranthus retroflexus* L.
A *Amaranthus × soproniensis* Priszter & Kárpáti
E *Amaranthus spinosus* L.
A *Amaranthus standleyanus* Parodi ex Covas
A *Amaranthus × thevenaei* Degen & Thell.
A *Amaranthus thunbergii* Moq.
A *Amaranthus tricolor* L.
A *Amaranthus tuberculatus* (Moq.) Sauer
A *Amaranthus × turicensis* Thell.
E *Amaranthus viridis* L.
A *Amaranthus × zobelii* Thell.
A *Atriplex aucheri* Moq.
A *Atriplex centralasiatica* Iljin
A *Atriplex dimorphostegia* Kar. & Kir.
A *Atriplex eardleyae* Aell.
A *Atriplex fera* (L.) Bunge
E *Atriplex halimus* L.
A *Atriplex holocarpa* F. Muell.
E *Atriplex hortensis* L.
E *Atriplex laciniata* Schkuhr
A *Atriplex leptocarpa* F. Müll.
E *Atriplex littoralis* L.
A *Atriplex megalotheca* Popov
E *Atriplex micrantha* Ledeb.
A *Atriplex muelleri* Benth.
E *Atriplex nitens* Schkuhr
A *Atriplex × northusiana* Wein.
A *Atriplex nummularia* Lindl.
E *Atriplex oblongifolia* Waldst. & Kit.
A *Atriplex patens* (Litv.) Iljin
E *Atriplex patula* L.
E *Atriplex polonica* Zapal.
E *Atriplex prostrata* Boucher ex DC.

- A *Atriplex pseudocampanulata* Aell.
- E *Atriplex rosea* L.
- A *Atriplex schugnanica* Iljin
- A *Atriplex semibaccata* R. Br.
- A *Atriplex semil*unaris Aellen
- E *Atriplex sibi*rica L.
- A *Atriplex suber*ecta Verd.
- E *Axyris amaranthoides* L.
- A *Bassia eriantha* (Fischer & C.A. Mey.) O. Kuntze
- E *Bassia hirsuta* (L.) Aschers.
- E *Bassia hyssopifolia* (Pall.) Kuntze
- E *Bassia laniflora* (S.G. Gmel.) A.J. Scott
- E *Bassia prostrata* (L.) A.J. Scott
- E *Bassia sedoides* (Pallas) Aschers.
- A *Bassia tricuspis* F. Mueller
- E *Beta macrocarpa* Guss.
- E *Beta trigyna* Waldst. & Kit.
- E *Beta vulgaris* L.
- A *Blitum nuttalianum* Schultes
- A *Celosia argentea* L.
- E *Ceratocarpus arenarius* L.
- E *Chenopodium acuminatum* Willd.
- E *Chenopodium album* L.
- A *Chenopodium album amaranticolor* Coste & Reyn.
- A *Chenopodium aristatum* L.
- A *Chenopodium atriplicinum* (F. Müll.) F. Müll.
- A *Chenopodium auricomiforme* J. Murr & Thell.
- A *Chenopodium berlandieri* Moq.
- A *Chenopodium* × *bontei* Aellen
- E *Chenopodium bonus-henricus* L.
- A *Chenopodium borbasioides* Ludwig
- A *Chenopodium botrys* L.
- A *Chenopodium bushianum* Aellen
- E *Chenopodium capitatum* (L.) Asch.
- A *Chenopodium carinatum* R. Br.
- A *Chenopodium carnosulum* Moq. in DC.
- E *Chenopodium chenopodioides* (L.) Aellen
- A *Chenopodium cristatum* (F. Muell.) F. Muell.
- A *Chenopodium desiccatum* A. Nelson
- E *Chenopodium ficifolium* Sm.
- E *Chenopodium foliosum* (Moench) Asch.
- E *Chenopodium glaucum* L.
- A *Chenopodium hircinum* Schrader
- E *Chenopodium hybridum* L.
- E *Chenopodium integrifolium* Vorosch.
- A *Chenopodium leptophyllum* (Nutt. ex Moq.) S. Watson
- A *Chenopodium macrospermum* Hook. f.
- A *Chenopodium missouriense* Aellen
- A *Chenopodium multifidum* L.
- E *Chenopodium murale* L.
- A *Chenopodium nitrariaceum* (F. Muell.) F. Muell. ex Benth.
- E *Chenopodium opulifolium* Schrader ex Koch & Ziz
- A *Chenopodium* × *pelgrimsianum* Aell.
- E *Chenopodium polyspermum* L.
- A *Chenopodium* × *preissmannii* J. Murr
- A *Chenopodium probs*tii Aellen
- A *Chenopodium prostratum* Herder
- A *Chenopodium pumilio* R. Br.
- A *Chenopodium quinoa* Willd.
- E *Chenopodium rubrum* L.
- E *Chenopodium schraderianum* Schultes
- A *Chenopodium simplex* (Torr.) Raf.
- E *Chenopodium strictum* Roth.
- E *Chenopodium strictum striatiforme* (J. Murr) Uotila
- E *Chenopodium suecicum* Murr
- E *Chenopodium urbicum* L.
- A *Chenopodium vachelii* Hook. & Arn.
- A *Chenopodium* × *variabile* Aellen
- E *Chenopodium vulvaria* L.
- A *Chenopodium wolffii* Simk.
- E *Corispermum algidum* Iljin
- E *Corispermum declinatum* Steph. ex Steven
- E *Corispermum intermedium* Schweigger
- E *Corispermum leptopterum* (Aschers.) Iljin
- E *Corispermum marschallii* Steven
- A *Corispermum membranaceum* (Bischoff) Iljin
- E *Corispermum orientale* Lam.
- A *Corispermum pallasii* Steven
- A *Dysphania glomulifera* (Nees) P.G. Wilson
- A *Einadia polygonoides* (J. Murr) P.G. Wilson
- A *Enchylaena tomentosa* R. Br.
- A *Froelichia floridana* (Nutt.) Moq. in DC.
- A *Froelichia gracilis* Moq.
- A *Gomphrena celosioides* Mart.
- A *Gomphrena globosa* L.
- A *Hablizia thamnoides* Bieb.
- E *Halimione portulacoides* (L.) Aellen
- A *Iresine herbstii* Hook. ex Lindl.
- A *Kochia indica* Wight
- E *Kochia scoparia* (L.) Schrader
- E *Kochia scoparia densiflora* (Moq.) Aellen

- E *Polycnemum arvense* L.
- E *Polycnemum heuffelii* A.F. Láng
- E *Polycnemum majus* A. Braun
- A *Rhagodia nutans* R. Br.
- E *Salicornia europaea* L.
- E *Salsola acutifolia* (Bunge) Botsch
- E *Salsola atriplicifolia* C.P.J. Sprengel
- E *Salsola collina* Pallas
- A *Sclerolaena muricata* (Moq.) Domin
- A *Spinacia oleracea* L.
- A *Spinacia turkestanica* Iljin
- E *Suaeda altissima* (L.) Pallas
- E *Suaeda baccifera* Pallas
- E *Suaeda corniculata* (C.A. Mey.) Bunge
- E *Suaeda maritima* (L.) Dumort.
- E *Suaeda maritima salsa* Soó ex Soó & Jáv.

Basellaceae
- A *Basella alba* L.
- A *Boussingaultia cordifolia* Ten.

Cactaceae
- A *Austrocylindropuntia cylindrica* (Lam.) Backeb.
- A *Austrocylindropuntia subulata* (Mühlenpfordt) Backeb.
- A *Cereus peruvianus* (L.) Miller
- A *Cylindropuntia imbricata* (Haw.) F.M. Knuth
- A *Cylindropuntia spinosior* (Engelm.) F.M. Kunth in Backeb.
- A *Hylocereus undatus* (Haw.) Britton & Rose
- A *Opuntia ammophila* Small
- A *Opuntia bernichiana* hort.
- A *Opuntia caespitosa* Poepp.
- A *Opuntia crassa* Haw.
- A *Opuntia dejecta* Salm-Dyck
- A *Opuntia dillenii* (Ker-Gawler) Haw.
- A *Opuntia engelmannii* Salm-Dyck ex Engelm.
- A *Opuntia ficus-indica* (L.) Mill.
- A *Opuntia huajuapensis* H. Bravo
- A *Opuntia imbricata* (Haw.) DC.
- A **Opuntia maxima** Mill.
- A *Opuntia microdasys* (Lehmann) Pfeiffer.
- A *Opuntia monacantha* (Willd.) Haw
- A *Opuntia phaeacantha* Engelm.
- A *Opuntia robusta* H.L. Wendland
- A *Opuntia rosea* DC.
- A *Opuntia stricta* (Haw.) Haw.
- A *Opuntia subulata* (Mühlenpfordt) Engelm.
- A *Opuntia tomentosa* Salm-Dyck
- A *Opuntia tortispina* Engelm. & Bigelow
- A *Opuntia tuna* (L.) Mill.
- A *Opuntia vulgaris* Mill.

Caryophyllaceae
- E *Agrostemma githago* L.
- E *Agrostemma gracile* Boiss.
- A *Agrostemma linicola* Terech.
- E *Arenaria balearica* L.
- A *Arenaria holosteoides* (C.A. Meyer) Edgew.
- E *Arenaria montana* L.
- E *Arenaria serpyllifolia* L.
- A *Atocion armeria* (L.) Raf.
- E *Bufonia paniculata* Dubois
- E *Bufonia willkommiana* Boiss.
- A *Cerastium arve*nse L.
- E *Cerastium biebersteinii* DC.
- E *Cerastium brachypetalum* Desp. ex Pers.
- E *Cerastium comatum* Desv.
- E *Cerastium dichotomum* L.
- E *Cerastium dubium* (Bastard) Géupin
- E *Cerastium fontanum* Baumg.
- E *Cerastium glomeratum* Thuill.
- E *Cerastium ligusticum* Viviani
- A *Cerastium × maureri* M. Schulze
- E *Cerastium perfoliatum* L.
- E *Cerastium siculum* Guss.
- E *Cerastium tomentosum* L.
- E *Corrigiola litoralis* L.
- E *Cucubalus baccifer* L.
- A *Dianthus × allwoodii* hort.
- E *Dianthus armeria* L.
- E *Dianthus barbatus* L.
- E *Dianthus campestris* M. Bieb.
- E *Dianthus carthusianorum* L.
- E *Dianthus caryophyllus* L.
- A *Dianthus chinensis* L.
- E *Dianthus cruentus* Griseb.
- E *Dianthus deltoides* L.
- E *Dianthus fisheri* Spreng.
- E *Dianthus gallicus* Pers.
- E *Dianthus giganteus* D'Urv.
- E *Dianthus gratianopolitanus* Vill.
- E *Dianthus plumarius* L.
- E *Dianthus subacaulis* Vill.
- E *Dianthus superbus* L.
- E *Dianthus tripunctatus* Sm.
- E *Dianthus versicolor* Fisch. ex Link
- E *Gypsophila acutifolia* Steven ex Spreng.
- E *Gypsophila altissima* L.
- E *Gypsophila elegans* M. Bieb.
- E *Gypsophila muralis* L.
- E *Gypsophila paniculata* L.
- E *Gypsophila perfoliata* L.
- E *Gypsophila pilosa* Huds.
- E *Gypsophila repens* L.
- E *Gypsophila scorzonerifolia* Ser.

- A *Gypsophila tubulosa* (Jaub. & Spach) Boiss.
- A *Gypsophila viscosa* Murray
- E *Herniaria glabra* L.
- E *Herniaria hirsuta* L.
- E *Herniaria hirsuta cinerea* (DC.) Coutinho
- A *Herniaria polygama* J. Gay
- E *Holosteum umbellatum* L.
- E *Illecebrum verticillatum* L.
- E *Lychnis chalcedonica* L.
- E *Lychnis flos-jovis* (L.) Desr.
- E *Lychnis viscaria* L.
- A *Minuartia glomerata* (M. Bieb.) Degen
- A *Minuartia hamata* (Hausskn. & Bornm. ex Hausskn.) Mattf.
- E *Minuartia hybrida* (Villars) Schischkin
- E *Minuartia laricifolia* (L.) Schinz & Thell.
- E *Minuartia mediterranea* (Link) K. Mal
- A *Minuartia viscosa* (Schreb.) Schinz & Thell
- E *Moenchia erecta* (L.) Gaertn., B. Mey. & Schreb.
- A *Moenchia mantica* (L.) Bartl.
- A *Paronychia brasiliana* DC.
- A *Paronychia kapela* (Hacq.) A. Kern.
- E *Petrorhagia nanteuilii* (Burnat) P.W. Ball & Heywood
- E *Petrorhagia prolifera* (L.) P. Ball & Heyw.
- E *Petrorhagia saxifraga* (L.) Link
- E *Petrorhagia velutina* (Guss.) Ball & Heywood
- E *Polycarpon diphyllum* Cav.
- E *Polycarpon tetraphyllum* (L.) L.
- E *Sagina apetala* Ard.
- E *Sagina apetala erecta* (Hornem.) F. Hermann
- E *Sagina procumbens* L.
- A *Sagina sabuletorum* (Gay) Lange
- E *Sagina saginoides* (L.) Karsten
- E *Sagina subulata* (Swartz) C. Presl
- E *Saponaria ocymoides* L.
- E *Saponaria officinalis* L.
- A *Saponaria orientalis* L.
- E *Scleranthus annuus* L.
- E *Scleranthus perennis* L.
- E *Silene antirrhina* L.
- E *Silene armeria* L.
- E *Silene behen* L.
- E *Silene bellidifolia* Juss. ex Jacq.
- E *Silene bupleuroides* L.
- A *Silene cinerea* Desf.
- E *Silene colorata* Poir.
- E *Silene conica* L.
- E *Silene conoidea* L.
- E *Silene coronaria* (L.) Clairv.
- A *Silene crassipes* Fenzl
- E *Silene cretica* L.
- E *Silene csereii* Baumg.
- E *Silene dichotoma* Ehrh.
- E *Silene dioica* (L.) Clairv.
- E *Silene disticha* Willd.
- A *Silene fimbriata* Sims
- E *Silene fuscata* Link ex Brot.
- E *Silene gallica* L.
- A *Silene* × *grecescui* Gusul.
- A *Silene* × *hampeana* Meusel & K. Werner
- E *Silene italica* (L.) Pers.
- E *Silene italica nemoralis* (Waldst. & Kit.) Nyman
- E *Silene latifolia* Poiret
- E *Silene linicola* C.C. Gmel.
- E *Silene lydia* Boiss.
- A *Silene multicaulis* Guss.
- E *Silene multiflora* (Waldst. & Kit.) Pers.
- E *Silene muscipula* L.
- E *Silene noctiflora* L.
- E *Silene nocturna* L.
- E *Silene otites* (L.) Wibel
- E *Silene pendula* L.
- E *Silene portensis* L.
- E *Silene procumbens* Murray
- E *Silene psammitis* Link ex Spreng.
- E *Silene pusilla* Waldst. & Kit.
- E *Silene repens* Patrin ex Pers.
- E *Silene scabriflora* Brot.
- A *Silene schafta* S.G. Gmel. ex Hohen.
- E *Silene sibirica* (L.) Pers.
- E *Silene coeli-rosa* (L.) Godr.
- E *Silene stricta* L.
- E *Silene subconica* Friv.
- A *Silene tatarica* (L.) Pers.
- A *Silene viridiflora* L.
- E *Silene viscosa* (L.) Pers.
- E *Silene wolgensis* (Hornem.) Otth in DC.
- E *Spergula arvensis* L.
- E *Spergula morisonii* Boreau
- E *Spergula pentandra* L.
- E *Spergularia bocconei* (Scheele) Graebner
- E *Spergularia marina* (L.) Griseb.
- E *Spergularia rubra* (L.) J. Presl & C. Presl
- E *Spergularia segetalis* (L.) G. Don

E *Stellaria cupaniana* (Jord. & Fourr.) Beguin.
E *Stellaria graminea* L.
A *Stellaria hebecalyx* Fenzl
E *Stellaria media* (L.) Vill.
E *Stellaria pallida* (Dumort.) Piré
E *Vaccaria hispanica* (Mill.) Rauschert
E *Velezia rigida* L.
Droseraceae
 A *Aldrovanda vesiculosa* L.
 A *Drosera capensis* L.
Frankeniaceae
 E *Frankenia laevis* L.
 E *Frankenia pulverulenta* L.
Molluginaceae
 E *Mollugo cerviana* (L.) Ser.
 A *Mollugo verticillata* L.
Nyctaginaceae
 A *Abronia fragrans* Nutt.
 A *Boerhavia repens* L.
 A *Bougainvillea glabra* Choisy
 A *Commicarpus africanus* (Lour.) Dandy in F.W. Andrews
 A *Commicarpus helenae* (Schult.) Meikle
 A *Mirabilis jalapa* L.
 A *Mirabilis longiflora* L.
 A *Oxybaphus nyctagineus* (Michx.) Sweet
Phytolaccaceae
 A *Phytolacca acinosa* Roxb.
 A *Phytolacca americana* L.
 A *Phytolacca dioica* L.
 A *Phytolacca esculenta* Van Houtte
 A *Phytolacca heterotepala* H. Walter
 A *Phytolacca polyandra* Batalin
 A *Rivina humilis* L.
 A *Rivina laevis* L.
Plumbaginaceae
 E *Armeria arenaria* (Pers.) Schult.
 E *Armeria maritima* (Mill.) Willd.
 A *Ceratostigma plumbaginoides* Bunge
 A *Goniolimon tataricum* (L.) Boiss.
 E *Limoniastrum monopetalum* (L.) Boiss.
 E *Limonium gerberi* Soldano
 E *Limonium hyblaeum* Brullo
 E *Limonium sinuatum* (L.) Miller
 E *Plumbago auriculata* Lam.
Polygonaceae
 A *Aconogonon coriarium* (Grig.) Sojk
 A *Antigonon leptopus* Hook. & Arn.
 A *Emex australis* Steinh.
 E *Emex spinosa* (L.) Campd.
 A *Fagopyrum dibotrys* (D. Don) H. Hara
 A *Fagopyrum esculentum* Moench

A *Fagopyrum tataricum* (L.) Gaertn.
A *Fallopia baldschuanica* (Regel) J. Holub
A *Fallopia* × *bohemica* (Chrtek & Chrtkov) J. Bailey
A *Fallopia* × *conollyana* J.P. Bailey
E *Fallopia convolvulus* (L.) Á. Löve
E *Fallopia dumetorum* (L.) J. Holub
A **Fallopia japonica** (Houtt.) Ronse Decr.
A *Fallopia sachalinensis* (F. Schmidt ex Maxim.) Ronse Decr.
A *Muehlenbeckia complexa* (A. Cunn.) Meissner
A *Muehlenbeckia sagittifolia* (Gómez Ortega) Meissner
E *Persicaria alpina* (All.) Gross
E *Persicaria amphibia* (L.) Gray
A *Persicaria amplexicaulis* (D. Don) Ronse Decraene
A *Persicaria campanulata* (Hook. f.) Ronse Decraene
E *Persicaria capitata* (Buch-Hamilton ex D. Don) H.Gross
A *Persicaria* × *fennica* (Reiersen) Stace
A *Persicaria filiformis* (Thunb.) Nakai
E *Persicaria hydropiper* (L.) Spach
A *Persicaria hydropiperoides* (Michx.) Small.
A *Persicaria lanigera* (R. Br.) Sojk
E *Persicaria lapathifolia* (L.) Gray
A *Persicaria longiseta* (Bruijn) Kitagawa
E *Persicaria maculosa* Gray
E *Persicaria minor* (Huds.) Opiz
E *Persicaria mitis* (Schrank) Assenov
A *Persicaria mollis* (D. Don) Gross
A *Persicaria nepalensis* (Meissner) Gross
E *Persicaria orientalis* (L.) Vilm.
A *Persicaria pensylvanica* (L.) M. Gómez
A *Persicaria sagittata* (L.) Gross ex Nakai
A *Persicaria salicifolia* (Brouss. ex Willd.) Assenov.
A *Persicaria senegalensis* (Meisner) Sojk
A *Persicaria virginiana* (L.) Gaertn.
A *Persicaria wallichii* Greuter & Burdet
A *Persicaria weyrichii* (F. Schmidt ex Maxim.) Ronse Decraene
A *Polygonum achoreum* S.F. Blake
E *Polygonum arenarium* Waldst. & Kit.
E *Polygonum arenastrum* Boreau
E *Polygonum argyrocoleon* Steud. ex Kunze
E *Polygonum aviculare* L.
E *Polygonum bellardii* All.
E *Polygonum bistorta* L.
E *Polygonum cognatum* Meisn.

- A *Polygonum divaricatum* L.
- E *Polygonum equisetiforme* Sm.
- E *Polygonum minus* Hudson
- A *Polygonum multiflorum* Thunb.
- E *Polygonum oxyspermum* C.A. Mey. & Bunge ex Ledeb.
- E *Polygonum patulum* M. Bieb.
- A *Polygonum plebejum* R. Br.
- E *Polygonum rurivagum* Jordan ex Boreau
- E *Rheum* × *hybridum (R. palmatum x rhaponticum)* Murray
- E *Rheum palmatum* L.
- E *Rheum rhaponticum* L.
- E *Rumex acetosa* L.
- E *Rumex acetosa ambiguus* (Gren.) Á. Löve
- A *Rumex acetosa thyrsiflorus* (Fingerhuth) Hayek
- E *Rumex acetosella* L.
- A *Rumex altissimus* Wood
- A *Rumex brownii* Campdera
- E *Rumex bucephalophorus* L.
- E *Rumex confertus* Willd.
- A *Rumex* × *confusus* Simonkai
- E *Rumex conglomeratus* Murray
- E *Rumex* × *corconticus* Kubát
- A *Rumex* × *cornubiensis* D.T. Holyoak
- E *Rumex crispus* L.
- E *Rumex cristatus* DC.
- E *Rumex dentatus* L.
- A *Rumex* × *dimidiatus* Hausskn.
- A *Rumex* × *erubescens* Simonk.
- A *Rumex frutescens* Thouars
- A *Rumex* × *hybridus* Kindberg
- E *Rumex hydrolapathum* Huds.
- E *Rumex kerneri* Borbás
- E *Rumex longifolius* DC.
- A *Rumex* × *lousleyi* D.H. Kent
- E *Rumex lunaria* L.
- E *Rumex maritimus* L.
- A *Rumex* × *mezii* Haussknecht
- A *Rumex obovatus* Danser
- E *Rumex obtusifolius* L.
- E *Rumex obtusifolius sylvestris* (Wallr.) Rech. f.
- E *Rumex obtusifolius transiens* (Simonk.) Rech. f.
- E *Rumex palustris* Sm.
- E *Rumex patientia* L.
- A *Rumex* × *propinquus* Aresch.
- E *Rumex pseudoalpinus* Höfft
- E *Rumex pulcher* L.
- A *Rumex salicifolius* Weinm.
- E *Rumex sanguineus* L.
- E *Rumex scutatus* L.
- A *Rumex* × *skofitzii* Blocki
- E *Rumex stenophyllus* Ledeb.
- A *Rumex vesicarius* L.
- A *Rumex violascens* Rech. f.

Portulacaceae
- A *Calandrinia compressa* Schrad. ex DC.
- A *Calandrinia menziesii* (Hook.) Torrey & A. Gray
- A *Claytonia parvifolia* Mocino ex DC.
- A *Claytonia perfoliata* Donn ex Willd.
- A *Claytonia sibirica* L.
- E *Montia fontana* L.
- A *Portulaca grandiflora* Hook.
- E *Portulaca oleracea* L.
- E *Portulaca oleracea granulostellulata* (Poelln.) Danin & Baker
- E *Portulaca oleracea nitida* Danin & H.G. Baker
- E *Portulaca oleracea papillatostellulata* Danin & H.G. Baker
- E *Portulaca oleracea stellata* Danin & H.G. Baker
- E *Portulaca pilosa* L.
- A *Talinum paniculatum* (Jacq.) Gaertn.

Simmondsiaceae
- A *Simmondsia chinensis* (Link) C.K. Schneid.

Tamaricaceae
- E *Myricaria germanica* (L.) Desv.
- E *Tamarix africana* Poiret
- E *Tamarix canariensis* Willd.
- A *Tamarix chinensis* Lour.
- E *Tamarix dalmatica* Baum
- E *Tamarix gallica* L.
- E *Tamarix parviflora* DC.
- E *Tamarix ramosissima* Ledeb.
- A *Tamarix tetrandra* Pall. ex M. Bieb.

Tetragoniaceae
- A *Tetragonia tetragonioides* (Pallas) O. Kuntze

Magnoliophytina, Dicotyledonae, Celestrales

Celastraceae
- A *Celastrus orbiculatus* Thunb. ex Murray
- A *Celastrus scandens* L.
- E *Euonymus europaeus* L.
- A *Euonymus fortunei* (Turcz.) Hand.-Mazz.
- A *Euonymus hederaceus* Champ. ex Benth.
- A *Euonymus japonicus* L. f.
- E *Euonymus latifolius* (L.) Miller
- E *Euonymus nanus* M. Bieb.
- A *Euonymus planipes* (Koehne) Koehne

Magnoliophytina, Dicotyledonae, Ceratophyllales

Ceratophyllaceae
- E *Ceratophyllum demersum* L.
- A *Ceratophyllum pentacanthum* Haynald
- E *Ceratophyllum submersum* L.

Magnoliophytina, Dicotyledonae, Cornales

Cornaceae
- E *Aucuba japonica* Thunb.
- E *Cornus alba* L.
- A *Cornus capitata* Wall.
- E *Cornus mas* L.
- E *Cornus sanguinea* L.
- A *Cornus sericea* L.
- A *Griselinia littoralis* (Raoul) Raoul

Hydrangeaceae
- A *Deutzia gracilis* Siebold & Zucc.
- A *Deutzia scabra* Thunb.
- A *Hydrangea arborescens* L.
- E *Hydrangea macrophylla* (Thunb.) Ser.
- E *Philadelphus coronarius* L.
- A *Philadelphus inodorus* L.
- A *Philadelphus* × *lemoinei* Lemoine
- A *Philadelphus microphyllus* Gray
- A *Philadelphus pubescens* Loisel.
- A *Philadelphus* × *virginalis* Rehder

Loasaceae
- A *Blumenbachia hieronymi* Urban
- A *Blumenbachia insignis* Schrad.

Magnoliophytina, Dicotyledonae, Crossosmatales

Staphyleaceae
- E *Staphylea pinnata* L.

Magnoliophytina, Dicotyledonae, Cucurbitales

Begoniaceae
- A *Begonia foliosa* Kunth.
- A *Begonia minor* Jacq.
- A *Begonia semperflorens* Link & Otto

Coriariaceae
- E *Coriaria myrtifolia* L.

Corynocarpaceae
- A *Corynocarpus laevigata* J.R. & G. Forst.

Cucurbitaceae
- E *Bryonia alba* L.
- E *Bryonia cretica* L.
- E *Bryonia dioica* Jacq.
- E *Citrullus colocynthis* (L.) Schrad.
- A *Citrullus lanatus* (Thunb.) Matsum. & Nakai
- E *Cucumis melo* L.
- A *Cucumis myriocarpus* E. Mey. ex Naud.
- A *Cucumis sativus* L.
- A *Cucurbita ficifolia* C.D. Bouché
- A *Cucurbita foetidissima* Kunth.
- A *Cucurbita maxima* Duchesne
- A *Cucurbita moschata* (Duchesne ex Lam.) Duchesne ex Poir.
- E *Cucurbita pepo* L.
- A *Cyclanthera pedata* L.
- E *Ecballium elaterium* (L.) A.Rich.
- A ***Echinocystis lobata*** (Michx.) Torr. & A. Gray
- A *Lagenaria siceraria* (Molina) Standley
- A *Luffa aegyptica* Mill.
- A *Momordica charantia* L.
- A *Sechium edule* (Jacq.) Sw.
- A *Sicyos angulatus* L.
- A *Thladiantha calcarata* (Wall.) C.B. Clarke
- E *Thladiantha dubia* Bunge

Datiscaceae
- A *Datisca cannabina* L.

Magnoliophytina, Dicotyledonae, Dipsacales

Adoxaceae
- E *Adoxa moschatellina* L.

Caprifoliaceae
- A *Abelia* × *grandiflora* (Rovelli ex André) Rehder
- A *Diervilla lonicera* Mill.
- A *Leycesteria formosa* Wallich
- E *Linnaea borealis* L.
- E *Lonicera alpigena* L.
- E *Lonicera caerulea* L.
- E *Lonicera caprifolium* L.
- E *Lonicera etrusca* Santi
- A *Lonicera henryi* Hemsley
- E *Lonicera involucrata* (Richardson) Banks ex Spreng.
- A *Lonicera* × *italica* Schmidt ex Tausch
- A *Lonicera japonica* Thunb. ex Murray
- A *Lonicera maackii* (Rupr.) Maxim.
- E *Lonicera nitida* E.H. Wilson
- E *Lonicera periclymenum* L.
- A *Lonicera pileata* Oliver
- A *Lonicera* × *purpusii* Rehd.
- E *Lonicera ruprechtiana* Regel
- A *Lonicera sempervirens* L.
- E *Lonicera tatarica* L.
- E *Lonicera* × *ylosteum* L.
- A *Sambucus canadensis* L.
- E *Sambucus ebulus* L.

- E *Sambucus nigra* L.
- E *Sambucus racemosa* L.
- A *Symphoricarpos albus* (L.) S.F. Blake
- A *Symphoricarpos* × *chenaultii* Rehder
- A *Symphoricarpos orbiculatus* Moench
- E *Viburnum lantana* L.
- A *Viburnum* × *rhytidophylloides* J.V. Suringar
- A *Viburnum rhytidophyllum* Hemsley ex Forbes & Hemsley
- E *Viburnum tinus* L.
- A *Weigela floribunda* (Siebold & Zucc.) K. Koch
- A *Weigela florida* (Bunge) A. DC.
- A *Weigela japonica* Thunb.

Dipsacaceae
- E *Cephalaria alpina* (L.) Schrad. ex Roem. & Schult.
- E *Cephalaria gigantea* (Ledeb.) Bobrov.
- E *Cephalaria radiata* Griseb. & Schenk
- E *Cephalaria syriaca* (L.) Roemer & Schultes
- E *Cephalaria transsylvanica* (L.) Schrad. ex Roem. & Schultz
- E *Dipsacus fullonum* L.
- E *Dipsacus laciniatus* L.
- A *Dipsacus* × *pseudosilvester* Schur
- E *Dipsacus sativus* (L.) Honck.
- E *Dipsacus strigosus* Willd.
- E *Knautia arvensis* (L.) Coult.
- A *Knautia degenii* Borbs
- E *Knautia integrifolia* (L.) Bertol.
- E *Knautia macedonica* Gris.
- A *Scabiosa ambigua* Ten.
- A *Scabiosa caucasica* M. Bieb.
- E *Scabiosa columbaria* L.
- E *Scabiosa ochroleuca* L.
- E *Sixalix atropurpurea* L.
- A *Sixalix atropurpurea maritima* (L.) Greuter & Burdet
- E *Succisella inflexa* (Kluk) Beck

Valerianaceae
- E *Centranthus calcitrapae* (L.) Dufr.
- A *Centranthus longiflorus* Steven
- E *Centranthus macrosiphon* Boiss.
- E *Centranthus ruber* (L.) DC.
- E *Valeriana officinalis* J.C. Mikan
- E *Valeriana phu* L.
- E *Valeriana pyrenaica* L.
- E *Valerianella carinata* Loisel.
- E *Valerianella coronata* (L.) DC.
- E *Valerianella dentata* (L.) Pollich
- E *Valerianella eriocarpa* Desv.
- E *Valerianella locusta* (L.) Laterr.
- E *Valerianella rimosa* Bast.
- E *Valerianella vesicaria* (L.) Moench

Magnoliophytina, Dicotyledonae, Ericales
Actinidiaceae
- A *Actinidia chinensis* Planchon
- A *Actinidia deliciosa* (Chevalier) Liang & Ferguson

Balsaminaceae
- E *Impatiens balfourii* Hook. f.
- A *Impatiens balsamina* L.
- A *Impatiens capensis* Meerb.
- A ***Impatiens glandulifera*** Royle
- E *Impatiens noli-tangere* L.
- A *Impatiens parviflora* DC.
- A *Impatiens scabrida* DC.
- A *Impatiens sodenii* Engl. & Warb. ex Engl.
- A *Impatiens walleriana* Hook. f. ex Oliver

Clethraceae
- A *Clethra alnifolia* L.
- A *Clethra arborea* Ait.

Ebenaceae
- A *Diospyros kaki* L. fil.
- A *Diospyros lotus* L.
- A *Diospyros virginiana* L.

Ericaceae
- E *Arbutus andrachne* L.
- E *Arbutus unedo* L.
- E *Calluna vulgaris* (L.) Hull
- A *Chimaphila maculata* (L.) Pursh
- E *Chimaphila umbellata* (L.) W.P.C. Barton
- A *Corema album* (L.) D. Don ex Sweet
- E *Daboecia cantabrica* (Hudson) C. Koch
- E *Erica arborea* L.
- E *Erica carnea* L.
- E *Erica ciliaris* L.
- E *Erica cinerea* L.
- A *Erica* × *darleyensis* Bean
- E *Erica lusitanica* Rudolphi
- E *Erica scoparia* L.
- E *Erica terminalis* Salisb.
- E *Erica tetralix* L.
- E *Erica vagans* L.
- A *Gaulnettya* × *wisleyensis* Marchant
- A *Gaultheria mucronata* (L. f.) Hook. & Arn.
- A *Gaultheria procumbens* L.
- A *Gaultheria shallon* Pursh
- A *Gaultheria* × *wisleyensis* Marchant ex D.J. Middleton
- A *Kalmia angustifolia* L.
- A *Kalmia latifolia* L.

- A *Kalmia polifolia* Wangenh.
- E *Ledum groenlandicum* Oeder
- E *Ledum palustre* L.
- E *Moneses uniflora* (L.) A. Gray
- E *Orthilia secunda* (L.) House
- A *Pieris floribunda* (Pursh) Benth. & Hook. f.
- A *Rhododendron arboreum* Sm.
- A *Rhododendron brachycarpum* D. Don ex G. Don
- E *Rhododendron ferrugineum* L.
- A *Rhododendron indicum* (L.) Sweet
- A *Rhododendron* × *intermedium* Tausch
- E *Rhododendron luteum* Sweet
- A *Rhododendron maximum* L.
- A *Rhododendron mucronatum* G. Don
- E **Rhododendron ponticum** L.
- A *Rhododendron sutchuense* D. Don ex G. Don
- A *Vaccinium corymbosum* L.
- A *Vaccinium macrocarpon* Aiton

Myrsinaceae
- A *Myrsine africana* L.

Polemoniaceae
- A *Collomia grandiflora* Dougl. ex Lindl.
- A *Collomia linearis* Nutt.
- A *Gilia achilleifolia* Bentham
- A *Gilia capitata* Sims
- A *Gilia tricolor* Benth.
- A *Navarretia squarrosa* (Eschsch.) Hook. & Arnott
- A *Phlox drummondii* Hook.
- A *Phlox paniculata* L.
- A *Phlox subulata* L.
- E *Polemonium caeruleum* L.

Primulaceae
- E *Anagallis arvensis* L.
- A *Anagallis* × *doerfleri* Ronniger
- E *Anagallis foemina* Mill.
- E *Anagallis monelli* L.
- A *Androsace elongata* L.
- E *Androsace filiformis* Retz.
- E *Androsace maxima* L.
- E *Cyclamen coum* Mill.
- E *Cyclamen creticum* Hildebr.
- E *Cyclamen hederifolium* Aiton
- E *Cyclamen persicum* Mill.
- E *Cyclamen purpurascens* Mill.
- E *Cyclamen repandum* Sibth. & Smith
- E *Hottonia palustris* L.
- A *Lysimachia atropurpurea* L.
- A *Lysimachia ciliata* L.
- E *Lysimachia nummularia* L.
- E *Lysimachia punctata* L.
- A *Lysimachia terrestris* (L.) Britton, Sterns & Pogg.
- E *Primula auricula* L.
- E *Primula elatior* (L.) Hill.
- A *Primula florindae* Kingdon-Ward
- A *Primula japonica* A. Gray
- A *Primula sikkimensis* Hook. f.
- E *Primula veris* L.
- E *Primula vulgaris* Huds.
- E *Soldanella montana* Willd.
- E *Trientalis europaea* L.

Sapotaceae
- A *Argania spinosa* (L.) Skeels

Sarraceniaceae
- A *Sarracenia purpurea* L.

Styracaceae
- E *Styrax officinalis* L.

Theaceae
- A *Camellia japonica* L.
- A *Camellia sinesis* (L) O. Kuntze

Magnoliophytina, Dicotyledonae, Fabales

Betulaceae
- E *Alnus cordata* (Loisel.) Duby
- E *Alnus glutinosa* (L.) Gaertn.
- E *Alnus* × *hybrida* A. Braun ex Reichb.
- E *Alnus incana* (L.) Moench
- A *Alnus japonica* (Thunb.) Steud.
- A *Alnus maritima* (Marshall) Muhl. ex Nutt.
- A *Alnus rubra* Bong.
- E *Alnus viridis* (Chaix) DC.
- E *Betula nana* L.
- E *Betula pendula* Roth
- E *Betula pubescens* Ehrh.
- E *Carpinus betulus* L.
- E *Corylus avellana* L.
- E *Corylus colurna* L.
- E *Corylus maxima* Miller

Casuarinaceae
- A *Casuarina cunninghamiana* Miq.
- A *Casuarina verticillata* Lam.

Fabaceae – Caesalpinioideae
- A *Bauhinia acuminata* L.
- A *Bauhinia variegata* L.
- A *Caesalpinia decapetala* (Roth) Alston
- A *Caesalpinia gilliesii* (Hooker) Dietr.
- A *Caesalpinia spinosa* (Molina) Kuntze
- E *Ceratonia siliqua* L.
- A *Chamaecrista fasciculata* (Michx.) Greene
- A *Clianthus puniceus* (G. Don) Soland. ex Lindl.
- A *Delonix regia* (Bojer ex Hook.) Raf.
- A *Erythrina abyssinica* Lam.

- A *Erythrina crista-galli* L.
- A *Erythrina falcata* Benth.
- A *Erythrina lysistemon* Hutchins.
- A *Erythrina speciosa* Andrews
- A *Gleditsia triacanthos* L.
- A *Gymnocladus dioica* (L.) K. Koch
- A *Indigofera herantha* Wall. ex Brandis
- E *Parkinsonia aculeata* L.
- A *Schotia brachypetala* Sond.
- A *Senna artemisioides* Isely
- A *Senna bicapsularis* (L.) Roxb.
- A *Senna corymbosa* (Lam.) H.S. Irwin & Barneby
- A *Senna didymobotrya* (Fresen.) H.S. Irwin & Barneby
- A *Senna multiglandulosa* (Jacq.) H.S. Irwin & Barneby
- A *Senna multijuga* (L.C. Rich.) Irwin & Barneby
- A *Senna obtusifolia* (L.) Irwin & Barneby
- A *Senna occidentalis* (L.) Link
- A *Senna pendula* Humb. & Bonpl. ex Willd.
- A *Senna septentrionalis* (Viv.) Irwin & Barneby
- A *Solandra nitida* Zucc.

Fabaceae – Cercideae
- E *Cercis siliquastrum* L.

Fabaceae – Faboideae
- A *Aeschynomene americana* L.
- A *Aeschynomene indica* L.
- A *Alhagi maurorum* Medik.
- A *Amorpha fruticosa* L.
- A *Apios americana* Medick.
- A *Arachis hypogaea* L.
- A *Astragalus alopecuroides* L.
- E *Astragalus boeticus* L.
- E *Astragalus cicer* L.
- A *Astragalus corrugatus* Bertol.
- E *Astragalus cymbicarpos* Brot.
- A *Astragalus falcatus* Lam.
- A *Astragalus galegiformis* L.
- A *Astragalus glycyphylloides* DC.
- E *Astragalus glycyphyllos* L.
- E *Astragalus hamosus* L.
- E *Astragalus odoratus* Lam.
- E *Astragalus onobrychis* L.
- E *Astragalus oxyglottis* Steven ex M. Bieb.
- E *Astragalus penduliflorus* Lam.
- A *Astragalus rytilobus* Bunge
- E *Astragalus stella* Gouan
- E *Biserrula pelecinus* L.
- A *Caragana arborescens* Lam.
- E *Caragana frutex* (L.) K. Koch
- E *Chamaecytisus hirsutus* (L.) Link
- E *Chamaecytisus purpureus* Scop. (Link)
- E *Chamaecytisus ratisbonensis* (Schaeff.) Rothm.
- A *Chamaecytisus* × *versicolor* Dippel
- E *Cicer arietinum* L.
- E *Colutea arborescens* L.
- A *Colutea* × *media* Willd.
- E *Colutea orientalis* Mill.
- E *Coronilla scorpioides* (L.) Koch
- E *Coronilla valentina* L.
- A *Cullen americanum* (L.) Rydb.
- A *Cullen cinereum* (Lindl.) J.W. Grimes
- E *Cytisus decumbens* (Durande) Spach
- A *Cytisus elongatus* Waldst. & Kit.
- E *Cytisus galianoi* Talavera & P.E. Gibbs
- E *Cytisus multiflorus* (L–Hér. ex Aiton) Sweet
- E *Cytisus sessilifolius* L.
- E *Cytisus striatus* (Hill) Rothm.
- E *Dorycnium hirsutum* (L.) Ser.
- E *Dorycnium pentaphyllum* Scop.
- E *Galega officinalis* L.
- A *Galega orientalis* Lam.
- E *Genista aetnensis* (Biv.) DC.
- E *Genista hispanica* L.
- E *Genista januensis* Viv.
- E *Genista linifolia* L.
- E *Genista monspessulana* (L.) L. Johnson
- A *Genista numidica* Spach
- E *Genista pilosa* L.
- E *Genista sagittalis* L.
- A *Genista stenopetala* Webb & Berthel.
- E *Genista tinctoria* L.
- A *Glycine max* (L.) Merrill
- E *Glycyrrhiza glabra* L.
- E *Halimodendron halodendron* (Pall.) Voss
- E *Hedysarum coronarium* L.
- E *Hippocrepis ciliata* Willd.
- E *Hippocrepis comosa* L.
- E *Hippocrepis emerus* (L.) Lassen
- A *Hymenocarpus circinatus* (L.) Savi
- A *Lablab purpureus* (L.) Sweet.
- E *Laburnum alpinum* (Mill.) Bercht. & J. Presl
- E *Laburnum anagyroides* Medicus
- A *Laburnum* × *watereri* (Wettst.) Dippel
- E *Lathyrus angulatus* L.
- E *Lathyrus annuus* L.
- E *Lathyrus aphaca* L.
- E *Lathyrus cicera* L.
- E *Lathyrus clymenum* L.
- E *Lathyrus grandiflorus* Sibth. & Sm.
- E *Lathyrus heterophyllus* L.

11 List of Species Alien in Europe and to Europe

- E *Lathyrus hirsutus* L.
- E *Lathyrus inconspicuus* L.
- E *Lathyrus incurvus* (Roth) Willd.
- E *Lathyrus latifolius* L.
- E *Lathyrus niger* (L.) Bernh
- E *Lathyrus nissolia* L.
- E *Lathyrus ochrus* (L.) DC.
- E *Lathyrus odoratus* L.
- E *Lathyrus palustris* L.
- E *Lathyrus pannonicus* (Jacq.) Garcke
- E *Lathyrus pratensis* L.
- E *Lathyrus sativus* L.
- E *Lathyrus setifolius* L.
- E *Lathyrus sphaericus* Retz.
- E *Lathyrus sylvestris* C. Presl
- E *Lathyrus tingitanus* L.
- E *Lathyrus tuberosus* L.
- E *Lathyrus vernus* (L.) Bernh.
- E *Lembotropis nigricans* (L.) Griseb.
- E *Lens culinaris* Medik.
- E *Lens ervoides* (Brign.) Grande
- E *Lens nigricans* (M. Bieb.) Godr.
- E *Lotus angustissimus* L.
- E *Lotus conimbricensis* Brot. ex Willd.
- E *Lotus corniculatus* L.
- E *Lotus cytisoides* L.
- E *Lotus drepanocarpus* Durieu
- E *Lotus maritimus* L.
- E *Lotus ornithopodioides* L.
- E *Lotus parviflorus* Desf.
- E *Lotus pedunculatus* Cav.
- E *Lotus suaveolens* Pers.
- E *Lupinus albus* L.
- E *Lupinus angustifolius* L.
- A *Lupinus arboreus* Sims
- A *Lupinus cosentinii* Guss.
- E *Lupinus hispanicus* Boiss. & Reut.
- E *Lupinus luteus* L.
- E *Lupinus micranthus* Guss.
- A *Lupinus nootkatensis* Sims
- A *Lupinus perennis* L.
- A *Lupinus polyphyllus* Lindley
- A *Lupinus* × *pseudopolyphyllus* C.P. Sm.
- A *Lupinus* × *regalis* (Hort.) Bergmans
- E *Medicago arabica* (L.) Hudson
- E *Medicago arborea* L.
- E *Medicago blancheana* Boiss.
- E *Medicago caerulea* Less. ex Ledeb.
- E *Medicago cancellata* M. Bieb.
- E *Medicago carstiensis* Wulfen
- E *Medicago ciliaris* (L.) All.
- E *Medicago coronata* (L.) Bart.
- E *Medicago disciformis* DC.
- E *Medicago falcata* L.
- A *Medicago glandulosa* (Mert. & W.D.J. Koch) Davidov
- E *Medicago intertexta* (L.) Mill.
- E *Medicago laciniata* (L.) Miller
- E *Medicago littoralis* Rohde ex Loisel.
- E *Medicago lupulina* L.
- E *Medicago minima* (L.) L.
- E *Medicago monantha* (C.A. Mey.) Trautv.
- E *Medicago murex* Willd.
- E *Medicago orbicularis* (L.) Bartal.
- E *Medicago polymorpha* L.
- E *Medicago praecox* DC.
- E *Medicago prostrata* Jacq.
- E *Medicago rigidula* (L.) Desr.
- E *Medicago romanica* Prod.
- E *Medicago rugosa* Desr.
- E *Medicago sativa* L.
- A *Medicago sativa microcarpa* Urb.
- E *Medicago scutellata* (L.) Mill.
- E *Medicago soleirolii* Duby
- E *Medicago tornata* (L.) Mill.
- E *Medicago truncatula* Gaertn.
- E *Medicago turbinata* (L.) All.
- A *Medicago* × *varia* Martyn
- E *Melilotus albus* Medik.
- E *Melilotus altissimus* Thuill.
- E *Melilotus dentatus* (Waldst. & Kit.) Desf.
- E *Melilotus indicus* (L.) All.
- E *Melilotus infestus* Guss.
- E *Melilotus italicus* (L.) Lam.
- E *Melilotus messanensis* (L.) All.
- E *Melilotus neapolitanus* Ten. ex Guss.
- E *Melilotus officinalis* (L.) Pall.
- E *Melilotus segetalis* (Brot.) Ser.
- E *Melilotus sulcatus* Desf.
- E *Melilotus wolgicus* Poir.
- E *Onobrychis aequidentata* (Sibth. & Sm.) d'Urv.
- E *Onobrychis caput-galli* (L.) Lam.
- E *Onobrychis crista-galli* Lam.
- C *Onobrychis tournefortii* (Willd.) Desv.
- E *Onobrychis viciifolia* Scop.
- E *Ononis alopecuroides* L.
- A *Ononis diffusa* Ten.
- E *Ononis mitissima* L.
- E *Ononis natrix* L.
- E *Ononis repens* L.
- E *Ononis spinosa* L.
- E *Ononis subspicata* Lag.
- E *Ornithopus compressus* L.
- E *Ornithopus perpusillus* L.
- E *Ornithopus sativus* Brot.

E *Oxytropis pilosa* (L.) DC
A *Phaseolus coccineus* L.
A *Phaseolus lunatus* L.
A *Phaseolus vulgaris* L.
E *Pisum sativum* L.
E *Psoralea bituminosa* L.
A *Psoralea pinnata* L.
A *Pueraria lobata* (Willd.) Ohwi
E *Retama monosperma* (L.) Boiss.
E *Retama sphaerocarpa* (L.) Boiss.
A *Rhynchosia caribaea* (Jacq.) DC.
A *Robinia* × *ambigua* Poir.
A *Robinia hispida* L.
A *Robinia neomexicana* Gray
A **Robinia pseudoacacia** L.
A *Robinia viscosa* Vent.
E *Scorpiurus muricatus* L.
E *Securigera securidaca* (L.) Degen & Dörfler
E *Securigera varia* (L.) Lassen
A *Sesbania cannabina* (Retz.) Pers.
A *Sesbania herbacea* (Mill.) McVaugh
A *Sesbania punicea* (Cav.) Benth.
A *Sesbania sesban* (L.) Merr.
A *Sophora jaubertii* Spach
E *Spartium junceum* L.
E *Styphnolobium japonicum* (L.) Schott
E *Tetragonolobus biflorus* (Desr.) Ser.
E *Tetragonolobus purpureus* Moench
A *Thermopsis montana* Nutt.
A *Tipuana tipu* (Benth.) Kuntze
E *Trifolium alexandrinum* L.
E *Trifolium angulatum* Waldst. & Kit.
E *Trifolium angustifolium* L.
E *Trifolium apertum* Bobrov
E *Trifolium argutum* Banks & Sol.
E *Trifolium arvense* L.
E *Trifolium aureum* Pollich
E *Trifolium campestre* Schreb.
E *Trifolium cernuum* Brot.
E *Trifolium constantinopolitanum* Ser.
E *Trifolium diffusum* Ehrh.
E *Trifolium dubium* Sibth.
E *Trifolium echinatum* M. Bieb.
E *Trifolium fragiferum* L.
E *Trifolium glomeratum* L.
E *Trifolium hirtum* All.
E *Trifolium hybridum* L.
E *Trifolium hybridum elegans* (Savi) Aschers. & Graebn.
E *Trifolium incarnatum* L.
E *Trifolium isthmocarpum* Brot.
E *Trifolium lappaceum* L.
E *Trifolium ligusticum* Balb. ex Loisel.
E *Trifolium lupinaster* L.
E *Trifolium micranthum* Viv.
E *Trifolium montanum* L.
E *Trifolium mutabile* Port.
E *Trifolium nigrescens* Viv.
A *Trifolium ornithopodioides* (L.) Sm.
E *Trifolium pallidum* Waldst. & Kit.
E *Trifolium pannonicum* Jacq.
E *Trifolium patens* Schreb.
E *Trifolium phleoides* Pourr. ex Willd.
E *Trifolium pratense* L.
E *Trifolium purpureum* Loisel.
E *Trifolium repens* L.
E *Trifolium resupinatum* L.
E *Trifolium retusum* L.
E *Trifolium rubens* L.
E *Trifolium scabrum* L.
E *Trifolium spadiceum* L.
E *Trifolium spumosum* L.
E *Trifolium squarrosum* L.
E *Trifolium stellatum* L.
E *Trifolium strictum* L.
E *Trifolium subterraneum* L.
E *Trifolium suffocatum* L.
E *Trifolium tomentosum* L.
E *Trifolium vesiculosum* Savi
A *Trigonella arabica* Delile
E *Trigonella balansae* Boiss. & Reut.
A *Trigonella caelesyriaca* Boiss.
E *Trigonella caerulea* (L.) Ser.
A *Trigonella calliceras* Fisch.
E *Trigonella fischeriana* Ser.
E *Trigonella foenum-graecum* L.
A *Trigonella grandiflora* Bunge
A *Trigonella hierosolymitana* Boiss.
A *Trigonella kotschyi* Benth.
E *Trigonella maritima* Delile
E *Trigonella monspeliaca* L.
A *Trigonella noeana* Boiss.
E *Trigonella orthoceras* Kar. & Kir.
E *Trigonella procumbens* (Bess.) Reichenb.
E *Trigonella spinosa* L.
A *Trigonella stellata* Forssk.
E *Ulex europaeus* L.
E *Ulex gallii* Planch.
E *Ulex minor* Roth
E *Vicia articulata* Hornem.
E *Vicia assyriaca* Boiss.
E *Vicia benghalensis* L.
E *Vicia bithynica* (L.) L.
E *Vicia ciliatula* Lipsky
E *Vicia cracca* L.
E *Vicia disperma* DC.

- E *Vicia dumetorum* L.
- E *Vicia ervilia* (L.) Willd.
- E *Vicia faba* L.
- E *Vicia grandiflora* Scop.
- E *Vicia hirsuta* (L.) Gray
- E *Vicia hybrida* L.
- A *Vicia johannis* Tamamsch.
- E *Vicia leucantha* Biv.
- E *Vicia lutea* L.
- E *Vicia melanops* Sibth. & Sm.
- E *Vicia monantha* Retz.
- E *Vicia narbonensis* L.
- E *Vicia onobrychioides* L.
- E *Vicia pannonica* Crantz
- E *Vicia parviflora* Cav.
- E *Vicia peregrina* L.
- E *Vicia pisiformis* L.
- A *Vicia* × *poechhackeri* J. Murr
- E *Vicia sativa* L.
- E *Vicia sativa nigra* (L.) Ehrh.
- E *Vicia sepium* L.
- E *Vicia serratifolia* Jacq.
- E *Vicia sicula* (Raf.) Guss.
- E *Vicia sylvatica* L.
- E *Vicia tenuifolia* Roth
- E *Vicia tetrasperma* (L.) Schreber
- E *Vicia villosa* Roth
- E *Vicia villosa eriocarpa* (Hausskn.) P.W. Ball
- E *Vicia villosa varia* (Host) Corb.
- A *Vigna radiata* (L.) Wilczek
- A *Vigna unguiculata* (L) Walp.
- A *Wisteria floribunda* (Willd.) DC.
- A *Wisteria sinensis* (Sims) Sweet

Fabaceae - Mimosoideae
- A *Acacia baileyana* F. Müll.
- A *Acacia cultriformis* A. Cunn. ex G. Don
- A *Acacia cyclops* A. Cunn. ex G. Don fil.
- A **Acacia dealbata** Link
- A *Acacia decurrens* (J.C. Wendl.) Willd.
- A *Acacia farnesiana* (L.) Willd.
- A *Acacia karoo* Hayne
- A *Acacia ligulata* Bentham
- A *Acacia longifolia* (Andrews) Willd.
- A *Acacia mearnsii* De Wild.
- A *Acacia melanoxylon* R. Br.
- A *Acacia neriifolia* A. Cunn. ex Benth.
- A *Acacia paradoxa* DC.
- A *Acacia pycnantha* Benth.
- A *Acacia retinodes* Schlecht.
- A *Acacia salicina* Lindley
- A *Acacia saligna* (Labill.) H.L. Wendl.
- A *Acacia terminalis* (Salisb.) J.F. Macbr.
- A *Acacia verticillata* (L'Hér.) Willd.
- A *Albizia julibrissin* Durazz.
- A *Albizia saman* (Jacq.) F. Müll.
- A *Calliandra tweedii* Benth.
- A *Desmanthus illinoensis* (Michx.) McMillan ex Robinson & Fernald
- A *Desmanthus virgatus* (L.) Willd.
- A *Faidherbia albida* (Delile) A. Cheval.
- A *Leucaena leucocephala* (Lam.) de Wit
- A *Mimosa pudica* L.
- A *Paraserianthes lophantha* (Willd.) I.C. Nielsen

Fagaceae
- A *Castanea crenata* Siebold & Zucc.
- E *Castanea sativa* Mill.
- E *Fagus sylvatica* L.
- E *Quercus canariensis* Willd.
- E *Quercus cerrioides* Willk. & Costa
- E *Quercus cerris* L.
- A *Quercus crenata* Lam.
- E *Quercus humilis* Mill.
- E *Quercus ilex* L.
- A *Quercus palustris* Münchh.
- E *Quercus robur* L.
- A *Quercus rubra* L.
- E *Quercus suber* L.
- A *Quercus velutina* Lam.

Juglandaceae
- A *Carya cordiformis* (Wangerin) K. Koch
- A *Carya ovata* (Mill.) K. Koch
- A *Juglans ailanthifolia* Carr.
- A *Juglans cinerea* L.
- A *Juglans nigra* L.
- E *Juglans regia* L.
- A *Pterocarya fraxinifolia* (Poiret) Spach
- A *Pterocarya* × *rehderiana* C.K. Schneider
- A *Pterocarya stenoptera* DC.

Myricaceae
- A *Comptonia peregrina* (L.) Coult.
- A *Myrica aeth*iopica L.
- A *Myrica pensylvanica* Lois. ex Duhamel

Nothofagaceae
- A *Nothofagus nervosa* (Phil.) Krasser
- A *Nothofagus obliqua* (Mirbel) Blume

Polygalaceae
- A *Polygala curtissii* A. Gray
- A *Polygala myrtifolia* L.

Magnoliophytina, Dicotyledonae, Gentianales

Apocynaceae
- A *Allamanda cathartica* L.
- A *Amsonia tabernaemontana* Walter
- A *Araujia sericifera* Brot.
- A *Asclepias physocarpa* (E. Mey.) Schlechter

- E *Asclepias curassavica* L.
- A *Asclepias syriaca* L.
- A *Catharanthus roseus* (L.) G. Don
- E *Cynanchum rossicum* (Kleopov) Borhidi
- A *Gomphocarpus fruticosus* (L.) Aiton f.
- E *Nerium oleander* L.
- E *Periploca graeca* L.
- A *Plumeria rubra* L.
- A *Thevetia peruviana* (Pers.) K. Schum.
- A *Trachelospermum jasminoides* (Lind.) Lem.
- A *Trachomitum venetum* (L.) Woodson
- E *Vinca difformis* Pourret
- E *Vinca major* L.
- E *Vinca major major* L.
- E *Vinca minor* L.
- E *Vincetoxicum nigrum* (L.) Moench

Gentianaceae
- E *Blackstonia perfoliata* (L.) Huds.
- E *Centaurium maritimum* (L.) Fritsch
- E *Centaurium pulchellum* (Sw.) Druce
- E *Centaurium tenuiflorum* (Hoffmanns. & Link) Fritsch ex Janch.
- E *Gentiana acaulis* L.
- E *Gentiana asclepiadea* L.
- E *Gentiana clusii* Perrier & Song.
- E *Gentiana lutea* L.
- A *Gentiana septemfida* Pall.
- A *Gentianella aurea* (L.) H. Sm.

Rubiaceae
- E *Asperula arvensis* L.
- E *Asperula laevigata* L.
- A *Asperula orientalis* Boiss. & Hohen.
- E *Asperula taurina* L.
- E *Asperula tinctoria* L.
- A *Coffea arabica* L.
- A *Coprosma repens* A. Rich.
- E *Crucianella angustifolia* L.
- E *Crucianella latifolia* L.
- E *Cruciata glabra* (L.) Ehrend.
- E *Cruciata laevipes* Opiz
- E *Galium aparine* L.
- E *Galium divaricatum* Pourr. ex Lam.
- E *Galium glaucum* L.
- A *Galium humifusum* M. Bieb.
- E *Galium mollugo* L.
- E *Galium murale* (L.) All.
- A *Galium* × *ochroleucum* Wolf. ex Schweigg. & Koert.
- E *Galium palustre* L.
- E *Galium parisiense* L.
- E *Galium* × *pomeranicum* Retz.
- E *Galium pumilum* Murray
- E *Galium rotundifolium* L.
- E *Galium rubioides* L.
- E *Galium rubrum* L.
- E *Galium saxatile* L.
- E *Galium sylvaticum* L.
- E *Galium tenuissimum* M. Bieb.
- E *Galium tricornutum* Dandy
- E *Galium uliginosum* L.
- E *Galium verrucosum* Hudson
- A *Galium verticillatum* Danthoine ex Lam.
- E *Galium verum* L.
- A *Nertera granadensis* (Mutis ex L. f.) Druce
- E *Phuopsis stylosa* (Trin.) Benth. & Hook. f. ex B.D. Jackson
- E *Rubia tinctorum* L.
- E *Sherardia arvensis* L.
- E *Valantia hispida* L.

Magnoliophytina, Dicotyledonae, Geraniales

Geraniaceae
- A *Erodium alnifolium* Guss.
- E *Erodium botrys* (Cav.) Bertol.
- E *Erodium brachycarpum* (Godron) Thell.
- E *Erodium chium* (L.) Willd.
- E *Erodium ciconium* (L.) L'Hér.
- E *Erodium cicutarium* (L.) L'Hér.
- A *Erodium crinitum* Carolin
- A *Erodium cygnorum* Nees
- A *Erodium glaucophyllum* (L.) L'Hér.
- A *Erodium gruinum* (L.) L'Hér.
- A *Erodium laciniatum* (Cav.) Willd.
- E *Erodium malacoides* (L.) L'Hér.
- E *Erodium manescavii* Cosson
- E *Erodium moschatum* (L.) L'Hér.
- A *Erodium neuradifolium* Delile
- E *Erodium salzmannii* Delile
- A *Erodium stephanianum* Willd.
- E *Geranium bohemicum* L.
- A *Geranium carolinianum* L.
- E *Geranium columbinum* L.
- E *Geranium dissectum* L.
- E *Geranium divaricatum* Ehrh.
- E *Geranium endressii* J. Gay
- A *Geranium himalayense* Klotzsch
- E *Geranium ibericum* Cav.
- E *Geranium lucidum* L.
- E *Geranium macrorrhizum* L.
- A *Geranium maderense* Yeo
- A *Geranium* × *magnificum* N. Hylander
- E *Geranium molle* L.
- A *Geranium* × *monacense* Harz
- A *Geranium nepalense* Sweet.

- E *Geranium nodosum* L.
- A *Geranium* × *oxonianum* Yeo
- E *Geranium phaeum* L.
- A *Geranium platypetalum* Fischer & C. Meyer
- E *Geranium pratense* L.
- E *Geranium psilostemon* Lebed.
- E *Geranium purpureum* Vill.
- E *Geranium pusillum* L.
- E *Geranium pyrenaicum* Burm. fil.
- A *Geranium reflexum* L.
- E *Geranium rotundifolium* L.
- A *Geranium rubescens* Yeo
- E *Geranium sanguineum* L.
- E *Geranium sibiricum* L.
- E *Geranium submolle* Steudel
- E *Geranium sylvaticum* L.
- E *Geranium versicolor* L.
- A *Monsonia angustifolia* E. Mey. ex A. Rich.
- A *Monsonia brevirostrata* Knuth
- A *Pelargonium capitatum* (L.) L'Hér. ex Aiton
- A *Pelargonium cordatum* L'Hér.
- A *Pelargonium glutinosum* (Jacq.) L'Hér
- A *Pelargonium* × *hybridum* (L.) Aiton
- A *Pelargonium inquinans* (L.) Aiton
- A *Pelargonium odoratissimum* (L.) L'Hér ex Aiton
- A *Pelargonium peltatum* (L.) L'Hér. ex Aiton
- A *Pelargonium quercifolium* (L. f.) L'Hér. ex Aiton
- A *Pelargonium radens* H.E. Moore
- A *Pelargonium tomentosum* Jacq.
- A *Pelargonium vitifolium* (L.) L'Hér ex Aiton
- A *Pelargonium zonale* (L.) L'Hér. ex Aiton

Magnoliophytina, Dicotyledonae, Gunnerales
Gunneraceae
- A *Gunnera manicata* Linden ex Andre
- A *Gunnera tinctoria* (Molina) Mirbel

Magnoliophytina, Dicotyledonae, Lamiales
Acanthaceae
- E *Acanthus mollis* L.
- E *Acanthus spinosus* L.
- A *Asystasia gangetica* (L.) T. Anders
- A *Hygrophila polysperma* (Roxb.) T. Anders.
- A *Hypoestes phyllostachya* Baker
- A *Jacobinia carnea* (Lindl.) G. Nicholson
- A *Justicia adhatoda* L.
- A *Ruellia brevifolia* (Pohl) Ezcurra
- A *Strobilanthes maculatus* (Wall.) Nees.
- A *Thunbergia alata* Bojer ex Sims
- A *Thunbergia grandiflora* Roxb.
- A *Thunbergia gregorii* S. Moore

Bignoniaceae
- A *Catalpa bignonioides* Walter
- A *Clytostoma callistegioides* (Cham.) Bureau ex Griseb.
- A *Jacaranda mimosifolia* D. Don
- A *Kigelia africana* (Lam.) Benth.
- A *Macfadyena unguis-cati* (L.) A. Gentry
- A *Markhamia platycalyx* (Bak.) Sprague
- A *Pandorea pandorana* (Andr.) Steenis
- A *Paulownia tomentosa* (Thunb.) Sieb. & Zucc. ex Steud.
- A *Phaedranthus buccinatorius* (DC.) Miers
- A *Podranea ricasoliana* (Tanfani) Sprague
- A *Pyrostegia venusta* (Ker-Gawl.) Miers
- A *Spathodea campanulata* P. Beauv.
- A *Tecoma radicans* (L.) Juss.
- A *Tecoma stans* (L.) Juss. ex Humb., Bonpl. & Kunth
- A *Tecomaria capensis* (Thunb.) Spach

Boraginaceae
- A *Alkanna lycopsoides* (Lehm.) Lehm.
- A *Amsinckia eastwoodiae* Macbr.
- A *Amsinckia intermedia* Fischer & C.A. Meyer
- A *Amsinckia lycopsoides* Lehm.
- A *Amsinckia menziesii* (Lehm.) A. Nels. & J.F. Macbr.
- A *Amsinckia tesselata* A. Gray
- E *Anchusa arvensis* (L.) M. Bieb.
- E *Anchusa arvensis orientalis* (L.) Nordh.
- E *Anchusa azurea* Mill.
- E *Anchusa barrelieri* (All.) Vitman
- A *Anchusa* × *baumgarteni* Nyman
- A *Anchusa gmelinii* Ledeb.
- E *Anchusa ochroleuca* M. Bieb.
- E *Anchusa officinalis* L.
- A *Anchusa procera* Besser ex Link
- E *Anchusa stylosa* M. Bieb.
- A *Anchusa thessala* Boiss. & Sprun.
- E *Anchusa undulata* L.
- E *Argusia sibirica* (L.) Dandy
- E *Arnebia decumbens* (Vent.) Coss. & Kralik
- E *Asperugo procumbens* L.
- E *Borago officinalis* L.
- E *Borago pygmaea* (DC.) Chater & Greuter

- E *Brunnera macrophylla* (Adams) I.M. Johnston
- E *Buglossoides arvensis* (L.) I.M. Johnston
- E *Cerinthe major* L.
- E *Cerinthe minor* L.
- A *Cordia myxa* L.
- E *Cynoglossum clandestinum* Desf.
- E *Cynoglossum columnae* Ten.
- E *Cynoglossum creticum* Mill.
- E *Cynoglossum glochidiatum* Wall. ex Benth.
- A *Cynoglossum mathezii* Greuter & Burdet
- E *Cynoglossum officinale* L.
- E *Echium arenarium* Guss.
- E *Echium italicum* L.
- A *Echium pininana* Webb & Bernh.
- E *Echium plantagineum* L.
- A *Echium rauwolfii* Delile
- A *Echium rosulatum* Lange
- E *Echium vulgare* L.
- A *Heliotropium amplexicaule* Vahl
- A *Heliotropium arborescens* L.
- A *Heliotropium curassavicum* L.
- E *Heliotropium dolosum* De Not.
- E *Heliotropium ellipticum* Ledeb.
- E *Heliotropium europaeum* L.
- E *Heliotropium suaveolens* M. Bieb.
- E *Heliotropium supinum* L.
- E *Lappula barbata* (M. Bieb.) Grke
- E *Lappula deflexa* (Wahlenb.) Grke
- E *Lappula marginata* (M. Bieb.) Grcke
- E *Lappula squarrosa* (Retz.) Dumort.
- A *Limnanthes douglasii* R. Br.
- E *Lithospermum officinale* L.
- A *Lithospermum sibthorpianum* Griseb.
- A *Mertensia denticulata* Ledeb.
- E *Myosotis alpestris* L.
- E *Myosotis arvensis* (L.) Hill
- E *Myosotis discolor* Pers.
- A *Myosotis* × *krajinae* Domin
- E *Myosotis latifolia* Poiret
- A *Myosotis* × *pseudohispida* Domin
- E *Myosotis ramosissima* Rochel
- E *Myosotis scorpioides* L.
- E *Myosotis secunda* A. Murray
- E *Myosotis sparsiflora* J.G. Mikan ex Pohl
- E *Myosotis stricta* Link ex Roem. & Schult.
- E *Myosotis sylvatica* Ehrh. ex Hoffm.
- E *Nonea erecta* Bernhardi
- E *Nonea lutea* (Desr.) DC.
- E *Nonea pallens* Petrovic
- A *Nonea rosea* (M. Bieb.) Link
- E *Nonea versicolor* (Steven) Sweet
- A *Nonea* × *popovii* Gusuleac & Tarnavschi
- E *Nonea pulla* (L.) DC. ex Lam & DC.
- E *Omphalodes linifolia* (L.) Moench
- E *Omphalodes verna* Moench
- E *Pentaglottis sempervirens* (L.) Tausch ex L.H. Bailey
- A *Plagiobothrys scouleri* (Hook. & Arn.) I.M. Johnston
- E *Pulmonaria affinis* Jord. ex F.W. Schultz
- E *Pulmonaria longifolia* (Bast.) Boreau
- E *Pulmonaria mollis* Wulfen ex Hornem.
- E *Pulmonaria montana* Lej.
- E *Pulmonaria officinalis* L.
- E *Pulmonaria rubra* Schott.
- E *Pulmonaria saccharata* Mill.
- A *Pulmonaria sibirica* L.
- A *Rochelia disperma* (L. f.) K. Koch
- E *Symphytum asperum* Lepech.
- E *Symphytum bulbosum* C. Schimper
- E *Symphytum caucasicum* M. Bieb.
- A *Symphytum* × *coeruleum* Petitm. ex Thell.
- E *Symphytum ibericum* Steven
- E *Symphytum officinale* L.
- E *Symphytum orientale* L.
- A *Symphytum peregrinum* auct. non Ledeb.
- E *Symphytum tauricum* Willd.
- E *Symphytum tuberosum* L.
- E *Symphytum* × *uplandicum* Nyman
- E *Trachystemon orientalis* (L.) G. Don f.

Calceolariaceae
- A *Calceolaria chelidonioides* Humb., Bonpl. & Kunth
- A *Calceolaria pinnata* L.
- A *Calceolaria tripartita* Ruiz & Pav.

Gesneriaceae
- E *Ramonda myconi* (L.) Reichb.

Hydrophyllaceae
- A *Ellisia nyctelea* L.
- A *Nemophila menziesii* Hook. & Arnott
- A *Phacelia campanularia* A. Gray
- A *Phacelia ciliata* Benth.
- A *Phacelia congesta* Hook.
- A *Phacelia purshii* Buckl.
- A *Phacelia tanacetifolia* Bentham
- A *Wigandia caracasana* Humb., Bonpl. & Kunth

Lamiaceae
- E *Acinos arvensis* (Lam.) Dandy
- A *Agastache foeniculum* (Pursh) Kuntze
- E *Ajuga chamaepitys* (L.) Schreber

11 List of Species Alien in Europe and to Europe

E *Ajuga genevensis* L.
A *Ajuga glabra* C. Presl
E *Ajuga pyramidalis* L.
E *Ajuga reptans* L.
A *Amethystea caeruea* L.
E *Ballota acetabulosa* (L.) Bentham
E *Ballota hirsuta* Benth.
E *Ballota hispanica* (L.) Benth.
E *Ballota nigra* L.
E *Ballota nigra meridionalis* (Béguinot) Béguinot
A *Ballota nigra nigra* L.
A *Ballota nigra uncinata* (Fiori & Bég.) Patzak
E *Ballota pseudodictamnus* (L.) Benth.
E *Calamintha grandiflora* (L.) Moench
E *Calamintha sylvatica* Bromf.
E *Cedronella canariensis* (L.) Webb & Berthel.
E *Clinopodium vulgare* L.
A *Coleus blumei* Benth.
E *Dracocephalum moldavica* L.
A *Dracocephalum parviflorum* Nutt.
A *Dracocephalum ruyschiana* L.
A *Dracocephalum sibiricum* (L.) L.
E *Dracocephalum thymiflorum* L.
A *Dracocephalum triflorum* L.
A *Elsholtzia ciliata* (Thunb.) Hyl.
A *Elsholtzia stauntonii* Benth.
E *Galeopsis angustifolia* Ehrh. ex Hoffm.
E *Galeopsis ladanum* L.
E *Galeopsis pubescens* Besser
E *Galeopsis segetum* Necker
E *Galeopsis speciosa* Mill.
E *Galeopsis tetrahit* L.
E *Glechoma hederacea* L.
E *Hyssopus officinalis* L.
A *Iboza riparia* (Hochst.) N.E. Br.
E *Lallemantia iberica* (M. Bieb.) Fisch. & C.A. Mey.
E *Lamiastrum galeobdolon* (L.) Ehrend. & Polatschek
E *Lamium album* L.
E *Lamium amplexicaule* L.
E *Lamium hybridum* Villars
E *Lamium maculatum* L.
A *Lamium moluccellifolium* Fries
E *Lamium orvala* L.
E *Lamium purpureum* L.
E *Lavandula angustifolia* Mill.
E *Lavandula dentata* L.
E *Lavandula* × *intermedia* Emeric ex Lois.
E *Lavandula latifolia* Medik.

E *Lavandula stoechas* L.
E *Lavandula viridis* L'Hér.
A *Leonitis leonorus* (L.) R. Br.
E *Leonurus cardiaca* L.
E *Leonurus cardiaca villosus* (Dum.-d'Urv.) Hyl.
A *Leonurus japonicus* Houtt.
E *Leonurus marrubiastrum* L.
A *Leonurus sibiricus* L.
E *Lycopus europaeus* L.
E *Lycopus exaltatus* L. f.
E *Marrubium alternidens* Rech. f.
A *Marrubium* × *paniculatum* Desr.
E *Marrubium peregrinum* L.
E *Marrubium vulgare* L.
E *Melissa officinalis* L.
E *Mentha arvensis* L.
A *Mentha* × *dalmatica* Tausch
A *Mentha* × *gentilis* L.
A *Mentha* × *gracilis* Sole
E *Mentha longifolia* (L.) Huds.
E *Mentha* × *niliaca* Jacq.
E *Mentha* × *piperita* L.
E *Mentha pulegium* L.
E *Mentha requienii* Bentham
E *Mentha* × *smithiana* R.A. Graham
E *Mentha spicata* L.
E *Mentha suaveolens* Ehrh.
A *Mentha* × *verticillata* L.
A *Mentha* × *villosa* Huds.
A *Mentha* × *villosonervata* Opiz
A *Molucella spinosa* L.
A *Monarda didyma* L.
A *Monarda punctata* L.
A *Nepeta amoena* Stapf.
E *Nepeta cataria* L.
A *Nepeta* × *faasenii* Stearn
A *Nepeta grandiflora* M. Bieb.
A *Nepeta mussinii* Spreng. ex Henckel
A *Nepeta sibirica* L.
A *Ocimum basilicum* L.
E *Origanum majorana* L.
E *Origanum onites* L.
E *Origanum vulgare* L.
A *Perilla frutescens* (L.) Britton
A *Perovskia atriplicifolia* Benth. ex DC.
E *Phlomis fruticosa* L.
E *Phlomis russeliana* (Sims) Benth.
E *Phlomis tuberosa* L.
A *Physostegia virginiana* (L.) Benth.
A *Plectranthus fruticosus* L'Hér
E *Prunella* × *intermedia* Link
E *Prunella laciniata* (L.) L.
E *Rosmarinus officinalis* L.

E *Salvia aethiopis* L.
E *Salvia amplexicaulis* Lam.
E *Salvia austriaca* Jacq.
A *Salvia coccinea* Juss. ex Murray
E *Salvia dumetorum* Andrz. ex Besser
A *Salvia farinacea* Benth.
E *Salvia fruticosa* Mill.
E *Salvia glutinosa* L.
A *Salvia leucantha* Cav.
A *Salvia microphylla* Kunth
E *Salvia nemorosa* L.
E *Salvia nutans* L.
E *Salvia officinalis* L.
E *Salvia pinnata* L.
E *Salvia pratensis* L.
A *Salvia reflexa* Hornem.
E *Salvia sclarea* L.
A *Salvia sessei* Benth.
A *Salvia spinosa* L.
A *Salvia splendens* Ker-Gawl.
E *Salvia* × *sylvestris* L.
A *Salvia tomentosa* Mill.
E *Salvia triloba* L. fil.
E *Salvia verbenaca* L.
E *Salvia verticillata* L.
E *Salvia viridis* L.
E *Satureja hortensis* L.
E *Satureja montana* L.
E *Scutellaria altissima* L.
E *Scutellaria columnae* All.
E *Scutellaria hastifolia* L.
E *Scutellaria minor* Huds.
E *Sideritis lanata* L.
E *Sideritis montana* L.
E *Stachys alpina* L.
E *Stachys annua* (L.) L.
E *Stachys arvensis* (L.) L.
E *Stachys atherocalyx* K. Koch
E *Stachys byzantina* C. Koch
E *Stachys cretica* L.
E *Stachys germanica* L.
A *Stachys macrantha* (K. Koch) Stearn
E *Stachys officinalis* (L.) Trévis.
E *Stachys recta* L.
A *Stachys sieboldii* Miq.
E *Teucrium botrys* L.
E *Teucrium chamaedrys* L.
E *Teucrium fruticans* L.
E *Teucrium marum* L.
E *Teucrium polium* L.
E *Teucrium resupinatum* Desf.
E *Teucrium scorodonia* L.
A *Thymus* × *citriodorus* Schreb.
E *Thymus odoratissimus* Mill.

E *Thymus pannonicus* All.
E *Thymus praecox* Opiz
E *Thymus pulegioides* L.
E *Thymus serpyllum* L.
E *Thymus vulgaris* L.
A *Ziziphora capitata* L.
Lentibulariaceae
E *Pinguicula crystallina* Sibth. & Sm.
E *Pinguicula grandiflora* Lam.
E *Utricularia gibba* L.
E *Utricularia stygia* G. Thor
Linderniaceae
A *Lindernia dubia* (L.) Pennell
A *Lindernia palustris* Hartmann
E *Lindernia procumbens* (Krocker) Borbs
Martyniaceae
A *Ibicella lutea* (Lindl.) van Eseltine
A *Proboscidea louisianica* (Mill.) Thell.
Oleaceae
A *Forsythia* × *intermedia* Zabel
A *Forsythia suspensa* (Thunb.) Vahl
A *Forsythia viridissima* Lindl.
A *Fraxinus americana* L.
E *Fraxinus angustifolia* Vahl.
E *Fraxinus excelsior* L.
E *Fraxinus ornus* L.
A *Fraxinus pennsylvanica* Marshall
A *Jasminum beesianum* Forrest
A *Jasminum fruticans* L.
A *Jasminum mesnyi* Hance
A *Jasminum nudiflorum* Lindley
A *Jasminum odoratissimum* L.
E *Jasminum officinale* L.
A *Jasminum polyanthum* Franch.
A *Ligustrum henryi* Hemsl.
A *Ligustrum japonicum* Thunb.
A *Ligustrum lucidum* Aiton
E *Ligustrum ovalifolium* Hassk.
A *Ligustrum sinense* Lour.
E *Ligustrum vulgare* L.
E *Olea europaea* L.
A *Syringa chinensis* Willd.
A *Syringa persica* L.
E *Syringa vulgaris* L.
Orobanchaceae
E *Bellardia trixago* (L.) All.
E *Cistanche phelypaea* (L.) Cout.
E *Euphrasia fennica* Kihml.
E *Euphrasia nemorosa* (Pers.) Wallr.
E *Euphrasia stricta* D. Wolff ex
   J.F. Lehm.
A *Euphrasia vernalis* List
E *Lathraea clandestina* L.
E *Melampyrum arvense* L.

- E *Melampyrum barbatum* Willd.
- E *Melampyrum cristatum* L.
- E *Melampyrum nemorosum* L.
- E *Odontites glutinosus* (M. Bieb.) Benth.
- E *Odontites jaubertianus* (Boreau) D. Dietr. ex Walp.
- E *Odontites luteus* (L.) Clairv.
- E *Odontites vernus* (Bellardi) Dumort.
- A *Odontites vulgaris* Moench
- E *Orobanche amethystea* Thuill.
- E *Orobanche coerulescens* Stephan ex Willd.
- E *Orobanche crenata* Forsskl
- E *Orobanche cumana* Wallr.
- E *Orobanche flava* C.F.P. Mart. ex F.W. Schultz
- A *Orobanche gracilis* Sm.
- E *Orobanche hederae* Duby
- E *Orobanche lavandulacea* Reichb.
- E *Orobanche lucorum* A. Braun
- E *Orobanche minor* Sm.
- E *Orobanche purpurea* Jacq.
- E *Orobanche ramosa* L.
- A *Orobanche ramosa nana* (Reuter) Coutinho
- E *Orobanche sanguinea* C. Presl
- A *Orobanche versicolor* F.G. Schultz
- E *Parentucellia viscosa* (L.) Caruel
- E *Rhinanthus alectorolophus* (Scop.) Pollich.
- E *Rhinanthus angustifolius* C. Gmelin
- E *Rhinanthus minor* L.

Pedaliaceae
- A *Sesamum indicum* L.

Phrymaceae
- A *Mazus japonicus* (Thunb.) O. Kuntze
- A *Mazus miquelii* Makino
- A *Mimulus* × *burnetii* hort.
- A *Mimulus cupreus* Regel
- A *Mimulus guttatus* DC.
- A *Mimulus luteus* L.
- A *Mimulus* × *maculosus* hort.
- A *Mimulus moschatus* Douglas ex Lindl.
- A *Mimulus* × *polymaculus* hort.
- A *Mimulus* × *robertsii* A.J. Silverside

Plantaginaceae
- E *Anarrhinum bellidifolium* (L.) Willd.
- E *Antirrhinum granaticum* Rothm.
- E *Antirrhinum majus* L.
- E *Antirrhinum siculum* Mill.
- E *Asarina procumbens* Mill.
- A *Bacopa monnieri* (L.) Pennell
- A *Bacopa rotundifolia* (Michaux) Wettst.
- E *Callitriche brutia* Petagna
- E *Callitriche deflexa* Hegelm.
- E *Chaenorrhinum minus* (L.) Lange
- E *Chaenorrhinum minus litorale* (Willd.) Hayek
- E *Chaenorrhinum origanifolium* (L.) Kostel.
- A *Chelone glabra* L.
- A *Collinsia heterophylla* Buist ex Graham
- E *Cymbalaria hepaticifolia* (Poir.) Wettst.
- E *Cymbalaria muralis* P. Gaertn., B. Mey. & Scherb.
- E *Cymbalaria muralis pubescens* (C.Presl) D.A. Webb
- E *Cymbalaria muralis visianii* (Kuemm. ex Jav.) D. Webb
- E *Cymbalaria pallida* (Ten.) Wettst.
- A *Diascia barberae* Hook. f.
- E *Digitalis ferruginea* L.
- E *Digitalis grandiflora* Mill.
- E *Digitalis lanata* Ehrh.
- E *Digitalis lutea* L.
- E *Digitalis purpurea* L.
- E *Erinus alpinus* L.
- E *Globularia punctata* Lapeyr.
- A *Gratiola neglecta* Torr.
- E *Gratiola officinalis* L.
- A *Gratiola virginiana* L.
- A *Hebe andersonii* (Lindl. & Paxton) Cockayne
- A *Hebe barkeri* (Cockayne) Wall
- A *Hebe brachysiphon* Summerh.
- A *Hebe dieffenbachii* (Benth.) Cockayne & Allen
- A *Hebe* × *franciscana* (Eastwood) Souster
- A *Hebe* × *lewisii* (J. Armstrong) Wall
- A *Hebe salicifolia* (G. Forster) Pennell
- E *Hippuris vulgaris* L.
- E *Kickxia cirrhosa* (L.) Fritsch
- E *Kickxia commutata* (Bernh. ex Rchb.) Fritsch
- E *Kickxia elatine* (L.) Dumort.
- E *Kickxia elatine crinita* (Mabille) Greuter
- E *Kickxia lanigera* (Desf.) Hand.-Mazz.
- E *Kickxia spuria* (L.) Dumort.
- A *Limnophila* × *ludoviciana* Thieret
- E *Limosella aquatica* L.
- E *Linaria angustissima* (Loisel.) Borb.
- E *Linaria arenaria* DC.
- E *Linaria arvensis* (L.) Desf.
- E *Linaria caesia* (Pers.) DC. ex Chav.
- E *Linaria chalepensis* (L.) Mill.
- A *Linaria* × *cornubiensis* Druce
- A *Linaria* × *dominii* Druce

- E *Linaria genistifolia* (L.) Mill.
- E *Linaria hirta* (L.) Moench
- E *Linaria incarnata* (Vent.) Spreng.
- A *Linaria* × *oligotricha* Borbs
- E *Linaria pelisseriana* (L.) Mill.
- E *Linaria purpurea* (L.) Mill.
- A *Linaria reflexa* (L.) Desf.
- E *Linaria repens* (L.) Mill.
- A *Linaria* × *sepium* Allman
- E *Linaria simplex* Desf.
- E *Linaria spartea* (L.) Willd.
- E *Linaria supina* (L.) Chazelles
- E *Linaria triornithophora* (L.) Willd.
- E *Linaria vulgaris* Mill.
- E *Littorella uniflora* (L.) Ascherson
- A *Lophospermum erubescens* D. Don ex Sweet
- A *Maurandya scandens* (Cav.) Pers.
- E *Misopates calycinum* Rothm.
- E *Misopates orontium* (L.) Raf.
- E *Plantago afra* L.
- A *Plantago alpina* L.
- E *Plantago arenaria* Waldst. & Kit.
- A *Plantago aristata* A. Michaux
- E *Plantago bellardii* All.
- E *Plantago coronopus* L.
- A *Plantago gentianoides* Sibth. & Sm.
- E *Plantago heterophylla* Nutt.
- E *Plantago holosteum* Scop.
- E *Plantago lagopus* L.
- E *Plantago lanceolata* L.
- E *Plantago loeflingii* L.
- E *Plantago major* L.
- E *Plantago maxima* Juss. ex Jacq.
- E *Plantago media* L.
- A *Plantago* × *mixta* Domin
- E *Plantago* × *moravica* Chrtek
- A *Plantago myosuros* Lam.
- E *Plantago ovata* Forssk.
- E *Plantago sempervirens* Crantz
- E *Plantago serpentina* All.
- A *Plantago virginica* L.
- E *Pseudolysimachion incanum* (L.) Holub
- A *Pseudolysimachion* × *neglectum* (Vahl) Trávníček
- A *Russelia equisetiformis* Schlecht. & Cham.
- A *Sibthorpia europaea* L.
- A *Sibthorpia peregrina* L.
- A *Sutera cordata* (Thunb.) O. Kuntze
- E *Veronica acinifolia* L.
- E *Veronica agrestis* L.
- E *Veronica anagallis-aquatica* L.
- E *Veronica arvensis* L.
- E *Veronica austriaca* L.
- E *Veronica austriaca jacquinii* (Baumg.) Eb. Fischer
- E *Veronica austriaca teucrium* (L.) D.A. Webb
- E *Veronica beccabunga* L.
- A *Veronica bornmuelleri* Hausskn.
- A *Veronica campylopoda* Boiss.
- E *Veronica catenata* Pennell
- A *Veronica ceratocarpa* C.A. Meyer
- E *Veronica chamaedrys* L.
- E *Veronica crista-galli* Steven
- E *Veronica cymbalaria* Bodard
- E *Veronica filiformis* Sm.
- E *Veronica fruticans* Jacq.
- E *Veronica fruticulosa* L.
- A *Veronica gentianoides* Vahl
- E *Veronica glauca* Sibth. & Smith
- E *Veronica hederifolia* L.
- E *Veronica longifolia* L.
- A *Veronica* × *macrosperma* Schust.
- E *Veronica officinalis* L.
- E *Veronica opaca* Fries
- E *Veronica peregrina* L.
- E *Veronica persica* Poiret
- E *Veronica polita* Fries
- E *Veronica praecox* All.
- E *Veronica prostrata* L.
- E *Veronica repens* Clarion ex DC.
- E *Veronica serpyllifolia* L.
- A *Veronica* × *sooiana* Borsos
- E *Veronica spicata* L.
- E *Veronica triphyllos* L.
- E *Veronica verna* L.

Scrophulariaceae
- A *Buddleja albiflora* Hemsl.
- A *Buddleja alternifolia* Maxim.
- A *Buddleja davidii* Franchet
- A *Buddleja globosa* J. Hope
- A *Buddleja japonica* Hemsl.
- A *Buddleja* × *weyeriana* Hort.
- A *Hebenstretia dentata* L.
- A *Myoporum insulare* R. Br.
- A *Myoporum laetum* G. Forster
- A *Myoporum tenuifolium* G. Forster
- A *Myoporum tetrandrum* (Labil.) Domin
- A *Phygelius capensis* E. Meyer ex Benth.
- E *Scrophularia auriculata* L.
- E *Scrophularia canina* L.
- A *Scrophularia lanceolata* Pursh
- E *Scrophularia scopolii* Hoppe
- E *Scrophularia scorodonia* L.
- E *Scrophularia vernalis* L.
- E *Verbascum banaticum* Schrad.

11 List of Species Alien in Europe and to Europe

- E *Verbascum blattaria* L.
- E *Verbascum bombyciferum* Boiss.
- E *Verbascum chaixii* Villars
- E *Verbascum creticum* (L.) Kuntze
- E *Verbascum densiflorum* Bertol.
- A *Verbascum* × *duernsteinense* Teyber
- A *Verbascum* × *interjectum* Pfund
- A *Verbascum* × *intermedium* Rupr. ex Bercht. & Pfund
- A *Verbascum* × *kerneri* Fritsch
- A *Verbascum* × *lemaitrei* Boreau
- E *Verbascum levanticum* I.K. Ferguson
- E *Verbascum lychnitis* L.
- E *Verbascum nigrum* L.
- A *Verbascum orientale* (L.) All.
- E *Verbascum phlomoides* L.
- E *Verbascum phoeniceum* L.
- E *Verbascum pulverulentum* Vill.
- E *Verbascum pyramidatum* M. Bieb.
- E *Verbascum sinuatum* L.
- E *Verbascum speciosum* Schrader
- E *Verbascum thapsus* L.
- E *Verbascum virgatum* Stokes

Verbenaceae
- A *Aloysia citrodora* Ortega ex Pers.
- A *Caryopteris* × *clandonensis* A. Simmonds
- A *Clerodendrum trichotomum* Thunb.
- A *Duranta erecta* L.
- A *Glandularia canadensis* (L.) Nutt.
- A *Holmskioldia sanguinea* Retz.
- A *Lantana camara* L.
- A *Lantana montevidensis* (Sprengel) Briq.
- E *Lippia canescens* Kunth
- A *Lippia triphylla* (L'Hér.) Kuntze
- A *Petrea volubilis* L.
- A *Verbena aristigera* S. Moore
- A *Verbena bonariensis* L.
- A *Verbena bracteata* Lag. & Rodr.
- A *Verbena canadensis* (L.) Britt.
- A *Verbena halei* Small
- A *Verbena hastata* L.
- A *Verbena* × *hybrida* hort.
- A *Verbena litoralis* Kunth
- E *Verbena officinalis* L.
- A *Verbena peruviana* (L.) Druce
- A *Verbena rigida* Sprengel
- E *Verbena supina* L.
- A *Verbena urticifolia* L.
- E *Vitex agnus-castus* L.

Magnoliophytina, Dicotyledonae, Laurales
Calycanthaceae
- A *Calycanthus floridus* L.
- A *Chimonanthus fragrans* Lindl.

Lauraceae
- A *Cinnamomun zeylanicum* Nees
- E *Laurus nobilis* L.
- E *Ocotea foetens* (Ait.) Benth. & Hook. f.
- A *Persea americana* Mill.
- A *Persea indica* (L.) Spreng.

Magnoliophytina, Dicotyledonae, Magnoliales
Magnoliaceae
- A *Liriodendron tulipifera* L.
- A *Magnolia grandiflora* L.
- A *Magnolia liliiflora* Desr.
- A *Magnolia stellata* (Siebold & Zucc.) Maxim.

Magnoliophytina, Dicotyledonae, Malpighiales
Elatinaceae
- A *Bergia capensis* L.
- E *Elatine alsinastrum* L.
- A *Elatine ambigua* Wight
- E *Elatine hungarica* Moesz.
- E *Elatine macropoda* Guss. Fl. Sic. Prodr (1827)

Euphorbiaceae
- A *Acalypha australis* L.
- A *Acalypha indica* L.
- A *Acalypha virginica* L.
- E *Andrachne telephioides* L.
- A *Breynia disticha* J.R. & G. Forst.
- A *Chamaesyce glomerifera* Millsp.
- E *Chamaesyce glyptosperma* (Engelm.) Small
- E *Chamaesyce humifusa* (Willd.) Prokh.
- A *Chamaesyce humistrata* (Engelm.) Small
- A *Chamaesyce maculata* (L.) Small
- E *Chamaesyce nutans* (Lag.) Small
- A *Chamaesyce polygonifolia* (L.) Small.
- E *Chamaesyce prostrata* (Ait.) Small
- A *Chamaesyce serpens* (Kunth) Small
- E *Chrozophora tinctoria* (L.) Raf.
- E *Euphorbia acuminata* Lam.
- A *Euphorbia agraria* M.Bieb.
- A *Euphorbia aleppica* L.
- E *Euphorbia amygdaloides* L.
- A *Euphorbia candelabrum* Kotschy
- E *Euphorbia carpatica* Wol.
- E *Euphorbia chamaesyce* L.
- E *Euphorbia characias* L.
- E *Euphorbia corallioides* L.
- A *Euphorbia cotinifolia* L.
- E *Euphorbia cyparissias* L.
- A *Euphorbia dentata* Michaux

E *Euphorbia dulcis* L.
E *Euphorbia epithymoides* L.
E *Euphorbia esula* L.
E *Euphorbia exigua* L.
E *Euphorbia falcata* L.
E *Euphorbia* × *gayeri* Boros & Soo
E *Euphorbia helioscopia* L.
A *Euphorbia ingens* E. Mey
E *Euphorbia lagascae* Sprengel
A *Euphorbia lasiocarpa* K. Koch Klotzsch
E *Euphorbia lathyris* L.
A *Euphorbia leptocaula* Boiss.
A *Euphorbia marginata* Pursh.
A *Euphorbia milii* Des Moul. ex Boiss.
E *Euphorbia myrsinites* L.
E *Euphorbia oblongata* Griseb.
E *Euphorbia palustris* L.
E *Euphorbia peplis* L.
E *Euphorbia peplus* L.
E *Euphorbia platyphyllos* L.
E *Euphorbia* × *pseudoesula* Schur
A *Euphorbia pseudograntii* Bruyns
E *Euphorbia* × *pseudovirgata* (Schur) Soó
A *Euphorbia pulcherrima* Willd. ex Klotzsch
E *Euphorbia rigida* M.Bieb. Loisel.
E *Euphorbia robbiae* Turrill
E *Euphorbia segetalis* L.
E *Euphorbia seguieriana* Neck.
A *Euphorbia serrulata* Thuill.
E *Euphorbia taurinensis* All.
A *Euphorbia trigona* Haw.
E *Euphorbia verrucosa* L. (1759)
E *Euphorbia virgata* Waldst. & Kit.
E *Mercurialis annua* L.
E *Mercurialis ovata* Sternb. & Hoppe
A *Phyllanthus niruri* L.
A *Phyllanthus rotundifolius* Willd.
E *Phyllanthus tenellus* Roxb.
A *Poinsettia heterophylla* (L.) Klotzsch & Carcke
A *Ricinus communis* L.
Hypericaceae
E *Hypericum androsaemum* L.
E *Hypericum atomarium* Boiss.
E *Hypericum calycinum* L.
A *Hypericum canadense* L.
A *Hypericum forrestii* (Chitt.) N. Robson
A *Hypericum gentianoides* (L.) Britton, Sterns & Poggenb.
A *Hypericum gymnanthum* Engelm. & A. Gray
E *Hypericum* × *hidcoteense* Hilling ex Geerinck

E *Hypericum hircinum* L.
E *Hypericum* × *inodorum* Miller
E *Hypericum maculatum* Cratz
A *Hypericum* × *medium* Peterm.
E *Hypericum mutilum* L.
E *Hypericum nummularium* L.
E *Hypericum perforatum* L.
A *Hypericum pseudohenryi* N. Robson
E *Hypericum pulchrum* L.
E *Hypericum triquetrifolium* Turra
E *Hypericum* × *ylosteifolium* (Spach) N. Robson
Linaceae
E *Asterolinon linum-stellatum* (L.) Duby
E *Linum austriacum* L.
E *Linum bienne* Miller
A *Linum grandiflorum* Desf.
E *Linum narbonense* L.
E *Linum perenne* L.
E *Linum trigynum* L.
E *Linum usitatissimum* L.
A *Reinwardtia indica* Dum.
Ochnaceae
A *Ochna atropurpurea* DC.
A *Ochna serrulata* (Hochst.) Walp.
Passifloraceae
A *Passiflora antioquiensis* Karst.
A *Passiflora caerulea* L.
A *Passiflora edulis* Sims.
A *Passiflora mollissima* (Humb., Bonpl. & Kunth) Bailey
A *Passiflora morifolia* Masters
A *Passiflora suberosa* L.
A *Passiflora subpeltata* Ortega
Salicaceae
E *Populus alba* L.
A *Populus balsamifera* L.
A *Populus* × *berolinensis* Dippel
A *Populus* × *canadensis* Moench
E *Populus canescens* (Ait.) Smith
A *Populus deltoides* Bartr. ex Marsh.
A *Populus euphratica* Olivier
A *Populus* × *generosa* A. Henry
A *Populus* × *jackii* Sarg.
A *Populus laurifolia* Ledeb.
E *Populus nigra* L.
A *Populus simonii* Carr.
A *Populus suaveolens* Fisch.
E *Populus tremula* L.
A *Populus trichocarpa* Torr. & A. Gray
E *Populus* × *canescens* (Aiton) Sm.
A *Populus yunnanensis* Dode
E *Salix alba* L.
E *Salix* × *alopecuroides* Tausch

- E *Salix × angusensis* Rech. f.
- A *Salix babylonica* L.
- E *Salix × basfordiana* Scaling ex J. Salter
- E *Salix × chrysocoma* Dode
- E *Salix cinerea* L.
- E *Salix daphnoides* Villars
- E *Salix × dasyclados* Wimm.
- E *Salix × ehrhartiana* Smith
- E *Salix elaeagnos* Scop.
- E *Salix elaeagnos angustifolia* (Cariot) Rech. f.
- A *Salix eriocephala* Michx.
- E *Salix × forbyana* Smith
- E *Salix fragilis* L.
- E *Salix × friesiana* Anderss.
- A *Salix × fruticosa* Doell
- A *Salix × hirtei* Strähler
- A *Salix irrorata* Anderss.
- E *Salix × meyeriana* Rostk. ex Willd.
- E *Salix × mollissima* Ehrh.
- E *Salix myrsinifolia* Salisb.
- A *Salix × pendulina* Wender.
- E *Salix pentandra* L.
- E *Salix purpurea* L.
- E *Salix × rubens* Schrank
- A *Salix × rubra* Hudson
- A *Salix sachalinensis* Schmidt
- A *Salix × seminigricans* E.G. Camus & A. Camus
- A *Salix × sepulcralis* Simonk.
- E *Salix × sericans* Tausch ex A. Kerner
- A *Salix × smithiana* Willd.
- E *Salix × stipularis* Smith
- A *Salix × taylorii* Rech. f.
- E *Salix triandra* L.
- E *Salix viminalis* L.

Violaceae
- E *Viola arvensis* Murr.
- E *Viola cornuta* L.
- A *Viola curtisii* E. Forster ex Sm.
- A *Viola × haynaldii* Wiesb.
- E *Viola hirta* L.
- A *Viola × hungarica* Degen & Sabr.
- E *Viola jooi* Janka
- A *Viola × kerneri* Wiesb.
- A *Viola obliqua* Hill
- E *Viola odorata* L.
- A *Viola × pluricaulis* Borbás
- A *Viola × poelliana* J. Murr.
- A *Viola × porphyrea* Uechtr.
- A *Viola × scabra* F. Braun
- E *Viola selkirkii* Pursh ex Goldie
- A *Viola sororia* Willd.
- A *Viola × sourekii* Procházka
- E *Viola suavis* M. Bieb.
- E *Viola tricolor* L.
- A *Viola × vindobonensis* Wiesb.
- A *Viola × wittrockiana* Gams

Magnoliophytina, Dicotyledonae, Malvales
Cistaceae
- E *Cistus albidus* L.
- E *Cistus ladanifer* L.
- E *Cistus laurifolius* L.
- E *Cistus monspeliensis* L.
- E *Cistus psilosepalus* Sweet
- E *Cistus salviifolius* L.
- A *Helianthemum syriacum* (Jacq.) Dum.-Cours.
- E *Tuberaria guttata* (L.) Fourr.

Malvaceae
- A *Abutilon grandifolium* (Willd.) Sweet
- E *Abutilon × hybridum* hort. ex Siebert & Voss
- A *Abutilon sonneratianum* (Cav.) Sweet
- A *Abutilon striatum* Dickson ex Lindl.
- A *Abutilon theophrasti* Medik.
- A *Alcea biennis* Winterl.
- A *Alcea rosea* L.
- A *Althaea armeniaca* Ten.
- E *Althaea cannabina* L.
- E *Althaea hirsuta* L.
- A *Althaea longiflora* Boiss. & Reut.
- E *Althaea officinalis* L.
- A *Althaea × taurinensis* DC.
- E *Anoda cristata* (L.) Schlecht.
- A *Brachychiton acerifolius* (A. Cunn.) MacArthur
- A *Brachychiton populneus* (Schott. & Endl.) R. Br.
- A *Chorisia speciosa* St. Hil.
- A *Dombeya wallichii* (Lindl.) Benth. & Hook. f.
- A *Gossypium arboreum* L.
- A *Gossypium barbadense* L.
- A *Gossypium herbaceum* L.
- A *Gossypium hirsutum* L.
- A *Hibiscus elatus* Swartz
- A *Hibiscus palustris* L.
- A *Hibiscus rosa-sinensis* L.
- A *Hibiscus syriacus* L.
- E *Hibiscus trionum* L.
- A *Hoheria populnea* Cunn.
- A *Kitaibela vitifolia* Willd.
- A *Lagunaria patersonii* (Andrews) G. Don f.
- E *Lavatera arborea* L.
- A *Lavatera × clementii* Cheek

E *Lavatera cretica* L.
E *Lavatera maroccana* (Batt. & Trabut) Maire
A *Lavatera plebeia* Sims
E *Lavatera punctata* All.
E *Lavatera thuringiaca* L.
E *Lavatera trimestris* L.
A *Lawrencia glomerata* Hooker
E *Malope malacoides* L.
E *Malope trifida* Cav.
E *Malva alcea* L.
E *Malva cretica* Cav.
E *Malva cretica althaeoides* (Cav.) Dalby
A *Malva × hybrida* Celak.
E *Malva moschata* L.
E *Malva neglecta* Wallr.
E *Malva nicaeensis* All.
E *Malva parviflora* L.
E *Malva pusilla* Sm.
E *Malva sylvestris* L.
E *Malva verticillata* L.
A *Malva × zoernigii* Fleischmann
A *Malvastrum americanum* (L.) Torr.
A *Malvastrum coromandelianum* (L.) Garcke
A *Malvella leprosa* (Gómez Ortega) Krapov.
E *Malvella sherardiana* (L.) Jaub. & Spach
A *Modiola caroliniana* (L.) G. Don fil.
A *Sida acuta* Burm. f.
A *Sida hermaphrodita* (L.) Rusby
A *Sida rhombifolia* L.
A *Sida spinosa* L.
A *Sidalcea candida* A. Gray
A *Sidalcea malviflora* (DC.) A. Gray ex Benth.
A *Sidalcea oregana* (Nutt.) A. Gray
A *Sidastrum paniculatum* (L.) Fryxell
A *Sphaeralcea bonariensis* (Cav.) Griseb.
A *Sphaeralcea miniata* (Cav.) Spach
A *Waltheria indica* L.
Thymelaeaceae
E *Daphne laureola* L.
A *Gnidia polystachya* P.J. Bergius
E *Thymelaea gussonei* Boreau
Tiliaceae
A *Corchorus depressus* (L.) Stocks
A *Corchorus olitorius* L.
A *Sparmannia africana* L. f.
A *Tilia americana* L.
E *Tilia × europaea* L.
E *Tilia platyphyllos* Scop.
E *Tilia tomentosa* Moench

Magnoliophytina, Dicotyledonae, Myrtales
Combretaceae
A *Quisqualis indica* L.
Lythraceae
A *Ammania baccifera* L.
A *Ammania senegalensis* Lam.
A *Ammania verticillata* (Ard.) Lam.
A *Ammannia auriculata* Willd.
A *Ammannia coccinea* Rottb.
A *Ammannia robusta* Heer & Regel
A *Cuphea lanceolata* Aiton ex Aiton
A *Cuphea viscosissima* Jacq.
A *Lagerstroemia indica* L.
A *Lawsonia inermis* L.
E *Lythrum acutangulatum* Lag.
A *Lythrum borysthenicum* (Schrank) Litv.
E *Lythrum hyssopifolium* L.
E *Lythrum junceum* Banks & Sol.
E *Lythrum portula* (L.) D.A. Webb
A *Lythrum thymifolia* L.
E *Lythrum virgatum* L.
E *Lythrum volgense* D.A. Webb
A *Nesaea myrtifolia* Desf. ex A. St.-Hil
A *Rotala densiflora* (Roem. & Schult.) Koehne
A *Rotala filiformis* (Bellardi) Hiern
A *Rotala indica* (Willd.) Koehne
A *Rotala macrandra* L.
A *Rotala ramosior* (L.) Koehne
E *Trapa natans* L.
Melastomataceae
A *Tibouchina urvilleana* (DC.) Cogn.
Myrtaceae
A *Callistemon citrinus* (Curtis) Stapf
A *Callistemon rigidus* R. Br.
A *Callistemon viminalis* (Gaertner) G. Don
A *Eucalyptus amygdalinus* Labil.
A *Eucalyptus camaldulensis* Dehnh.
A *Eucalyptus ficifolia* F. Müll.
A *Eucalyptus globulus* Labill.
A *Eucalyptus gomphocephalus* DC.
A *Eucalyptus gunnii* Hook. f.
A *Eucalyptus johnstonii* Maiden
A *Eucalyptus pulchella* Desf.
A *Eucalyptus robustus* Sm.
A *Eucalyptus sideroxylon* A. Cunn. ex Woolls
A *Eucalyptus tereticornis* Sm.
A *Eugenia uniflora* L.
A *Feijoa sellowiana* (O.Berg) O. Berg
A *Leptospermum lanigerum* (Aiton) Smith
A *Leptospermum scoparium* Forster & G. Forster

- A *Metrosideros excelsa* Sol. ex Gaertn.
- E *Myrtus communis* L.
- A *Myrtus luma* Molina
- A *Psidium guajava* L.
- A *Psidium guineense* Swartz
- A *Syzygium jambos* (L.) Alston

Onagraceae
- E *Chamaenerion angustifolium* (L.) Scop.
- A *Clarkia amoena* (Lehm.) Nelson & J.F. Macbr.
- A *Clarkia pulchella* Pursh
- A *Clarkia unguiculata* Lindl.
- A *Epilobium adenocaulon* Hausskn.
- E *Epilobium brachycarpum* C. Presl
- A *Epilobium* × *brunnatum* Kitchener & McKean
- A *Epilobium brunnescens* (Cockayne) Raven & Engelhorn
- A *Epilobium* × *chateri* Kitchener & McKean
- A *Epilobium ciliatum* Raf.
- A *Epilobium* × *confusilobum* Kitchener & McKean
- E *Epilobium dodonaei* Vill.
- A *Epilobium* × *floridulum* Smejkal
- A *Epilobium* × *fossicola* Smejkal
- A *Epilobium glandulosum* Lehm.
- E *Epilobium hirsutum* L.
- A *Epilobium* × *iglaviense* Smejkal
- A *Epilobium* × *interjectum* Smejkal
- A *Epilobium* × *josefi-holubi* Krahulec
- A *Epilobium komarovianum* Lév.
- A *Epilobium* × *novae-civitatis* Smejkal
- A *Epilobium* × *nutantiflorum* Smejkal
- A *Epilobium* × *obscurescens* Kitchener & McKean
- E *Epilobium obscurum* Schreb.
- E *Epilobium parviflorum* Schreb.
- A *Epilobium pedunculare* Cunn.
- A *Epilobium* × *prochazkae* Krahulec
- A *Epilobium pseudorubescens* A. Skvorts.
- E *Epilobium roseum* Schreb.
- E *Epilobium tetragonum* L.
- A *Epilobium* × *vicinum* Smejkal
- A *Epilobium watsonii* Barbey
- A *Fuchsia arborescens* Sims
- A *Fuchsia boliviana* Carrire
- A *Fuchsia coccinea* Aiton
- A *Fuchsia 'Corallina'* (*F. cordifolia* x *globosa*) hort. ex Lynch
- A *Fuchsia magellanica* Lam.
- A *Gaura biennis* L.
- A *Gaura lindheimeri* Engelm. & A.Gray
- A *Lopezia coronata* Andrews
- A *Lopezia racemosa* Cav.
- A *Ludwigia alternifolia* L.
- A *Ludwigia grandiflora* (M. Micheli) Greuter & Burdet
- A *Ludwigia peploides* (Kunth) P.H. Raven
- A *Ludwigia repens* J.R. Forst.
- C *Oenothera* × *acerviphila* Rostanski
- A *Oenothera* × *acutifolia* Rostanski
- A *Oenothera adriatica* Soldano
- A *Oenothera affinis* Cambess.
- C *Oenothera* × *albipercurva* Renner
- A *Oenothera ammophila* Focke
- A *Oenothera biennis* L.
- A *Oenothera* × *braunii* Döll
- A *Oenothera brevispicata* Hudziok
- A *Oenothera cambrica* Rostanski
- A *Oenothera canovirens* Steele
- A *Oenothera carinthiaca* Rostanski
- A *Oenothera* × *clavifera* Hudziok
- A *Oenothera compacta* Hudziok
- A *Oenothera coronifera* Renner
- A *Oenothera cruciata* Nutt. ex G. Don
- A *Oenothera deflexa* Gates
- A *Oenothera* × *drawertii* Renner ex Rostanski
- A *Oenothera drummondii* Hook.
- A *Oenothera elata* Humb., Bonpl. & Kunth
- A *Oenothera ersteinensis* R. Linder & R. Jean
- A *Oenothera fallacoides* Soldano & Rostanski
- A *Oenothera* × *fallax* Renner
- C *Oenothera* × *flaemingina* Hudziok
- A *Oenothera flava* (A. Nels.) Garrett
- A *Oenothera fruticosa* L.
- A *Oenothera glazioviana* Micheli
- A *Oenothera grandiflora* L'Hér. ex Aiton
- A *Oenothera hazelae* Gates
- A *Oenothera* × *heiniana* Teyber
- A *Oenothera humifusa* Nutt.
- A *Oenothera indecora* Camb.
- A *Oenothera indivisa* Hudziok
- A *Oenothera issleri* Renner ex Rostanski
- A *Oenothera italica* Rostanski & Soldano
- A *Oenothera laciniata* Hill
- A *Oenothera longiflora* L.
- A *Oenothera macrocarpa* Nutt.
- A *Oenothera moravica* Jehlík & Rostanski
- A *Oenothera nuda* Renner
- A *Oenothera oakesiana* (A. Gray) Robbins ex S. Watson & Coulter
- A *Oenothera* × *oehlkersii* Kappus
- A *Oenothera paradoxa* Hudziok

- A *Oenothera parviflora* L.
- A *Oenothera pedemontana* Soldano
- A *Oenothera perangusta* R.R. Gates
- A *Oenothera* × *pseudocernua* Hudziok
- C *Oenothera* × *pseudochicaginensis* Rostanski
- A *Oenothera* × *punctulata* Rostanski & Gutte
- A *Oenothera pycnocarpa* G.F. Atkinson & Bartlett
- A *Oenothera rosea* L'Hér. ex Aiton
- A *Oenothera rubricaulis* Klebahn
- A *Oenothera sesitensis* Soldano
- A *Oenothera speciosa* Nutt.
- A *Oenothera stricta* Ledeb. ex Link
- A *Oenothera stucchii* Soldano
- A *Oenothera suaveolens* Desf. ex Pers.
- A *Oenothera subterminalis* Gates
- A *Oenothera tetraptera* Cav.
- A *Oenothera turoviensis* Rostanski
- A *Oenothera victorini* Gates & Catcheside
- A *Oenothera villosa* Thunb.
- A *Oenothera* × *wienii* Renner ex Rostanski

Punicaceae
- E *Punica granatum* L.

Magnoliophytina, Dicotyledonae, Nymphaeales

Cabombaceae
- A *Cabomba caroliniana* A. Gray

Nymphaeaceae
- A *Nuphar advena* (Aiton) Aiton f.
- A *Nuphar japonica* DC.
- E *Nuphar pumila* (Timm.) DC
- E *Nymphaea alba* L.
- A *Nymphaea* × *marliacea* hort.
- A *Nymphaea mexicana* Zucc.
- A *Nymphaea rubra* Roxb. ex Salisb.

Magnoliophytina, Dicotyledonae, Oxidales

Oxalidaceae
- A *Oxalis articulata* Savigny
- A *Oxalis bowiei* Lindl.
- E *Oxalis corniculata* L.
- A *Oxalis debilis* Humb., Bonpl. & Kunth
- A *Oxalis decaphylla* Kunth
- A *Oxalis dillenii* Jacq.
- A *Oxalis exilis* Cunn.
- A *Oxalis incarnata* L.
- E *Oxalis latifolia* Kunth
- E *Oxalis parvifolia* DC.
- A **Oxalis pes-caprae** L.
- A *Oxalis purpurata* Jacq.
- A *Oxalis rosea* Jacq.
- E *Oxalis stricta* L.
- A *Oxalis tetraphylla* Cav.
- A *Oxalis triangularis* A. St. Hil.
- A *Oxalis tuberosa* Mol.
- A *Oxalis* × *uittienii* J. Jansen
- A *Oxalis valdiviensis* Barnéoud
- A *Oxalis variabilis* Jacq.

Magnoliophytina, Dicotyledonae, Piperales

Aristolochiaceae
- E *Aristolochia clematitis* L.
- A *Aristolochia macrophylla* Lam.
- E *Aristolochia rotunda* L.
- E *Aristolochia sempervirens* L.
- E *Asarum europaeum* L.

Saururaceae
- A *Saururus cernuus* L.

Magnoliophytina, Dicotyledonae, Proteales

Nelumbonaceae
- A *Nelumbo nucifera* Gaertn.

Platanaceae
- A *Platanus* × *hispanica* Mill. ex Münchh.
- A *Platanus occidentalis* L.
- E *Platanus orientalis* L.
- A *Banksia integrifolia* L. f.
- A *Grevillea robusta* A. Cunn. ex R. Br.
- A *Hakea salicifolia* (Vent.) B.L. Burtt
- A *Hakea sericea* Schrad.
- A *Protea cynaroides* L.

Magnoliophytina, Dicotyledonae, Ranuncales

Berberidaceae
- A *Berberis aggregata* C. Schneider
- A *Berberis buxifolia* Lam.
- A *Berberis darwinii* Hook.
- A *Berberis gagnepainii* C. Schneider
- A *Berberis julianae* C. Schneider
- A *Berberis* × *ottawensis* C.K. Schneider
- A *Berberis* × *stenophylla* Lindley
- E *Berberis thunbergii* L.
- E *Berberis vulgaris* L.
- A *Berberis wilsoniae* Hemsley
- E *Epimedium alpinum* L.
- E *Leontice leontopetalum* L.
- A *Mahonia aquifolium* (Pursh) Nutt.
- A *Mahonia* × *decumbens* Stace
- A *Mahonia* × *domestica* Ambrózy
- A *Mahonia repens* (Lindl.) G. Don

Fumariaceae
- E *Ceratocapnos claviculata* L.
- E *Corydalis alba* Mansfeld
- E *Corydalis cava* (L.) Schweigger & Koerte
- A *Corydalis cheilanthifolia* Hemsley

A *Corydalis nobilis* (L.) Pers.
E *Corydalis solida* (L.) Clairv.
A *Dicentra formosa* (Haw.) Walp.
A *Dicentra spectabilis* (L.) Lemaire
E *Fumaria agraria* Lag.
E *Fumaria bastardii* Boreau
E *Fumaria capreolata* L.
E *Fumaria densiflora* DC.
E *Fumaria kralikii* Jord.
E *Fumaria martinii* Clavaud
E *Fumaria muralis* Sond. ex W.D.J. Koch
E *Fumaria officinalis* L.
A *Fumaria* × *painteri* Pugsley
E *Fumaria parviflora* Lam.
E *Fumaria petteri* Reichb.
E *Fumaria rostellata* Knaf
E *Fumaria rupestris* Boiss. & Reut.
E *Fumaria schleicheri* Soy.-Will.
A *Fumaria schrammii* (Aschers.) Velen.
E *Fumaria sepium* Boiss.
E *Fumaria vaillantii* Loisel.
E *Pseudofumaria alba* (Miller) Lidén.
E *Pseudofumaria lutea* (L.) Borkh.
Lardizabalaceae
A *Akebia quinata* (Houtt.) Decne.
Papaveraceae
A *Argemone mexicana* L.
A *Capnoides sempervirens* (L.) Borkh.
E *Chelidonium majus* L.
A *Eschscholzia californica* Cham.
E *Glaucium corniculatum* (L.) J.H. Rudolph
E *Glaucium flavum* Crantz
A *Hunnemannia fumariifolia* Sweet
A *Hypecoum imberbe* Sm.
E *Hypecoum pendulum* L.
A *Lamprocapnos spectabilis* (L.) Endl.
A *Macleaya cordata* (Willd.) R. Br.
A *Macleaya* × *kewensis* Turrill
E *Meconopsis cambrica* (L.) Vig.
E *Papaver apulum* Tenore
E *Papaver argemone* L.
E *Papaver atlanticum* (Ball) Cosson
A *Papaver commutatum* Fisch. & C.A. Mey.
E *Papaver dubium* L.
A *Papaver* × *hungaricum* Borbás
E *Papaver hybridum* L.
A *Papaver orientale* L.
A *Papaver pilosum* Sibth. & Sm.
A *Papaver pinnatifidum* Moris
E *Papaver pseudoorientale* (Fedde) Medw.
E *Papaver rhoeas* L.
E *Papaver somniferum* L.
E *Papaver somniferum setigerum* (DC.) Arcang.

A *Papaver strigosum* (Boen.) Schur
E *Roemeria hybrida* (L.) DC.
E *Roemeria refracta* DC.
Ranunculaceae
A *Aconitum* × *cammarum* L.
E *Aconitum napellus* L.
A *Aconitum* × *stoerkianum* L.
E *Aconitum variegatum* L.
E *Adonis aestivalis* L.
E *Adonis annua* L.
E *Adonis flammea* Jacq.
E *Adonis microcarpa* DC.
E *Anemone apennina* L.
E *Anemone blanda* Schott & Kotschy
E *Anemone coronaria* L.
A *Anemone dichotoma* L.
A *Anemone hupehensis* Lemoine
A *Anemone* × *hybrida* Paxton
E *Anemone nemorosa* L.
E *Anemone ranunculoides* L.
E *Anemone sylvestris* L.
A *Anemone tomentosa* (Maxim.) C. Pi
A *Anemonella thalictroides* (L.) Spach
E *Aquilegia atrata* Koch
E *Aquilegia pyrenaica* DC.
E *Aquilegia vulgaris* L.
E *Ceratocephala testiculata* (Crantz) Roth
E *Clematis flammula* L.
A *Clematis glauca* Willd.
E *Clematis integrifolia* L.
A *Clematis* × *jackmanii* T. Moore
A *Clematis montana* Buch.-Ham. ex DC.
A *Clematis orientalis* L.
E *Clematis recta* L.
E *Clematis vitalba* L.
E *Clematis viticella* L.
E *Consolida ajacis* (L.) Schur
E *Consolida hispanica* (Costa) Greuter & Burdet
E *Consolida regalis* S.F. Gray
A *Consolida uechtritziana* (Panc.) Soó
A *Delphinium* × *cultorum* Voss
E *Delphinium elatum* L.
E *Eranthis hyemalis* (L.) Salisb.
E *Helleborus argutifolius* Viv.
E *Helleborus foetidus* L.
A *Helleborus* × *hybridus*
E *Helleborus niger* L.
A *Helleborus odorus* Waldst. & Kit.
E *Helleborus orientalis* Lam.
E *Helleborus viridis* L.
E *Hepatica nobilis* Schreber
E *Myosurus minimus* L.
E *Nigella agrestis* J. & C. Presl
E *Nigella damascena* L.

- E *Nigella hispanica* L.
- E *Nigella sativa* L.
- E *Pulsatilla vulgaris* Mill.
- E *Ranunculus aconitifolius* L.
- E *Ranunculus acris* L.
- E *Ranunculus arvensis* L.
- E *Ranunculus bulbosus* L.
- A *Ranunculus cymbalaria* Pursh
- E *Ranunculus flammula* L.
- E *Ranunculus gramineus* L.
- E *Ranunculus hederaceus* L.
- E *Ranunculus illyricus* L.
- E *Ranunculus lanuginosus* L.
- A *Ranunculus lateriflorus* DC.
- E *Ranunculus macrophyllus* Desf.
- E *Ranunculus marginatus* Urv.
- E *Ranunculus muricatus* L.
- E *Ranunculus parviflorus* L.
- E *Ranunculus pedatus* Waldst. & Kit.
- E *Ranunculus polyanthemos* L.
- A *Ranunculus psilostachys* Griseb.
- E *Ranunculus repens* L.
- E *Ranunculus sardous* Crantz
- E *Ranunculus sceleratus* L.
- E *Ranunculus trilobus* Desf.
- E *Thalictrum aquilegiifolium* L.
- A *Thalictrum delavayi* Franchet
- E *Thalictrum lucidum* L.
- E *Thalictrum minus* L.
- E *Trollius europaeus* L.

Magnoliophytina, Dicotyledonae, Rosales
Cannabaceae
- A *Cannabis* × *intersita* Sojk
- E *Cannabis sativa* L.
- E *Humulus lupulus* L.
- E *Humulus scandens* (Lour.) Merrill

Elaeagnaceae
- E *Elaeagnus angustifolia* L.
- A *Elaeagnus commutata* Bernh. ex Rydb.
- A *Elaeagnus macrophylla* Thunb.
- A *Elaeagnus multiflora* Thunb.
- A *Elaeagnus pungens* Thunb.
- A *Elaeagnus* × *submacrophylla* Servettaz
- A *Elaeagnus umbellata* Thunb.
- E *Hippophaë rhamnoides* L.

Moraceae
- A *Broussonetia papyrifera* (L.) Vent.
- A *Fatoua villosa* (Thunb.) Nakai
- A *Ficus bengalensis* L.
- A *Ficus benjamina* L.
- E *Ficus carica* L.
- A *Ficus microcarpa* L.
- A *Ficus pumila* L.
- A *Ficus radicans* Desf.
- A *Ficus religiosa* L.
- A *Ficus watkinsiana* F.M. Bailey
- A *Maclura pomifera* (Raf.) C.K. Schneider
- A *Morus alba* L.
- A *Morus nigra* L.
- A *Morus rubra* L.

Rhamnaceae
- A *Ceanothus* × *delilianus* Spach
- E *Paliurus spina-christi* Mill.
- E *Rhamnus alaternus* L.
- A *Ziziphus jujuba* Mill.

Rosaceae
- A *Acaena anserinifolia* (J.R. & G. Forst.) Druce
- A *Acaena inermis* Hook. f.
- A *Acaena novae-zelandiae* Kirk
- A *Acaena ovalifolia* Ruiz Lopez & Pavon
- A *Agrimonia pilosa* Ledeb.
- E *Agrimonia procera* Wallr.
- E *Alchemilla acutiloba* Opiz
- E *Alchemilla baltica* Sam. ex Juz.
- E *Alchemilla conjuncta* Bab.
- E *Alchemilla cymatophylla* Juz.
- A *Alchemilla gibberulosa* H. Lindb.
- E *Alchemilla glabricaulis* H. Lindb.
- E *Alchemilla glaucescens* Wallr.
- A *Alchemilla heptagona* Juz.
- A *Alchemilla hirsuticaulis* H. Lindb.
- A *Alchemilla leiophylla* Juz.
- E *Alchemilla micans* Buser
- E *Alchemilla mollis* (Buser) Rothm.
- E *Alchemilla monticola* Opiz
- E *Alchemilla plicata* Buser
- E *Alchemilla propinqua* H. Lindb. ex. Juz.
- A *Alchemilla sarmatica* Juz.
- A *Alchemilla semilunaris* Alechin
- E *Alchemilla sericata* Reichenb.
- A *Alchemilla speciosa* Buser
- A *Alchemilla splendens* Christ ex Gremli
- E *Alchemilla subcrenata* Buser
- E *Alchemilla tytthantha* Juz.
- E *Alchemilla vulgaris* L.
- A *Alchemilla* × *anthochlora* Rothm.
- A *Amelanchier alnifolia* (Nutt.) Nutt.
- A *Amelanchier canadensis* (L.) Medik.
- A *Amelanchier confusa* Hyl.
- A *Amelanchier laevis* Wiegand
- A *Amelanchier lamarckii* F.G. Schroed.
- E *Amelanchier ovalis* Medikus
- A *Amelanchier spicata* (Lam.) K. Koch
- E *Aphanes arvensis* L.
- E *Aphanes australis* Rydb.
- E *Aremonia agrimonioides* (L.) DC.

11 List of Species Alien in Europe and to Europe

A *Aronia arbutifolia* (L.) Pers.
A *Aronia melanocarpa* (Michaux) Elliott
E *Aruncus dioicus* (Walter) Fernald
E *Bencomia caudata* (Ait.) W. & B.
A *Chaenomeles japonica* (Thunb.) Lindl. ex Spach
A *Chaenomeles speciosa* (Sweet) Nakai
A *Cotoneaster acuminatus* Lindl.
A *Cotoneaster acutifolius* Turcz.
A *Cotoneaster adpressus* Bois
A *Cotoneaster affinis* Lindley
A *Cotoneaster amoenus* E. Wilson
A *Cotoneaster apiculatus* Rehder & E. Wilson
A *Cotoneaster ascendens* Flinck & Hylmö
A *Cotoneaster astrophoros* J. Fryer & E.C. Nelson
A *Cotoneaster atropurpureus* Flinck & Hylmö
A *Cotoneaster boisianus* G. Klotz
A *Cotoneaster bullatus* Bois
E *Cotoneaster cambricus* J. Fryer & Hylmö
A *Cotoneaster cashmiriensis* G. Klotz
A *Cotoneaster cochleatus* (Franchet) G. Klotz
A *Cotoneaster congestus* Baker
A *Cotoneaster conspicuus* Marquand
A *Cotoneaster cooperi* Marquand
A *Cotoneaster dammeri* C. Schneider
A *Cotoneaster dielsianus* E. Pritzel ex Diels
A *Cotoneaster divaricatus* Rehder & E.H. Wilson
A *Cotoneaster ellipticus* (Lindley) Loudon
A *Cotoneaster fangianus* T.T. Yu
A *Cotoneaster franchetii* Bois
A *Cotoneaster frigidus* Wallich ex Lindley
A *Cotoneaster henryanus* (C.K. Schneider) Rehder & E.H. Wilson
A *Cotoneaster hissaricus* Pojark.
A *Cotoneaster hjelmqvistii* Flinck & Hylmö
E *Cotoneaster horizontalis* Decne.
A *Cotoneaster hsingshangensis* J. Fryer & Hylmö
A *Cotoneaster hummelii* J. Fryer & Hylmö
A *Cotoneaster hylmoei* Flinck & J. Fryer
A *Cotoneaster ignotus* G. Klotz
A *Cotoneaster insculptus* Diels
E *Cotoneaster integerrimus* Medik.
A *Cotoneaster integrifolius* (Roxb.) G. Klotz
A *Cotoneaster lacteus* W.W. Smith
A *Cotoneaster laetevirens* (Rehder & E.H. Wilson) G. Klotz
A *Cotoneaster linearifolius* (G. Klotz) G. Klotz
A *Cotoneaster lucidus* Schltdl.
A *Cotoneaster mairei* H. Lév.
A *Cotoneaster marginatus* (Loud.) Schltdl.
E *Cotoneaster melanocarpus* G. Lodd.
A *Cotoneaster microphyllus* Wall. ex Lindl.
A *Cotoneaster monopyrenus* (W.W. Smith) Flinck & Hylmö
A *Cotoneaster moupinensis* Franchet
A *Cotoneaster mucronatus* Franchet
A *Cotoneaster multiflorus* Bunge
A *Cotoneaster nanshan* M. Vilm. ex Mottet
E *Cotoneaster nebrodensis* (Guss.) K. Koch
A *Cotoneaster nitens* Rehder & E.H. Wilson
A *Cotoneaster nitidus* Jacques
A *Cotoneaster obscurus* Rehder & E.H. Wilson
A *Cotoneaster obtusus* Wallich ex Lindley
A *Cotoneaster pannosus* Franch.
A *Cotoneaster prostratus* Baker
A *Cotoneaster przewalskii* Pojark.
A *Cotoneaster pseudoambiguus* J. Fryer & Hylmö
A *Cotoneaster rehderi* Pojark.
A *Cotoneaster roseus* Edgew.
A *Cotoneaster rotundifolius* Wallich ex Lindley
A *Cotoneaster salicifolius* Franchet
A *Cotoneaster sherriffii* Klotz
A *Cotoneaster simonsii* Baker
A *Cotoneaster splendens* Flink & Hylmö
A *Cotoneaster sternianus* (Turrill) Boom
A *Cotoneaster* × *suecicus* G. Klotz
A *Cotoneaster tengyuehensis* J. Fryer & Hylmö
A *Cotoneaster tomentellus* Pojark.
A *Cotoneaster transens* G. Klotz
A *Cotoneaster villosulus* (Rehder & E.H. Wilson) Flinck & B. Hylmö
A *Cotoneaster vilmorinianus* G. Klotz
A *Cotoneaster wardii* W.W. Smith
A *Cotoneaster* × *watereri* Exell
A *Cotoneaster zabelii* C. Schneider
E *Crataegus azarolus* L.
A *Crataegus coccinioides* Ashe
E *Crataegus crus-galli* L.
A *Crataegus douglasii* Lindl.

- A *Crataegus flabellata* (Spach) Kirchn.
- E *Crataegus heterophylla* Flgge
- E *Crataegus laevigata* (Poiret) DC.
- A *Crataegus* × *media* Bechst.
- A *Crataegus mollis* (Torrey & A. Gray) Scheele
- E *Crataegus monogyna* Jacq.
- E *Crataegus orientalis* Pallas ex M. Bieb.
- A *Crataegus pedicellata* Sarg.
- A *Crataegus persimilis* Sarg.
- E *Crataegus sanguinea* Pall.
- A *Crataegus sinaica* Boiss.
- A *Crataegus submollis* Sarg.
- A *Crataegus succulenta* Schrader
- A *Crataegus* × *uhrovae* Soó
- A × *Crataemespilus gillotii* Beck
- A × *Crataemespilus grandiflora* (Sm.) E.G. Camus
- E *Cydonia oblonga* Miller
- A *Duchesnea indica* (Andrews) Focke
- A *Eriobotrya japonica* (Thunb.) Lindl.
- A *Exochorda racemosa* (Lindl.) Rehder
- A *Filipendula kamtschatica* (Pallas) Maxim.
- A *Filipendula rubra* (Hill) Robinson
- E *Filipendula vulgaris* Moench
- A *Fragaria* × *ananassa* (Duchesne) Duchesne
- A *Fragaria chiloensis* (L.) Duchesne
- E *Fragaria moschata* (Duchesne) Weston
- A *Fragaria virginiana* Mill.
- E *Fragaria viridis* L.
- E *Geum aleppicum* Jacq.
- E *Geum coccineum* Sibth. & Sm.
- A *Geum* × *gajewskii* Smejkal
- A *Geum humifusum* M. B.
- A *Geum macrophyllum* Willd.
- A *Geum quellyon* Sweet
- A *Geum* × *teszlense* Simonk.
- A *Geum vernum* (Raf.) Torr. & A. Gray
- A *Geum verrucosum* Huds.
- A *Holodiscus discolor* (Pursh) Maxim.
- A *Kerria japonica* (L.) DC.
- A *Malus baccata* (L.) Borkh.
- E *Malus domestica* Borkh.
- A *Malus floribunda* Siebold ex van Houtte
- A *Malus* × *purpurea* (E. Barbier) Rehder
- A *Malus sachalinensis* Juzepczuk
- E *Malus sylvestris* (L.) Mill.
- A *Malus toringo* (Siebold) Siebold ex K. Koch
- E *Malus trilobata* (Labill. ex Poir.) C.K. Schneid.
- E *Mespilus germanica* L.
- A *Oemleria cerasiformis* (Torr. & Gray ex Hook. & Arn.) Landon
- A *Photinia davidiana* (Decne.) Cardot
- A *Photinia floribunda* (Lindl.) Robertson & Phipps
- A *Photinia serratifolia* (Desf.) Kalkm.
- A *Physocarpus opulifolius* (L.) Maxim.
- E *Potentilla alba* L.
- E *Potentilla anglica* Laichard.
- E *Potentilla argentea* L.
- A *Potentilla atrosanguinea* Raf.
- E *Potentilla bifurca* L.
- A *Potentilla davurica* Nestl.
- E *Potentilla erecta* (L.) Räusch.
- E *Potentilla fruticosa* L.
- E *Potentilla inclinata* Villars
- E *Potentilla intermedia* L.
- E *Potentilla longifolia* Willd. ex Schlecht.
- E *Potentilla micrantha* Ramond ex DC.
- E *Potentilla montana* Brot.
- E *Potentilla multifida* L.
- E *Potentilla neumanniana* Reichb.
- E *Potentilla norvegica* L.
- A *Potentilla pensylvanica* A. Gray
- E *Potentilla recta* L.
- E *Potentilla reptans* L.
- A *Potentilla rivalis* Nutt. ex Torrey & A. Gray
- E *Potentilla rupestris* L.
- A *Potentilla* × *semiargentea* Borbs ex Zimmeter
- A *Potentilla subarenaria* Borbs ex Zimmeter
- E *Potentilla supina* L.
- E *Potentilla thuringiaca* Bernh. ex Link
- A *Prunus armeniaca* L.
- E *Prunus avium* L. (L.)
- E *Prunus cerasifera* Ehrh.
- E *Prunus cerasus* L.
- E *Prunus domestica* L.
- E *Prunus domestica domestica* L.
- E *Prunus domestica insititia* (L.) Bonnier & Layens
- E *Prunus dulcis* (Miller) D.A. Webb
- A *Prunus* × *eminens* G. Beck
- A *Prunus* × *fruticans* Weihe
- E *Prunus fruticosa* Pall.
- A *Prunus incisa* Thunb. ex Murray
- A *Prunus insititia* L.
- E *Prunus laurocerasus* L.
- E *Prunus lusitanica* L.
- E *Prunus mahaleb* L.
- A *Prunus pensylvanica* L. f.
- E *Prunus persica* (L.) Batsch

11 List of Species Alien in Europe and to Europe

A *Prunus* × *persicoides* (Ser.) Vilmorin & Bois
A ***Prunus serotina*** Ehrh.
A *Prunus serrulata* Lindley
E *Prunus spinosa* L.
A *Prunus tenella* Batsch
A *Prunus tomentosa* Thunb.
A *Prunus triloba* Lindl.
A *Prunus virginiana* L.
A *Pyracantha angustifolia* (Franch.) C.K. Schneid.
E *Pyracantha coccinea* M. Roemer
A *Pyracantha crenato-serrata* (Hance.) Rehd.
A *Pyracantha crenulata* (D. Don) Roemer
A *Pyracantha rogersiana* (A.B. Jackson) Coltman-Rogers
A *Pyrus* × *amphigenea* Domin & J. Dostál ex J. Dostálek
E *Pyrus communis* L.
A *Pyrus elaeagnifolia* Pall.
E *Pyrus nivalis* Jacq.
E *Pyrus pyraster* (L.) Burgst.
E *Pyrus spinosa* Forsskl
E *Pyrus syriaca* Boiss.
A *Rhaphiolepis umbellata* (Thunb.) Makino
A *Rhodotypos scandens* (Thunb.) Makino
E *Rosa acicularis* Lindl.
A *Rosa* × *alba* L.
E *Rosa arvensis* L.
A *Rosa blanda* Ait.
A *Rosa bracteata* Wendl.
A *Rosa centifolia* L.
A *Rosa chinensis* Jacq.
A *Rosa damascena* Mill.
A *Rosa davurica* Pall.
E *Rosa ferruginea* Vill.
A *Rosa foetida* J. Herrmann
E *Rosa gallica* L.
E *Rosa glabrifolia* C.A. Mey. ex Rupr.
E *Rosa gorenkensis* Besser
E *Rosa* × *hybrida* Vill.
A *Rosa laevigata* Michx.
E *Rosa luciae* Franchet & Rochebr.
E *Rosa majalis* Herrm.
E *Rosa moschata* J. Herrmann
A *Rosa multiflora* Thunb. ex Murray
A *Rosa* × *paulii* Rehder
E *Rosa pendulina* L.
E *Rosa pimpinellifolia* L.
A *Rosa* × *praegeri* Wolley-Dod
A *Rosa pratorum* Sukatch.
A ***Rosa rugosa*** Thunb.
E *Rosa setigera* Michx.
A *Rosa* × *turbinata* Ait.
E *Rosa villosa* L.
A *Rosa virginiana* J. Herrmann
A *Rubus allegheniensis* Porter
E *Rubus armeniacus* Focke
A *Rubus canadensis* L.
E *Rubus chlorothyrsos* Focke
A *Rubus cockburnianus* Hemsley
E *Rubus elegantispinosus* (A.Schumach.) H.E. Weber
E *Rubus fabrimontanus* (Sprib.) Sprib.
A *Rubus fissipetalus* P.J. Müll.
A *Rubus flagellaris* Willd.
A *Rubus* × *fraseri* Rehder
E *Rubus gratus* Focke
E *Rubus idaeus* L.
A *Rubus illecebrosus* Focke
A *Rubus imbricatus* hort.
A *Rubus laciniatus* Willd.
A *Rubus leptothyrsos* G. Braun
A *Rubus loehrii* Wirtg.
A *Rubus loganobaccus* L. Bailey
A *Rubus moschus* Juz.
E *Rubus nemoralis* P.J. Müll
A *Rubus nemorosus* Hayne ex Willd.
A *Rubus occidentalis* L.
A *Rubus odoratus* L.
A *Rubus parviflorus* Nutt.
A *Rubus* × *paxii* W.O. Focke
E *Rubus phoenicolasius* Maxim.
A *Rubus pinnatus* Willd.
E *Rubus plicatus* Weihe & Nees
E *Rubus senticosus* J. Kohler ex Weihe
A *Rubus siekensis* G. Braun ex Foche
A *Rubus spectabilis* Pursh
A *Rubus sylvaticus* Weihe & Nees
A *Rubus tricolor* Focke
E *Rubus ulmifolius* Schott
A *Rubus* × *anthocarpus* Bureau & Franchet
A *Sanguisorba alpina* Bunge
A *Sanguisorba canadensis* L.
E *Sanguisorba dodecandra* Moretti
E *Sanguisorba minor* Scop.
E *Sanguisorba minor balearica* (Bourgeau ex Nyman) Muñoz Garmendia & C. Navarro
E *Sanguisorba minor muricata* (Gremli) Briq.
E *Sanguisorba minor verrucosa* (Ehrenb. ex Decne.) Cout.
E *Sanguisorba officinalis* L.
A *Sanguisorba tenuifolia* Fisc. ex Link
A *Sorbaria kirilowii* (Regel) Maxim.

- A *Sorbaria sorbifolia* (L.) A. Braun
- A *Sorbaria tomentosa* (Lindley) Rehder
- E *Sorbus aria* (L.) Crantz
- A *Sorbus croceocarpa* Sell
- E *Sorbus decipiens* (Bechst.) Irmisch.
- E *Sorbus domestica* L.
- E *Sorbus graeca* (Spach) Kotschy
- E *Sorbus hybrida* L.
- E *Sorbus intermedia* (Ehrh.) Pers.
- E *Sorbus latifolia* (Lam.) Pers.
- E *Sorbus porrigentiformis* E. Warb.
- A *Sorbus × rotundifolia* (Bechst.) Hedl.
- E *Sorbus teodorii* Liljef.
- E *Sorbus × thuringiaca* (Ilse) Fritsch
- A *Sorbus torminalis* (L.) Crantz
- E *Spiraea alba* Du Roi
- A *Spiraea × arguta* Zabel
- A *Spiraea betulifolia* Pallas
- E *Spiraea × billardii* Hérincq
- A *Spiraea × brachybotrys* Lange
- A *Spiraea × bumalda* Burvenich
- A *Spiraea canescens* D. Don
- A *Spiraea cantoniensis* Lour.
- A *Spiraea chamaedryfolia* L.
- A *Spiraea crenifolia* C.A. Meyer
- A *Spiraea douglasii* Hook.
- A *Spiraea douglasii douglasii* Hook.
- E *Spiraea hypericifolia* L.
- A *Spiraea japonica* L. f.
- E *Spiraea media* F. Schmidt
- A *Spiraea prunifolia* Sieb. & Zucc.
- A *Spiraea × pseudosalicifolia* Silverside
- A *Spiraea × rosalba* Dippel
- A *Spiraea salicifolia* L.
- A *Spiraea × semperflorens* Zabel
- A *Spiraea × vanhouttei* (Briot) Carrire
- A *Stephanandra incisa* (Thunb.) Zabel
- E *Waldsteinia geoides* Willd.
- E *Waldsteinia trifolia* Rochel ex Koch

Ulmaceae
- E *Celtis australis* L.
- A *Celtis occidentalis* L.
- E *Ulmus glabra* Huds.
- E *Ulmus laevis* Pallas
- E *Ulmus minor* Mill.
- E *Ulmus minor minor* Mill.
- E *Ulmus procera* Salisb.
- A *Ulmus pumila* L.
- A *Zelkova serrata* (Thunb.) Makino

Urticaceae
- A *Boehmeria nivea* (L.) Gaudich.
- A *Laportea aestuans* (L.) Chew
- E *Parietaria debilis* G. Forst.
- E *Parietaria judaica* L.
- E *Parietaria lusitanica* L.
- E *Parietaria microphylla* L.
- E *Parietaria officinalis* L.
- A *Parietaria pensylvanica* Muhl. ex Willd.
- A *Pilea hyalina* Fenzl
- A *Pilea peperomioides* Diels
- E *Soleirolia soleirolii* (Req.) Dandy
- A *Urtica chamaedryoides* Pursh
- E *Urtica dioica* L.
- A *Urtica incisa* Poiret
- E *Urtica membranacea* Poir.
- E *Urtica morifolia* Poir.
- E *Urtica pilulifera* L.
- E *Urtica urens* L.

Magnoliophytina, Dicotyledonae, Sapindales

Anacardiaceae
- E *Cotinus coggygria* Scop.
- A *Mangifera indica* L.
- A *Pistacia vera* L.
- E *Rhus coriaria* L.
- A *Rhus glabra* L.
- A *Rhus × pulvinata* E. Greene
- E *Rhus typhina* L.
- A *Schinus molle* L.
- A *Schinus terebenthifolia* Raddi
- A *Toxicodendron radicans* (L.) O. Kuntze.

Meliaceae
- A *Melia azedarach* L.

Rutaceae
- A *Calodendrum capense* (L. f.) Thunb.
- A *Choisya ternata* Kunth
- A *Citrus × aurantium* L.
- A *Citrus deliciosa* Ten.
- A *Citrus limon* (L.) Burm.f.
- A *Citrus medica* L.
- A *Citrus reticulata* Blanco
- A *Citrus sinensis* (L.) Osbeck
- A *Correa backhousiana* Hook.
- A *Haplophyllum buxbaumii* (Poir.) G. Don f.
- A *Haplophyllum linifolium* (L.) G. Don f.
- A *Haplophyllum villosum* (M. Bieb.) G. Don f.
- A *Poncirus trifoliata* (L.) Raf.
- A *Ptelea trifoliata* L.
- E *Ruta chalepensis* L.
- E *Ruta graveolens* L.
- A *Tetradium daniellii* (Benn.) T.G. Hartley

Santalaceae
- E *Thesium arvense* Horvatovszky
- A *Thesium refractum* C.A. Mey.

Sapindaceae
- E *Acer campestre* L.
- E *Acer cappadocicum* Gled.
- A *Acer ginnala* Maxim.
- E *Acer monspessulanum* L.
- A *Acer negundo* L.
- A *Acer obtusatum* Waldst. & Kit. ex Willd.
- A *Acer palmatum* Thunb.
- E *Acer platanoides* L.
- E *Acer pseudoplatanus* L.
- A *Acer saccharinum* L.
- A *Acer saccharum* Marshall
- E *Acer tataricum* L.
- A *Aesculus* × *carnea* Hayne
- A *Aesculus flava* Sol. ex Hope
- E *Aesculus hippocastanum* L.
- A *Aesculus indica* (Cambess.) Hook.
- A *Aesculus parviflora* Walter
- A *Aesculus pavia* L.
- A *Cardiospermum grandiflorum* Swartz
- A *Cardiospermum halicacabum* L.
- A *Dodonaea viscosa* (L.) Jacq.
- A *Koelreuteria paniculata* Laxm.
- A *Xanthoceras sorbifolia* Bunge

Simaroubaceae
- A **Ailanthus altissima** (Miller) Swingle

Magnoliophytina, Dicotyledonae, Saxifragales

Crassulaceae
- E *Aeonium arboreum* (L.) Webb & Berth.
- E *Aeonium cuneatum* Webb & Berth.
- E *Aeonium decorum* Webb ex Bolle
- E *Aeonium glutinosum* (Ait.) Webb & Berth.
- E *Aeonium haworthii* Webb & Berth.
- E *Aeonium simsii* (Sweet) Stearn
- E *Aichryson dichotomum* (DC.) Webb & Berth.
- A *Bryophyllum fedtschenkoi* (Raym.-Hamet & H. Perrier) Lauz.-March.
- A *Bryophyllum proliferum* Bowie ex Curtis
- A *Cotyledon macracantha* erger
- A *Cotyledon orbiculata* L.
- E *Crassula aquatica* (L.) Schönl.
- A *Crassula arborescens* (Mill.) Willd.
- A *Crassula argentea* L. f.
- A *Crassula campestris* (Ecklon & Zeyher) Walpers
- A *Crassula decumbens* Thunb.
- A **Crassula helmsii** (Kirk) Cockayne
- A *Crassula lycopodioides* Lam.
- A *Crassula multicava* Lemaire
- A *Crassula ovata* (P. Mill.) Druce
- A *Crassula peduncularis* (Smith) Meigen
- A *Crassula pubescens* Thunb.
- A *Crassula rupestris* Thunb.
- A *Crassula tetragona* L.
- E *Crassula tillaea* Lester - Garland
- E *Crassula vaillantii* (Willd.) Roth
- A *Graptopetalum paraguayense* (N.E. Br.) Walther
- E *Hylotelephium carpaticum* (G. Reuss) Sojk
- E *Jovibarba globifera* (L.) J. Parn.
- A *Kalanchoë daigremontiana* R. Hamet & H. Perrier
- A *Kalanchoë longiflora* Schlechter.
- A *Kalanchoë pinnata* (Lam.) Pers.
- A *Kalanchoë tubiflora* (Harvey) Raym.-Hamet
- E *Sedum acre* L.
- A *Sedum aizoon* L.
- E *Sedum album* L.
- E *Sedum anacampseros* L.
- A *Sedum annuum* L.
- E *Sedum anopetalum* DC.
- E *Sedum arboroseum* Bak.
- E *Sedum cepaea* L.
- A *Sedum confusum* Hemsley
- E *Sedum dasyphyllum* L.
- A *Sedum dendroideum* (Moq. & Sessé) ex DC.
- A *Sedum ewersii* Ledeb.
- E *Sedum forsterianum* Smith
- E *Sedum hispanicum* L.
- A *Sedum hybridum* L.
- A *Sedum kamtschaticum* Fisch. & C.A. Mey.
- E *Sedum lydium* Boiss.
- A *Sedum mexicanum* Britton
- E *Sedum montanum* Perr. & Song.
- E *Sedum multiceps* Coss. & Durieu
- E *Sedum ochroleucum* Chaix in Vill.
- E *Sedum pallidum* M. Bieb.
- E *Sedum praealtum* A. DC.
- A *Sedum rubrotinctum* R.T. Clausen
- E *Sedum rupestre* L.
- A *Sedum sarmentosum* Bunge
- E *Sedum sediforme* (Jacq.) Pau
- E *Sedum sexangulare* L.
- A *Sedum sieboldii* hort. ex D. Don
- A *Sedum spathulifolium* Hook.
- E *Sedum spectabile* Boreau
- E *Sedum spurium* L.
- E *Sedum stoloniferum* S. Gmelin
- E *Sedum telephium* L.

- E *Sedum thartii* L.P. Hbert
- E *Sempervivum arachnoideum* L.
- A *Sempervivum × barbulatum* Schott
- A *Sempervivum funckii* Braun ex W.D.J. Koch
- E *Sempervivum tectorum* L.
- E *Umbilicus rupestris* (Salisb.) Dandy

Grossulariaceae
- A *Escallonia bifida* Link & Otto
- A *Escallonia rubra* (Ruiz & Pavón) Pers.
- E *Ribes alpinum* L.
- A *Ribes aureum* Pursh
- A *Ribes divaricatum* Douglas
- E *Ribes nigrum* L.
- A *Ribes odoratum* H.L. Wendl.
- E *Ribes rubrum* L.
- A *Ribes sanguineum* Pursh
- E *Ribes spicatum* Robson
- E *Ribes uva-crispa* L.
- A *Ribes × virginalis*

Haloragaceae
- A *Haloragis micrantha* (Thunb.) R. Br. ex Siebold & Zucc.
- A *Myriophyllum aquaticum* (Velloso) Verdc.
- A *Myriophyllum heterophyllum* Michx.

Hamamelidaceae
- A *Hamamelis mollis* Oliver ex Forb. & Hemsl.
- A *Liquidambar styraciflua* L.

Paeoniaceae
- E *Paeonia mascula* (L.) Miller
- E *Paeonia officinalis* L.

Saxifragaceae
- A *Astilbe × arendsii* Arends
- A *Astilbe japonica* (Morren & Decne.) A. Gray
- A *Astilbe rivularis* Buch.-Ham. ex D. Don
- A *Bergenia cordifolia* (Haw.) Sternb.
- A *Bergenia crassifolia* (L.) Fritsch
- A *Bergenia × schmidtii* (Regel) Silva Tar.
- A *Darmera peltata* (Torrey ex Benth.) Voss ex Post & Kuntze
- A *Heuchera americana* L.
- A *Heuchera micrantha* Dougl. ex Lindl.
- A *Heuchera sanguinea* Engelm.
- A *Rodgersia pinnata* Franchet
- A *Rodgersia podophylla* A. Gray
- A *Saxifraga × arendsii* hort.
- E *Saxifraga cuneifolia* L.
- E *Saxifraga cymbalaria* L.
- E *Saxifraga × geum* L.
- E *Saxifraga granulata* L.
- E *Saxifraga hirsuta* L.
- E *Saxifraga hostii* Tausch
- E *Saxifraga hypnoides* L.
- E *Saxifraga paniculata* Miller
- E *Saxifraga × polita* (Haw.) Link
- E *Saxifraga rosacea* Moench
- E *Saxifraga rotundifolia* L.
- E *Saxifraga spathularis* Brot.
- A *Saxifraga stolonifera* Curtis
- E *Saxifraga umbrosa* L.
- E *Saxifraga × urbium* D.A. Webb
- A *Tellima grandiflora* (Pursh) Douglas ex Lindl.
- A *Tiarella cordifolia* L.
- A *Tolmiea menziesii* (Pursh) Torrey & A. Gray

Magnoliophytina, Dicotyledonae, Solanales
Convolvulaceae
- A *Calystegia × howittiorum* Brummitt
- A *Calystegia × lucana* (Ten.) G. Don
- E *Calystegia pulchra* Brummitt & Heywood
- A *Calystegia × scanica* Brummitt
- E *Calystegia sepium* (L.) R. Br.
- E *Calystegia silvatica* (Kit.) Griseb.
- E *Calystegia soldanella* (L.) R. Br. ex Roemer & J.A. Schultes
- E *Convolvulus arvensis* L.
- A *Convolvulus arvensis crispatus* Franco
- E *Convolvulus betonicifolius* Miller
- E *Convolvulus elegantissimus* Mill.
- E *Convolvulus farinosus* L.
- A *Convolvulus persicus* L.
- A *Convolvulus sabatius* Viv.
- E *Convolvulus tricolor* L.
- E *Convolvulus tricolor cupanianus* (Tod.) Cavara & Grande
- A *Cuscuta approximata* Bab.
- E *Cuscuta campestris* Yuncker
- E *Cuscuta epilinum* Weihe
- E *Cuscuta epithymum* (L.) L.
- E *Cuscuta epithymum epithymum* L.
- E *Cuscuta europaea* L.
- E *Cuscuta gronovii* Willd. ex Roem. & Schult.
- A *Cuscuta hyalina* Roth (1820)
- E *Cuscuta lupuliformis* Krocker
- E *Cuscuta monogyna* Vahl
- E *Cuscuta palaestina* Boiss.
- E *Cuscuta planiflora* Ten.
- E *Cuscuta scandens* Brot.
- E *Cuscuta suaveolens* Ser.
- A *Dichondra micrantha* Urban
- A *Ipomoea arenaria* Roem. & Schult

- A *Ipomoea batatas* (L.) Lam.
- A *Ipomoea cairica* (L.) Sweet
- A *Ipomoea hederacea* (L.) Jacq.
- A *Ipomoea hederifolia* L.
- A *Ipomoea imperati* (Vahl) Griseb.
- A *Ipomoea indica* (Burm.) Merr.
- A *Ipomoea lacunosa* L.
- A *Ipomoea ochracea* (Lindl.) G. Don
- A *Ipomoea pes-caprae* (L.) R. Br.
- A *Ipomoea purpurea* Roth
- A *Ipomoea sagittata* Poiret
- A *Ipomoea tricolor* Cav.
- A *Ipomoea triloba* L.
- A *Ipomoea wrightii* Gray
- A *Quamoclit coccinea* (L.) Moench
- A *Turbina corymbosa* (L.) Raf.

Hydrophyllaceae
- A *Hydrophyllum virginianum* L.

Solanaceae
- E *Atropa bella-dona* L.
- A *Brugmansia suaveolens* (Willd.) Bercht. & J. Presl
- A *Capsicum annuum* L.
- A *Cestrum foetidissimum* Jacq.
- A *Cestrum nocturnum* L.
- A *Cestrum parqui* L. Hér.
- A *Cestrum purpureum* (Lindl.) Standl.
- A *Cyphomandra betacea* (Cav.) Sendtn.
- A *Datura arborea* L.
- A *Datura ferox* L.
- A *Datura innoxia* Mill.
- A *Datura metel* L.
- A *Datura stramonium* L.
- A *Datura wrightii* Regel
- E *Hyoscyamus albus* L.
- E *Hyoscyamus niger* L.
- E *Hyoscyamus pusillus* L.
- A *Iochroma australe* Griseb.
- A *Lycium afrum* L.
- E *Lycium barbarum* L.
- A *Lycium chinense* Mill.
- E *Lycium europaeum* L.
- A *Lycium ferocissimum* Miers
- E *Lycium intricatum* Boiss.
- A *Lycopersicon esculentum* Mill.
- A *Nicandra physalodes* (L.) P. Gaertn.
- A *Nicotiana alata* Link & Otto
- A *Nicotiana forgetiana* Hemsley
- A *Nicotiana glauca* R.C. Graham
- A *Nicotiana langsdorffii* J.A. Weinm.
- A *Nicotiana longiflora* Cav.
- A *Nicotiana paniculata* L.
- A *Nicotiana rustica* L.
- A *Nicotiana* × *sanderae* W. Watson
- A *Nicotiana sylvestris* Spegazz. & Comes
- A *Nicotiana tabacum* L.
- A *Nicotiana velutina* H. Wheeler
- A *Nicotiana wigandioides* K. Koch & Fint.
- A *Petunia* × *hybrida* hort. ex Vilmorin
- A *Petunia integrifolia* (Hooker) Schinz & Thell.
- A *Petunia* × *punctata* Paxt.
- E *Physalis alkekengi* L.
- A *Physalis angulata* L.
- A *Physalis cinerascens* (Dun.) A.S. Hitchc.
- A *Physalis fusco-maculata* Rouville ex Dunal
- A *Physalis grisea* (Waterfall) M. Martínez
- A *Physalis heterophylla* Nees.
- A *Physalis ixocarpa* Brot. ex Hornem.
- A *Physalis lanceolata* Michx.
- A *Physalis minima* L.
- A *Physalis peruviana* L.
- A *Physalis philadelphica* Lam.
- A *Physalis pubescens* L.
- A *Physalis virginiana* Mill.
- A *Physalis viscosa* L.
- A *Salpichroa origanifolia* (Lam.) Thell.
- A *Salpiglossis sinuata* Ruiz & Pavón
- A *Schizanthus pinnatus* Ruiz & Pavón
- E *Scopolia carniolica* Jacq.
- A *Solanum aethiopicum* L.
- A *Solanum americanum* Mill.
- A *Solanum bonariense* L.
- A *Solanum burbankii* Bitter
- A *Solanum capsicastrum* Link ex Schauer
- A *Solanum capsicoides* All.
- A *Solanum carolinense* L.
- A *Solanum cornutum* Lam.
- E *Solanum dulcamara* L.
- A *Solanum elaeagnifolium* Cav.
- A *Solanum fastigiatum* Willd.
- A *Solanum fontanesianum* Dunal
- A *Solanum giganteum* Jacq.
- A *Solanum heterodoxum* Dunal
- A *Solanum hispidum* Pers.
- A *Solanum jasminoides* Paxton
- A *Solanum juvenale* Thell.
- A *Solanum laciniatum* Aiton
- A *Solanum mammosum* L.
- A *Solanum marginatum* L. fil.
- A *Solanum mauritianum* Scop.
- A *Solanum melongena* L.
- A *Solanum microcarpum* (Pers.) Vahl
- E *Solanum nigrum* L.
- E *Solanum nigrum nigrum* L.

- E *Solanum nigrum schultesii* (Opiz) Wessely
- A *Solanum nitidibaccatum* Bitter
- A *Solanum physalifolium* Rusby
- A *Solanum × procurrens* A.C. Leslie
- A *Solanum pseudocapsicum* L.
- A *Solanum pygmaeum* Cav.
- A *Solanum pyracanthos* Lam.
- A *Solanum rantonnetii* Lesc.
- A *Solanum robustum* H.L. Wendl.
- A *Solanum sarachoides* Sendtner
- A *Solanum sisymbriifolium* Lam.
- A *Solanum sodomaeum* L.
- A *Solanum sublobatum* Willd. ex Roemer & Schultes
- A *Solanum tenuifolium* Dunal
- A *Solanum torvum* Sw.
- A *Solanum triflorum* Nutt.
- A *Solanum tuberosum* L.
- A *Solanum vernei* Bitter & Wittm.
- A *Solanum viarum* Dun.
- E *Solanum villosum* Mill.
- E *Solanum villosum miniatum* (Bernh. ex Willd.) Edmonds
- E *Solanum villosum villosum* Mill.
- A *Solanum wendlandii* Hook. f.
- A *Streptosolen jamesonii* (Benth.) Miers
- A *Withania somnifera* (L.) Dunal

Magnoliophytina, Dicotyledonae, Vitales
Vitaceae
- A *Parthenocissus inserta* (A. Kerner) Fritsch
- A *Parthenocissus quinquefolia* (L.) Planchon
- A *Parthenocissus tricuspidata* (Siebold & Zucc.) Planchon
- A *Vitis aestivalis* Michx.
- E *Vitis labrusca* L.
- A *Vitis rotundifolia* Michx.
- A *Vitis rupestris* Scheele
- E *Vitis vinifera* L.
- A *Vitis vulpina* L.

Magnoliophytina, Dicotyledonae, Zygophyllales
Zygophyllaceae
- A *Peganum harmala* L.
- E *Tribulus terrestris* L.
- E *Zygophyllum fabago* L.

Magnoliophytina, Monocotyledonae
Acoraceae
- E *Acorus calamus* L.
- A *Acorus gramineus* Sol. ex Aiton

Agapanthaceae
- A *Agapanthus praecox* Willd.

Agavaceae
- A *Agave americana* L.
- A *Agave angustifolia* Haw.
- A *Agave atrovirens* Karwinski ex Salm-Dyck
- A *Agave attenuata* Salm-Dyck
- A *Agave fourcroydes* Lem.
- A *Agave salmiana* Otto
- A *Agave sisalana* Perrine
- A *Chlorophytum comosum* (Thunb.) Jacques
- A *Cordyline australis* (G. Forst.) Endl.
- A *Doryanthes palmeri* W. Hill. ex Beth.
- E *Dracaena draco* (L.) L.
- A *Furcraea foetida* (L.) Haw.
- A *Hosta lancifolia* (Thunb.) Engelm.
- A *Hosta plantaginea* (Lam.) Aschers.
- A *Nolina recurvata* (Lem.) Hemsl.
- A *Phormium cookianum* Le Jol.
- A *Phormium tenax* J.R. & G. Forst.
- A *Polianthes tuberosa* L.
- A *Yucca aloifolia* L.
- A *Yucca filamentosa* L.
- A *Yucca gloriosa* L.
- A *Yucca recurvifolia* Salisb.

Alismataceae
- E *Alisma lanceolatum* With.
- A *Baldellia ranunculoides* (L.) Parl.
- E *Luronium natans* (L.) Rafn.
- E *Sagittaria graminea* Michx.
- A *Sagittaria lancifolia* L.
- E *Sagittaria latifolia* Willd.
- A *Sagittaria platyphylla* (Engelm.) J.G. Sm.
- A *Sagittaria rigida* Pursh
- E *Sagittaria sagittifolia* L.
- A *Sagittaria subulata* (L.) Buchenau
- A *Sagittaria trifolia* L.

Alliaceae
- E *Allium ampeloprasum* L.
- E *Allium angulosum* L.
- E *Allium atropurpureum* Waldst. & Kit.
- E *Allium carinatum* L.
- A *Allium cepa* L.
- A *Allium christophii* Trautv.
- A *Allium cyrilii* Ten.
- E *Allium fistulosum* L.
- E *Allium flavum* L.
- E *Allium moly* L.
- E *Allium neapolitanum* Cirillo
- E *Allium nigrum* L.
- A *Allium odorum* L. non Ten.

- E *Allium oleraceum* L.
- E *Allium paniculatum* L.
- E *Allium paradoxum* (M. Bieb.) G. Don
- E *Allium pendulinum* Ten.
- E *Allium porrum* L.
- E *Allium rose*um L.
- E *Allium sativum* L.
- E *Allium schoenoprasum* L.
- E *Allium scorodoprasum* L.
- E *Allium sphaerocephalon* L.
- E *Allium subhirsutum* L.
- E *Allium subvillosum* Salzm. ex Schult. & Schult.f.
- E *Allium triquetrum* L.
- A *Allium unifolium* Kellogg
- E *Allium ursinum* L.
- E *Allium vineale* L.
- E *Nectaroscordum siculum* (Ucria) Lindley ex Edwards
- A *Nothoscordum bivalve* (L.) Britt.
- A *Nothoscordum borbonicum* Kunth
- A *Nothoscordum gracile* (Aiton) Stearn
- A *Tristagma uniflorum* (Lindl.) Traub

Alstroemeriaceae
- A *Alstroemeria aurea* Graham
- A *Alstroemeria pulchella* L. f.

Amaryllidaceae
- A *Amaryllis belladonna* L.
- A *Clivia miniata* Regel
- A *Clivia nobilis* Lindl.
- A *Crinum bulbispermum* (Burm.) Milne-Redh. & Schweick.
- A *Crinum moonieri* Hook. f.
- A *Crinum × powellii* Baker
- E *Galanthus caucasicus* (Baker) Grossh.
- E *Galanthus elwesii* Hook. f.
- E *Galanthus ikariae* Baker
- E *Galanthus nivalis* L.
- E *Galanthus plicatus* M. Bieb.
- E *Galanthus reginae-olgae* Orph.
- A *Galanthus × warei* J. Allen
- E *Leucojum aestivum* L.
- E *Leucojum vernum* L.
- E *Narcissus assoanus* Dufour ex Schultes & Schultes f.
- E *Narcissus bulbocodium* L.
- E *Narcissus calcicola* Mendona
- E *Narcissus cyclamineus* DC.
- E *Narcissus × incomparabilis* Mill.
- A *Narcissus × intermedius* Hort. Dammann
- E *Narcissus jonquilla* L.
- E *Narcissus × medioluteus* Mill.
- E *Narcissus minor* L.
- A *Narcissus × monochromus* P.D. Sell
- A *Narcissus × odorus* L.
- E *Narcissus papyraceus* Ker-Gawl.
- E *Narcissus poeticus* L.
- E *Narcissus pseudonarcissus* L.
- E *Narcissus tazetta* L.
- E *Narcissus triandrus* L.
- A *Nerine sarniensis* (L.) Herb.
- E *Pancratium maritimum* L.
- E *Sternbergia lutea* (L.) Ker-Gawler ex Spreng.

Aponogetonaceae
- A *Aponogeton distachyos* L. f.

Araceae
- E *Arisarum proboscideum* (L.) Savi
- E *Arum italicum* Mill.
- E *Arum maculatum* L.
- E *Calla palustris* L.
- A *Colocasia esculenta* (L.) Schott
- A *Cryptocoryne crispatula* Engl.
- A *Cryptocoryne wendtii* de Wit
- E *Dracunculus vulgaris* Schott
- A *Lysichiton americanus* Hultén & St John
- A *Lysichiton camtschatcensis* (L.) Schott
- A *Monstera deliciosa* Liebm.
- A *Orontium aquaticum* L.
- A *Pinellia ternata* (Thunb.) Makino ex Breitenbach
- A *Spirodela oligorrhiza* (Kurz) Hegelm.
- E *Spirodela polyrhiza* (L.) Schleiden
- A *Spirodela punctata* (G.F.W. Meyer) Thompsom
- E *Wolffia arrhiza* (L.) Horkel ex Wimm.
- A *Zantedeschia aethiopica* (L.) Spreng.

Arecaceae
- A *Archontophoenix cunninghamiana* (H.A. Wendl.) H.A. Wendl. & Drude
- A *Arecastrum romanzoffianum* (Cham.) Becc.
- A *Brahea armata* S. Wats.
- A *Chamaedorea elegans* C. Martius
- E *Chamaerops humilis* L.
- A *Chambeyronia macrocarpa* (Brongn.) Vieill. ex Becc.
- A *Kentia belmoreana* F.J. Müll.
- A *Kentia forsteriana* F. Müll. ex H. Wendl.
- A *Phoenix canariensis* Chabaud
- A *Phoenix dactylifera* L.
- A *Roystonea regia* (Kunth) O.F. Cook
- A *Trachycarpus fortunei* (Hooker) H.A. Wendl.
- A *Washingtonia filifera* (J.A. Linden) H.A. Wendl.
- A *Washingtonia robusta* Wendland

Asparagaceae
  A *Asparagus asparagoides* (L.) Druce
  A *Asparagus densiflorus* (Kunth) Jessop
  E *Asparagus officinalis* L.
  A *Asparagus setaceus* (Kunth) Jessop
Asphodelaceae
  A *Aloë arborescens* Mill.
  A *Aloë brevifolia* Mill.
  A *Aloë ciliaris* Haw.
  A *Aloë ferox* Mill.
  A *Aloë latifolia* (Haw.) Haw.
  A *Aloë maculata* All.
  A *Aloë plicatilis* Mill.
  A *Aloë succotrina* All.
  A *Aloë variegata* L.
  A *Aloë vera* (L.) Burm. f.
  E *Asphodelus albus* L.
  E *Asphodelus fistulosus* L.
  A *Kniphofia* × *praecox* Baker
  A *Kniphofia uvaria* (L.) Oken
Bromeliaceae
  A *Bromelia carnea* Beer
  A *Fascicularia bicolor* (Ruiz & Pavon) Mez
  A *Fascicularia pitcairniifolia* (Verlot) Mez
Butomaceae
  E *Butomus umbellatus* L.
Cannaceae
  A *Canna flaccida* Salisb.
  A *Canna* × *generalis* L.H. Bailey
  A *Canna indica* L.
  A *Canna variabilis* Willd.
Colchicaceae
  E *Colchicum autumnale* L.
Commelinaceae
  A *Commelina benghalensis* L.
  A *Commelina coelestis* Willd.
  A *Commelina communis* L.
  A *Commelina diffusa* Burm.f.
  A *Commelina obliqua* Vahl
  A *Commelina virginica* L.
  A *Dichorisandra thyrsiflora* Mikan f.
  A *Murdannia keisak* (Hassk.) Hand.-Mazz.
  A *Tinantia erecta* (Jacq.) Schltdl.
  A *Tinantia fugax* Scheidw.
  A *Tradescantia cerinthoides* Kunth
  A *Tradescantia fluminensis* Velloso
  A *Tradescantia pallida* (Rose) D.R. Hunt
  A *Tradescantia virginiana* L.
  A *Zebrina pendula* Schnizlein
Cyperaceae
  A *Bolboschoenus glaucus* (Lam.) S.G. Sm.
  A *Bolboschoenus planiculmis* (F. Schmidt) T.V. Egorova

E *Carex acuta* L.
E *Carex bebbii* Olney ex Fern.
E *Carex bohemica* Schreb.
E *Carex buchananii* Berggr.
E *Carex caryophyllea* Latourr.
A *Carex crawfordii* Fernald
A *Carex crus-corvi* Shuttlw. ex Kunze
E *Carex curta* Gooden.
A *Carex debilis* Michx.
E *Carex hirta* L.
A *Carex livida* (Wahlenb.) Willd.
A *Carex macloviana* d'Urv.
A *Carex morrowii* Boott
A *Carex muskingumensis* Schwein.
E *Carex nigra* (L.) Reichard
E *Carex ovalis* Gooden.
E *Carex panicea* L.
E *Carex pendula* Huds.
A *Carex praticola* Rydb.
A *Carex scoparia* Schkuhr
A *Carex straminea* Willd.
E *Carex viridula* Michx.
E *Carex vulpina* L.
A *Carex vulpinoidea* Michx.
A *Cyperus aggregatus* (Willd.) Endl.
A *Cyperus alternifolius* L.
A *Cyperus amuricus* Maxim.
A *Cyperus auricomus* Sieber ex Spreng.
A *Cyperus brevifolius* (Rottb.) Hassk.
A *Cyperus conglomeratus* Rottb.
A *Cyperus cyperoides* (L.) Kuntze
E *Cyperus difformis* L.
E *Cyperus eragrostis* Lam.
E *Cyperus esculentus* L.
A *Cyperus exaltatus* Retz.
E *Cyperus flavescens* L.
E *Cyperus flavidus* Retz.
E *Cyperus fuscus* L.
E *Cyperus glomeratus* L.
E *Cyperus hamulosus* M. Bieb.
A *Cyperus imbricatus* Retz.
E *Cyperus longus* L.
E *Cyperus longus badius* (Desf.) Murb.
E *Cyperus longus longus* L.
E *Cyperus longus tenuiflorus* (Rottb.) Kük.
A *Cyperus lupulinus* (Spreng.) Marcks
E *Cyperus michelianus* (L.) Link
A *Cyperus odoratus* L.
A *Cyperus ovularis* (Michx.) Torr.
A *Cyperus papyrus* L.
A *Cyperus polystachyos* Rottb.
A *Cyperus reflexus* Vahl
A *Cyperus rigens* J. Presl & C. Presl

- E *Cyperus rotundus* L.
- A *Cyperus schweinitzii* Torr.
- E *Cyperus serotinus* Rottb.
- A *Cyperus squarrosus* L.
- A *Cyperus strigosus* L.
- A *Cyperus textilis* Thunb.
- E *Cyperus virens* Michx.
- E *Eleocharis atropurpurea* (Retz.) C. Presl
- A *Eleocharis bonariensis* Nees
- E *Eleocharis flavescens* (Poir.) Urban
- A *Eleocharis obtusa* (Willd.) Schultes
- A *Eleocharis parvula* (Roem. & Schult.) Bluff & al.
- A *Fimbristylis annua* (All.) Roemer & Schultes
- E *Fimbristylis bisumbellata* (Forsskl) Bubani
- A *Fimbristylis dichotoma* (L.) Vahl
- A *Fimbristylis ferruginea* (L.) Vahl
- A *Kyllinga gracillima* Miq.
- A *Mariscus congestus* (Vahl) C.B. Clarke
- A *Pycreus propinquus* Nees
- A *Schoenoplectus juncoides* (Roxb.) Palla
- E *Schoenoplectus litoralis* (Schrad.) Palla
- E *Schoenoplectus mucronatus* (L.) A. Kerner
- E *Schoenus ferrugineus* L.
- A *Scirpus atrovirens* Willd.
- A *Scirpus cyperinus* (L.) Kunth
- A *Scirpus georgianus* Harper
- A *Scirpus pendulus* Muhl.
- E *Scirpus pungens* Vahl
- E *Scirpus supinus* L.
- A *Scirpus uninodis* (Gray) Beetle

Dioscoreaceae
- E *Tamus communis* L.

Hemerocallidaceae
- A *Hemerocallis fulva* (L.) L.
- E *Hemerocallis lilioasphodelus* L.
- E *Simethis planifolia* (L.) Gren.

Hyacinthaceae
- A *Bellevalia ciliata* (Cirillo) Nees
- E *Bellevalia romana* (L.) Reichb.
- E *Bellevalia trifoliata* (Ten.) Kunth
- A *Chionodoxa forbesii* Baker
- A *Chionodoxa luciliae* Boiss.
- A *Chionodoxa sardensis* Whittall ex Barr & Sugden
- E *Hyacinthella leucophaea* (K. Koch) Schur
- A *Hyacinthella rumelica* Velen.
- E *Hyacinthoides hispanica* (Mill.) Rothm.
- E *Hyacinthoides italica* (L.) Rothm.
- A *Hyacinthoides* × *massartiana* Geerinck pro parte
- E *Hyacinthoides non-scripta* (L.) Chouard ex Rothm.
- A *Hyacinthus orientalis* L.
- A *Muscari armeniacum* Leichtlin ex Baker
- A *Muscari azureum* Fenzl
- E *Muscari botryoides* (L.) Mill
- E *Muscari comosum* (L.) Miller
- E *Muscari neglectum* Guss. ex Ten.
- E *Ornithogalum angustifolium* Boreau
- E *Ornithogalum arabicum* L.
- E *Ornithogalum narbonense* L.
- E *Ornithogalum nutans* L.
- E *Ornithogalum pyramidale* L.
- E *Ornithogalum pyrenaicum* L.
- A *Ornithogalum thyrsoides* Jacq.
- E *Ornithogalum umbellatum* L.
- A *Puschkinia scilloides* Adams
- E *Scilla amoena* L.
- E *Scilla bifolia* L.
- E *Scilla bithynica* Boiss.
- E *Scilla hyacinthoides* L.
- E *Scilla lilio-hyacinthus* L.
- E *Scilla messeniaca* Boiss.
- E *Scilla peruviana* L.
- E *Scilla siberica* Haw.

Hydrocharitaceae
- A *Blyxa japonica* (Miq.) Maxim. ex Aschers & Grcke
- A *Egeria densa* Planch.
- A *Elodea callitrichoides* (Rich.) Casp.
- A **Elodea canadensis** Michx.
- A *Elodea nuttallii* (Planch.) St. John
- A **Halophila stipulacea** (Forsskal) Ascherson
- C *Hydrilla verticillata* (L. f.) Royle
- E *Hydrocharis morsus-ranae* L.
- A *Lagarosiphon major* (Ridley) Moss
- E *Najas flexilis* Rostk. & W.L.E. Schmidt
- A *Najas gracillima* (A. Braun ex Engelm.) Magnus
- A *Najas graminea* Delile
- A *Najas guadalupensis* (Spreng.) Magnus
- E *Najas marina* L.
- A *Najas orientalis* Triest & Uotila
- A *Ottelia alismoides* (L.) Pers.
- E *Stratiotes aloides* L.
- A *Vallisneria spiralis* L.

Iridaceae
- A *Anomatheca laxa* (Thunb.) Goldblatt
- A *Aristea ecklonii* Baker
- A *Chasmanthe aethiopica* (L.) N.E. Br.
- A *Chasmanthe bicolor* (Gasp. ex Ten.) N.E. Br.

- A *Chasmanthe floribunda* (Salisb.) N.E. Br.
- A *Crocosmia aurea* (Pappe ex Hook.) Planch.
- A *Crocosmia* × *crocosmiiflora* (V. Lemoine) N.E. Br.
- A *Crocosmia masoniorum* (L. Bolus) N.E. Br.
- A *Crocosmia paniculata* (Klatt) Gloldblatt
- A *Crocosmia pottsii* (Macnab ex Baker) N.E. Br.
- E *Crocus ancyrensis* (Herbert) Maw
- E *Crocus biflorus* Mill.
- E *Crocus chrysanthus* Herb.
- E *Crocus flavus* West.
- A *Crocus* × *hybridus* Petrovic
- E *Crocus kotschyanus* K. Koch
- E *Crocus ligusticus* M.G. Mariotti
- E *Crocus longiflorus* Raf.
- E *Crocus nudiflorus* Sm.
- E *Crocus pulchellus* Herb.
- A *Crocus sativus* L.
- E *Crocus serotinus* Salisb.
- E *Crocus sieberi* J. Gay
- E *Crocus speciosus* M. Bieb.
- A *Crocus* × *stellaris* Haw.
- E *Crocus tommasinianus* Herb.
- E *Crocus vernus* (L.) Hill
- A *Ferraria crispa* Burm.
- A *Freesia alba* (Baker) Gumbl.
- A *Freesia* × *hybrida* L.H. Bailey
- A *Freesia refracta* (Jacq.) Ecklon ex Klatt
- A *Gladiolus blandus* Aiton
- A *Gladiolus cardinalis* Curtis
- E *Gladiolus communis* L.
- E *Gladiolus illyricus* W.D.J. Koch
- E *Gladiolus imbricatus* L.
- E *Gladiolus italicus* Mill.
- A *Gladiolus natalensis* (Eckl.) Hook. f.
- E *Gladiolus palustris* Gaud.
- A *Gladiolus undulatus* L.
- E *Hermodactylus tuberosus* (L.) Mill.
- E *Iris albicans* Lange
- E *Iris cretensis* Janka
- A *Iris ensata* Thunb.
- E *Iris foetidissima* L.
- E *Iris germanica* L.
- E *Iris graminea* L.
- A *Iris* × *hollandica* hort.
- A *Iris japonica* Thunb
- E *Iris latifolia* (Miller) Voss ex Sieber & Voss
- E *Iris* × *lurida* Aiton
- E *Iris lutescens* Lam.
- E *Iris pallida* Lam.
- E *Iris pseudacorus* L.
- E *Iris pumila* L.
- A *Iris* × *robusta* E.S. Anderson
- A *Iris setosa* Pall. ex Link
- E *Iris sibirica* L.
- E *Iris spuria* L.
- A *Iris susiana* L.
- A *Iris tectorum* Maxim.
- E *Iris unguicularis* Poiret
- A *Iris versicolor* L.
- E *Iris* × *iphium* L.
- A *Ixia campanulata* Houtt.
- A *Ixia dubia* Vent.
- A *Ixia maculata* L.
- A *Ixia paniculata* Delaroche
- A *Lapeyrousia cruenta* (Lindley) Baker
- A *Libertia caerulescens* Kunth
- A *Libertia elegans* oepp.
- A *Libertia formosa* Graham
- A *Moraea iridioides* L.
- E *Romulea columnae* Sebast. & Mauri
- A *Romulea rosea* (L.) Ecklon
- A *Sisyrinchium angustifolium* Mill.
- E *Sisyrinchium bermudiana* L.
- A *Sisyrinchium californicum* (Ker-Gawl.) Aiton
- A *Sisyrinchium laxum* Otto ex Sims
- E *Sisyrinchium montanum* E.L. Greene
- A *Sisyrinchium platense* I.M. Johnst.
- A *Sisyrinchium septentrionale* Bicknell
- A *Sisyrinchium striatum* Sm.
- A *Sparaxis bulbifera* (L.) Ker-Gawler
- A *Sparaxis tricolor* (Schneev.) Ker-Gawler
- A *Tigridia pavonia* (L. f.) DC.
- A *Tritonia crocata* (L.) Ker-Gawl.
- A *Tritonia lineata* (Salisb.) Ker-Gawl.
- A *Watsonia borbonica* (Pourr.) Goldblatt

Juncaceae
- A *Juncus anthelatus* (Wiegand) R.E. Brooks & Whittem.
- E *Juncus articulatus* L.
- E *Juncus bracteosus* Jacq.
- E *Juncus bufonius* L.
- A *Juncus canadensis* J. Gay ex Laharpe
- E *Juncus compressus* Jacq. Humb.
- E *Juncus conglomeratus* L.
- A *Juncus dudleyi* Wiegand
- A *Juncus ensifolius* Wikstr.
- E *Juncus foliosus* Desf.
- E *Juncus gerardii* Loisel.
- E *Juncus hybridus* Brot.
- A *Juncus imbricatus* Laharpe
- A *Juncus pallescens* Lam.

- A *Juncus pallidus* R. Br.
- E *Juncus planifolius* R. Br.
- E *Juncus pygmaeus* Rich. ex Thuill.
- A *Juncus radula* Buchen.
- E *Juncus squarrosus* L.
- A *Juncus striatus* Schousb. ex E. Mey.
- A *Juncus subsecundus* N.A. Wakef.
- E *Juncus subulatus* Forsskl
- A *Juncus tenuis* Willd.
- E *Luzula campestris* (L.) DC.
- E *Luzula congesta* (Thuill.) Lej.
- E *Luzula forsteri* (Sm.) DC.
- E *Luzula luzuloides* (Lam.) Dandy & Wilmott
- E *Luzula multiflora* (Retz.) Lej.
- E *Luzula nivea* (Nathh.) DC.
- A *Luzula pallidula* Kirschner

Juncaginaceae
- A *Lilaea scilloides* (Poiret) Hauman
- A *Triglochin striata* Ruiz & Pavón

Lemnaceae
- A *Lemna aequinoctialis* Welw.
- E *Lemna minor* L.
- C *Lemna minuscola* Herter
- A *Lemna minuta* Kunth
- A *Lemna perpusilla* Torr.
- A *Lemna turionifera* Landolt

Liliaceae
- E *Erythronium dens-canis* L.
- A *Fritillaria imperialis* L.
- E *Fritillaria meleagris* L.
- E *Gagea arvensis* (Pers.) Dumort
- E *Gagea lutea* (L.) Ker-Gawl.
- A *Gagea minima* (L.) Ker Gawl.
- E *Gagea pratensis* (Pers.) Dumort
- E *Lilium bulbiferum* L.
- A *Lilium candidum* L.
- A *Lilium lancifolium* Thunb.
- E *Lilium martagon* L.
- E *Lilium pyrenaicum* Gouan
- E *Tulipa agenensis* DC.
- E *Tulipa clusiana* DC.
- A *Tulipa didieri* Jordan
- A *Tulipa gesneriana* L.
- A *Tulipa grengiolensis* Thommen
- E *Tulipa raddii* Reboul
- E *Tulipa saxatilis* Sieber ex Spreng.
- E *Tulipa sylvestris* L.

Melanthiaceae
- E *Veratrum album* L.

Musaceae
- A *Ensete ventricosum* (Welw.) Cheeseman
- A *Musa × cavendishii* Lambert ex Paxton

Orchidaceae
- A *Cymbidium insigne* Rolfe
- A *Cypripedium reginae* Walt.
- E *Dactylorhiza lapponica* (Hartm.) Soó
- E *Goodyera repens* (L.) R. Br.
- E *Ophrys insectifera* L.
- E *Orchis laxiflora* Lam.
- A *Paphiopedilum insigne* (Wall. ex. Lindl.) Pfitz.
- A *Sobralia macrantha* Lindl.
- E *Traunsteinera globosa* (L.) Reichb.

Pistiaceae
- A *Pistia stratiotes* L.

Poaceae
- A *Achnatherum brachychaetum* (Godr.) Barkworth
- E *Aegilops columnaris* Zhuk.
- E *Aegilops cylindrica* Host
- E *Aegilops geniculata* Roth
- A *Aegilops juvenalis* (Thell.) Eig
- E *Aegilops markgrafii* (Greuter) Hamm.
- E *Aegilops ovata* L. nom ambig
- E *Aegilops speltoides* Tausch
- E *Aegilops tauschii* Coss.
- E *Aegilops triuncialis* L.
- E *Aegilops* ventricosa Tausch
- E *Aeluropus littoralis* (Gouan) Parl.
- A × *Agropogon robinsonii* (Druce) Melderis & D.C. McClint.
- E *Agropyron dasyanthum* Ledeb.
- E *Agropyron desertorum* (Fisch. ex Link) Schult.
- E *Agropyron fragile* (Roth) P. Candargy
- E *Agropyron pectinatum* (Bieb.) Beauv.
- A *Agropyron tenerum* Vasey
- A *Agrostis avenacea* J.F. Gmel.
- A *Agrostis billardierei* R. Br.
- A *Agrostis × bjoerkmanii* Widén
- E *Agrostis capillaris* L.
- E *Agrostis castellana* Boiss. & Reut.
- A *Agrostis eriantha* Hack.
- A *Agrostis exarata* rin.
- A *Agrostis × fouilladei* P. Fourn.
- E *Agrostis gigantea* Roth
- A *Agrostis hyemalis* (Walter) Britton, Sterns & Pogg.
- A *Agrostis lachnantha* Nees
- E *Agrostis nebulosa* Boiss. & Reut.
- A *Agrostis scabra* Willd.
- E *Agrostis stolonifera* L.
- E *Aira caryophyllea* L.
- E *Aira elegantissima* Schur
- E *Alopecurus aequalis* Sobol.
- A *Alopecurus carolinianus* Walter

- E *Alopecurus geniculatus* L.
- E *Alopecurus myosuroides* Huds.
- E *Alopecurus pratensis* L.
- E *Alopecurus rendlei* Eig
- E *Alopecurus utriculatus* Sol.
- A *Ammophila breviligulata* Fernald
- E *Ampelodesmos mauritanicus* (Poir.) Th. Dur. & Schinz
- A *Amphibromus recurvatus* W. Swallen
- A *Andropogon distachyos* L.
- E *Anthoxanthum aristatum* Boiss.
- E *Anthoxanthum odoratum* Boiss.
- E *Apera intermedia* Hack.
- E *Apera interrupta* (L.) P. Beauv.
- E *Apera spica-venti* (L.) P. Beauv.
- A *Aristida adscensionis* L.
- A *Aristida congesta* Roem. & Schult.
- A *Aristida hordeacea* Kunth
- A *Aristida longespica* Poiret
- E *Arrhenatherum elatius* (L.) J. Presl & C. Presl
- A *Arundinaria chino* (Franch. & Sav.) Makino
- A *Arundinaria pygmaea* (Miq.) Makino
- A *Arundinaria simonii* (Carr.) Rivire & C.Rivire
- A *Arundo donax* L.
- A *Austrodanthonia setacea* (R. Br.) H.P. Linder
- A *Austrodanthonia tenuior* (Steud.) H.P. Linder
- E *Avena barbata* Pott ex Link
- E *Avena brevis* Roth
- E *Avena byzantina* C. Koch
- E *Avena clauda* Durieu
- E *Avena eriantha* Durieu
- E *Avena fatua* L.
- A *Avena* × *marquandii* Druce
- E *Avena matritensis* Baum
- E *Avena nuda* L.
- E *Avena sativa* L.
- E *Avena sterilis* L.
- E *Avena strigosa* Schreb.
- E *Avenula bromoides* (Gouan) H. Scholz
- A *Axonopus fissifolius* (Raddi) Kuhlm.
- E *Beckmannia eruciformis* (L.) Host
- A *Beckmannia syzigachne* (Steud.) Fernald
- A *Bothriochloa barbinodis* (Lag.) Herter
- A *Bothriochloa decipiens* (Hack.) C.E. Hubbard
- E *Bothriochloa insculpta* (A. Rich.) A. Camus
- E *Bothriochloa ischaemum* (L.) Keng
- A *Bothriochloa macra* (Steud.) S.T. Blake
- A *Bothriochloa saccharoides* (Sw.) Rydb.
- A *Brachiaria eruciformis* (Sibth. & Sm.) Griseb.
- A *Brachiaria mutica* (Forssk.) Stapf
- E *Brachiaria platyphylla* (Munro ex Wright) Nash
- A *Brachiaria texana* (Buckley) S.T. Blake
- A *Brachyachne tenella* (R. Br.) C.E. Hubb.
- E *Brachypodium distachyon* (L.) P. Beauv.
- E *Brachypodium pinnatum* (L.) P. Beauv.
- E *Briza maxima* L.
- E *Briza media* L.
- E *Briza minor* L.
- E *Bromus alopecuros* Poir.
- E *Bromus arvensis* L.
- A *Bromus brachystachys* Hornung
- A *Bromus brevis* Steud.
- E *Bromus briziformis* Fisch. & Mey.
- E *Bromus carinatus* Hook. & Arn.
- E *Bromus catharticus* Vahl
- A *Bromus cebadilla* Steud.
- E *Bromus commutatus* Schrad.
- A *Bromus danthoniae* Trin.
- E *Bromus diandrus* Roth
- E *Bromus erectus* Huds.
- E *Bromus grossus* Desf. ex DC.
- E *Bromus hordeaceus* L.
- E *Bromus inermis* Leyss.
- E *Bromus interruptus* G.C. Druce
- E *Bromus japonicus* Thunb.
- A *Bromus kalmii* A. Gray
- E *Bromus lanceolatus* Roth
- E *Bromus lepidus* Holmb.
- E *Bromus madritensis* L.
- A *Bromus marginatus* Steud.
- A *Bromus pectinatus* Thunb.
- A *Bromus pseudosecalinus* P.M. Sm.
- A *Bromus* × *pseudothominii* P.M. Sm.
- A *Bromus pumpellianus* Scribn.
- E *Bromus racemosus* L.
- E *Bromus rigidus* Roth
- E *Bromus riparius* Rehmann
- E *Bromus rubens* L.
- E *Bromus scoparius* L.
- E *Bromus secalinus* L.
- A *Bromus sitchensis* (Trin.) T.A. Cope & T.B. Ryves
- E *Bromus squarrosus* L.
- E *Bromus sterilis* L.
- E *Bromus tectorum* L.
- E *Calamagrostis arundinacea* (L.) Roth
- E *Catapodium marinum* (L.) C.E. Hubb.
- E *Catapodium rigidum* (L.) C.E. Hubb.
- E *Catapodium siculum* (Jacq.) Link

- A *Cenchrus biflorus* Roxb.
- E *Cenchrus ciliaris* L.
- E *Cenchrus echinatus* L.
- E *Cenchrus incertus* M.A. Curtis
- A *Cenchrus longispinus* (Hack.) Fernald
- A *Chasmanthium latifolium* (Michx.) Yates
- A *Chimonobambusa quadrangularis* (Fenzi) Makino
- A *Chloris barbata* Sw.
- A *Chloris divaricata* R. Br.
- A *Chloris gayana* Kunth
- A *Chloris pectinata* Benth.
- A *Chloris pycnothrix* Trin.
- A *Chloris radiata* (L.) Sw.
- A *Chloris truncata* R. Br.
- A *Chloris ventricosa* R. Br.
- A *Chloris virgata* Sw.
- A *Coix lacryma-jobi* L.
- E *Cornucopiae cucullatum* L.
- A *Cortaderia richardii* (Endl.) Zotov
- E **Cortaderia selloana** (Schult. & Schult. f.) Aschers. & Graebn.
- E *Corynephorus divaricatus* (Pourr.) Breistr.
- E *Crypsis aculeata* (L.) Ait.
- E *Crypsis alopecuroides* (Pill. & Mitterp.) Schrad.
- E *Crypsis schoenoides* (L.) Lam.
- A *Ctenopsis cynosuroides* (Desf.) Paunero ex Romero García
- E *Ctenopsis pectinella* (Delile) De Not.
- E *Cutandia memphitica* (Spreng.) Benth
- E *Cynodon dactylon* (L.) Pers.
- A *Cynodon incompletus* Nees
- E *Cynosurus cristatus* L.
- E *Cynosurus echinatus* L.
- E *Dactylis glomerata* L.
- A *Dactyloctenium aegyptinum* (L.) Willd.
- A *Dactyloctenium australe* Steud.
- A *Dactyloctenium radulans* (R. Br.) P. Beauv.
- E *Danthonia decumbens* (L.) DC.
- E *Danthonia gracilis* J.D. Hook.
- A *Danthonia spicata* (L.) Beauv. ex Roemer & Schultes
- E *Dasypyrum villosum* (L.) P. Candargy
- E *Deschampsia cespitosa* (L.) P.Beauv.
- A *Dichanthelium acuminatum* (Sw.) Gould & C.A. Clark
- A *Dichanthium annulatum* (Forssk.) Stapf
- A *Dichanthium intermedium* (R.Br.) de Wet & J.R. Harlan
- A *Dichanthium sericeum* (R. Br.) A. Camus
- A *Digitaria acuminatissima* Stapf
- A *Digitaria aequiglumis* (Hack. & Arech.) Parodi
- A *Digitaria brownii* (Roem. & Schult.) Hughes
- E *Digitaria ciliaris* (Retz.) Koeler
- E *Digitaria debilis* (Desf.) Willd.
- E *Digitaria ischaemum* (Schreb.) Muhl
- A *Digitaria longiflora* (Retz.) Pers.
- E *Digitaria sanguinalis* (L.) Scop.
- A *Digitaria violascens* Link
- A *Dinebra retroflexa* (Vahl) Panz.
- A *Diplachne fusca* (L.) P. Beauv.
- E *Echinaria capitata* (L.) Desf.
- E *Echinochloa colona* (L.) Link
- E *Echinochloa crus-galli* (L.) P. Beauv.
- A *Echinochloa elliptica* Michael & Vickery
- A *Echinochloa esculenta* (A. Braun) H. Scholz
- A *Echinochloa frumentacea* Link
- A *Echinochloa inundata* Michael & Vickery
- A *Echinochloa muricata* (Beauv.) Fernald
- E *Echinochloa oryzoides* (Ard.) Fritsch
- A *Echinochloa turneriana* (Domin) J.M. Black
- A *Ehrharta calycina* Sm.
- E *Ehrharta erecta* Lam.
- A *Ehrharta longiflora* Sm.
- A *Ehrharta stipoides* Labill.
- A *Eleusine africana* Kenn.-O'Byrne
- A *Eleusine coracana* (L.) Gaertn.
- A *Eleusine indica* (L.) Gaertn.
- A *Eleusine multiflora* Hochst. ex A. Rich.
- A *Eleusine tristachya* (Lam.) Lam.
- A *Elymus athericus* (Link) Kerguélen
- A *Elymus canadensis* L.
- E *Elymus dahuricus* Turcz. ex Griseb.
- E *Elymus elongatus* (Host) Runemark
- E *Elymus fibrosus* (Schrenk) Tzvelev
- A *Elymus hystrix* L.
- E *Elymus junceus* Fisch.
- A *Elymus scabrus* (R. Br.) Á. Löve
- A *Elymus sibiricus* L.
- E *Elytrigia intermedia* (Host) Nevski
- E *Elytrigia repens* (L.) Desv. ex. Nevski
- A *Enneapogon desvauxii* P. Beauv.
- A *Enneapogon scaber* Lehm.
- E *Enteropogon acicularis* (Lindl.) Lazarides
- A *Eragrostis advena* (Stapf) S.M. Phillips
- E *Eragrostis albensis* H. Scholz
- A *Eragrostis alveiformis* Lazarides
- A *Eragrostis atrovirens* (Desf.) Trin. ex Steud.

A *Eragrostis bahiensis* Schrad. ex Schult.
E *Eragrostis barrelieri* Daveau
A *Eragrostis bicolor* Nees
A *Eragrostis brownii* (Kunth) Nees ex Wight
A *Eragrostis capillaris* (L.) Nees
A *Eragrostis caroliniana* (Biehler) Scribn.
E *Eragrostis cilianensis* (All.) Janchen
E *Eragrostis ciliaris* (L.) R. Br.
A *Eragrostis curvula* (Schrad.) Nees
A *Eragrostis dielsii* Pilger
A *Eragrostis echinochloidea* Stapf
A *Eragrostis elongata* (Willd.) Jacq.
A *Eragrostis flaccida* Lindm.
A *Eragrostis frankii* C.A. Mey. ex Steud.
A *Eragrostis japonica* (Thunb.) Trin.
A *Eragrostis lac*unaria F. Müll. ex Benth.
A *Eragrostis leptocarpa* Benth.
A *Eragrostis leptostachya* (R. Br.) Steud.
A *Eragrostis lugens* Nees
A *Eragrostis mexicana* (Hornnm.) Link
E *Eragrostis minor* Becker ex Claus
E *Eragrostis multicaulis* Steud.
A *Eragrostis neesii* Trin.
A *Eragrostis nigra* Nees ex Steud
A *Eragrostis obtusa* Munro ex Ficalho & Hiern
A *Eragrostis palmeri* S. Wats.
E *Eragrostis papposa* (Duf.) Steud
A *Eragrostis parviflora* (R. Br.) Trin.
A *Eragrostis pectinacea* (Michx.) Nees
E *Eragrostis pilosa* (L.) Beauv.
A *Eragrostis plana* Nees
A *Eragrostis planiculmis* Nees
A *Eragrostis procumbens* Nees
A *Eragrostis prolifera* (Sw.) Steud.
A *Eragrostis rotifer* Rendle
A *Eragrostis sarmentosa* (Thunb.) Trin.
A *Eragrostis schimperi* Benth.
A *Eragrostis tef* (Zucc.) Trotter
A *Eragrostis trichodes* Wood
A *Eragrostis trichophora* Coss. & Durieu
A *Eremopoa persica* (Trin.) Roshev.
A *Eremopyrum bonaepartis* (Spreng.) Nevski
E *Eremopyrum orientale* (L.) Jaub. & Spach
E *Eremopyrum triticeum* (Gaertn.) Nevski
A *Eriochloa australiensis* Stapf ex Thell.
A *Eriochloa crebra* S.T. Blake
A *Eriochloa procera* (Retz.) C.E. Hubb.
A *Eriochloa pseudoacrotricha* (Stapf ex Thell.) J.M. Black
E *Eriochloa succinta* (Trin.) Kunth
A *Eriochloa villosa* (Thunb.) Kunth
A *Eustachys neglecta* (Nash) Nash
A *Eustachys retusa* (Lag.) Kunth
A *Fargesia spathacea* Franch.
E *Festuca arundinacea* Schreb.
E *Festuca filiformis* Pourr.
E *Festuca gautieri* (Hack.) K. Richter
E *Festuca glauca* Vill.
E *Festuca heterophylla* Lam.
A *Festuca nigrescens* Lam.
E *Festuca ovina* L.
E *Festuca pratensis* Huds.
E *Festuca rubra* L.
E *Festuca rupicaprina* (Hack.) A. Kern.
E *Festuca trachyphylla* (Hack.) Kroj.
A × *Festulolium braunii* (K. Rich.) A. Camus
A × *Festulolium loliaceum* (Huds.) F. Fourn.
E *Gastridium phleoides* (Nees & Meyen) C.E. Hubb.
E *Gastridium ventricosum* (Gouan) Schinz & Thellung
E *Gaudinia fragilis* (L.) Beauv.
E *Glinus lotoides* L.
A *Glyceria canadensis* (Michx.) Trin.
E *Glyceria declinata* Breb.
E *Glyceria fluitans* (L.) R.Br.
A *Glyceria grandis* Wats.
E *Glyceria maxima* (Hartm.) O.R. Holmberg
A *Glyceria striata* (Lam.) Hitchc.
E *Hainardia cylindrica* (Willd.) Greuter
A *Helictotrichon neesii* (Steudel) C.A. Stace
E *Helictotrichon pratense* (L.) Pilger
E *Helictotrichon pubescens* (Huds.) Schult.
E *Helictotrichon sempervirens* (Vill.) Pilger
E *Hemarthria altissima* (Poiret) Stapf & C.E. Hubbard
A *Hemarthria compressa* (L. f.) R.Br.
E *Heteropogon contortus* (L.) Beauv. ex Roem. & Schult.
E *Hierochloë odorata* (L.) P. Beauv.
E *Holcus lanatus* L.
E *Holcus mollis* L.
E *Hordeum bulbosum* L.
A *Hordeum distichon* L.
A *Hordeum euclaston* Steud.
A *Hordeum flexuosum* Nees ex Steud.
A *Hordeum jubatum* L.
E *Hordeum marinum* Huds.
E *Hordeum marinum gussoneanum* (Parl.) Thell.

- E *Hordeum murinum* L.
- E *Hordeum murinum glaucum* (Steud.) Tzvelev
- E *Hordeum murinum leporinum* (Link) Arcang.
- A *Hordeum pubiflorum* Hook. f.
- A *Hordeum pusillum* Nutt.
- E *Hordeum secalinum* Schreb.
- A *Hordeum turkestanicum* Nevski
- E *Hordeum vulgare* L.
- E *Hordeum vulgare agriocrithon* (Aberg) A. & D. Love
- E *Hyparrhenia hirta* (L.) Stapf
- A *Hyparrhenia rufa* (Ness) Stapf
- E *Imperata cylindrica* (L.) Raeusch.
- E *Koeleria macrantha* (Ledeb.) Schult.
- E *Koeleria pyramidata* (Lam.) Beauv.
- E *Lagurus ovatus* L.
- E *Lamarckia aurea* (L.) Moench
- E *Leersia oryzoides* (L.) Sw.
- A *Leptochloa brachiata* Steud.
- A *Leptochloa chinensis* (L.) Nees
- A *Leptochloa decipiens* (R. Br.) Stapf ex Maiden
- A *Leptochloa divaricatissima* S.T. Blake
- A *Leptochloa dubia* (Kunth) Nees
- A *Leptochloa mucronata* (Michx.) Kunth.
- A *Leptochloa parviflora* (R. Br.) Verloove & Lambinon
- A *Leptochloa peacockii* (Maiden & Betche) Domin
- E *Leymus arenarius* (L.) Hochst.
- A *Leymus innovatus* (Beal) Pilger
- E *Leymus racemosus* (Lam.) Tzvel.
- E *Lolium × boucheanum* Kunth
- E *Lolium multiflorum* Lam.
- E *Lolium perenne* L.
- E *Lolium persicum* Boiss. & Hohen.
- E *Lolium remotum* Schrank
- E *Lolium rigidum* Gaud.
- E *Lolium temulentum* L.
- E *Melica altissima* L.
- A *Melinis minutiflora* P. Beauv.
- E *Mibora minima* (L.) Desv.
- E *Micropyrum tenellum* (L.) Link
- A *Microstegium vimineum* (Trin.) A. Camus
- E *Milium vernale* Bieb.
- A *Miscanthus sacchariflorus* (Maxim.) Hack.
- A *Miscanthus sinensis* Anderss.
- A *Muhlenbergia frondosa* (Poir.) Fernald
- A *Muhlenbergia mexicana* (L.) Trin.
- A *Muhlenbergia schreberi* J.F. Gmel.
- A *Muhlenbergia vaginiflora* (Torr. ex A. Gray) Jogan
- A *Nasella neesiana* (Trin. & Rupr.) Barkworth
- A *Nassella mucronata* (Kunth) R.W. Pohl
- A *Nassella poeppigiana* (Trin. & Rupr.) Barkworth
- A *Nassella tenuissima* (Trin.) Barkworth
- A *Nassella trichotoma* (Nees) Hackel ex Arech.
- A *Oplismenus hirtellus* (L.) P. Beauv.
- E *Oplismenus undulatifolius* (Ard.) Roemer & Schultes
- A *Oryza rufipogon* Griff.
- A *Oryza sativa* L.
- A *Oryzopsis hymenoides* (Roem. & Schult.) Ricker ex Piper
- A *Oxychloris scariosa* (F. Müll.) Lazarides
- A *Panicum antidotale* Retz.
- A *Panicum bergii* Arechav.
- A *Panicum bulbosum* H. B. & K.
- A *Panicum capillare* L.
- A *Panicum capillare barbipulvinatum* (Nash) Tzvelev
- A *Panicum clandestinum* L.
- A *Panicum coloratum* L.
- A *Panicum decompositum* R. Brown
- A *Panicum dichotomiflorum* Michx.
- A *Panicum dichotomum* L.
- A *Panicum hillmanii* Chase
- A *Panicum hirticaule* J. Presl
- A *Panicum laevinode* Lindl.
- A *Panicum maximum* Jacq.
- E *Panicum miliaceum* L.
- A *Panicum obtusum* H. B. & K.
- A *Panicum oligosanthes* Schult.
- A *Panicum philadelphicum* Bernh. ex Trin.
- E *Panicum repens* L.
- A *Panicum repentellum* Napper
- A *Panicum schinzii* Hack. ex Schinz
- A *Panicum virgatum* L.
- E *Parapholis incurva* (L.) C.E. Hubb.
- E *Parapholis strigosa* (Dum.) C.E. Hubb.
- A *Parvotrisetum myrianthum* (Bertol.) Chrtek
- E *Paspalum dilatatum* Poiret
- E *Paspalum distichum* L.
- A *Paspalum notatum* Flüggé
- A *Paspalum paniculatum* L.
- A **Paspalum paspalodes** (Michx) Scribner
- A *Paspalum pubiflorum* Rupr. ex E. Fourn.
- A *Paspalum quadrifarium* Lam.
- A *Paspalum saurae* (L. Parodi) L. Parodi
- A *Paspalum thunbergii* Kunth ex Steud.

- A *Paspalum urvillei* Steud.
- A *Paspalum vaginatum* Swartz
- A *Pennisetum clandestinum* Chiov.
- A *Pennisetum flaccidum* Griseb.
- A *Pennisetum orientale* Rich.
- A *Pennisetum petiolare* (Hochst.) Chiov.
- A *Pennisetum purpureum* Schum.
- A *Pennisetum setaceum* (Forsk.) Chiov.
- A *Pennisetum villosum* R. Br. ex Fresen.
- A *Pentaschistis airoides* (Nees) Stapf
- A *Phalaris angusta* Nees ex Trin.
- E *Phalaris aquatica* L.
- E *Phalaris arundinacea* L.
- E *Phalaris brachystachys* Link
- E *Phalaris canariensis* L.
- E *Phalaris coerulescens* Desf.
- E *Phalaris minor* Retz.
- E *Phalaris paradoxa* L.
- A *Phalaris platensis* Henrard ex Heukels
- E *Phleum arenarium* L.
- E *Phleum echinatum* Host
- E *Phleum exaratum* Griseb.
- A *Phleum graecum* Boiss. & Heldr.
- E *Phleum hirsutum* Honck.
- E *Phleum paniculatum* Huds.
- E *Phleum phleoides* (L.) H. Karst.
- E *Phleum pratense* L.
- E *Phleum rhaeticum* (Humphries) Rauschert
- E *Phleum subulatum* (Savi) Ascherson & Graebner
- A *Phyllostachys aurea* Rivire & C. Rivire
- A *Phyllostachys bambusoides* Sieb. & Zucc.
- A *Phyllostachys nigra* (Lodd.) Munro
- A *Phyllostachys pubescens* Mazel ex Houz.
- A *Phyllostachys viridiglaucescens* (Carrire) A. & C. Rivire
- A *Phyllostachys viridis* (Young) McClure
- E *Piptatherum miliaceum* (L.) Cosson
- A *Pleioblastus argenteo-striatus* (Regel) Nakai
- A *Pleuraphis jamesii* Torr.
- E *Poa alpina* L.
- E *Poa angustifolia* L.
- E *Poa annua* L.
- E *Poa bulbosa* L.
- E *Poa chaixii* Vill.
- E *Poa compressa* L.
- E *Poa crispa* Thuill.
- A *Poa flabellata* (Lam.) Raspail
- E *Poa infirma* H. B. & K.
- E *Poa jubata* A. Kern.
- E *Poa nemoralis* L.
- E *Poa palustris* L.
- E *Poa pratensis* L.
- A *Poa sieberiana* Spreng.
- E *Poa supina* Schrad.
- E *Poa trivialis* L.
- A *Polypogon chilensis* (Kunth) Pilger
- A *Polypogon elongatus* Humb., Bonpl. & Kunth
- E *Polypogon fugax* Nees ex Steud.
- A *Polypogon maritimus* Willd.
- E *Polypogon monspeliensis* (L.) Desf.
- A *Polypogon rioplatensis* Herter
- E *Polypogon viridis* (Gouan) Breistr.
- A *Pseudosasa japonica* (Steud.) Makino
- C *Puccinellia ciliata* Bor
- E *Puccinellia distans* (Jacq.) Parl.
- A *Puccinellia fominii* Bilyk
- E *Puccinellia gigantea* (Grossh.) Grossh.
- E *Puccinellia hauptiana* Krecz.
- A *Puccinellia nuttaliana* (Schult.) Hitchc.
- A *Puccinellia stricta* (Hook. f.) Blom
- E *Rostraria cristata* (L.) Tzvelev
- A *Rytidosperma racemosum* (R.Br.) H.E. Connor & E. Edgar
- A *Rytidosperma tenuius* (Steud.) Connor & Edgar
- A *Rytidosperma thomsonii* (Buchanan) Connor & Edgar
- A *Saccharum officinarum* L.
- E *Saccharum ravennae* (L.) Murr.
- E *Saccharum spontaneum* L.
- A *Sasa palmata* (Burb.) E.G. Camus
- A *Sasa ramosa* (Makino) Makino & Shibata
- A *Sasa veitchii* (Carr.) Rehder
- E *Schismus barbatus* (L.) Thell.
- A *Schmidtia kalahariensis* Stent
- A *Schoenefeldia gracilis* Kunth
- A *Sclerochloa dura* (L.) Beauv.
- E *Secale cereale* L.
- E *Secale montanum* Guss.
- A *Semiarundinaria fastuosa* (Mitf.) Makino
- A *Seslaria dactyloides* Nutt.
- E *Sesleria caerulea* (L.) Arduin.
- A *Setaria adhaerens* (Forssk.) Chiov.
- A *Setaria dielsii* R.A.W. Herrmann
- E *Setaria faberi* Herrm.
- A *Setaria globulifera* (Steud.) Griseb.
- A *Setaria grisebachii* Fourn.
- E *Setaria italica* (L.) Beauv.
- A *Setaria megaphylla* (Steud.) Dur. & Schinz
- E *Setaria pachystachys* (Franch. & Sav.) Matsum.

- E *Setaria parviflora* (Poiret) M. Kerguelen
- E *Setaria pumila* (Poiret) Schultes
- A *Setaria pycnocoma* (Steud.) Henrard ex Nakai
- A *Setaria sphacelata* (Schumach.) Stapf & C.E. Hubbard ex M.B. Moss
- E *Setaria verticillata* (L.) P. Beauv.
- E *Setaria viridis* (L.) P. Beauv.
- A *Sorghum bicolor* (L.) Moench
- E *Sorghum halepense* (L.) Pers.
- A *Sorghum saccharatum* (L.) Moench
- A *Spartina alterniflora* Loisel.
- A **Spartina anglica** C.E. Hubb.
- A *Spartina densiflora* Brongn.
- A *Spartina juncea* auct., non (Michx) Willd.
- E *Spartina maritima* (Curtis) Fernald
- A *Spartina patens* (Ait.) Muhl.
- A *Spartina pectinata* Link
- E *Spartina* × *townsendii* H. & J. Groves
- A *Spartina versicolor* Fabre
- E *Sporobolus africanus* (Poir.) Robyns & Tournay
- A *Sporobolus carolii* Mez
- A *Sporobolus creber* De Nardi
- A *Sporobolus cryptandrus* (Torr.) A. Gray
- A *Sporobolus elongatus* R. Br.
- A *Sporobolus fertilis* (Steud.) W.D. Clayton
- E *Sporobolus indicus* (L.) R. Br.
- A *Sporobolus neglectus* Nash
- A *Sporobolus pyramidatus* (Lam.) Hitchcock
- A *Sporobolus vaginiflorus* (Gray) Wood
- A *Sporobolus virginicus* (L.) Kunth
- E *Stenotaphrum secundatum* (Walter) O. Kuntze
- E *Stipa bromoides* (L.) Dörfl.
- E *Stipa calamagrostis* (L.) Wahlenb.
- E *Stipa capensis* Thunb.
- E *Stipa capillata* L.
- A *Stipa filiculmis* Delile
- A *Stipa formicarum* Delile
- A *Stipa hyalina* Nees
- A *Stipa nitida* Summerh. & C.E. Hubbard
- A *Stipa papposa* Nees
- E *Stipa pennata* L.
- A *Stipa poeppigiana* Trin. & Rupr.
- E *Stipa splendens* Trin.
- E *Stipa tenacissima* L.
- A *Stipa verticillata* Nees ex Spreng.
- A *Stipagrostis pungens* (Desf.) De Winter
- E *Taeniatherum caput-medusae* (L.) Nevski
- E *Taeniatherum crinitum* (Schreb.) Nevski
- A *Thamnocalamus tesselatus* (Nees) Soderstr. & R.P. Ellis
- A *Tragus australianus* S.T. Blake
- A *Tragus berteronianus* Schult.
- E *Tragus racemosus* (L.) All.
- A *Triraphis mollis* R. Br.
- E *Trisetum flavescens* (L.) Beauv.
- E *Trisetum paniceum* (Lam.) Pers.
- E *Triticum aestivum* L.
- E *Triticum baeoticum* Boiss.
- A *Triticum carthlicum* Nevski
- E *Triticum compactum* Host
- E *Triticum dicoccon* Schrank
- E *Triticum durum* Desf.
- E *Triticum monococcum* L.
- A *Triticum polonicum* L.
- E *Triticum spelta* L.
- E *Triticum turgidum* L.
- E *Urochloa panicoides* Beauv.
- E *Vulpia alopecuros* (Schousb.) Link
- A *Vulpia bromoides* (L.) S.F. Gray
- E *Vulpia ciliata* Dum.
- E *Vulpia delicatula* (Lag.) Dumort
- A *Vulpia eriolepis* (Desv.) C. Blom
- E *Vulpia geniculata* (L.) Link
- E *Vulpia ligustica* (All.) Link
- E *Vulpia muralis* (Kunth) Nees
- E *Vulpia myuros* (L.) C.C. Gmel.
- A *Vulpia octoflora* (Walter) Rydberg
- E *Vulpia unilateralis* (L.) C.A. Stace
- A *Yushania anceps* (Mitford) W.C. Lin
- A *Zea mays* L.
- A *Zizania aquatica* L.
- A *Zizania latifolia* (Griseb.) Turcz. ex Stapf

Pontederiaceae
- A *Eichhornia crassipes* (Mart.) Solms
- A *Heteranthera limosa* (Sw.) Willd.
- A *Heteranthera reniformis* Ruiz & Pav.
- A *Heteranthera rotundifolia* (Kunth) Griseb.
- A *Heteranthera zosterifolia* Martius
- A *Monochoria korsakowii* Regel & Maack
- A *Pontederia cordata* L.
- A *Pontederia renniformis* Larraaga
- A *Pontederia rotundifolia* L.f.

Potamogetonaceae
- E *Groenlandia densa* (L.) Fourr.
- E *Potamogeton compressus* L.
- E *Potamogeton epihydrus* Rafn.
- E *Potamogeton nodosus* Poiret
- E *Potamogeton trichoides* Cham. & Schlecht.

Ruscaceae
- E *Convallaria majalis* L.
- A *Danae racemosa* (L.) Moench

E *Maianthemum bifolium* (L.) F.W. Schmidt
A *Maianthemum stellatum* (L.) Link
A *Ophiopogon japonicus* (L. f.) Ker-Gawl.
A *Polygonatum* × *hybridum* Brügger
E *Polygonatum latifolium* (Jacq.) Desf.
E *Polygonatum multiflorum* (L.) All.
A *Reineckea carnea* (Andrews) Kunth
E *Ruscus aculeatus* L.
E *Ruscus hypoglossum* L.
E *Ruscus hypo*phyllum L.
A *Sansevieria trifasciata* Prain
E *Semele androgyna* (L.) Kunth
Smilacaceae
 E *Smilax aspera* L.
 E *Smilax canariensis* Brouss. ex Willd.
 E *Smilax excelsa* L.
Typhaceae
 A *Typha domingensis* (Pers.) Steud.
 E *Typha laxmannii* Lepechin
 E *Typha minima* Funck
Zingiberaceae
 A *Alpinia zerumbet* (Pers.) Burtt & R.M. Sm.
 A *Hedychium flavescens* Carey ex Roscoe
 A **Hedychium gardnerianum** Shepard ex Ker-Gawl.

Cnidaria
Aiptasiidae
 A *Aiptasia pulchella* Carlgren, 1943
Bougainvilliidae
 A *Bimeria franciscana* Torrey, 1902
 A *Bougainvillia macloviana* Lesson, 1836
 C *Bougainvillia megas* (Kinne, 1896)
 A *Bougainvillia rugosa* Clarke, 1882
 A *Clavopsella navis* (Millard, 1959)
 A *Garveia franciscana* Torrey, 1902
 A *Nemopsis bachei* L. Agassiz, 1849
 A *Thieliana navis* (Millard, 1959)
Campanulariidae
 A *Clytia hummelincki* (Leloup, 1934)
 A *Clytia linearis* (Thorneley, 1900)
 C *Blackfordia virginica* Mayer, 1910
Cassiopeidae
 A *Cassiopeia andromeda* (Forsskål, 1775)
Clavidae
 C **Cordylophora caspia** (Pallas, 1766)
 A *Rhizogeton nudum* Broch, 1909
Corynidae
 A *Coryne pintneri* Schneider, 1897
Diadumenidae
 A *Diadumene cincta* Stephenson, 1925

Diphyidae
 C *Muggiaea atlantica* Cunningham, 1892
Haliplanellidae
 A *Haliplanella lineata* (Verrill, 1870)
Kirchenpaueriidae
 C *Ventromma halecioides* (Alder, 1859)
Lovenellidae
 A *Eucheilota paradoxica* Mayer, 1900
Mastigiidae
 A *Phyllorhiza punctata* von Lendenfeld, 1884
Melithaeidae
 A *Acabaria erythraea* (Ehrenberg, 1834)
Moerisiidae
 A *Moerisia maeotica* (Ostroumov, 1896)
 A *Ostroumovia inkermanica* (Paltschikowa-Ostroumova, 1925)
Oculinidae
 A *Oculina patagonica* de Angelis, 1908
Olindiidae
 E *Craspedacusta sowerbyi* Lankester, 1880
 A *Gonionemus vertens* A. Agassiz, 1862
 E *Maeotias marginata* Modeer, 1791
Plumulariidae
 A *Macrorhynchia philippina* (Kirchenpauer, 1872)
Polypodiidae
 E *Polypodium hydriforme* Ussov, 1885
Rathkeidae
 C *Rathkea octopunctata* (Sars, 1835)
Rhizostomatidae
 A **Rhopilema nomadica** Galil, 1990
Sagartiidae
 E *Cereus pedunculatus* (Pennant, 1777)
Tubulariidae
 A *Tubularia crocea* (L. Agassiz, 1862)
 E *Tubularia indivisa* Linnaeus, 1758
Ulmaridae
 E *Aurelia aurita* Linnaeus, 1758

Ctenophora
Beroidae
 A *Beroe ovata* Bruguiere, 1789
Bolinopsidae
 A **Mnemiopsis leidyi** A. Agassiz, 1865

Porifera
Chalinidae
 C *Chalinula loosanoffi* (Hartmann, 1958)
 C *Haliclona rosea* (Bowerbank, 1866)
 C *Haliclona simplex* (Czerniavsky, 1880)

Halichondriidae
    C *Hymeniacidon perlevis* (Montagu, 1818)
Mycalidae
    C *Mycale micracanthoxea* Buizer & Van Soest, 1977
Spongillidae
    A *Eunapius carteri* (Bowerbank, 1863)
Suberitidae
    C *Suberites massa* Nardo, 1847
Sycettidae
    C *Scypha scaldiensis* Van Koolwijk, 1982
Tetillidae
    A *Cinachyrella alloclada* (Uliczka, 1929)

Platyhelminthes
Anoplocephalidae
    E *Andrya cuniculi* (Blanchard, 1891)
    E *Mosgovoyia ctenoides* (Railliet, 1890)
Bipaliidae
    A *Bipalium kewense* Moseley, 1878
Bothriocephalidae
    A *Bothriocephalus acheilognathi* Yamaguti, 1934
    E *Bothriocephalus claviceps* (Goeze, 1782)
Callioplanidae
    A *Pseudostylochus ostreophagus* Hyman, 1955
Dactylogyridae
    A *Dactylogyrus vastator* Nybelin, 1924
    E *Dactylogyrus yinwenyingae* Gusev, 1962
    A *Glyphidohaptor plectocirra* (Paperna, 1972)
    A *Pseudodactylogyrus anguillae* (Yin & Sproston, 1948)
    A *Pseudodactylogyrus bini* (Kikuchi, 1929)
    C *Tetrancistrum polymorphus* (Paperna, 1972)
    A *Tetrancistrum strophosolenum* Kritsky, Galli & Yang, 2007
    A *Tetrancistrum suezicus* (Paperna, 1972)
    A *Urocleidus dispar* (Müller, 1936)
    A *Urocleidus similis* (Müller, 1936)
Davaineidae
    A *Raillietina friedbergeri* (von Linstow, 1878)
Dendrocoelidae
    C *Bdellocephala punctata* (Pallas, 1774)
Dendrocoelidae
    E *Dendrocoelum romanodanubiale* (Codreanu, 1949)
Didymozoidae
    C *Monilicaecum ventricosum* Yamaguti, 1942

Diphyllobothriidae
    C *Diphyllobothrium latum* (Linnaeus, 1758)
    C *Ligula intestinalis* (Linnaeus, 1758)
Diplostomatidae
    C *Diplostomum helveticum* (Dubois, 1929)
    C *Diplostomum nordmanni* Shigin & Shapirov, 1986
    C *Diplostomum paraspathaceum* Shigin, 1965
    C *Diplostomum p*hoxini (Faust, 1918)
    C *Diplostomum volvens* Nordmann, 1832
    C *Tetracotyle percafluviatilis* Linstow, 1856
    C *Posthodiplostomum cuticola* (Nordmann, 1832)
    C *Tylodelphys clavata* (von Nordmann, 1832)
    C *Tylodelphys podicipina* Kozicka & Niewiadomska, 1960
Diplozoidae
    C *Diplozoon paradoxum* von Nordmann, 1832
Fasciolidae
    A *Fascioloides magna* (Bassi, 1875)
Hymenolepididae
    A *Echinolepis carioca* (Magelhaes, 1898)
Faustulidae
    C *Pseudobacciger harengulae* (Yamaguti, 1938)
Geoplanidae
    A *Arthurdendyus triangulata* (Dendy, 1895)
    A *Australopacifica coxii* (Fletcher & Hamilton, 1888)
Gyrodactylidae
    A *Gyrodactylus gasterostei* Glaser, 1974
    E ***Gyrodactylus salaris*** Malmberg, 1957
Hemiuridae
    A *Hysterolecitha sigani* Manter, 1969
    A *Lecithochirium magnicaudatum* Fischthal & Kuntz, 1963
Heterophyidae
    C *Parorchis avitus* (Linton, 1914)
Leptocreadiidae
    A *Allolepidapedon fistularia* Yamaguti, 1940
Leptoplanidae
    A *Euplana gracilis* (Girard, 1850)
Lytocestidae
    C *Caryophyllaeides fennica* (Schneider, 1902)
    C *Caryophyllueus laticeps* (Pallas, 1781)

Lytocestidae
  A *Khawia sinensis* Hsü, 1935
Monocelididae
  A *Pseudomonocelis cetinae* Meixner, 1943
Notocotylidae
  A *Quinqueserialis quinqueserialis* Barker & Laughlin, 1911
Opisthorchiidae
  A *Clonorchis sinensis* (Cobbold, 1875)
  A *Opisthorchis viverrini* (Poirier, 1886)
Planariidae
  A *Dugesia tigrina* (Girard, 1850)
Promesostomidae
  A *Promesostoma bilineatum* Pereyaslawzevw, 1892
Prosthogonimidae
  C *Prosthogonimus pellucidus* (von Linstow, 1873)
Rhynchodemidae
  A *Dolichoplana fieldeni* von Graff, 1899
  A *Rhynchodemus sylvaticus* (Leidy, 1851)
Stylochidae
  A *Imogine necopinata* Sluys, Faubel, Rajagopal, & van der Velde, 2005
  A *Stylochus flevensis* (Hofker, 1930)
Taeniidae
  A *Taenia multiceps* Leske, 1780

Mollusca
Aeolidiidae
  A *Aeolidiella indica* (Bergh, 1888)
Aglajidae
  A *Chelidonura fulvipunctata* Baba, 1938
Agriolimacidae
  E *Deroceras agreste* (Linnaeus, 1758)
  E *Deroceras caruanae* (Pollonera, 1891)
  E *Deroceras lothari* Giusti, 1973
  E *Deroceras oertzeni* (Simroth, 1889)
  E *Deroceras panormitanum* (Lessona & Pollonera, 1882)
  E *Deroceras reticulatum* (O.F. Müller, 1774)
  E *Deroceras sturanyi* (Simroth, 1894)
  E *Krynickillus melanocephalus* Kaleniczenko, 1851
Amathinidae
  A *Amathina tricarinata* (Linnaeus, 1767)
Ampullariidae
  A *Pomacea bridgesi* (Reeve, 1856)
Ancylidae
  E *Ferrissia clessiniana* (Jickeli, 1882)
  A *Ferrissia wautieri* (Mirolli, 1960)
Anomiidae
  A *Anomia chinensis* Philippi, 1849

Aplysiidae
  A *Aplysia dactylomela* Rang, 1828
  A *Bursatella leachi* De Blainville, 1817
  A *Syphonota geographica* (Adams & Reeve, 1850)
Arcidae
  A *Acar plicata* (Dillwyn, 1817)
  A *Anadara demiri* (Piani, 1981)
  A *Anadara inaequivalvis* (Bruguière, 1789)
  A *Anadara natalensis* (Krauss, 1848)
  A *Barbatia trapezina* (Lamarck, 1819)
Arionidae
  E *Arion distinctus* J. Mabille, 1868
  E *Arion fasciatus* (Nilsson, 1823)
  E *Arion flagellus* Collinge, 1893
  E *Arion hortensis* A. Férussac, 1819
  E *Arion rufus* (Linnaeus, 1758)
  E *Arion silvaticus* Lohmander, 1937
  E **Arion vulgaris** (Moquin-Tandon, 1855)
Bithyniidae
  C *Bithynia leachii* (Sheppard, 1823)
Boettgerillidae
  A *Boettgerilla pallens* Simroth, 1912
Buccinidae
  A *Cantharus tranqubaricus* Gmelin, 1791
  A *Engina mendicaria* (Linnaeus, 1758)
  A *Pollia dorbignyi* (Payraudeau, 1826)
Bullidae
  A *Bulla ampulla* Linnaeus, 1758
Calyptraeidae
  E *Calyptraea chinensis* Linnaeus, 1758
  A *Crepidula aculeata* (Gmelin, 1791)
  A **Crepidula fornicata** (Linnaeus, 1758)
Cardiidae
  A *Afrocardium richardi* (Audouin, 1826)
  C *Cerastoderma edule* (Linnaeus, 1758)
  A *Fulvia australis* (Sowerby G.B., 1834)
  A *Fulvia fragilis* (Forsskål in Niehbur, 1775)
  A *Monodacna colorata* (Eichwald, 1829)
Carditidae
  A *Cardites akabana* (Sturany, 1899)
Cerithiidae
  A *Cerithium columna* Soweby, 1834
  A *Cerithium egenum* Gould, 1849
  A *Cerithium litteratum* (Born, 1778)
  A *Cerithium nesioticum* Pilsbry & Vanatta, 1906
  A *Cerithium scabridum* Philippi, 1848
  A *Clypeomorus bifasciatus* (Sowerby G.B. II, 1855)
  A *Diala varia* Adams A., 1860
  A *Rhinoclavis kochi* (Philippi, 1848)
  A *Rhinoclavis sinensis* (Gmelin, 1791)

Cerithiopsidae
- A *Cerithiopsis pulvis* (Issel, 1869)
- A *Cerithiopsis tenthrenois* (Melville, 1896)

Chamidae
- A *Chama asperella* Lamarck, 1819
- A *Chama aspersa* Reeve, 1846
- A *Chama brassica* Reeve, 1846
- A *Chama pacifica* Broderip, 1834
- A *Pseudochama corbieri* (Jonas, 1846)

Chitonidae
- A *Chiton hululensis* (Smith E.A. in Gardiner, 1903)

Chondrinidae
- E *Solatopupa similis* (Bruguière, 1792)

Chromodorididae
- A *Chromodoris quadricolor* (Rüppell & Leuckart, 1830)
- A *Hypselodoris infucata* (Rüppell & Leuckart, 1830)

Clausiliidae
- E *Bulgarica thessalonica* (Rossmässler, 1839)
- E *Charpentieria itala braunii* (Rossmässler, 1836)
- E *Herilla bosniensis bosniensis* (L. Pfeiffer, 1868)
- E *Leucostigma candidescens* (Rossmässler, 1835)
- E *Medora almissana* (Küster, 1847)
- E *Medora macascarensis* (Küster, 1854)
- E *Papillifera bidens* (Philippi, 1836)

Cochlicellidae
- E *Cochlicella acuta* (O.F. Müller, 1774)
- E *Cochlicella barbara* (Linnaeus, 1758)

Cochlostomatidae
- E *Cochlostoma septemspirale* (Razoumowsky, 1789)

Columbellidae
- A *Zafra savignyi* (Moazzo, 1939)
- A *Zafra selasphora* (Melville & Standen, 1901)

Conidae
- A *Conus fumigatus* Hwass in Bruguiere, 1792

Corambidae
- A *Corambe obscura* (A.E. Verrill, 1870)
- A *Doridella obscura* A.E. Verrill, 1870

Corbiculidae
- A *Corbicula fluminalis* (O.F. Müller, 1774)
- A **Corbicula fluminea** (O.F. Müller, 1774)

Corbulidae
- E *Corbula gibba* (Olivi, 1792)

Costellariidae
- A *Vexillum depexum* Deshayes in Laborde, 1834

Cylichnidae
- A *Acteocina crithodes* Melville & Standen, 1907
- A *Acteocina mucronata* (Philippi, 1849)

Cypraeidae
- A *Erosaria turdus* (Lamarck, 1810)
- A *Palmadusta lentiginosa* (Gray, 1825)
- A *Purpuradusta gracilis notata* (Gill, 1858)

Dendrodorididae
- A *Dendrodoris fumata* (Rüppell & Leuckart, 1830)

Dentaliidae
- A *Dentalium octangulatum* Donovan, 1804

Dialidae
- A *Cerithidium perparvulum* (Watson, 1886)

Discodorididae
- A *Discodoris lilacina* (Gould, 1852)
- A *Geitodoris planata* (Alder & Hancock, 1846)

Dreissenidae
- A *Congeria leucophaeta* (Conrad, 1831)
- E *Dreissena bugensis* Andrusov, 1897
- E **Dreissena polymorpha** (Pallas, 1771)
- A *Mytilopsis leucophaeta* (Conrad, 1831)

Ebalidae
- A *Murchisonella columna* (Hedley, 1907)

Elysiidae
- A *Elysia grandifolia* Kelaart, 1858
- A *Elysia tomentosa* Jensen, A.S., 1997
- E *Elysia viridis* (Montagu, 1804)

Epitoniidae
- A *Cycloscala hyalina* (Sowerby, 1844)

Facelinidae
- A *Caloria indica* (Bergh, 1896)
- A *Favorinus ghanensis* Edmunds, 1968

Fasciolariidae
- A *Fusinus verrucosus* (Gmelin, 1791)

Fissurellidae
- A *Diodora funiculata* Reeve, 1850
- A *Diodora ruppellii* (Sowerby, 1834)

Flabellinidae
- A *Flabellina rubrolineata* (O'Donoghue, 1929)

Gastrochaenidae
- A *Gastrochaena cymbium* (Spengler, 1783)

Gastrodontidae
- A *Zonitoides arboreus* (Say, 1817)
- A *Zonitoides excavatus* (Alder, 1830)

Glycymerididae
A *Glycymeris arabicus* (Adams H., 1871)
Gryphaeidae
A *Hyotissa hyotis* (Linnaeus, 1758)
Gryphaeidae
A *Parahyotissa imbricata* (Lamarck, 1819)
Halgerdidae
A *Halgerda willeyi* Eliot, 1904
Haliotidae
A *Haliotis discus hannai* Ino, 1952
A *Haliotis pustulata cruenta* Reeve, 1846
E *Haliotis tuberculata* Linnaeus, 1758
Haminoeidae
A *Atys angustatus* Smith, 1872
A *Atys cylindricus* (Helbling, 1779)
E *Haminoea callidegenita* Gibson & Chia, 1989
A *Haminoea cyanomarginata* Heller & Thompson, 1983
Helicidae
E *Arianta arbustorum* (Linnaeus, 1758)
E *Cantareus apertus* (Born, 1778)
E *Cepaea nemoralis* (Linnaeus, 1758)
E *Chilostoma banaticum* (Rossmässler, 1838)
E *Chilostoma cingulatum* (S. Studer, 1820)
E *Chilostoma cingulatum gobanzi* (Frauenfeld, 1867)
E *Chilostoma illyricum* (Stabile, 1864)
E *Chilostoma planospira* (Lamarck, 1822)
E *Chilostoma planospira erjaveci* (Lamarck, 1822)
E *Cornu aspersum* (O.F. Müller, 1774)
E *Eobania vermiculata* (O.F. Müller, 1774)
E *Helix cincta* O.F. Müller, 1774
E *Helix lucorum* Linnaeus, 1758
A *Helix nucula* Pfeiffer, 1859
E *Helix pomatia* L., 1758
E *Marmorana serpentina serpentina* (A. Férussac, 1821)
E *Theba pisana* (O.F. Müller, 1774)
Helicodiscidae
A *Helicodiscus parallelus* (Say, 1821)
E *Helicodiscus riparbellii* Giusti, 1976
A *Lucilla singleyana* (Pilsbry, 1889)
Hiatellidae
C *Hiatella arctica* (Linnaeus, 1767)
Hipponicidae
A *Sabia conica* (Schumacher, 1817)
Hydrobiidae
E *Emmericia patula* (Brumati, 1938)
E *Lithoglyphus naticoides* (C. Pfeiffer, 1828)
A *Potamopyrgus antipodarum* (J.E. Gray, 1843)

Hygromiidae
E *Candidula gigaxii* (L. Pfeiffer, 1847)
E *Candidula intersecta* (Poiret, 1801)
E *Cernuella cisalpina* (Rossmässler, 1837)
E *Cernuella ionica* (Mousson, 1854)
E *Cernuella neglecta* (Draparnaud, 1805)
E *Cernuella virgata* (Da Costa, 1778)
E *Helicella bolenensis* (Locard, 1882)
E *Helicella itala* (Linnaeus, 1758)
E *Hygromia cinctella* (Draparnaud, 1801)
E *Microxeromagna armillata* (Lowe, 1852)
E *Monacha cantiana* (Montagu, 1803)
E *Monacha cartusiana* (O.F. Müller, 1774)
E *Trichia hispida* (Linnaeus, 1758)
E *Trochoidea elegans* (Gmelin, 1791)
E *Trochoidea pyramidata* (Draparnaud, 1805)
E *Xerolenta obvia* (Menke, 1828)
E *Xeromunda candiota* (Mousson, 1854)
E *Xeropicta derbentina* (Krynicki, 1836)
C *Xeropicta krynickii* (Krynicki, 1833)
E *Xerotricha conspurcata* (Draparnaud, 1801)
Isognomonidae
A *Isognomon ephippium* (Linnaeus, 1758)
Laternulidae
A *Laternula anatina* (Linnaeus, 1758)
Lauriidae
E *Lauria cylindracea* (Da Costa, 1778)
Leptochitonidae
A *Lepidopleurus cancellatus* (G.B. Sowerby II, 1840)
Limacidae
E *Lehmannia nyctelia* (Bourguignat, 1861)
E *Lehmannia valentiana* (A. Férussac, 1822)
E *Limacus flavus* (Linnaeus, 1758)
E *Limax albipes* Dumont & Mortillet, 1853
E *Limax maximus* Linnaeus, 1758
Limopsidae
A *Limopsis multistriata* (Forsskål, 1775)
Litiopidae
A *Gibborissoa virgata* (Philippi, 1849)
Loliginidae
A *Sepioteuthis lessoniana* Lesson, 1830
Lucinidae
A *Divalinga arabica* Dekker & Goud, 1994

Lymnaeidae
  A *Pseudosuccinea columella* (Say, 1817)
Mactridae
  A *Mactra lilacea* Lamarck, 1818
  A *Mactra olorina* Philippi, 1846
  A *Spisula solidissima* (Dillwyn, 1817)
Malleidae
  A *Malvufundus regulus* (Forsskål, 1775)
Mesodesmatidae
  A *Atactodea glabrata* Linnaeus, 1767
Milacidae
  E *Milax gagates* (Draparnaud, 1801)
  E *Milax nigricans* (Philippi, 1836)
  E *Tandonia budapestensis* (Hazay, 1880)
  E *Tandonia rustica* (Millet, 1843)
  E *Tandonia sowerbyi* (A. Férussac, 1823)
Muricidae
  C *Bedeva paivae* (Crosse, 1864)
  A *Ergalatax contracta* (Reeve, 1846)
  A *Ergalatax obscura* Houart, 1996
  E *Hexaplex trunculus* Linnaeus, 1758
  A *Murex forskoehlii* Röding, 1798
  E *Ocenebra erinacea* Linnaeus, 1758
  A *Ocinebrellus inornatus* (Récluz, 1851)
  A ***Rapana venosa*** (Valenciennes, 1846)
  A *Thais lacera* (Born, 1778)
  A *Thais sacellum* (Gmelin, 1791)
  A *Trunculariopsis trunculus* (Linnaeus, 1758)
  A *Urosalpinx cinerea* (Say, 1822)
Myidae
  C *Mya arenaria* (Linnaeus, 1758)
  A *Sphenia ruppellii* Adams A., 1850
Mytilidae
  A *Aulocomya ater* (Molina, 1782)
  A ***Brachidontes pharaonis*** (Fischer P., 1870)
  A *Leiosolenus hanleyanus* Reeve, 1857
  A *Modiolus auriculatus* (Krauss, 1848)
  A *Musculista perfragilis* (Dunker, 1857)
  A ***Musculista senhousia*** (Benson in Cantor, 1842)
  E *Mytilaster lineatus* (Gmelin, 1791)
  A *Septifer bilocularis* Linnaeus, 1758
  A *Septifer forskali* Dunker, 1855
  A *Xenostrobus securis* (Lamarck, 1819)
Nacellidae
  A *Cellana rota* (Gmelin, 1791)
Nassariidae
  E *Cyclope neritea* (Linnaeus, 1758)
  A *Nassarius arcularia pilcatus* (Roding, 1798)
Naticidae
  A *Natica gualteriana* Récluz, 1844
Neritidae
  A *Nerita sanguinolenta* Menke, 1829
  A *Smaragdia souverbiana* (Montrouzier, 1863)
  E *Theodoxus fluviatilis* Linnaeus, 1758
Obtortionidae
  A *Ceritidium diplax* (Watson, 1886)
  A *Finella pupoides* Adams A., 1860
Octopodidae
  A *Octopus aegina* Gray, 1849
  A *Octopus cyaneus* Gray, 1849
Ostreidae
  E *Crassostrea angulata* Lamarck, 1819
  A ***Crassostrea gigas*** (Thunberg, 1793)
  A *Crassostrea rhizophorae* (Guilding, 1828)
  A *Crassostrea rivularis* (Gould, 1861)
  A *Crassostrea sikamea* (Amemiya, 1928)
  A *Crassostrea virginica* (Gmelin, 1791)
  A *Dendrostrea frons* Linnaeus, 1758
  A *Nanostrea exigua* Harry, 1985
  A *Ostrea angasi* Sowerby, 1871
  A *Ostrea denselamellosa* Linnaeus, 1758
  A *Ostrea puelchana* dOrbigny, 1842
  A *Planostrea pestigris* Hanley, 1846
  A *Saccostrea commercialis* (Iredale & Roughley, 1933)
  A *Saccostrea cucullata* (Born, 1778)
  A *Tiostrea chilensis* (Philippi, 1845)
Oxychilidae
  E *Aegopinella nitens* (Michaud, 1831)
  E *Aegopinella nitidula* (Draparnaud, 1805)
  E *Aegopinella pura* (Alder, 1830)
  E *Nesovitrea hammonis* (Ström, 1765)
  C *Oxychilus camelinus* (Bourguignat, 1852)
  E *Oxychilus cellarius* (O.F. Müller, 1774)
  E *Oxychilus cyprius* (Pfeiffer, 1847)
  E *Oxychilus draparnaudi* (H. Beck, 1837)
  A *Oxychilus translucidus* (Mortillet, 1853)
Oxynoidae
  A *Oxynoe viridis* Pease, 1863
Partulidae
  A *Synanodonta woodiana* (Lea, 1834)
Patulidae
  E *Discus rotundatus* (O.F. Müller, 1774)

Pectinidae
- A *Chlamys lischkei* (Dunker, 1850)
- A *Patinopecten yessoensis* Jay, 1857
- E *Pecten maximus* (Linnaeus, 1758)
- A *Semipallium coruscans* Hinds, 1845

Petricolidae
- A *Petricola hemprichi* Issel, 1869
- A *Petricola pholadiformis* Lamarck, 1818

Pharidae
- A ***Ensis americanus*** (Gould, 1870)

Physidae
- E *Physella acuta* (Draparnaud, 1805)
- A *Physella gyrina* (Say, 1821)
- A *Physella heterostropha* (Say, 1817)
- A *Physella integra* (Haldeman, 1841)

Pisidiidae
- C *Musculum transversum* (Say, 1829)

Planaxidae
- A *Angiola punctostriata* (Smith E.A., 1872)
- A *Planaxis griseus* (Brocchi, 1821)

Planorbidae
- A *Biomphalaria glabrata* (Say, 1818)
- A *Gyraulus chinensis* Dunker, 1848
- A *Gyraulus parvus* (Say, 1817)
- A *Helisoma anceps* (Menke, 1830)
- A *Helisoma duryi* (Wetherby, 1879)
- A *Helisoma tenue* (Dunker, 1850)
- A *Menetus dilatatus* Gould, 1841
- C *Planorbarius corneus* Linnaeus, 1758
- A *Planorbella anceps anceps* (Menke, 1830)
- A *Planorbella duryi* (Wetherby, 1879)

Pleurobranchidae
- A *Pleurobranchus forskalii* Rüppell & Leuckart, 1828

Pleurodiscidae
- E *Pleurodiscus balmei* (Potiez & Michaud, 1835)

Plicatulidae
- A *Plicatula chinensis* Mörch, 1853

Polyceridae
- A *Polycera hedgpethi* Marcus Er., 1964
- A *Polycerella emertoni* Verrill, 1881

Pomatiidae
- E *Pomatias elegans* (O.F. Müller, 1774)

Pristilomatidae
- A *Hawaiia minuscula* (Binney, 1841)
- E *Vitrea contracta* (Westerlund, 1871)

Psammobiidae
- A *Hiatula ruppelliana* (Reeve, 1857)

Pteriidae
- A *Electroma vexillum* (Reeve, 1857)
- A *Pinctada margaritifera* (Linnaeus, 1758)
- A ***Pinctada radiata*** (Leach, 1814)
- E *Pteria hirundo* (Linnaeus, 1758)

Punctidae
- C *Hebetodiscus inermis* (Backer, 1929)
- E *Paralaoma servilis* (Shuttleworth, 1852)

Pyramidellidae
- A *Chrysallida fischeri* (Hornung & Mermod, 1925)
- A *Chrysallida maiae* (Hornung & Mermod, 1924)
- A *Chrysallida micronana* (Hornung & Mermod, 1924)
- A *Chrysallida pirinthella* (Melvill, 1910)
- A *Cingulina isseli* (Tryon, 1886)
- A *Hinemoa cylindrica* (de Folin, 1879)
- A *Iolaea neofelixoides* (Nomura, 1936)
- A *Leucotina natalensis* (Adams A., 1851)
- A *Monotygma lauta* (Adams A., 1851)
- A *Oscilla jocosa* Melvill, 1904
- A *Pyramidella dolabrata* (Linnaeus, 1758)
- A *Odostomia lorioli* (Hornung & Mermod, 1924)
- A *Syrnola cinctella* Adams A., 1860
- A *Syrnola fasciata* (Jickeli, 1882)
- A *Syrnola lendix* A. Adams, 1863
- A *Turbonilla edgarii* (Melvill, 1896)

Retusidae
- A *Cylichnina girardi* (Audouin, 1826)
- A *Pyrunculus fourierii* (Audouin, 1826)
- A *Retusa desgenettii* (Audouin, 1826)

Rissoidae
- A *Rissoina ambigua* (Gould, 1849)
- A *Rissoina bertholleti* Issel, 1869
- A *Rissoina siprata* Sowerby, 1825
- A *Voorwindia tiberiana* (Issel, 1869)

Semelidae
- E *Abra ovata* Philippi, 1836
- A *Theora lubrica* Gould, 1861

Sepiidae
- A *Sepia pharaonis* Ehrenberg, 1831

Siphonariidae
- A *Siphonaria crenata* Blainville, 1827

Spondylidae
- A *Spondylus groschi* Lamprell & Kilburn, 1995
- A *Spondylus nicobaricus* Schreibers, 1793
- A *Spondylus spinosus* Schreibers, 1793

Streptaxidae
- A *Gulella io* Verdcourt, 1974

Strombidae
  A *Canarium mutabilis* (Swainson, 1821)
  A *Conomurex persicus* (Swainson, 1821)
Subulinidae
  A *Lamellaxis clavulinus* (Potiez & Michaud, 1838)
  A *Lamellaxis gracilis* (Hutton, 1834)
  A *Lamellaxis micra* (Orbigny, 1835)
  A *Opeas pumilum* (Pfeiffer, 1840)
  A *Rumina decollata* (Linnaeus, 1758)
  A *Subulina octona* Bruguiere, 1789
  A *Subulina striatella* Rang, 1831
Tellinidae
  A *Angulus flacca* (Römer, 1871)
  A *Psammotreta praerupta* (Salisbury, 1934)
  A *Tellina valtonis* Hanley, 1844
Teredinidae
  C **Teredo navalis** Linnaeus, 1758
Tergipedidae
  A *Cuthona perca* (Marcus, 1958)
  E *Tenellia adspersa* (Nordmann, 1845)
Testacellidae
  E *Testacella haliotidea* Draparnaud, 1801
Tethyidae
  A *Melibe viridis* Alder & Hancock, 1864
Thiaridae
  A *Melanoides tuberculatus* (Muller, 1774)
Trapeziidae
  A *Trapezium oblongum* (Linnaeus, 1758)
Tremoctopodidae
  A *Tremoctopus gracilis* (Eydoux & Souleyet, 1852)
Triophidae
  A *Plocamopherus ocellatus* Rüppell & Leuckart, 1828
Triphoridae
  A *Metaxia bacillum* (Issel, 1869)
Trissexodontidae
  E *Caracollina lenticula* (Michaud, 1831)
Trochidae
  C *Gibbula adansoni* (Payraudeau, 1826)
  E *Gibbula albida* (Gmelin, 1791)
  E *Gibbula cineraria* (Linnaeus, 1758)
  A *Pseudominolia nedyma* (Melvill, 1897)
  A *Stomatella impertusa* (Burrow, 1815)
  A *Trochus erithreus* Brocchi, 1821
Truncatellidae
  E *Truncatella subcylindrica* Linnaeus, 1758
Turridae
  A *Lienardia mighelsi* Iredale & Tomlin, 1917
Ungulinidae
  A *Diplodonta bogii* van Aartsen, 2004

Unionidae
  C *Anodonta cellensis* Pfeiffer, 1821
  A *Sinanodonta woodiana* (Lea, 1834)
  A *Unio mancus* Lamarck, 1819
Urocoptidae
  A *Holospira fusca* von Martens, 1897
  A *Microceramus gossei* Pfeiffer, 1846
Valloniidae
  E *Vallonia excentrica* Sterki, 1893
Veneridae
  A *Antigona lamellaris* Schumacher, 1817
  A *Callista florida* (Lamarck, 1818)
  A *Cirenita callipyga* (von Born, 1778)
  A *Clementia papyracea* (Gray, 1825)
  A *Dosinia erythraea* Römer, 1860
  A *Gafrarium pectinatum* (Linnaeus, 1758)
  A *Mercenaria mercenaria* (Linnaeus, 1758)
  A *Paphia textile* (Gmelin, 1791)
  A *Redicirce sulcata* Gray, 1838
  E *Ruditapes decussatus* (Linnaeus, 1758)
  A *Ruditapes philippinarum* (Adams & Reeve, 1850)
  A *Timoclea marica* (Issel, 1869)
Veronicellidae
  A *Semperula maculata* (Templeton, 1858)
Vertiginidae
  A *Gastrocopta pellucida* (Pfeiffer, 1841)
Viviparidae
  C *Viviparus viviparus* (Linnaeus, 1758)
Zonitidae
  E *Aegopis verticillus* (Lamarck, 1822)
  E *Zonites algirus* (Linnaeus, 1758)

Annelida, Hirudinea
Cambarincolidae
  A *Cambarincola mesochoreus* Hoffman, 1963
Erpobdellidae
  C *Erpobdella testacea* (Savigny, 1820)
Hirudinidae
  E *Hirudo medicinalis* Linnaeus, 1758
  A *Piscicola haranti* Jarry, 1960
Piscicolidae
  E *Piscicola geometra* (Linnaeus, 1761)
  E *Acipenserobdella volgensis* (Zykoff, 1903)
  E *Caspiobdella fadejewi* (Selensky, 1915)
Salifidae
  A *Barbronia weberi* (Blanchard, 1897)

Annelida, Oligochaeta
Acanthodrilidae
  A *Microscolex phosphoreus* (Dugès, 1837)

Aelosomatidae
   A *Aelosoma bengalese* Steph.
Branchiobdellidae
   A *Isochaetides michaelseni* (Lastockin, 1937)
   A *Xironogiton instabilis* (Moore, 1894)
   A *Xironogiton victoriensis* Gelder & Hall, 1990
Eudrilidae
   A *Eudrilus eugeniae* (Kinberg, 1867)
Glossoscolecidae
   A *Pontoscolex corethrurus* (Müller, 1857)
Lumbricidae
   E *Dendrobaena attemsi* (Michaelsen, 1902)
   E *Dendrobaena hortensis* (Michaelsen, 1890)
   E *Dendrobaena veneta* (Rosa, 1886)
   E *Eisenia andrei* Bouché, 1972
   C *Eisenia carolinensis* (Michaelsen, 1910)
   A *Eisenia japonica* (Michaelsen, 1891)
   A *Eisenia parva* (Eisen, 1874)
   A *Helodrilus tenuis* Eisen, 1874
   E *Satchellius mammalis* (Savigny, 1826)
Megascolecidae
   A *Amynthas gracilis* (Kinberg, 1867)
   A *Amynthas morrisi* (Beddard, 1892)
   A *Amynthas rodericensis* (Grube, 1879)
   C *Metaphire californica* (Kinberg, 1867)
   A *Perionyx excavatus* Perrier, 1872
   A *Pithemera bicincta* (Perrier, 1875)
Naididae
   C *Chaetogaster limnaei* Von Baer, 1827
   E *Paranais botniensis* Sperber, 1948
   E *Paranais frici* Hrabe, 1941
Ocnerodrilidae
   A *Eukerria saltensis* (Beddard, 1895)
   A *Ocnerodrilus occidentalis* Eisen, 1878
Octochaetidae
   A *Dichogaster modiglianii* (Rosa, 1896)
Tubificidae
   A *Branchiura sowerbyi* Beddard, 1892
   E *Potamothrix bavaricus* (Oschman, 1913)
   E *Potamothrix bedoti* (Piquet, 1913)
   E *Potamothrix heuscheri* (Bretscher, 1900)
   E *Potamothrix moldaviensis* Vejdovsky & Mrazek, 1902
   E *Potamothrix vejdovskyi* (Hrabe, 1941)
   E *Psammoryctides moravicus* (Hrabe, 1934)
   E *Tubifex newaensis* (Michaelsen, 1903)
   C *Tubificoides heterochaetus* (Michaelsen, 1926)
   A *Tubificoides pseudogaster* (Dahl, 1960)

Annelida, Polychaeta
Ampharetidae
   C *Alkmaria rominji* Korst, 1919
Amphinomidae
   A *Linopherus incarunculata* (Peters, 1858)
Capitellidae
   A *Notomastus mossambicus* (Thomassin, 1970)
Cirratulidae
   C *Aphelochaeta marioni* (Saint-Joseph, 1894)
   E *Hypania invalida* (Grube, 1860)
   A *Pseudeurythoe acarunculata* Monro, 1937
   A *Tharyx dorsobranchialis* (Kirkegaard, 1959)
   A *Tharyx killariensis* (Southern, 1914)
Cossuridae
   A *Cossura coasta* Litamori, 1960
Dorvilleidae
   A *Ophryotrocha japonica* (Pleijel & Eide, 1996)
Eunicidae
   A *Lysidice collaris* Grube, 1870
   A *Lysidice ninetta* Audouin & Milne-Edwards, 1833
Goniadidae
   A *Goniadella gracilis* (Verrill, 1873)
Hesionidae
   A *Syllidia armata* Quatrefages, 1865
   C *Microphthalmus similis* Bobretzky, 1870
Lumbrineridae
   A *Lumbrineris inflata* Moore, 1911
Maldanidae
   A *Metasychis gotoi* (Izuka, 1902)
   A *Clymenella torquata* (Leidy, 1855)
Nereididae
   A *Ceratonereis mirabilis* Kinberg, 1866
   A *Leonnates decipiens* Fauvel, 1929
   A *Leonnates indicus* Kinberg, 1865
   A *Leonnates persica* Wesenberg-Lund, 1948
   C *Neanthes succinea* Frey & Leuckart, 1847
   A *Neanthes willeyi* (Day, 1934)
   E *Nereis diversicolor* O.F. Müller, 1776
   A *Nereis persica* Fauvel, 1911
   E *Nereis virens* Sars, 1835
   A *Perinereis nuntia* (Savigny, 1818)
   A *Pseudonereis anomala* Gravier, 1901
Polynoidae
   A *Lepidonotus tenuisetosus* (Gravier, 1901)

Sabellariidae
  E *Sabellaria alveolata* (Linnaeus, 1767)
  E *Sabellaria spinulosa* Leuckart, 1849
Sabellidae
  A *Branchiomma boholense* (Grube, 1878)
  E *Branchiomma bombyx* (Koelliker, 1858)
  A *Branchiomma luctuosum* (Grube, 1869)
  C *Desdemona ornata* Banse, 1957
  C *Sabella spallanzani* Viviani, 1805
Serpulidae
  A ***Ficopomatus enigmaticus*** (Fauvel, 1923)
  E *Filogranula calyculata* (Costa, 1961)
  A *Hydroides dianthus* (Verrill, 1871)
  A *Hydroides dirampha* Mörch, 1863
  A *Hydroides elegans* (Haswell, 1883)
  A *Hydroides heterocerus* (Grube, 1868)
  A *Hydroides homocerus* Pixell, 1913
  A *Hydroides minax* (Grube, 1878)
  A *Hydroides operculatus* (Treadwell, 1929)
  A *Hydroides ezoensis* Okuda, 1934
  A *Janua brasiliensis* (Grube, 1872)
  A *Laonome elegans* (Fauvel, 1923)
  C *Metavermilia multicristata* (Philippi, 1844)
  C *Paralaeospira malardi* Caullery & Mesnil, 1897
  A *Pileolaria berkeleyana* Rioja, 1942
  C *Pileolaria militaris* Claparede, 1868
  A *Pomatoleios kraussii* (Baird, 1865)
  A *Spirobranchus tetraceros* (Schmarda, 1861)
  A *Spirorbis marioni* Caullery & Mesnil, 1897
  C *Vermiliopsis straticeps* Bianchi, 1981
Spionidae
  C *Boccardia ligerica* Ferronière, 1898
  A *Boccardia polybranchia* (Haswell, 1885)
  E *Boccardia redeki* (Horst, 1920)
  E *Boccardia semibranchiata* Guerin, 1990
  A ***Marenzelleria neglecta*** (Sikorski & Bick, 2004)
  A *Polydora ciliata* (Johnston, 1838)
  A *Polydora cornuta* Bosc, 1802
  A *Polydora hoplura* Claparede, 1870
  C *Polydora ligerica* (Ferronière, 1898)
  C *Polydora redeki* Horst, 1920
  C *Polydora websteri* Hartman, 1943
  A *Prionospio saccifera* Mackie & Hartley, 1990
  A *Streblospio benedicti* Webster, 1879
  A *Streblospio gynobranchiata* Rice & Levin, 1998

Syllidae
  A *Branchiosyllis exilis* (Gravier, 1900)
  A *Eusyllis kupfferi* Langerhans, 1879
  A *Opisthosyllis brunnea* Langerhans, 1879
  A *Proceraea cornuta* (Agassiz, 1862)
  A *Sphaerosyllis longipapillata* Hartmann-Schröder, 1979
  A *Syllis gracilis* Grube, 1840
Terebellidae
  A *Pista unibranchia* Day, 1963
  A *Terebella ehrenbergi* Grube, 1870
  E *Terebella lapidaria* (Linnaeus, 1767)

Chaetognatha
Sagittidae
  A *Parasagitta setosa* (Mueller, 1847)

Ectoprocta/Bryozoa
Adeonidae
  C *Reptadeonella violacea* (Julien, 1903)
Aeteidae
  C *Aetea anguina* (Linnaeus, 1758)
  C *Aetea ligulata* (Busk, 1852)
  C *Aetea longicollis* Jullien in Jullien & Calvet, 1903
  C *Aetea sica* (Couch, 1844)
  C *Aetea truncata* (Landsborough, 1852)
Barentsiidae
  E *Barentsia benedeni* (Foetinger, 1878)
  A *Urnatella gracilis* (Leidyi, 1851)
Beaniidae
  C *Beania mirabilis* Johnston, 1840
Bugulidae
  C *Bugula fulva* Ryland, 1960
  A *Bugula neritina* (Linnaeus, 1758)
  C *Bugula simplex* Hincks, 1886
  C *Bugula stolonifera* Ryland, 1960
Chorizoporidae
  C *Chorizopora brogniartii* (Audouin, 1826)
Cribrilinidae
  C *Puellina innominata* (Couch, 1844)
Electridae
  C *Electra pilosa* (Linnaeus, 1768)
  A *Electra tenella* Hincks, 1880
Hippothoidae
  A *Celleporella carolinensis* Ryland, 1979
Lepraliellidae
  A *Celleporaria aperta* Hincks, 1882
  A *Celleporaria pilaefera* Canu & Bassler, 1929
Lophopodidae
  C *Lophopus crystallinus* (Pallas, 1768)

Loxosomatidae
   A *Loxosomella antedonis* Mortensen, 1911
   A *Loxosomella kefersteinii* (Claparède, 1867)
Mambraniporidae
   C *Membranipora tuberculata* (Bosc, 1802)
Membraniporidae
   E *Conopeum seurati* (Canu, 1928)
Microporellidae
   C *Fenestrulina malusii* (Audouin, 1826)
   C *Microporella ciliata* (Pallas, 1766)
Pectinatellidae
   A *Pectinatella magnifica* (Leidy, 1851)
Schizoporellidae
   C *Escharina vulgaris* (Moll, 1803)
   C *Schizoporella errata* Waters, 1878
   C *Schizoporella unicornis* (Johnston, 1874)
Scrupariidae
   C *Scruparia ambigua* (D'orbigny, 1841)
Scrupocellariidae
   A **Tricellaria inopinata** DHondt & Occhipinti Ambrogi, 1985
Smittinidae
   A *Smittoidea prolifica* Osburn, 1952
Victorellidae
   A *Victorella pavida* Saville Kent, 1870
Walkeriidae
   E *Walkeria uva* (Linnaeus, 1758)
Watersiporidae
   A *Watersipora aterrima* (Ortmann, 1890)

Nematoda
Ancylostomatidae
   C *Ancylostoma duodenale* (Dubini, 1843)
Anguillicolidae
   A **Anguillicola crassus** Kuwahara, Niimi & Itagaki, 1974
Ascarididae
   A *Baylisascaris procyonis* (Stefanski & Zarnowski, 1951)
   A *Ascaridia dissimilis* Vigueras, 1931
Capillariidae
   A *Aonchotheca annulosa* (Dujardin, 1845)
   A *Eucoleus gastricus* (Baylis, 1926)
Chabertiidae
   A *Oesophagostomum columbianum* Curtice, 1890
Dictyocaulidae
   A *Dictyocaulus filaria* (Rudolphi, 1809)
Gongylonematidae
   A *Gongylonema neoplasticum* (Fibiger & Ditlevsen, 1914)
Heteroderidae
   A *Cactodera cacti* (Filipjev & Schuurmans Stekhoven, 1941)
   A *Globodera pallida* (Stone, 1973)
   A *Globodera rostochiensis* (Wollenweber, 1923)
   C *Heterodera avenae* Wollenweber, 1924
   C *Heterodera fici* Kirjanova, 1954
   C *Heterodera schachtii* Schmidt, 1871
Longidoridae
   E *Paralongidorus maximus* (Bütschli, 1874)
Meloidogynidae
   C *Meloidogyne arenaria* (Neal, 1889)
   A *Meloidogyne chitwoodi* Golden & al., 1980
   A *Meloidogyne incognita* (Kofoid & White, 1919)
   C *Meloidogyne javanica* (Treub, 1885)
Molineidae
   E *Nematodirus battus* Crofton & Thomas, 1951
Oxyuridae
   A *Syphacia muris* (Yamaguti, 1935)
Parasitaphelenchidale
   A **Bursaphelenchus xylophilus** (Steiner & Bührer, 1934)
Pratylenchidae
   A *Pratylenchus bolivianus* Corbett, 1984
   E *Pratylenchus penetrans* (Cobb, 1917)
   A *Pratylenchus vulnus* Allen & Jensen, 1951
   C *Radopholus similis* (Cobb, 1893)
Protostrongylidae
   A *Cystocaulus ocreatus* (Railliet & Henry, 1907)
   A *Muellerius capillaris* (Müller, 1889)
   E *Muellerius tenuispiculatus* Gebauer, 1932
   A *Neostrongylus linearis* (Marotel, 1913)
   A *Protostrongylus rufescens* (Leuckart, 1865)
Strongyloididae
   A *Strongyloides myopotami* Artigas & Pacheco, 1933
   A *Strongyloides ratti* Sandground, 1925
   C *Strongyloides stercoralis* (Bavay, 1876)
Telotylenchidae
   A *Tylenchorhynchus claytoni* Steiner, 1937
Thelastomatidae
   C *Blatticola blattae* (Graeffe, 1860)
   C *Hammerschmidtiella diesingi* (Hammerschmidt, 1838)
   A *Leidynema appendiculatum* (Leidy, 1850)

Trichinellidae
  E *Trichinella spiralis* (Owen, 1835)
Trichodoridae
  C *Paratrichodorus renifer* Siddiqi, 1974
  A *Trichosomoides crassicauda* (Bellingham, 1840)
  A *Ashworthius sidemi* Schulz, 1933
  A *Cooperia curticei* (Giles, 1892)
  A *Marshallagia marshalli* (Ransom, 1907)
  A *Ostertagia mossi* Dikmans, 1931
  A *Spiculopteragia asymmetrica* (Ware, 1925)
Trichuridae
  A *Trichuris myocastoris* Enigk, 1933
  A *Trichuris opaca* Barker & Noyes, 1915
Xiphinematidae
  A *Xiphinema coxi coxi* Tarjan, 1964
  A *Xiphinema diversicaudatum* (Micoletzky, 1927)
  A *Xiphinema index* Thorne & Allen, 1950
  E *Xiphinema pachtaicum* (Tulaganov, 1938)
  E *Xiphinema vuittenezi* Luc & al., 1964

Nemertea
Plectonemertidae
  A *Antiponemertes pantini* (Southgate, 1954)
  A *Argonemertes dendyi* (Dakin, 1915)
  A *Leptonemertes chalicophora* (Graff, 1879)

Acanthocephala
Echinorhynchidae
  C *Acanthocephalus anguillae* (Mueller, 1780)
Pomphorhynchidae
  C *Pomphorhynchus laevis* (Zoega, 1776)

Phoronida
Phoronidae
  A *Phoronis hippocrepia* Wright, 1856
  C *Phoronis psammophila* Cori, 1889
  A *Phoronopsis harmeri* Pixell, 1912

Rotifera
Brachionidae
  A *Brachionus plicatilis* Mueller, 1786
  A *Brachionus variabilis* Hampel, 1896

Arthropoda, Crustacea
Acartiidae
  A *Acartia centrura* Giesbrecht, 1889
  E *Acartia clausi* Giesbrecht, 1889
  C *Acartia margalefi* Alcaraz, 1976
  A *Acartia omorii* Bradford, 1976
  A *Acartia tonsa* J.D. Dana, 1852
  A *Paracartia grani* Sars, 1904
Alpheidae
  A *Alpheus audouini* Coutiere, 1905
  A *Alpheus inopinatus* Holthuis & Gottlieb, 1958
  A *Alpheus migrans* Lewinsohn & Holthuis, 1978
  A *Alpheus rapacida* de Man, 1908
Ameiridae
  A *Ameira divagans* Nicholls, 1939
Ammotheidae
  A *Ammothea hilgendorfi* Böhm, 1879
Ampithoidae
  A *Cymadusa filosa* Savigny, 1816
Aoridae
  A *Bemlos leptocheirus* Walker, 1909
  A *Grandidierella japonica* Stephensen, 1938
Archaeobalanidae
  C *Solidobalanus fallax* (Broch, 1927)
Armadillidiidae
  E *Armadillidium assimile* Budde-Lund, 1879
  E *Armadillidium kossuthi* Arcangeli, 1929
  E *Armadillidium nasatum* Budde-Lund, 1885
  C *Reductoniscus costulatus* Kessleyak, 1930
Artemiidae
  A *Artemia franciscana* Kellogg, 1906
Asellidae
  E *Asellus aquaticus* (Linnaeus, 1758)
  E *Proasellus coxalis* (Dollfus, 1892)
  E *Proasellus meridianus* (Racovitza, 1919)
Astacidae
  E *Astacus leptodactylus* (Eschscholtz, 1823)
  E *Austropotamobius pallipes* (Lereboullet, 1858)
  A *Niphargus hrabei* Karaman, 1932
  A *Pacifastacus leniusculus* (Dana, 1852)
Balanidae
  A *Balanus albicostatus* Pilsbry, 1916
  A *Balanus amphitrite* Darwin, 1854
  A *Balanus eburneus* Gould, 1841
  A ***Balanus improvisus*** Darwin, 1854
  A *Balanus reticulatus* Utinomi, 1967
  A *Balanus trigonus* Darwin, 1854
  A *Balanus variegatus* Darwin, 1854
  A *Elminius modestus* Darwin, 1854
  A *Megabalanus coccopoma* (Darwin, 1854)
  A *Megabalanus tintinnabulum* Linnaeus, 1758
  A *Megabalanus tulipiformis* (Ellis, 1758)

Bosminidae
　A　*Eubosmina longispina* (Leydig, 1860)
Buddelundiellidae
　E　*Buddelundiella cataractae* Verhoeff, 1930
Calanidae
　E　*Calanus euxinus* Hulsemann, 1991
Calappidae
　A　*Calappa hepatica* Linnaeus, 1758
　E　*Calappa pelii* Kerklots, 1851
Caligidae
　C　*Caligus pageti* Russell, 1925
Cambaridae
　A　*Orconectes immunis* (Hagen, 1870)
　A　*Orconectes limosus* (Rafinesque, 1817)
　A　*Orconectes rusticus* (Girard, 1852)
　A　*Orconectes virilis* (Hagen, 1870)
　A　**Procambarus clarkii** (Girard, 1852)
Caprellidae
　A　*Caprella mutica* Schurin, 1935
　A　*Caprella scaura* Templeton, 1836
Centropagidae
　E　*Boeckella triarticulata* (Thomson, 1883)
Cercopagididae
　E　*Bythotrephes longimanus* Leydig, 1860
　E　**Cercopagis pengoi** (Ostroumov, 1891)
Chthamalida
　E　*Chthamalus stellatus* (Poli, 1791)
Chydoridae
　C　*Alona protzi* Hartwig, 1900
　C　*Alona rustica* Scott, 1895
　C　*Camptocercus uncinatus* Smirnov, 1971
　A　*Disparalona hamata* (Birge, 1879)
　A　*Picripleuroxus denticulatus* (Birge, 1879)
Corophiidae
　C　*Corophium acherusicum* Costa, 1857
　E　*Corophium curvispinum* Sars, 1895
　C　*Corophium multisetosum* Stock, 1952
　A　*Corophium robustum* Sars, 1895
　C　*Corophium sextonae* Crawford, 1937
　E　*Corophium volutator* (Pallas, 1766)
　A　*Unciolella lunata* Chevreux, 1911
Crangonyctidae
　A　*Crangonyx pseudogracilis* Bousfield, 1958
　A　*Synurella ambulans* (F. Müller, 1846)
Cyclopidae
　E　*Cyclops kolensis* Lilljeborg, 1901
　E　*Cyclops vicinus* Uljanin, 1875
　A　*Mesocyclops ruttneri* Kiefer, 1981
Cyclopinidae
　C　*Muceddina multispinosa* Jaume & Boxshall, 1996

Cyprididae
　C　*Chlamydotheca incisa* Claus, 1892
　C　*Cypretta turgida* (Sars, 1896)
　C　*Dolerocypris sinensis* G.O. Sars, 1903
　C　*Hemicypris dentatomarginata* (Baird, 1859)
　C　*Isocypris beauchampi cicatricosa* Fox, 1963
　A　*Notodromas persica* Gurney, 1921
　C　*Stenocypris major* (Baird, 1859)
　C　*Strandesia spinulosa* Bronstein, 1958
　C　*Tanycypris pellucida* (Klie, 1932)
Daphniidae
　C　*Ceriodaphnia rotunda* Sars, 1862
　E　*Daphnia ambigua* Scourfield, 1946
　C　*Daphnia atkinsoni* Baird, 1859
　E　*Daphnia cristata* Sars, 1861
　E　*Daphnia longiremis* G.O. Sars, 1861
　C　*Daphnia parvula* Fordyce, 1901
　A　*Dasya sessilis* Yamada, 1928
　C　*Moina affinis* Birge, 1893
　A　*Moina weismanni* Ishikawa, 1897
　C　*Simocephalus hejlongjiangensis* Shi. & Shi., 1994
Diaptomidae
　C　*Diaptomus wierzejskii* Richard, 1888
　A　*Eudiaptomus gracilis* (G.O. Sars, 1862)
Diogenidae
　C　*Clibanarius erythropus* (Latreille, 1818)
　E　*Diogenes pugilator* (Roux, 1829)
Dorippidae
　A　*Dorippe quadridens* Fabricius, 1793
Dromiidae
　A　*Dromia spinirostris* Miers, 1881
Ectinosomatidae
　E　*Halectinosoma abrau* (Kricagin, 1878)
Epialtidae
　A　*Menaethius monoceros* (Latreille, 1825)
Ergasilidae
　C　*Ergasilus gibbus* von Nordmann, 1832
Euryplacidae
　A　*Eucrate crenata* de Haan, 1835
Gammaridae
　E　*Chaetogammarus ischnus* (Stebbing, 1899)
　E　*Chaetogammarus warpachowskyi* (Sars, 1894)
　E　*Dikerogammarus bispinosus* Martynov, 1925
　E　*Dikerogammarus haemobaphes* (Eichwald, 1841)
　E　**Dikerogammarus villosus** (Sovinskii, 1894)

E *Echinogammarus berilloni* (Catta, 1878)
E *Echinogammarus ischnus* (Stebbing, 1899)
E *Echinogammarus trichiatus* (Martynov, 1932)
A *Elasmopus pectenicrus* Bate, 1862
E *Gammarus pulex* (Linnaeus, 1758)
A *Gammarus roeseli* Gervais, 1835
A *Gammarus tigrinus* Sexton, 1939
A *Gmelinoides fasciatus* (Stebbing, 1899)
A *Incisocalliope aestuarius* (Watling & Maurer, 1973)
E *Iphigenella shablensis* (Carausu, 1943)
E *Obesogammarus crassus* (Sars, 1894)
E *Obesogammarus obesus* (Sars, 1894)
E *Orchestia cavimana* Heller, 1865
E *Pontogammarus robustoides* G.O. Sars, 1894

Grapsidae
E *Brachynotus sexdentatus* (Risso, 1827)
A *Plagusia depressa* (J.C. Fabricius, 1775)
A *Plagusia tuberculata* Lamarck, 1818

Hippolytidae
A *Bythocaris cosmetops* Holthuis, 1951

Idoteidae
A *Synidotea laevidorsalis* (Miers, 1881A)
A *Synidotea laticauda* J.E. Benedict, 1897

Isaeidae
A *Gammaropsis togoensis* Schellenberg, 1925
A *Photis lamellifera* Schellenberg, 1928

Janiridae
E *Jaera istri* Veuille, 1979

Lepadidae
A *Conchoderma auritum* (Linnaeus, 1758)

Leucosiidae
A *Ixa monodi* Holthuis & Gottlieb, 1956
A *Leucosia signata* Paulson, 1875
A *Myra subgranulata* Kossmann, 1877

Lichomolgidae
C *Lichomolgus actinae* D.V.

Ligiidae
A *Ligia italica* Fabricius, 1798
E *Ligia oceanica* (Linnaeus, 1767)

Limnoridae
C *Limnoria quadripunctata* (Holthuis, 1949)
A *Limnoria tripunctata* Menzies, 1951

Limulidae
A *Limulus polyphemus* Linnaeus, 1758

Lithodidae
A ***Paralithodes camtschaticus*** (Tilesius, 1815)

Luciferidae
A *Lucifer hanseni* Nobili, 1905

Macrothricidae
A *Drepanothrix dentata* (Euren, 1861)
C *Wlassicsia pannonica* Daday, 1904

Matutidae
A *Ashtoret lunaris* (Forsskål, 1775)

Melitidae
A *Melita nitida* S.I. Smith, 1873

Menippidae
A *Sphaerozius nitidus* Stimpson, 1858

Myicolidae
A *Myicola ostreae* Hoshina & Sugiura, 1953
A *Pseudomyicola spinosus* (Raffaele & Monticelli, 1885)

Mysidae
E *Diamysis bahirensis* (G.O. Sars, 1877)
E *Hemimysis anomala* G.O. Sars, 1907
E *Katamysis warpachowskyi* G.O. Sars, 1893
E *Limnomysis benedeni* Czerniavsky, 1882
E *Mysis relicta* Lovén, 1862
E *Paramysis baeri* Czerniavsky, 1882
E *Paramysis intermedia* (Czerniavsky, 1882)
E *Paramysis lacustris* (Czerniavsky, 1882)
E *Paramysis ullskyi* (Czerniavsky, 1882)

Mytilicolidae
C *Mytilicola intestinalis* Steuer, 1902
A *Mytilicola orientalis* Mori, 1935

Nephropidae
A *Homarus americanus* Herrick, 1895

Nucellicolidae
C *Nucellicola holmanae* Lamb, Boxshall, Mill & Grahame, 1996

Ocypodidae
A *Macrophthalmus graeffei* A. Milne-Edwards, 1873

Ogyrididae
A *Ogyrides mjoebergi* (Balss, 1921)

Oithonidae
A *Oithona similis* Claus, 1866

Oregoniidae
A *Chionoecetes opilio* (O. Fabricius, 1788)

Palaemonidae
E *Atyaephyra desmaresti* (Millet, 1831)
A *Macrobrachium nipponense* (De Haan, 1849)

E *Palaemon adspersus* Rathke, 1837
E *Palaemon elegans* Rathke, 1837
E *Palaemon longirostris* H. Milne-Edwards, 1837
A *Palaemon macrodactylus* Rathbun, 1902
A *Palaemonella rotumana* (Borradaile, 1898)
A *Periclimenes calmani* Tattersall, 1921
C *Tuleariocaris neglecta* Chace, 1969
Palinuridae
A *Jasus lalandii* (H. Milne-Edwards, 1837)
E *Palinurus elephas* (Fabricius, 1787)
A *Panulirus guttatus* (Latreille, 1804)
A *Panulirus ornatus* (Fabricius, 1798)
Panopeidae
A *Dyspanopeus sayi* (Smith, 1869)
A *Neopanope sayi* (S.I. Smith, 1869)
A *Rhithropanopeus harrisii* (Gould, 1841)
Parastacidae
A *Cherax destructor* Clark, 1936
Pasiphaeidae
A *Leptochela aculeocaudata* Paulson, 1875
A *Leptochela pugnax* de Man, 1916
Penaeidae
A **Marsupenaeus japonicus** (Bate, 1888)
A *Melicertus hathor* (Burkenroad, 1959)
A *Metapenaeopsis aegyptia* Galil & Golani, 1990
A *Metapenaeopsis mogiensis consobrina* (Nobili, 1904)
A *Metapenaeus monoceros* (Fabricius, 1798)
A *Metapenaeus stebbingi* (Nobili, 1904)
A *Penaeus semisulcatus* de Haan, 1844
A *Trachysalambria palaestinensis* (Steinitz, 1932)
Philosciidae
C *Anchiphiloscia balssi* (Verhoeff, 1928)
A *Benthana olfersii* (Brandt, 1833)
Phoxichilidiidae
A *Anoplodactylus portus* (Boehm, 1879)
Phyllodicolidae
C *Phyllodicola petiti* Delamare-Deboutteville & Laubier, 1960
Pilumnidae
A *Halimede tyche* (Herbst, 1801)
A *Heteropanope laevis* (Dana, 1852)
A *Pilumnopeus vauquelini* (Audouin, 1826)
C *Pilumnus spinifer* H. Milne-Edwards, 1834

Pisidae
E *Herbstia nitida* Manning & Holthuis, 1981
A *Hyastenus hilgendorfi* de Man, 1887
A *Libina dubia* H. Milne-Edwards, 1834
A *Micippa thalia* (Herbst, 1803)
Plagusiidae
A **Percnon gibbesi** (H. Milne-Edwards, 1853)
Platyarthridae
E *Platyarthrus schoebli* Budde-Lund, 1879
A *Trichorhina tomentosa* (Budde-Lund, 1893)
Podonidae
E *Cornigerius maeoticus* (Pengo, 1879)
E *Evadne anonyx* Sars, 1897
E *Pleopis polyphemoides* (Leuckart, 1859)
E *Podonevadne trigona* (Sars, 1897)
Pollicipedidae
C *Pollicipes pollicipes* (Gmelin, 1789)
Polyphemidae
E *Podon intermedius* Lilljeborg, 1853
Porcellidiidae
E *Agabiformius lentus* (Budde-Lund, 1885)
C *Heterolaophonte hammondi* Hicks, 1975
C *Porcellidium ovatum* Haller, 1879
E *Porcellio dilatatus* Brandt, 1833
E *Porcellio laevis* Latreille, 1804
C *Porcellionides pruinosus* (Brandt, 1833)
E *Proporcellio quadriseriatus* (Verhoeff, 1917)
Portunidae
A *Callinectes danae* Smith, 1869
A *Callinectes sapidus* M.J. Rathbun, 1896
A *Carupa tenuipes* Dana, 1851
A *Charybdis helleri* (A. Milne-Edwards, 1867)
A *Charybdis longicollis* Leene, 1938
E *Portumnus latipes* (Pennant, 1777)
A **Portunus pelagicus** Linnaeus, 1758
A *Thalamita gloriensis* Crosnier, 1962
A *Thalamita poissonii* (Audouin, 1826)
Potamidae
A *Potamobius leptodactylus* Esch., 1823
A *Potamon ibericum* (Bieberstein, 1809)
C *Potamon ibericum tauricium* (Czerniavsky, 1884)
Pseudocumatidae
A *Stenocuma graciloides* (G.O. Sars, 1894)

Pseudodiaptomidae
  E *Calanipeda aquaedulcis* Kritschagin, 1873
Raninidae
  A *Notopus dorsipes* Linnaeus, 1758
Sabelliphidae
  C *Herrmannella duggani* Holmes & Minchin, 1991
Sacculinidae
  A *Heterosaccus dollfusi* Boschma., 1960
Sarsiellidae
  A *Eusarsiella zostericola* (Cushman, 1906)
Scyllaridae
  A *Scyllarus caparti* Holthuis, 1952
Sididae
  C *Latonopsis australis* Sars, 1888
  A *Limnosida frontosa* Sars, 1862
Solenoceridae
  A *Solenocera crassicornis* (H. Milne-Edwards, 1837)
Sphaeromatidae
  A *Paracereis sculpta* Holmes, 1904
  A *Paradella dianae* Menzies, 1962
  C *Sphaeroma serratum* (Fabricius, 1787)
  A *Sphaeroma walkeri* Stebbing, 1905
Spiophanicolidae
  C *Spiophanicola spinosus* Ho, 1984
Squillidae
  A *Erugosquilla massavensis* (Kossmann, 1880)
Stenothoidae
  A *Stenothoe gallensis* Walker, 1904
Styloniscidae
  C *Cordioniscus stebbingi* (Patience, 1907)
  C *Cordioniscus stebbingi boettgeri* (Patience, 1907)
Talitridae
  E *Orchestia cavimana* Heller, 1865
  A *Platorchestia platensis* (Kr>yer, 1845)
Tanaidae
  A *Tanais dulongi* (Audouin, 1826)
  A *Zeuxo coralensis* Sieg, 1980
Temoridae
  A *Eurytemora americana* L.W. Williams, 1906
  A *Eurytemora pacifica* Sato, 1913
  E *Eurytemora velox* (Lilljeborg, 1853)
  E *Heterocope appendiculata* Sars, 1893
  E *Heterocope caspia* G.O. Sars, 1897
Tetraclitidae
  A *Tesseropora atlantica* Newman & Ross, 1976
Trachelipodidae
  A *Nagurus cristatus* (Dollfus, 1889)
  E *Protracheoniscus major* (Dollfus, 1903)
Trichoniscidae
  E *Androniscus dentiger* Verhoeff, 1908
  E *Haplophthalmus danicus* Budde-Lund, 1880
  E *Metatrichoniscoides leydigi* (Weber, 1880)
  A *Miktoniscus linearis* (Patience, 1908)
  E *Trichoniscus provisorius* Racovitza, 1908
  E *Trichoniscus pusillus* Brandt, 1833
Varunidae
  A **Eriocheir sinensis** H. Milne-Edwards, 1853
  A *Hemigrapsus penicillatus* (De Haan, 1835)
  A *Hemigrapsus sanguineus* (de Haan, 1835)
Verrucidae
  A *Verruca spengleri* Darwin, 1854
Xanthidae
  A *Atergatis roseus* (Rüppell, 1830)
  A *Daira perlata* (Herbst, 1790)
  A *Macromedaeus voeltzkowi* (Lenz, 1905)
  A *Parapilumnus malardi* (De Man, 1913)
  A *Pilumnoides perlatus* (Poeppig, 1836)
  A *Pilumnus hirsutus* Stimpson, 1858

Arthropoda, Myriapoda
Henicopidae
  A *Lamyctes coeculus* (Brölemann, 1889)
  A *Lamyctes emarginatus* (Newport, 1844)
Paradoxosomatidae
  A *Oxidus gracilis* (C.L. Koch, 1847)
Schizopetalidae
  A *Eurygyrus ochraceus* C.L. Koch, 1847
Spirobolellidae
  A *Sechellobolus dictyonotus* (Latzel, 1895)

Arthropoda, Insecta, Apterygota
Campodeidae
  E *Campodea lubbocki* Silvestri, 1912
  E *Campodea quilisi* Silvestri, 1932
  E *Campodea rhopalota* Denis, 1930
Entomobryidae
  E *Lepidocyrtus cyaneus* Tullberg, 1871
Isotomidae
  C *Cryptopygus thermophilus* (Axelson, 1900)
  C *Desoria trispinata* (MacGillivray, 1896)

C *Parisotoma notabilis* (Schäffer, 1896) Bagnall, 1940
C *Proisotoma minuta* (Tullberg, 1871)
Katiannidae
  C *Sminthurinus trinotatus* Axelson, 1905
Lepismatidae
  C *Ctenolepisma longicaudata* Escherisch, 1905
  C *Thermobia domestica* (Packard, 1837)
Neanuridae
  C *Friesea claviseta* Axelson, 1900
Sminthuridae
  C *Sphyrotheca multifasciata* (Reuter, 1881)
  C *Sphaeridia pumilis* (Krausbauer, 1898)
Tullbergiidae
  C *Mesaphorura krausbaueri* (Börner, 1901)

Arthropoda, Insecta, Ephemeroptera
Ametropodidae
  A *Ametropus fragilis* Albada, 1878
Baetidae
  E *Baetis liebenauae* Keffermüller, 1974
  A *Baetis tracheatus* Keffermüller & Machel, 1967

Arthropoda, Insecta, Blattodea
Blaberidae
  A *Blaberus atropos* (Stoll, 1813)
  A *Blaberus parabolicus* (Walker, 1868)
  A *Henschoutedenia flexivitta* (Walker, 1868)
  A *Panchlora peruana* Saussure, 1864
  A *Panchlora fraterna* Saussure & Zehntner, 1893
  A *Pycnoscelus surinamensis* (Linaeus, 1767)
Blattellidae
  C *Nyctibora laevigata* (Beauvois, 1805)
  A *Supella longipalpa* (Fabricius, 1798)
Blattidae
  C *Blatta orientalis* Linnaeus, 1758
  C *Neostylopyga rhombifolia* (Stall, 1861)
  A *Periplaneta americana* (Linnaeus, 1758)
  A *Periplaneta australasiae* (Fabricius, 1775)
  A *Periplaneta brunnea* Burmeister, 1838
Epilampridae
  A *Phoetalia pallida* (Brunner, 1865)
  A *Phoetalia circumvagans* (Burmeister, 1838)
Nauphoetidae
  A *Nauphoeta cinerea* (Olivier, 1789)
  A *Rhyparobia maderae* (Fabricius, 1781)

Arthropoda, Insecta, Isoptera
Kalotermitidae
  A *Cryptotermes brevis* (Walker, 1853)
  E *Kalotermes flavicollis* (Fabricius, 1793)
Rhinotermitidae
  A *Reticulitermes flavipes* (Kollar, 1837)
  E *Reticulitermes lucifugus* (Rossi, 1792)
Termitidae
  A *Macrotermes bellicosus* (Smeathman, 1781)

Arthropoda, Insecta, Orthoptera
Acrididae
  E *Anacridium aegyptium* (Linnaeus, 1764)
  A *Dociostaurus tartarus* Shchelkanovtsev, 1921
  A *Locusta migratoria* (Linnaeus, 1758)
  A *Notostaurus albicornis* (Eversmann, 1848)
  A *Ramburiella turcomana* (Fischer von Waldheim, 1846)
Bradyporidae
  A *Ephippigerida nigromarginata* (Lucas, 1849)
Gryllidae
  A *Gryllodes sigillatus* Walker, 1869
  C *Myrmecophilus americana* Saussure, 1877
Meconematidae
  E *Meconema meridionale* A. Costa, 1860
  A *Phlugiola dahlemica* Eichler, 1938
Phaneropteridae
  E *Leptophyes punctatissima* (Bosc, 1792)
  A *Topana cincticornis* (Stål, 1873)
Rhaphidophoridae
  E *Dolichopoda bormansi* Brunner von Wattenwyl, 1882
  A *Tachycines asynamorus* (Adelung, 1902)
  E *Troglophillus neglectus* (Kraus, 1879)
Tettigoniidae
  E *Antaxius spinibrachius* (Fischer, 1853)
  A *Copiphora brevirostris* Stål, 1873

Arthropoda, Insecta, Phasmatodea
Bacillidae
  E *Bacillius rossius* (Rossi, 1788)
  E *Clonopsis gallica* (Charpentier, 1825)
Phasmatidae
  A *Acanthoxyla geisovii* (Kaup, 1866)
  A *Acanthoxyla inermis* Salmon, 1955
  A *Carausius morosus* (Sinéty, 1901)
  A *Clitarchus hookeri* (White, 1846)

Arthropoda, Insecta, Dermaptera
Anisolabididae
  C *Anisolabis maritima* (Bonelli, 1832)
Carcinophoridae
  C *Euborellia annulipes* (Lucas, 1847)
  A *Euborellia stali* (Dohrn, 1864)
Labiduridae
  A *Nala lividipes* (Dufour, 1828)
Labiidae
  E *Forficula smyrnensis* Serville, 1838

Arthropoda, Insecta, Phtiraptera
Goniodidae
  A *Chelopistes meleagridis* (Linnaeus, 1758)
  A *Stenocrotaphus gigas* (Taschenberg, 1879)
  A *Zlotorzyckella colchici* (Denys, 1842)
Gyropidae
  A *Gliricola porcelli* (Schrank, 1781)
  A *Gyropus ovalis* Burmeister, 1838
  A *Pitrufquenia coypus* Marelli, 1932
Haematopinidae
  C *Linognathus stenopsis* (Burmeister, 1838)
  A *Polyplax spinulosa* (Burmeister, 1839)
Hoplopleuridae
  C *Haemodipsus lyriocephalus* (Burmeister, 1839)
  E *Haemodipsus ventricosus* (Denny, 1842)
Menoponidae
  C *Menopon gallinae* (Linneaus, 1758)
  C *Myrsidea quadrifasciata* (Piaget, 1880)
  C *Eomenacanthus stramineus* (Nitzsch, 1818)
  A *Hohorstiella gigantea* (Piaget, 1880)
  C *Neocolpocephalum turbinatum* (Denny, 1842)
  A *Uchida phasiani* (Modrzejewska & Zlotorzycka, 1977)
Philopteridae
  C *Cuclotogaster heterographa* (Nitzsch in Giebel, 1866)
  C *Goniocotes chrysocephalus* Giebel, 1874
  C *Goniocotes gallinae* (De Geer, 1778)
  C *Goniocotes rectangulatus* Nitzsch in Giebel, 1866
  C *Goniodes pavonis* (Linnaeus, 1758)
  A *Lagopoecus colchicus* Emerson, 1949
  A *Lipeurus maculosus* Clay, 1938
  C *Reticulipeurus polytrapezius* (Burmeister, 1838)
Trichodectidae
  E *Bovicola alpinus* Kéler, 1942
  E *Bovicola caprae* (Gurlt, 1843)
  C *Bovicola ovis* (Schrank, 1781)
  C *Trichodectes canis* (De Geer, 1778)
  A *Trichodectes octomaculatus* Paine, 1912
Trimenoponidae
  A *Trimenopon hispidum* (Burmeister, 1838)

Arthropoda, Insecta, Psocoptera
Caeciliidae
  E *Enderleinella obsoleta* Stephens, 1836
Caeciliusidae
  C *Lacroixiella martini* (Lacroix, 1919)
Ectopsocidae
  A *Ectopsocopsis cryptomeriae* (Enderlein, 1907)
  A *Ectopsocus axillaris* (Smithers, 1969)
  C *Ectopsocus briggsi* McLachlan, 1899
  A *Ectopsocus maindroni* Badonnel, 1935
  C *Ectopsocus meridionalis* Ribaga, 1903
  A *Ectopsocus pumilis* (Banks, 1920)
  A *Ectopsocus richardsi* (Pearman, 1929)
  A *Ectopsocus rileyae* Schmidt & Thornton, 1993
  A *Ectopsocus titschaki* Jentsch, 1939
  E *Ectopsocus vachoni* Badonnel, 1945
Lachesillidae
  E *Lachesilla greeni* (Pearman, 1933)
  A *Lachesilla pacifica* Chapman, 1930
Lepidopsocidae
  A *Echemeteryx madagascariensis* Kolbe, 1885
  A *Nepticulomima sakuntala* Enderlein, 1906
  A *Pteroxanium kelloggi* (Ribaga, 1905)
  A *Soa flaviterminata* Enderlein, 1906
Liposcelididae
  A *Embidopsocus minor* (Pearman, 1931)
  A *Liposcelis albothoracica* Broadhead, 1955
  C *Liposcelis arenicola* Günther, 1974
  C *Liposcelis bostrychophila* Badonnel, 1931
  C *Liposcelis brunnea* Motschulsky, 1852
  C *Liposcelis corrodens* (Heymons, 1909)
  C *Liposcelis decolor* (Pearman, 1925)
  C *Liposcelis entomophila* (Enderlein, 1907)
  A *Liposcelis mendax* Pearman, 1946
  A *Liposcelis obscura* Broadhead, 1954
  C *Liposcelis paeta* Pearman, 1942
  C *Liposcelis paetulus* Broadhead, 1950
  A *Liposcelis pearmani* Leinhard, 1990
  C *Liposcelis pubescens* Broadhead, 1947
  E *Liposcelis rufa* Broadhead, 1950

Pachytroctidae
  A *Nanopsocus oceanicus* Pearman, 1928
  A *Tapinella castanea* Pearman, 1932
Peripsocidae
  E *Peripsocus milleri* (Tillyard, 1923)
Peripsocidae
  E *Peripsocus parvulus* Kolbe, 1880
Philotarsidae
  A *Trichadenotecnum innuptum* Betz, 1983
Psoquilidae
  A *Psoquilla marginepunctata* (Hagen, 1865)
Psyllipsocidae
  A *Dorypteryx domestica* (Smithers, 1958)
  C *Dorypteryx longipennis* Smithers, 1991
  A *Dorypteryx pallida* Aaron, 1883
  A *Psocathropos lachlani* Ribaga, 1899
  C *Psyllipsocus ramburi* Sélys-Longs-champs, 1872
Trichopsocidae
  E *Trichopsocus clarus* (Banks, 1908)
  E *Trichopsocus dalii* (Mac Lachlan, 1867)
Trogiidae
  E *Cerobasis annulata* (Hagen, 1865)
  C *Lepinotus inquilinus* Heyden, 1850
  C *Lepinotus patruelis* Pearman, 1931
  C *Lepinotus reticulatus* Enderlein, 1905
  C *Trogium pulsatorium* (Linnaeus, 1758)

Arthropoda, Insecta, Thysanoptera
Aeolothripidae
  E *Aeolothrips fasciatus* (Linnaeus, 1758)
  A *Franklinothrips megalops* (Trybom, 1912)
  A *Franklinothrips vespiformis* (Crawford, 1909)
  E *Rhipidothrips gratiosus* Uzel, 1895
Merothripidae
  A *Merothrips floridensis* Watson, 1927
Phlaeothripidae
  C *Aleurodothrips fasciapennis* (Franklin, 1908)
  E *Apterygothrips pinicolus* Pelikan & Schliephake, 1994
  A *Bagnalliella yuccae* (Hinds, 1902)
  A *Eurythrips tristis* Hood, 1941
  A *Gynaikothrips ficorum* (Marchal, 1908)
  A *Haplothrips gowdeyi* (Franklin, 1908)
  A *Haplothrips rivnayi* Priesner, 1936
  E *Hoplandrothrips consobrinus* (Knechtel, 1951)
  C *Hoplothrips lichenis* Knechel, 1954
  E *Hoplothrips ulmi* (Fabricius, 1781)
  C *Hoplothrips unicolor* (Vuillet, 1914)
  A *Karnyothrips americanus* (Hood, 1912)
  A *Karnyothrips flavipes* (Jones, 1912)
  A *Karnyothrips melaleucus* (Bagnall, 1911)
  E *Liothrips vaneeckei* Priesner, 1920
  A *Nesothrips propinquus* (Bagnall, 1916)
  A *Podothrips semiflavus* Hood, 1913
  C *Suocerathrips linguis* Mound & Marullo, 1994
Thripidae
  A *Anaphothrips sudanensis* Trybom, 1911
  A *Anisopilothrips venustulus* (Priesner, 1923)
  E *Aptinothrips rufus* Haliday, 1836
  A *Aurantothrips orchidaceus* (Bagnall, 1909)
  A *Bradinothrips musae* Hood, 1956
  A *Caliothrips fasciatus* (Pergande, 1895)
  A *Chaetanaphothrips orchidii* (Moulton, 1908)
  E *Chirothrips manicatus* Halyday, 1836
  A *Copidothrips octarticulatus* (Schmutz, 1913)
  E *Dendrothrips eastopi* Pitkin & Palmer, 1975
  A *Dichromothrips corbetti* (Priesner, 1936)
  A *Dichromothrips phalaenopsidis* Sakimura, 1955
  A *Dorcadothrips billeni* Zur Strassen, 1995
  A *Echinothrips americanus* Morgan, 1913
  E *Euphysothrips minozzii* Bagnall, 1926
  A *Frankliniella fusca* (Hinds, 1902)
  A **Frankliniella occidentalis** (Pergande, 1895)
  C *Frankliniella schultzei* (Trybom, 1910)
  A *Heliothrips haemorrhoidalis* (Bouché, 1833)
  A *Hercinothrips bicinctus* (Bagnall, 1919)
  A *Hercinothrips femoralis* (Reuter, 1891)
  A *Isoneurothrips australis* Bagnall, 1915
  A *Leucothrips nigripennis* Reuter, 1904
  E *Limothrips cerealium* Halyday, 1836
  A *Microcephalothrips abdominalis* (Crawford, 1910)
  A *Neohydatothrips samayunkur* (Kudo, 1995)
  E *Odontothrips meliloti* Priesner, 1951
  A *Organothrips indicus* Bhatti, 1974
  A *Palmiothrips palmae* (Ramakrishna, 1934)

A *Parthenothrips dracaenae* (Heeger, 1854)
C *Pezothrips kellyanus* (Bagnall, 1916)
A *Phibalothrips peringueyi* (Faure, 1925)
A *Plesiothrips perplexus* (Beach, 1896)
A *Pseudodendrothrips mori* (Niwa, 1908)
A *Psydrothrips kewi* Palmer & Mound, 1985
A *Pteridothrips pteridicola* (Karny, 1914)
C *Scirtothrips longipennis* (Bagnall, 1909)
A *Stenchaetothrips biformis* (Bagnall, 1913)
A *Stenchaetothrips spinalis* Reyes, 1994
A *Thrips australis* (Bagnall, 1915)
A *Thrips palmi* Karny, 1925
A *Thrips simplex* (Morison, 1930)
E *Thrips tabaci* Lindeman, 1889

Arthropoda, Insecta, Hemiptera
Adelgidae
E *Adelges abietis* Linneaus, 1758
A *Adelges cooleyi* (Gillette, 1907)
A *Adelges coweni* (Gillette, 1907)
E *Adelges laricis* Vallot, 1836
A *Adelges nordmannianae* (Eckstein, 1890)
E *Adelges piceae* (Ratzeburg, 1844)
A *Adelges prelli* Grosmann, 1935
A *Adelges viridana* Cholodkovsky, 1896
E *Adelges viridis* Ratzeburg, 1843
A *Aphrastasia pectinatae* (Cholodkowsky, 1888)
A *Dreyfusia merkeri* Eichhorn, 1957
E *Pineus cembrae* (Cholodkovsky, 1888)
A *Pineus orientalis* (Dreyfus, 1889)
E *Pineus pineoides* Cholodkovsky, 1903
E *Pineus similis* (Gillette, 1907)
A *Pineus strobi* Hartig, 1837
Aleyrodidae
A *Aleuroclava guyavae* (Takahashi, 1932)
A *Aleurodicus dispersus* Russell, 1965
A *Aleurolobus marlatti* (Quaintance, 1903)
E *Aleurolobus olivinus* (Silvestri, 1911)
A *Aleuropteridis filicicola* (Newstead, 1911)
A *Aleurothrixus floccosus* (Maskell, 1896)
A *Aleurotrachelus atratus* Hempel, 1922
C *Aleurotulus nephrolepidis* (Quaintance, 1900)
A *Bemisia afer* (Priesner & Hosny, 1934)
A **Bemisia tabaci** (Gennadius, 1889)
A *Ceraleurodicus varus* (Bondar, 1928)

A *Crenidorsum aroidephagus* Martin & Aguiar, 2001
A *Dialeurodes chittendeni* Laing, 1928
A *Dialeurodes citri* (Ashmead, 1885)
A *Dialeurodes formosensis* Takahashi, 1933
C *Dialeurodes kirkaldy* (Kotinsky, 1907)
C *Filicaleyrodes williamsi* (Trehan, 1938)
A *Lecanoideus floccissimus* Martin et al., 1997
A *Parabemisia myricae* (Kuwana, 1927)
A *Paraleyrodes bondari* Peracchi, 1971
A *Paraleyrodes citricolus* Costa Lima, 1928
A *Paraleyrodes minei* Iccarino, 1990
A *Pealius azaleae* (Baker & Moles, 1920)
A *Singhiella citrifolii* (Morgan, 1893)
A *Trialeurodes packardi* (Morrill, 1903)
A *Trialeurodes vaporariorum* (Westwood, 1856)
Anthocoridae
A *Amphiareus obscuriceps* (Poppius, 1909)
E *Anthocoris butleri* Le Quesne, 1954
C *Buchananiella continua* (White, 1880)
C *Lyctocoris campestris* (Fabricius, 1794)
E *Orius laevigatus* (Fieber, 1860)
Aphididae
E *Acyrthosiphon auriculae* Martin, 1981
A *Acyrthosiphon caraganae* Cholodkovsky, 1908
A *Acyrthosiphon kondoi* Shinji, 1938
A *Acyrthosiphon primulae* (Theoblad, 1913)
A *Aloephagus myersi* Essig, 1950
E *Amphorophora tuberculata* Brown & Blackman, 1985
E *Aphis balloticola* Szelegiewicz, 1968
A *Aphis catalpae* Mamontova, 1953
E *Aphis craccivora* C.L. Kock, 1854
E *Aphis cytisorum* Hartig, 1841
A *Aphis forbesi* Weed, 1889
A **Aphis gossypii** Glover, 1877
A *Aphis illinoisensis* Shimer, 1866
A *Aphis oenotherae oenotherae* Oestlund 1887
E *Aphis salicariae* Koch, 1855
A *Aphis spiraecola* Patch, 1914
A *Aphis spiraephaga* F.P. Müller, 1961
A *Aphis spiraephila* Patch, 1914
E *Aphis thalictri* C.L. Kock, 1854
A *Appendiseta robiniae* (Gillette, 1907)
A *Brachycaudus rumexicolens* (Patch, 1917)

E *Brachycorynella asparagi* (Mordvilko, 1929)
A *Cerataphis brasiliensis* (Hempel, 1901)
A *Cerataphis lataniae* (Boisduval, 1867)
A *Cerataphis orchidearum* (Westwood, 1879)
A *Chaetosiphon fragaefolii* (T.D.A. Cockerell, 1901)
A *Chaitophorus populifolii* (Essig, 1912)
A *Chaitophorus saliapterus quinquemaculatus* Bozhko 1976
A *Chromaphis juglandicola* (Kaltenbach, 1843)
E *Cinara acutirostris* Hille Ris Lambers, 1956
E *Cinara brauni* Börner, 1940
A *Cinara cedri* Mimeur, 1936
E *Cinara confinis* (Koch, 1856)
E *Cinara costata* (Zetterstedt, 1828)
E *Cinara cuneomaculata* (Del Guercio, 1909)
E *Cinara cupressi* (Buckton, 1881)
A *Cinara curvipes* (Patch, 1912)
E *Cinara fresai* E.E. Blanchard, 1939
E *Cinara juniperi* (De Geer, 1773)
E *Cinara kochiana* (Börner, 1939)
A *Cinara laportei* (Remaudière, 1954)
E *Cinara laricis* (Hartig, 1839)
E *Cinara nuda* (Mordvilko, 1895)
E *Cinara pectinatae* (Nördlinger, 1880)
E *Cinara piceae* (Panzer, 1800)
E *Cinara pilicornis* (Hartig, 1841)
E *Cinara pinea* (Mordvilko, 1894)
E *Cinara pini* (Linnaeus, 1758)
E *Cinara pinihabitans* (Mordvilko, 1894)
E *Cinara pruinosa* (Hartig, 1841)
E *Cinara tujafilina* Del Guercio, 1909)
E *Cinara viridescens* (Cholodkovsky, 1898)
E *Crypturaphis grassii* Silvestri, 1935
E *Diuraphis noxia* (Kurdjumov, 1913)
A *Drepanaphis acerifoliae* (Thomas, 1878)
E *Drepanosiphum acerinum* (Walker, 1848)
E *Dysaphis tulipae* (Boyer de Fonscolombe, 1841)
E *Elatobium abietinum* (Walker, 1849)
A *Ericaphis scammelli* Mason, 1940
A *Ericaphis wakibae* (Hottes, 1934)
A *Eriosoma lanigerum* (Hausmann, 1802)
A *Essigella californica* (Essig, 1909)
E *Eulachnus agilis* Kaltenbach, 1843
E *Eulachnus bluncki* Börner, 1940

E *Eulachnus brevipilosus* Börner, 1940
A *Greenidea ficicola* Takahashi, 1921
E *Hysteroneura setariae* (Thomas, 1878)
A *Idiopterus nephrelepidis* Davis, 1909
A *Illinoia andromedae* (MacGillivray, 1958)
A *Illinoia azaleae* Mason, 1925
A *Illinoia goldamaryae* (Knowlton, 1938)
A *Illinoia lambersi* (MacGillivray, 1960)
A *Illinoia liriodendri* (Monell, 1879)
A *Illinoia morrisoni* (Swain, 1918)
A *Illinoia rhododendri* (Wilson, 1918)
A *Impatientinum asiaticum* Nevsky, 1929
A *Iziphya flabella* (Sanborn, 1904)
A *Macrosiphoniella sanborni* (Gillette, 1908)
A *Macrosiphum albifrons* Essig, 1911
A *Macrosiphum euphorbiae* (Thomas, 1878)
A *Macrosiphum ptericolens* Patch, 1919
A *Megoura lespedezae* (Essig & Kuwana, 1918)
A *Melanaphis bambusae* (Fullaway, 1910)
A *Melaphis rhois* (Fitch, 1866)
E *Mindarus abietinus* Koch, 1857
E *Mindarus obliquus* (Cholodkovsky, 1896)
A *Monellia caryella* (Fitch, 1855)
A *Monelliopsis caryae* (Monell, 1879)
A *Monelliopsis pecanis* Bissell, 1983
C *Myzaphis turanica* Nevsky, 1929
E *Myzocallis boerneri* Stroyan, 1957
E *Myzocallis schreiberi* Hille Ris Lambers & Stroyan, 1959
A *Myzocallis walshii* (Monell, 1879)
A *Myzus ascalonicus* Doncaster, 1946
C *Myzus cymbalariae* Stroyan, 1954
A *Myzus hemerocallis* Takahashi, 1921
A *Myzus ornatus* Laing, 1932
C *Myzus persicae* Sulzer, 1776
A *Myzus varians* Davidson, 1912
A *Nearctaphis bakeri* (Cowen, 1895)
A *Neomyzus circumflexus* Buckton, 1876
A *Neophyllaphis podocarpi* Takahashi, 1920
A *Neotoxoptera formosana* (Takahashi, 1921)
A *Neotoxoptera oliveri* (Essig, 1935)
A *Neotoxoptera violae* (Pergande, 1900)
A *Panaphis juglandis* (Goeze, 1778)
A *Paoliella eastopi* Hille Ris Lambers, 1973
E *Paracolopha morrisoni* Baker, 1919

A *Pemphigus populitransversus* Riley, 1879
A *Pentalonia nigronervosa* Coquerel, 1859
E *Periphyllus acericola* (Walker, 1848)
A *Periphyllus californiensis* (Shinji, 1917)
E *Periphyllus xanthomelas* Koch
E *Prociphilus fraxini* (Fabricius, 1777)
A *Prociphilus fraxinifolii* Riley, 1879
A *Pterochloroides persicae* (Cholodkovsky, 1899)
A *Pterocomma pseudopopuleum* Palmer, 1952
A *Reticulaphis distylii* der Goot, 1917
A *Rhodobium porosum* (Sanderson, 1900)
A *Rhopalosiphoninus latysiphon* (Davidson, 1912)
A *Rhopalosiphum insertum* (Walker, 1849)
A *Rhopalosiphum maidis* (Fitch, 1856)
A *Rhopalosiphum rufiabdominale* (Sasaki, 1899)
E *Schizolachnus pineti* (Fabricius, 1781)
A *Sipha flava* (Forbes, 1884)
A *Siphonatrophia cupressi* Swain, 1918
A *Sitobion alopecuri* (Takahashi, 1921)
C *Sitobion luteum* (Buckton, 1876)
E *Stagona pini* Burmeister, 1835
A *Stomaphis mordvilkoi* Hille Ris Lambers, 1933
A *Takecallis arundicolens* (Clarke, 1903)
A *Takecallis arundinariae* (Essig, 1917)
A *Takecallis taiwana* (Takahashi, 1926)
A *Tinocallis kahawaluokalani* (Kirkaldy, 1906)
A *Tinocallis nevskyi* Remaudière, Quednau & Heie, 1988
A *Tinocallis saltans* (Nevsky, 1929)
A *Tinocallis takachihoensis* Higuchi, 1972
A *Tinocallis ulmiparvifoliae* Matsumura, 1919
A *Tinocallis zelkowae* (Takahashi, 1919)
A *Toxoptera aurantii* Boyer de Fonscolombe, 1841
A *Toxoptera citricidus* Kirkaldy, 1906
A *Trichosiphonaphis polygonifoliae* (Shinji, 1944)
A *Tuberculatus kuricola* (Matsumura, 1917)
A *Uroleucon erigeronense* (Thomas, 1878)
A *Uroleucon pseudoambrosiae* (Olive, 1963)
E *Uroleucon telekiae* (Holman, 1965)
A *Utamphorophora humboldti* (Essig, 1941)
A *Wahlgreniella arbuti* (Davidson, 1910)
A *Wahlgreniella nervata* Gillette, 1908)

Aphrophoridae
E *Philaenus spumarius* (Linnaeus, 1758)

Asterolecaniidae
A *Asterolecanium epidendri* (Bouché, 1844)
A *Bambusaspis bambusae* (Boisduval, 1869)

Cicadellidae
E *Anoscopus albifrons* (Linnaeus, 1758)
A *Cicadulina bipunctata* (Melichar, 1904)
A *Edwardsiana ishidai* (Matsumura, 1932)
E *Edwardsiana platanicola* (Vidano, 1961)
E *Empoasca pteridis* (Dahlbom, 1850)
A *Empoasca punjabensis* Singh-Pruthi, 1940
A *Endria nebulosa* (Ball, 1900)
A *Erythroneura vulnerata* (Fitch, 1851)
E *Eupteryx decemnotata* Rey, 1891
E *Eupteryx melissae* Curtis, 1837
E *Eupteryx rostrata* Ribaut, 1936
E *Eupteryx salviae* Arzone & Vidano, 1994
A *Graphocephala fennahi* Young, 1977
E *Grypotes puncticollis* (Herrich-Schäffer, 1834)
E *Iassus scutellaris* (Fieber, 1868)
A *Jacobiasca lybica* (Bergevin & Zanon, 1922)
A *Japananus hyalinus* (Osborn, 1900)
A *Kyboasca bipunctata* (Oshanin, 1871)
A *Kyboasca maligna* (Walsh, 1862)
A *Macropsis elaeagni* Emeljanov, 1964
A *Melillaia desbrochersi* (Lethierry, 1889)
E *Opsius stactogalus* Fieber, 1866
A *Orientus ishidae* (Matsumura, 1902)
E *Placotettix taeniatifrons* (Kirschbaum, 1868)
A *Psammotettix saxatilis* Emeljanov, 1962
A *Scaphoideus titanus* Ball, 1932
A *Vilbasteana oculata* (Lindb.)
E *Wagneripteryx germari* (Zetterstedt, 1840)

Coccidae
- A *Ceroplastes ceriferus* (Fabricius, 1798)
- A *Ceroplastes floridensis* Comstock, 1881
- A *Ceroplastes japonicus* Green, 1921
- A *Ceroplastes sinensis* Del Guercio, 1900
- A *Coccus hesperidum* Linnaeus, 1758
- A *Coccus longulus* (Douglas, 1887)
- A *Coccus pseudohesperidum* (Cockerell, 1895)
- A *Coccus pseudomagnoliarum* (Kuwana, 1914)
- A *Eucalymnatus tessellatus* (Signoret, 1873)
- A *Eulecanium excrescens* Ferris, 1920
- E *Eulecanium tiliae* (Linnaeus, 1758)
- A *Inglisia lounsburyi* (Cockerell, 1900)
- A *Kilifia acuminata* (Signoret, 1873)
- A *Neopulvinaria innumerabilis* (Rathvon, 1880)
- A *Parasaissetia nigra* (Nietner, 1861)
- A *Parthenolecanium fletcheri* (Cockerel, 1893)
- E *Parthenolecanium persicae* (Fabricius, 1776)
- A *Protopulvinaria pyriformis* (Cockerell, 1894)
- A *Pulvinaria floccifera* (Westwood, 1870)
- A *Pulvinaria horii* Kuwana, 1902
- A *Pulvinaria hydrangeae* (Steinweden, 1946)
- A *Pulvinaria psidii* Maskell, 1893
- A *Pulvinaria regalis* Canard, 1968
- A *Pulvinariella mesembryanthemi* (Vallot, 1830)
- A *Saissetia coffeae* (Walker, 1852)
- A *Saissetia oleae* (Olivier, 1791)

Coreidae
- A *Leptoglossus occidentalis* Heidemann, 1910

Corixidae
- A *Trichocorixa verticalis* (Fieber, 1851)

Dactylopiidae
- A *Dactylopius coccus* Costa, 1829

Delphacidae
- A *Prokelisia marginata* (Van Duzee, 1897)

Diaspididae
- C *Abgrallaspis cyanophylli* (Signoret, 1869)
- A *Acutaspis perseae* (Comstock, 1881)
- E *Aonidia lauri* (Bouché, 1833)
- A *Aonidiella aurantii* (Maskell, 1879)
- A *Aonidiella citrina* (Coquillet, 1891)
- A *Aonidiella taxus* Leonardi, 1906
- A *Aonidiella tinerfensis* (Lindinger, 1911)
- A *Aspidiotus destructor* Signoret, 1869
- A *Aspidiotus elaeidis* Marchal, 1909
- A *Aspidiotus nerii* (Bouché, 1833)
- A *Aulacaspis rosae* (Bouché, 1833)
- A *Aulacaspis tubercularis* Newstead, 1906
- A *Chrysomphalus aonidum* (Linnaeus, 1758)
- A *Chrysomphalus dictyospermi* (Morgan, 1889)
- A *Comstockiella sabalis* (Comstock, 1883)
- A *Diaspidiotus osborni* (Newell & Cockerell, 1898)
- A *Diaspidiotus perniciosus* (Comstock, 1881)
- E *Diaspidiotus pyri* (Lichtenstein, 1881)
- A *Diaspidiotus uvae* (Comstock, 1881)
- A *Diaspis boisduvalii* Signoret, 1869
- A *Diaspis bromeliae* (Kerner, 1778)
- A *Diaspis echinocacti* (Bouché, 1833)
- E *Dynaspidiotus britannicus* (Newstead, 1898)
- A *Entaspidiotus lounsburyi* (Marlatt, 1908)
- A *Eulepidosaphes pyriformis* (Maskell, 1897)
- A *Fiorinia fioriniae* (Targioni Tozzeti, 1867)
- A *Fiorinia pinicola* Maskell, 1897
- A *Furchadaspis zamiae* (Morgan, 1890)
- A *Gymnaspis aechmeae* Newstead, 1898
- A *Hemiberlesia lataniae* (Signoret, 1869)
- A *Hemiberlesia palmae* (Cockerell, 1892)
- A *Hemiberlesia rapax* (Comstock, 1881)
- A *Howardia biclavis* (Comstock, 1883)
- A *Ischnaspis longirostris* (Signoret, 1882)
- A *Kuwanaspis bambusae* Kuwana, 1902
- A *Kuwanaspis pseudoleucaspis* (Kuwana, 1923)
- A *Lepidosaphes beckii* (Newman, 1869)
- A *Lepidosaphes gloverii* (Packard, 1869)
- A *Lepidosaphes pinnaeformis* (Bouché, 1851)
- A *Leucaspis podocarpi* (Green, 1929)
- A *Lindingaspis rossi* (Maskell, 1814)
- A *Lopholeucaspis cockerelli* (Grandpré & Charmoy, 1899)
- A *Lopholeucaspis japonica* (Cockerell, 1897)
- A *Mycetaspis personata* (Comstock, 1883)
- A *Oceanaspidiotus araucariae* Adachi & Fullaway, 1953
- A *Oceanaspidiotus spinosus* (Comstock, 1883)

- A *Odonaspis greeni* (Cockerell, 1902)
- A *Odonaspis secreta* (Cockerell, 1896)
- A *Opuntaspis philococcus* (Cockerell, 1893)
- A *Parlatoria blanchardi* Targioni Tozzeti, 1868
- A *Parlatoria camelliae* Comstock, 1883
- A *Parlatoria cinerea* Hadden, 1909
- A *Parlatoria crotonis* Douglas, 1867
- A *Parlatoria oleae* (Colvée, 1880)
- A *Parlatoria pergandii* Comstock, 1881
- A *Parlatoria proteus* (Curtis, 1843)
- A *Parlatoria theae* Cockerell, 1896
- A *Parlatoria ziziphi* (Lucas, 1853)
- A *Pinnaspis aspidistrae* (Signoret, 1869)
- A *Pinnaspis buxi* (Bouché, 1851)
- A *Pinnaspis strachani* (Cooley, 1899)
- A *Pseudaonidia paeoniae* (Cockerell, 1899)
- A *Pseudaulacaspis cockerelli* (Cooley, 1897)
- A *Pseudaulacaspis pentagona* (Targioni Tozzeti, 1886)
- A *Pseudauparlatoria parlatorioides* (Comstock, 1883)
- A *Pseudoparlatoria ostreata* Cockerell, 1892
- A *Rutherfordia major* (Cockerell, 1894)
- A *Selenaspidus albus* McKenzie, 1953
- A *Umbaspis regularis* (Newstead, 1911)
- A *Unaspis citri* (Comstock, 1846)
- A *Unaspis euonymi* (Comstock, 1881)
- A *Unaspis yanonensis* (Kuwana, 1923)

Eriococcidae
- A *Eriococcus araucariae* Maskell, 1879
- E *Eriococcus buxi* (Fonscolombe, 1834)
- A *Eriococcus occineus* Cockerell, 1894
- A *Ovaticoccus agavium* (Douglas, 1888)

Flatidae
- A *Metcalfa pruinosa* (Say, 1830)

Halimococcidae
- A *Colobopyga kewensis* (Newstead, 1901)

Issidae
- A *Acanalonia conica* (Say, 1830)

Lygaeidae
- E *Arocatus longiceps* Stål, 1872
- E *Gastrodes grossipes* (De Geer, 1773)
- A *Nysius huttoni* F.B. White, 1878
- E *Orsillus depressus* (Mulsant & Rey, 1852)
- E *Oxycarenus lavaterae* (Fabricius, 1787)

Margarodidae
- A *Icerya formicarum* Newstead, 1897
- A *Icerya purchasi* (Maskell, 1879)
- E *Matsucoccus feytaudi* Ducasse, 1941

Membracidae
- A *Stictocephala bisonia* Kopp & Yonke, 1977

Miridae
- E *Closterotomus trivialis* (A. Costa, 1853)
- E *Deraeocoris flavilinea* (A. Costa, 1862)
- E *Dichrooscytus gustavi* Josifov, 1981
- E *Dicyphus escalerae* Lindberg, 1934
- E *Macrolophus glaucescens* Fieber, 1858
- E *Macrolophus melanotoma* (A. Costa, 1853)
- C *Nesidiocoris tenuis* (Reuter, 1895)
- E *Orthotylus caprai* Wagner, 1955
- C *Taylorilygus apicalis* (Fieber, 1861)
- A *Tupiocoris rhododendri* (Dolling, 1972)
- E *Tuponia brevirostris* Reuter, 1883
- E *Tuponia elegans* (Jakovlev, 1867)
- E *Tuponia hippophaes* (Fieber, 1861)
- E *Tuponia macedonica* Wagner, 1957
- E *Tuponia mixticolor* (A. Costa, 1862)

Ortheziidae
- A *Insignorthezia insignis* Browne, 1887

Pentatomidae
- A *Halyomorpha halys* (Stål, 1855)
- A *Nezara viridula* (Linnaeus, 1758)
- A *Perillus bioculatus* (Fabricius, 1775)

Phoenicococcidae
- A *Phoenicococcus marlatti* (Cockerell, 1899)

Phylloxeridae
- C *Moritziella corticalis* (Kaltenbach, 1867)
- A *Viteus vitifoliae* (Fitch, 1855)

Pseudococcidae
- A *Antonina crawi* Cockerell, 1900
- A *Antonina graminis* (Maskell, 1897)
- A *Balanococcus diminutus* (Leonardi, 1918)
- A *Chaetococcus bambusae* (Maskell, 1892)
- A *Chorizococcus rostellum* (Lobdell, 1930)
- A *Delottococcus euphorbiae* (Ezzat & McConnell, 1956)
- A *Dysmicoccus brevipes* (Cockerell, 1893)
- A *Dysmicoccus grassii* (Leonardi, 1913)
- A *Dysmicoccus mackenziei* Beardsley, 1965
- A *Dysmicoccus neobrevipes* Beardsley, 1959
- A *Ferrisia virgata* (Cockerell, 1893)
- A *Geococcus coffeae* Green, 1933
- A *Hypogeococcus pungens* Granara de Willlink, 1981

A *Nipaecoccus nipae* (Maskell, 1893)
A *Palmicultor palmarum* (Ehrhorn, 1916)
A *Peliococcus multispinus* (Siraiwa, 1939)
A *Peliococcus serratus* (Ferris, 1925)
A *Phenacoccus gossypii* Townsend & Cockerell, 1898
A *Phenacoccus madeirensis* Green, 1923
A *Phenacoccus pumilus* Kiritchenko, 1931
A *Phenacoccus solani* Ferris, 1918
A *Planococcus citri* (Risso, 1813)
E *Planococcus ficus* (Signoret, 1875)
A *Planococcus halli* Ezzat & McConnell, 1956
E *Planococcus vovae* (Nasonov, 1908)
A *Pseudococcus calceolariae* (Maskell, 1879)
A *Pseudococcus comstocki* (Kuwana, 1902)
A *Pseudococcus longispinus* (Targioni-Tozzetti, 1867)
A *Pseudococcus maritimus* (Ehrhorn, 1900)
A *Pseudococcus microcirculus* McKenzie, 1960
A *Pseudococcus viburni* (Signoret, 1875)
A *Rhizoecus americanus* (Hambleton, 1946)
A *Rhizoecus cacticans* (Hambleton, 1946)
A *Rhizoecus dianthi* Green, 1926
E *Rhizoecus falcifer* Künckel d'Herculais, 1878
A *Rhizoecus hibisci* (Kawai & Takagi, 1971)
A *Rhizoecus latus* (Hambleton, 1946)
A *Spilococcus mamillariae* (Bouché, 1844)
A *Trionymus angustifrons* Hall, 1926
A *Trochiscococcus speciosus* (De Lotto, 1961)
A *Vryburgia amaryllidis* (Bouché, 1837)
A *Vryburgia brevicruris* (McKenzie, 1960)
A *Vryburgia rimariae* Tranfaglia, 1981
Psyllidae
A *Acizzia acaciaebaileyanae* (Froggatt, 1901)
A *Acizzia hollisi* Burckhardt, 1981
A *Acizzia jamatonica* (Kuwayama, 1908)
A *Acizzia uncatoides* (Ferris & Klyver, 1932)
A *Blastopsylla occidentalis* Taylor, 1985
A *Cacopsylla fulguralis* (Kuwayama, 1908)
E *Cacopsylla pulchella* (Löw, 1877)
E *Calophya rhois* (Löw, 1877)
A *Ctenarytaina eucalypti* (Maskell, 1890)
A *Ctenarytaina spatulata* Taylor, 1967
E *Homotoma ficus* (Linnaeus, 1758)
E *Livilla variegata* (Löw, 1881)
Reduviidae
C *Empicoris rubromaculatus* (Blackburn, 1889)
A *Ploiaria chilensis* (Philippi, 1862)
Saldidae
A *Pentacora sphacelata* (Uhler, 1877)
Tingidae
A *Corythucha arcuata* (Say, 1832)
A *Corythucha ciliata* (Say, 1832)
E *Stephanitis oberti* (Kolenati, 1857)
A *Stephanitis pyrioides* (Scott, 1874)
A *Stephanitis rhododendri* Horvath, 1905
A *Stephanitis takeyai* Drake & Maa, 1955
E *Tingis cardui* (Linnaeus, 1758)
Triozidae
A *Bactericera tremblayi* (Wagner, 1961)
E *Epitrioza neglecta* (Loginova, 1978)
E *Laurotrioza alacris* (Flor, 1861)
A *Trioza erythreae* (Del Gercio, 1918)
A *Trioza vitreoradiata* (Maskell, 1879)
Tropiduchidae
A *Ommatissus binotatus* Fieber, 1876

Arthropoda, Insecta, Coleoptera
Acanthocnemidae
A *Acanthocnemus nigricans* (Hope, 1845)
Anobiidae
E *Anobium punctatum* De Geer, 1774
A *Calymmaderus oblongus* (Gorham, 1883)
C *Epauloecus unicolor* (Piller & Mitterpacher, 1783)
C *Ernobius mollis* (Linnaeus, 1758)
A *Gibbium aequinoctiale* Boieldieu, 1854
C *Gibbium psylloides* (Czempinski, 1778)
A *Lasioderma serricorne* (Fabricius, 1792)
C *Mezium affine* Boieldieu, 1856
A *Mezium americanum* Laporte de Castelnau, 1840
C *Nicobium castaneum* (Olivier, 1790)
E *Oligomerus ptilinoides* (Wollaston, 1854)
A *Ozognathus cornutus* (Le Conte, 1859)
A *Pseudeurostus hilleri* (Reitter, 1877)
A *Ptilineurus marmoratus* (Reitter, 1877)
C *Ptinus bicinctus* Sturm, 1837
C *Ptinus clavipes* Panzer, 1792

11 List of Species Alien in Europe and to Europe

E *Ptinus dubius* Sturm, 1837
C *Ptinus fur* (Linnaeus, 1758)
C *Ptinus latro* Fabricius, 1775
A *Ptinus ocellus* Brown, 1929
E *Sphaericus gibboides* (Boieldieu, 1854)
A *Tricorynus tabaci* (Guérin-Méneville, 1850)
A *Trigonogenius globulus* Solier, 1849
Anthicidae
   A *Anthicus crinitus* La Ferte-Senectere, 1849
   A *Anthicus czernohorskyi* Pic, 1912
   E *Cordicomus instabilis* (Schmidt, 1842)
   E *Cyclodinus humilis* (Germar, 1824)
   A *Omonadus floralis* (Linnaeus, 1758)
   E *Omonadus formicarius* (Goeze, 1777)
   A *Stricticomus tobias* (Marseul, 1879)
Anthribidae
   A *Araecerus coffeae* (Fabricius, 1801)
   E *Bruchela rufipes* Olivier, 1790
   A *Tropideres dorsalis* (Thunberg, 1796)
Aphodiidae
   A *Aphodius gracilis* Boheman, 1857
   E *Calamosternus granarius* Linnaeus, 1767
   E *Pleurophorus caesus* Creutzer, 1796
   A *Saprosites mendax* Blackburn, 1892
   A *Saprosites natalensis* (Peringuey, 1901)
   A *Tesarius caelatus* (Laconte, 1857)
Apionidae
   A *Alocentron curvirostre* (Gyllenhal, 1833)
   E *Apion haematodes* W. Kirby, 1808
   A *Aspidapion validum* (Germar, 1817)
   E *Ixapion variegatum* (Wencker, 1864)
   A *Rhopalapion longirostre* (Olivier, 1807)
Bostrichidae
   A *Apate monachus* Fabricius, 1775
   A *Bostrychoplites cornutus* (Olivier, 1790)
   A *Dinoderus bifoveolatus* (Wollaston, 1858)
   A *Dinoderus minutus* (Fabricius, 1775)
   A *Heterobostrychus hamatipennis* (Lesne, 1895)
   A *Rhyzopertha dominica* (Fabricius, 1792)
   A *Sinoxylon senegalense* Karsch, 1831
Buprestidae
   E *Agrilus angustulus* (Illiger, 1803)
   A *Buprestis decora* Fabricius, 1775
   E *Buprestis novemmaculata* Linnaeus, 1758
   A *Chrysobothris dorsata* (Fabricius, 1787)
   E *Melanophila acuminata* (De Geer, 1774)

Byrrhidae
   E *Simplocaria semistriata* (Fabricius, 1794)
Carabidae
   E *Abax parallelus* Duftschmid, 1812
   E *Amara aenea* (De Geer, 1774)
   E *Amara anthobia* A. Villa & G.B. Villa, 1833
   E *Amara aulica* (Panzer, 1797)
   E *Amara montivaga* Sturm, 1825
   E *Anisodactylus binotatus* (Fabricius, 1787)
   E *Callistus lunatus* (Fabricius, 1775)
   E *Carabus auratus* Linnaeus, 1758
   E *Carabus cancellatus* Linnaeus, 1758
   E *Carabus convexus* Fabricius, 1775
   E *Carabus nemoralis* O.F. Müller, 1764
   E *Demetrias atricapillus* (Linnaeus, 1758)
   E *Epaphius secalis* (Paykull, 1790)
   E *Graniger femoralis* (Coquerel, 1858)
   E *Harpalus distinguendus* (Duftschmid, 1812)
   A *Laemostenus complanatus* (Dejean, 1828)
   A *Leistus nubivagus* Wollaston, 1864
   E *Leistus rufomarginatus* (Duftschmid, 1812)
   E *Leistus terminatus* (Panzer, 1793)
   E *Licinus punctatulus* (Fabricius, 1792)
   E *Lymnastis galilaeus* Piochard de la Brélerie, 1876
   E *Microlestes minutulus* (Goeze, 1777)
   E *Notaphus varius* (Olivier, 1795)
   A *Notiobia cupripennis* (Germar, 1824)
   E *Ocydromus tetracolus* (Say, 1823)
   E *Paranchus albipes* (Fabricius, 1796)
   E *Philochthus guttula* (Fabricius, 1792)
   A *Plochionus pallens* (Fabricius, 1775)
   E *Pterostichus angustatus* Duftschmid, 1812
   A *Pterostichus caspius* (Ménétriés, 1832)
   E *Pterostichus cristatus* Dufour, 1820
   A *Pterostichus quadrifoveolatus* Letzner, 1852
   E *Pterostichus vernalis* (Panzer, 1796)
   E *Scybalicus oblongiusculus* (Dejean, 1829)
   A *Somotrichus unifasciatus* (Dejean, 1831)
   E *Sphodrus leucophthalmus* (Linnaeus, 1758)
   E *Tachyta nana* (Gyllenhal, 1810)
   A *Trechicus nigriceps* (Dejean, 1831)
   E *Trechus subnotatus* Dejean, 1831
   E *Tschitscherinellus cordatus* (Dejean, 1825)

Cerambycidae
- A *Acanthoderes jaspidea* Germar, 1824
- A *Acrocinus longimanus* (Linnaeus, 1758)
- A **Anoplophora chinensis** (Förster, 1848)
- A **Anoplophora glabripennis** (Motschulsky, 1853)
- E *Arhopalus rusticus* (Linnaeus, 1758)
- E *Aromia moschata* (Linnaeus, 1758)
- A *Callidiellum rufipenne* (Motschulsky, 1860)
- E *Cerambyx carinatus* Küster, 1846
- E *Cerambyx nodulosus* Germar, 1817
- A *Chlorophorus annularis* (Fabricius, 1787)
- E *Clytus arietis* (Linnaeus, 1758)
- E *Cyrthognathus forficatus* (Fabricius, 1792)
- E *Derolus mauritanicus* Buquet, 1840
- A *Deroplia albida* (Brullé, 1838)
- E *Gracilia minuta* (Fabricius, 1781)
- E *Icosium tomentosum atticum* Ganglbauer, 1881
- A *Lucasianus levaillantii* (Lucas, 1846)
- E *Monochamus galloprovincialis* (Olivier, 1795)
- E *Monochamus sartor* (Fabricius, 1787)
- E *Monochamus sutor* (Linnaeus, 1758)
- E *Morimus asper funereus* Mulsant, 1863
- E *Nathrius brevipennis* (Mulsant, 1839)
- A *Neoclytus acuminatus* (Fabricius, 1775)
- A *Oxymerus aculeatus lebasi* Dupont, 1838
- A *Parandra brunnea* (Fabricius, 1789)
- A *Phoracantha recurva* Newman, 1840
- A *Phoracantha semipunctata* (Fabricius, 1775)
- A *Phryneta leprosa* (Fabricius, 1775)
- E *Phymatodes testaceus* (Linnaeus, 1758)
- E *Poecilium lividum* (Rossi, 1794)
- A *Psacothea hilaris* (Pascoe, 1847)
- E *Rhagium inquisitor* (Linnaeus, 1758)
- E *Rosalia alpina* (Linnaeus, 1758)
- E *Stictoleptura rubra* (Linnaeus, 1758)
- E *Stromatium unicolor* (Olivier, 1795)
- A *Taeniotes cayennensis* Thomson, 1859
- E *Trichoferus fasciculatus* (Faldermann, 1837)
- E *Trichoferus griseus* (Fabricius, 1792)
- A *Trinophylum cribratum* (Bates, 1878)
- E *Xylotrechus arvicola* (Olivier, 1795)
- A *Xylotrechus stebbingi* Gahan, 1906

Cerylonidae
- A *Murmidius ovalis* (Beck, 1817)
- A *Philothermus montandoni* Aube, 1843

Chrysomelidae
- A *Acanthoscelides obtectus* Say, 1831
- A *Acanthoscelides pallidipennis* (Motschulsky, 1874)
- E *Altica ampelophaga* Guérin-Méneville, 1858
- E *Altica carduorum* Guérin-Méneville, 1858
- A *Aspidomorpha fabricii* Sekerka, 2008
- E *Bruchidius foveolatus* (Gyllenhal, 1833)
- E *Bruchidius lividimanus* (Gyllenhal, 1833)
- E *Bruchus ervi* Fröhlich, 1799
- E *Bruchus lentis* Fröhlich, 1799
- A *Bruchus pisorum* (Linnaeus, 1758)
- A *Bruchus rufimanus* Bohemann, 1833
- E *Bruchus rufipes* Herbst, 1783
- E *Bruchus signaticornis* Gyllenhal, 1833
- A *Callosobruchus chinensis* (Linnaeus, 1758)
- A *Callosobruchus maculatus* (Fabricius, 1775)
- A *Callosobruchus phaseoli* (Gyllenhal, 1833)
- A *Caryedon serratus* (Olivier, 1790)
- E *Chaetocnema hortensis* (Geoffroy, 1785)
- E *Chrysolina americana* Linnaeus, 1758
- E *Chrysolina caerulans* Scriba, 1791
- E *Chrysolina viridana* Küster, 1844
- E *Crioceris asparagi* (Linnaeus, 1758)
- E *Cryptocephalus sulphureus* G.A. Olivier, 1808
- A **Diabrotica virgifera virgifera** LeConte, 1868
- A *Epitrix cucumeris* (Harris, 1851)
- A *Epitrix hirtipennis* (Melsheimer, 1847)
- E *Epitrix pubescens* (Koch, 1803)
- E *Gonioctena fornicata* (Bruggemann, 1873)
- A **Leptinotarsa decemlineata** (Say, 1824)
- E *Lilioceris lilii* (Scopoli, 1763)
- E *Longitarsus kutscherae* (Rye, 1872)
- E *Longitarsus lateripunctatus* (Rosenhauer, 1856)
- A *Luperomorpha xanthodera* Fairmaire, 1888
- A *Megabruchidius dorsalis* (Fahreus, 1839)
- A *Megabruchidius tonkineus* György, 2007
- A *Mimosestes mimose* (Fabricius, 1781)
- E *Neocrepidodera brevicollis* (J. Daniel, 1904)

E   *Neocrepidodera ferruginea* (Scopoli, 1763)
A   *Phaedon brassicae* Baly, 1874
A   *Pistosia dactyliferae* Maulik, 1919
C   *Polyspilla polyspilla* Germar, 1821
E   *Psylliodes chrysocephalus* (Linnaeus, 1758)
A   *Zabrotes subfasciatus* (Bohemann, 1833)
A   *Zygogramma suturalis* (Fabricius, 1775)

Ciidae
A   *Xylographus bostrychoides* (Dufour, 1843)

Clambidae
E   *Clambus pallidulus* Reitter, 1911
A   *Clambus simsoni* Blackburn, 1902

Cleridae
E   *Enoplium serraticorne* (Olivier, 1790)
C   *Necrobia ruficollis* (Fabricius, 1775)
A   *Necrobia rufipes* (De Geer, 1775)
C   *Necrobia violacea* (Linnaeus, 1758)
A   *Opetiopalpus scutellaris* (Panzer, 1797)
E   *Opilo domesticus* (Sturm, 1837)
E   *Opilo mollis* (Linnaeus, 1758)
A   *Paratillus carus* (Newmann, 1840)
C   *Tarsostenus univittatus* (Rossi, 1792)
A   *Thaneroclerus buqueti* (Lefebvre, 1835)

Coccinellidae
E   *Anatis ocellata* (Linnaeus, 1758)
E   *Aphidecta obliterata* (Linnaeus, 1758)
A   *Chilocorus kuwanae* Silvestri, 1909
A   *Chilocorus nigrita* (Fabricius, 1798)
A   *Cryptolaemus montrouzieri* Mulsant, 1853
A   *Delphastus catalinae* (Horn, 1895)
E   *Exochomus quadripustulatus* (Linnaeus, 1758)
E   *Harmonia quadrpunctata* (Pontoppidan, 1763)
A   **Harmonia axyridis** (Pallas, 1773)
E   *Henosepilachna argus* (Geoffroy, 1762)
A   *Hippodamia convergens* Guerin-Meneville, 1842
A   *Hyperaspis pantherina* Fürsch, 1975
E   *Myrrha 18-guttata* Linnaeus
E   *Myzia oblongogutatta* (Linnaeus, 1758)
A   *Nephus reunioni* Fürsch, 1974
A   *Rhyzobius forestieri* (Mulsant, 1853)
A   *Rhyzobius lophanthae* (Blaisdell, 1892)
A   *Rodolia cardinalis* (Mulsant, 1850)
E   *Scymnus nigrinus* Kugelann, 1794
E   *Scymnus pullus* Mulsant, 1850
A   *Serangium parcesetosum* Sicard, 1929

Colydiidae
C   *Aglenus brunneus* (Gyllenhall, 1813)

Corylophidae
A   *Orthoperus aequalis* Sharp, 1885
E   *Sericoderus lateralis* (Gyllenhal, 1827)

Cryptophagidae
E   *Atomaria apicalis* Erichson, 1846
E   *Atomaria bella* Reitter, 1875
E   *Atomaria fuscata* (Schönherr, 1808)
E   *Atomaria fuscipes* (Gyllenhal, 1808)
E   *Atomaria hislopi* Wollaston, 1857
A   *Atomaria lewisi* Reitter, 1877
E   *Atomaria lohsei* Johnson & Strand, 1968
E   *Atomaria munda* Erichson, 1846
E   *Atomaria nitidula* Marsham, 1802
E   *Atomaria punctithorax* Reitter, 1887
E   *Atomaria pusilla* (Paykull, 1798)
E   *Atomaria strandi* Johnson, 1967
E   *Atomaria testacea* Stephens, 1830
E   *Atomaria turgida* Erichson, 1846
A   *Caenoscelis subdeplanata* C. Brisout de Barneville, 1882
C   *Cryptophagus acutangulus* Gyllenhall, 1828
C   *Cryptophagus affinis* Sturm, 1845
C   *Cryptophagus cellaris* (Scopoli, 1763)
C   *Cryptophagus dentatus* (Herbst, 1793)
E   *Cryptophagus distinguendus* Sturm, 1845
C   *Cryptophagus fallax* Balfour-Browne, 1953
C   *Cryptophagus pilosus* Gyllenhal, 1828
E   *Cryptophagus saginatus* Sturm, 1845
E   *Cryptophagus scanicus* (Linnaeus, 1758)
E   *Cryptophagus schmidti* Sturm, 1845
C   *Cryptophagus subfumatus* Kraatz, 1856
E   *Ephistemus globulus* Paykull, 1798
A   *Henoticus californicus* (Mannhereim, 1843)

Curculionidae
A   *Asperogronops inaequalis* (Boheman, 1842)
A   *Asynonychus godmani* Crotch, 1867
E   *Barynotus squamosus* Germar, 1824
A   *Barypeithes pellucidus* (Boheman, 1834)
E   *Brachytemnus porcatus* (Germar, 1824)
E   *Cathormiocerus curvipes* (Wollaston, 1854)
A   *Caulophilus oryzae* (Gyllenhal, 1838)
E   *Ceutorhynchus assimilis* (Paykull, 1800)
A   *Demyrsus meleoides* Pascoe, 1872
A   *Euophryum confine* (Broun, 1881)
A   *Euophryum rufum* (Broun, 1880)

A *Gonipterus scutellatus* Gyllenhal, 1833
E *Hypera postica* (Gyllenhal, 1813)
A *Lignyodes bischoffi* (Blatchley, 1916)
E *Liparus glabrirostris* Küster, 1849
A *Lissorhoptrus oryzophilus* Kuschel, 1952
A *Listroderes costirostris* Schoenherr, 1826
E *Magdalis memnonia* (Gyllenhal, 1837)
E *Mecinus pascuorum* (Gyllenhal, 1813)
A *Micromimus osellai* Voss, 1968
E *Mogulones geographicus* (Goeze, 1777)
A *Naupactus leucoloma* Boheman, 1840
E *Neoderelomus piriformis* (Hoffmann, 1938)
E *Otiorhynchus armadillo* (Rossi, 1792)
E *Otiorhynchus armatus* Boheman, 1843
E *Otiorhynchus corruptor* (Host, 1789)
E *Otiorhynchus crataegi* Germar, 1824
E *Otiorhynchus cribricollis* Gyllenhal, 1834
E *Otiorhynchus dieckmanni* Magnano, 1979
E *Otiorhynchus singularis* (Linnaeus, 1767)
E *Otiorhynchus sulcatus* (Fabricius, 1775)
E *Otiorrhynchus salicicola* Heyden, 1908
A *Paradiophorus crenatus* (Billbarg, 1820)
A *Pentarthrum huttoni* Wollaston, 1854
E *Philopedon plagiatum* (Schaller, 1783)
E *Psallidium maxillosum* (Fabricius, 1792)
E *Pselactus spadix* (Herbst, 1795)
E *Rhinocyllus conicus* (Froelich, 1792)
E *Rhinoncus pericarpius* Stephens, 1829
E *Rhopalomesites tardyi* (Curtis, 1825)
A *Rhyephenes humeralis* (Guérin-Méneville, 1830)
E *Sitona cinnamomeus* Allard, 1863
E *Sitona discoideus* Gyllenhal, 1834
E *Sitona lepidus* Gyllenhal, 1834
E *Sitona puberulus* Reitter, 1903
E *Sitona puncticollis* Stephens, 1831
E *Strophosoma melanogrammum* (Forster, 1771)
A *Syagrius intrudens* Waterhouse, 1903
E *Tychius cuprifer* (Panzer, 1799)
E *Tychius picirostris* (Fabricius, 1787)
Cybocephalidae
A *Aglyptinus agathidioides* Blair, 1930
A *Cybocephalus nipponicus* Endrody-Younga, 1971
Dermestidae
A *Anthrenocerus australis* (Hope, 1843)
A *Anthrenus caucasicus* Reitter, 1881

E *Anthrenus coloratus* Reitter, 1881
E *Anthrenus festivus* Erichson, 1846
A *Anthrenus flavidus* Solsky, 1876
C *Anthrenus flavipes* LeConte, 1854
E *Anthrenus museorum* (Linnaeus, 1761)
A *Anthrenus oceanicus* Fauvel, 1903
E *Anthrenus olgae* Kalik, 1946
E *Attagenus bifasciatus* (Olivier, 1790)
A *Attagenus diversepubescens* Pic, 1936
C *Attagenus fasciatus* (Thunberg, 1795)
A *Attagenus gobicola* Frivaldszky, 1892
A *Attagenus lynx* (Mulsant & Rey, 1868)
E *Attagenus quadrimaculatus* Kraatz, 1858
E *Attagenus rossi* Ganglbauer, 1904
E *Attagenus simplex* Reitter, 1881
C *Attagenus smirnovi* Zhantiev, 1973
E *Attagenus trifasciatus* (Fabricius, 1787)
C *Attagenus unicolor* Reitter, 1877
E *Attagenus brunneus* Faldermann, 1835
E *Attagenus pellio* Linnaeus, 1758
C *Dermestes ater* De Geer, 1774
A *Dermestes bicolor bicolor* Fabricius, 1781
A *Dermestes carnivorus* Fabricius, 1775
A *Dermestes coronatus* Steven, 1808
E *Dermestes dermestinus undulatus* Brahm, 1790
C *Dermestes frischi* Kugelann, 1792
C *Dermestes lardarius* (Linnaeus, 1758)
A *Dermestes leechi* Kalik, 1952
C *Dermestes maculatus* De Geer, 1774
E *Dermestes murinus* Linnaeus, 1758
A *Dermestes peruvianus* Laporte de Castelnau, 1840
A *Dermestes vorax* Motschulsky, 1860
A *Novelsis horni* (Jayne, 1882)
A *Orphinus fulvipes* Guérin-Méneville, 1838
A *Phradonoma tricolor* (Arrow, 1915)
A *Reesa vespulae* (Milliron, 1939)
A *Sefrania bleusei* Pic, 1899
A *Telopes heydeni* Reitter, 1875
A *Thaumaglossa rufocapillata* Redtenbacher, 1867
A *Thylodrias contractus* Motschulsky, 1839
A *Trogoderma angustum* (Solier, 1849)
C *Trogoderma glabrum* (Herbst, 1783)
A *Trogoderma granarium* Everts, 1898
A *Trogoderma inclusum* LeConte, 1854
A *Trogoderma insulare* Chevrolat, 1863
A *Trogoderma longisetosum* Chao & Lee, 1966

- A *Trogoderma megatomoides* Reitter, 1881
- A *Trogoderma variabile* Ballion, 1878
- C *Trogoderma versicolor* (Creutzer, 1799)

Derodontidae
- E *Laricobius erichsonii* Rosenhauer, 1846

Dryophthoridae
- A *Cosmopolites sordidus* (Germar, 1824)
- A *Rhynchophorus ferrugineus* (Olivier, 1790)
- A *Sitophilus linearis* (Herbst, 1795)
- A *Sitophilus oryzae* (Linnaeus, 1763)
- C *Sitophilus zeamais* Motschulsky, 1855
- E *Sphenophorus meridionalis* Gyllenhal, 1838
- A *Sphenophorus venatus* (Say, 1831)

Dytiscidae
- A *Megadytes costalis* Fabricius, 1775

Elateridae
- E *Athous haemorrhoidalis* (Fabricius, 1801)
- A *Cardiophorus taylori* Cobos, 1970
- A *Conoderus posticus* (Eschscholtz)
- E *Melanotus dichrous* (Erichson, 1841)
- A *Panspaeus guttatus* Sharp, 1877

Endomychidae
- C *Holoparamecus caularum* Aubé, 1843
- C *Holoparamecus depressus* Curtis, 1833

Erirhinidae
- A *Stenopelmus rufinasus* Gyllenhal, 1835

Erotylidae
- A *Dacne picta* Crotch, 1873

Histeridae
- E *Acritus nigricornis* (Hoffmann, 1803)
- C *Carcinops pumilio* (Erichson, 1834)
- A *Carcinops troglodytes* (Paykull, 1811)
- A *Chalcionellus decemstriatus* Reichardt, 1932
- A *Diplostix mayeti* (Marseul, 1870)
- E *Halacritus punctum* (Aubé, 1843)
- A *Hister bipunctatus* Paykull, 1811
- C *Hypocaccus brasiliensis* (Paykull, 1811)
- E *Hypocaccus dimidiatus* (Illiger, 1807)
- E *Macrolister major* (Linnaeus, 1767)
- A *Paromalus luderti* Marseul, 1862
- E *Saprinus acuminatus* (Fabricius, 1798)
- E *Saprinus caerulescens* (Hoffmann, 1803)
- A *Saprinus lugens* Erichson, 1834
- E *Saprinus planiusculus* Motschulsky, 1849
- E *Saprinus semistriatus* (Scriba, 1790)
- E *Saprinus subnitescens* Bickhardt, 1909

Hydrophilidae
- E *Cercyon depressus* Stephens, 1829
- E *Cercyon haemorrhoidalis* (Fabricius, 1775)
- A *Cercyon inquinatus* Wollaston, 1854
- A *Cercyon laminatus* Sharp, 1873
- A *Cercyon nigriceps* (Marsham, 1802)
- E *Cercyon obsoletus* (Gyllenhal, 1808)
- E *Cercyon quisquilius* (Linnaeus, 1761)
- A *Cryptopleurum subtile* Sharp, 1884
- A *Dactylosternum abdominale* (Fabricius, 1792)
- E *Enochrus bicolor bicolor* (Fabricius, 1792)
- E *Helochares lividus* (Forster, 1771)
- A *Oosternum sharpi* Hansen, 1999
- A *Pachysternum capense* (Mulsant, 1894)
- A *Pelosoma lafertei* Mulsant, 1844
- E *Sphaeridium bipustulatum* Fabricius, 1781
- E *Sphaeridium scarabaeoides* (Linnaeus, 1758)

Kateretidae
- E *Brachypterolus antirrhini* (Murray, 1864)
- E *Brachypterolus vestitus* (Kiesenwetter, 1850)

Laemophloeidae
- E *Cryptolestes capensis* (Waltl, 1834)
- C *Cryptolestes duplicatus* (Waltl, 1834)
- C *Cryptolestes ferrugineus* (Stephens, 1831)
- C *Cryptolestes pusilloides* (Steel & Howe, 1952)
- A *Cryptolestes pusillus* (Schönherr, 1817)
- C *Cryptolestes spartii* (Curtis, 1834)
- C *Cryptolestes turcicus* (A. Grouvelle, 1876)

Languriidae
- C *Cryptophilus integer* (Heer, 1841)
- A *Cryptophilus obliteratus* Reitter, 1878
- C *Curelius japonicus* (Reitter, 1877)
- C *Pharaxonotha kirschii* Reitter, 1875

Latridiidae
- C *Adistemia watsoni* (Wollaston, 1871)
- A *Cartodere australicus* (Belon, 1887)
- A *Cartodere bifasciata* (Reitter, 1877)
- C *Cartodere constricta* (Gyllenhal, 1827)
- A *Cartodere delamarei* Dajoz, 1962
- A *Cartodere nodifer* (Westwood, 1839)
- E *Cartodere norvegica* (Strand, 1940)
- E *Corticaria abietorum* Motschulsky, 1867
- C *Corticaria elongata* (Gyllenhal, 1827)
- C *Corticaria fulva* (Comolli, 1837)
- C *Corticaria fenestralis* L.
- C *Corticaria pubescens* (Gyllenhal, 1827)
- C *Corticaria serrata* (Paykull, 1798)
- C *Dienerella argus* (Reitter, 1884)
- C *Dienerella costulata* (Reitter, 1877)

C *Dienerella filum* (Aubé, 1850)
E *Dienerella ruficollis* (Marsham, 1802)
C *Latridius minutus* (Linnaeus, 1767)
A *Metophthalmus serripennis* Broun, 1914
C *Migneauxia orientalis* Reitter, 1877
E *Thes bergrothi* (Reitter, 1880)
Leiodidae
E *Catops fuliginosus* Erichson, 1837
Lyctidae
A *Lyctus africanus* Lesne, 1907
A *Lyctus brunneus* (Stephens, 1830)
A *Lyctus cavicollis* Le Conte, 1805
A *Lyctus parallelocollis* Blackburn
A *Lyctus planicollis* Le Conte, 1858
A *Lyctus sinensis* Lesne, 1911
A *Minthea rugicollis* (Walker, 1858)
Meloidae
E *Mylabris variabilis* (Pallas, 1781)
Melyridae
A *Axinotarsus marginalis* (Laporte de Castelnau, 1840)
Monotomidae
E *Monotoma bicolor* A. Villa & G.B. Villa, 1835
E *Monotoma longicollis* (Gyllenhal, 1827)
E *Monotoma picipes* Herbst, 1793
E *Monotoma quadrifoveolata* Aubé, 1837
E *Monotoma spinicollis* Aubé, 1837
E *Rhizophagus grandis* Gyllenhal, 1827
Mordellidae
A *Mordellistena cattleyana* Champion, 1913
Mycetophagidae
E *Berginus tamarisci* Wollaston, 1854
E *Eulagius filicornis* (Reitter, 1887)
A *Litargus balteatus* Leconte, 1856
C *Typhaea stercorea* (Linnaeus, 1758)
Nemonychidae
E *Cimberis attelaboides* Fabricus, 1881
Nitidulidae
A *Brachypeplus mauli* Gardner & Classey, 1962
A *Carpophilus dimidiatus* (Fabricius, 1792)
A *Carpophilus freemani* Dobson, 1956
A *Carpophilus fumatus* Boheman, 1851
A *Carpophilus hemipterus* (Linnaeus, 1758)
A *Carpophilus ligneus* Murray, 1864
A *Carpophilus marginellus* Motschulsky, 1858
A *Carpophilus mutilatus* Erichson, 1843

A *Carpophilus nepos* Murray, 1864
A *Carpophilus obsoletus* Erichson, 1843
A *Carpophilus pilosellus* Motschulsky, 1858
E *Carpophilus quadrisignatus* Erichson, 1843
A *Carpophilus succisus* Erichson, 1843
A *Carpophilus tersus* Wollaston, 1865
A *Carpophilus zeaphilus* Dobson, 1969
E *Epuraea aestiva* (Linnaeus, 1758)
E *Epuraea biguttata* (Thunberg, 1784)
E *Epuraea longula* Erichson, 1845
A *Epuraea luteola* Erichson, 1843
A *Epuraea ocularis* Fairmaire, 1849
A *Glischrochilus fasciatus* (Olivier, 1790)
A *Glischrochilus quadrisignatus* (Say, 1835)
E *Meligethes aeneus* (Fabricius, 1775)
E *Meligethes incanus* Sturm, 1845
E *Meligethes ruficornis* (Marsham, 1802)
C *Nitidula carnaria* (Schaller, 1783)
E *Nitidula flavomaculata* Rossi, 1790
C *Omosita colon* (Linnaeus, 1758)
C *Omosita discoidea* (Fabricius, 1775)
A *Phenolia tibialis* (Boheman, 1851)
E *Pocadius adustus* Reitter, 1888
A *Stelidota geminata* (Say, 1825)
A *Urophorus humeralis* (Fabricius, 1798)
Oedemeridae
E *Nacerdes melanura* (Linnaeus, 1758)
Passandridae
A *Catogenus rufus* (Fabricius), 1798
Phalacridae
E *Phalacrus corruscus* (Panzer, 1797)
A *Phalacrus politus* Melsheimer, 1844
Platypodidae
A *Megaplatypus mutatus* (Chapuis, 1865)
A *Platypus parallelus* Fabricius, 1801
A *Treptoplatypus solidus* (Walker, 1859)
Ptiliidae
E *Acrotrichis cognata* (Matthews, 1877)
A *Acrotrichis henrici* (Matthews, 1872)
A *Acrotrichis insularis* (Maklin, 1852)
A *Acrotrichis josephi* (Matthews, 1872)
A *Acrotrichis sanctaehelenae* Johnson, 1972
E *Actinopteryx fucicola* (Allibert, 1844)
A *Baeocrara japonica* (Matthews, 1884)
A *Bambara contorta* (Dybas, 1066)
A *Bambara fusca* (Dybas, 1966)
E *Ptenidium pusillum* (Gyllenhal, 1808)
A *Ptinella cavelli* (Broun, 1893)
A *Ptinella errabunda* Johnson, 1975
A *Ptinella johnsoni* Rutanen, 1985

A *Ptinella simsoni* (Matthews, 1878)
A *Ptinella taylorae* Johnson, 1977
Ptilodactylidae
    A *Ptilodactyla exotica* Chapin, 1927
    C *Ptilodactyla luteipes* Pic
Ripiphoridae
    A *Ripidius pectinicornis* Thunberg, 1806
Rutelidae
    A *Popillia japonica* Newman, 1841
Scarabaeidae
    E *Onthophagus illyricus* (Scopoli, 1763)
    E *Onthophagus taurus* (Schreber, 1759)
    E *Onthophagus vacca* (Linnaeus, 1767)
    E *Oryctes nasicornis* Linnaeus, 1746
Scolytidae
    A *Coccotrypes carpophagus* Hornung, 1842
    A *Coccotrypes dactyliperda* (Fabricius, 1801)
    E *Cryphalus abietis* (Ratzeburg, 1837)
    E *Cryphalus piceae* (Ratzeburg, 1837)
    E *Crypturgus cinereus* (Herbst, 1793)
    A *Cyclorhipidion bodoanus* Reitter, 1913
    A *Dactylotrypes longicollis* (Wollaston, 1864)
    E *Dendroctonus micans* (Kugelann, 1794)
    A *Dryocoetes himalayensis* Strohmeyer, 1908
    A *Gnathotrichus materiarius* (Fitch, 1858)
    E *Hylastes ater* (Paykull, 1800)
    E *Hylastinus fankhauseri* Reitter, 1894
    A *Hypothenemus eruditus* Westwood, 1836
    A *Hypothenemus hampei* Ferrari, 1867
    E *Ips amitinus* Eichhoff, 1871
    E *Ips cembrae* Herr, 1836
    E *Ips duplicatus* (Sahlberg, 1836)
    E *Ips sexdentatus* Börner, 1776
    E *Phloeosinus armatus* Reitter, 1887
    A *Phloeosinus rudis* Blandford, 1894
    A *Phloeotribus caucasicus* Reitter, 1891
    A *Phloeotribus liminaris* (Harris, 1852)
    E *Pityogenes bidentatus* Herbst, 1784
    E *Pityogenes quadridens* (Hartig, 1834)
    A *Scolytogenes jalapae* (Letzner, 1848)
    E *Scolytus laevis* Chapuis, 1869
    E *Scolytus pygmaeus* (Fabricius, 1787)
    A *Trypodendron laeve* Eggers, 1939
    A *Xyleborinus alni* Niisima, 1909
    E *Xyleborinus saxesenii* Ratzeburg, 1837
    A *Xyleborus affinis* Eichhoff, 1868
    A *Xyleborus perforans* Wollaston, 1857
    A *Xyleborus volvulus* (Fabricius, 1775)
    A *Xylosandrus crassiusculus* (Motschulsky, 1866)
    A *Xylosandrus germanus* (Blandford, 1894)
    A *Xylosandrus morigenus* Blandford, 1894
Scydmaenidae
    E *Stenichnus collaris* (Müller & Kunze, 1822)
Silphidae
    E *Ablattaria laevigata* (Fabricius, 1775)
    E *Blitophaga opaca* (L., 1758)
Silvanidae
    C *Ahasverus advena* (Waltl, 1832)
    A *Cryptamorpha desjardinsi* (Guérin-Méneville, 1844)
    C *Nausibius clavicornis* (Kugelann, 1794)
    A *Oryzaephilus acuminatus* Halstead, 1980
    A *Oryzaephilus mercator* (Fauvel, 1889)
    C *Oryzaephilus surinamensis* (Linnaeus, 1758)
    A *Silvanus lateritius* (Broun, 1880)
    A *Silvanus lewisi* Reitter, 1876
    A *Silvanus recticollis* Reitter, 1876
    E *Silvanus unidentatus* (Olivier, 1790)
Sphindidae
    E *Sphindus dubius* (Gyllenhal, 1808)
Staphylinidae
    A *Acrotona pseudotenera* (Cameron, 1933)
    A *Adota maritima* Mannerheim, 1843
    E *Aleochara bipustulata* (Linnaeus, 1760)
    E *Aleochara clavicornis* Redtenbacher, L., 1849
    C *Aleochara puberula* Klug, 1833
    E *Aleochara sparsa* Heer, 1839
    E *Amischa analis* (Gravenhorst, 1802)
    C *Anotylus nitidifrons* (Wollaston, 1871)
    E *Anotylus nitidulus* (Gravenhorst, 1802)
    E *Anotylus speculifrons* (Kraatz, 1857)
    E *Atheta acuticollis* Fauvel
    E *Atheta amicula* (Stephens, 1832)
    E *Atheta atramentaria* (Gyllenhal, 1810)
    E *Atheta castanoptera* (Mannerheim, 1830)
    E *Atheta coriaria* (Kraatz, 1856)
    A *Atheta dilutipennis* (Motschulsky, 1858)
    E *Atheta divisa* (Maerkel, 1844)
    E *Atheta fungi* (Gravenhorst, 1806)
    E *Atheta gregaria* (Casey, 1910)
    E *Atheta harwoodi* Williams, 1930
    E *Atheta luridipennis* (Mannerheim, 1830)
    A *Atheta mucronata* (Kraatz, 1859)
    E *Atheta nigra* (Kraatz, 1856)
    E *Atheta nigricornis* (Thomson, 1852)
    E *Atheta oblita* (Erichson, 1839)

E *Atheta palustris* (Kiesenwetter, 1844)
E *Atheta sordida* Marsham, 1802
E *Atheta triangulum* (Kraatz, 1856)
E *Atheta trinotata* (Kraatz, 1856)
A *Bisnius palmi* (Smetana, 1955)
A *Bisnius parcus* (Sharp, 1874)
E *Bisnius sordidus* (Gravenhorst, 1802)
C *Bohemiellina flavipennis* (Cameron, 1921)
E *Brachygluta paludosa* (Peyron, 1858)
E *Cafius Xantholoma* (Gravenhorst, 1806)
C *Carpelimus bilineatus* Stephens, 1834
C *Carpelimus corticinus* (Gravenhorst, 1806)
C *Carpelimus gracilis* (Mannerheim, 1830)
C *Carpelimus pusillus* (Gravenhorst, 1802)
C *Carpelimus subtilis* (Erichson, 1839)
A *Carpelimus zealandicus* (Sharp, 1900)
C *Cilea silphoides* (Linnaeus, 1767)
A *Coproporus pulchellus* (Erichson, 1839)
E *Cordalia obscura* (Gravenhorst, 1802)
E *Creophilus max illosus* (Linnaeus, 1758)
E *Cypha pulicaria* (Erichson, 1839)
A *Diestota guadalupensis* Pace, 1987
E *Edaphus beszedesi* Reitter, 1914
E *Euplectus infirmus* Raffray, 1910
E *Gabrius nigritulus* (Gravenhorst, 1802)
E *Gabronthus thermarum* (Aubé, 1850)
E *Geostiba circellaris* Gravenhorst, 1806
E *Gyrophaena bihamata* Thomson, 1867
E *Gyrohypnus fracticornis* (O. Müller, 1776)
E *Hadrognathus longipalpis* (Mulsant & Rey, 1851)
E *Halobrecta flavipes* Thomson, 1861
E *Heterota plumbea* (Waterhouse, 1858)
E *Lathrobium fulvipenne* (Gravenhorst, 1806)
E *Leptacinus pusillus* (Stephens, 1833)
A *Leptoplectus remyi* (Jeannel, 1961)
A *Lithocharis nigriceps* (Kraatz, 1859)
E *Lithocharis ochracea* (Gravenhorst, 1802)
E *Micropeplus marietti* Jacquelin du Val, 1857
E *Mycetoporus nigricollis* (Stephens, 1832)
E *Myllaena brevicornis* (Matthews, 1838)
C *Myrmecocephalus concinna* (Erichson, 1840)
C *Myrmecopora brevipes* Butler, 1909
E *Myrmecopora sulcata* (Kiesenwetter, 1850)
E *Myrmecopora uvida* (Erichson, 1840)
A *Nacaeus impressicollis* (Motschulsky, 1857)

E *Neobisnius lathrobioides* (Baudi, 1848)
E *Neobisnius procerulus* (Gravenhorst, 1806)
E *Ocalea picata* (Stephens, 1832)
A *Oligota parva* Kraatz, 1862
E *Oligota pusillima* (Gravenhorst, 1806)
E *Olophrum fuscum* (Gravenhorst, 1806)
E *Omalium excavatum* Stephens, 1834
E *Omalium rivulare* (Paykull, 1789)
E *Oxypoda haemorrhoa* (Mannerheim, 1830)
A *Oxytelus migrator* Fauvel, 1904
E *Oxytelus sculptus* Gravenhorst, 1806
A *Paraphloeostiba gayndahensis* (Mac Leay, 1871)
E *Phacophallus parumpunctatus* (Gyllenhal, 1827)
E *Philonthus cephalotes* (Gravenhorst, 1802)
E *Philonthus concinnus* (Gravenhorst, 1802)
E *Philonthus discoideus* (Gravenhorst, 1802)
E *Philonthus fenestratus* Fauvel, 1872
E *Philonthus fimetarius* (Gravenhorst)
E *Philonthus longicornis* Stephens, 1832
E *Philonthus marginatus* (O. Müller, 1764)
E *Philonthus politus* (Linnaeus, 1758)
E *Philonthus quisquiliarius* (Gyllenhal, 1810)
A *Philonthus rectangulus* Sharp, 1874
A *Philonthus spinipes* Sharp, 1874
E *Philonthus umbratilis* (Gravenhorst, 1802)
E *Phloeopora angustiformis* Baudi, 1869
E *Phloeopora teres* (Gravenhorst, 1802)
E *Phloeopora testacea* (Mannerheim, 1830)
E *Proteinus brachypterus* (Fabricius, 1792)
E *Quedius mesomelinus* (Marsham, 1802)
E *Remus pruinosus* (Erichson, 1840)
E *Sunius propinquus* (Brisout de Barneville, 1867)
E *Tachinus laticollis* Gravenhorst, 1802
A *Tachinus sibiricus* Sharp, 1888
E *Tachinus signatus* Gravenhorst, 1802
E *Tachyporus chrysomelinus* (Linnaeus, 1758)
E *Tachyporus nitidulus* (Fabricius, 1781)
A *Teropalpus unicolor* (Sharp, 1900)
E *Thecturota marchii* (Dodero, 1922)
A *Trichiusa immigrata* Lohse, 1984

E *Xantholinus linearis* (Olivier, 1795)
E *Xantholinus longiventris* Heer, 1839
E *Xantholinus concinnus* (Marsham, 1802)
E *Xantholinus deplanatus* Gyllenhal, 1810
Tenebrionidae
A *Alphitobius diaperinus* (Panzer, 1797)
A *Alphitobius laevigatus* (Fabricius, 1781)
C *Alphitophagus bifasciatus* (Say, 1823)
E *Blaps gigas* (Linnaeus, 1758)
E *Blaps lethifera* Marsham, 1802
E *Blaps mortisaga* (Linnaeus, 1758)
E *Blaps mucronata* Latreille, 1804
E *Corticeus linearis* (Fabricus, 1790)
E *Corticeus pini* (Panzer, 1799)
A *Cynaeus angustus* (Leconte, 1851)
A *Cynaeus depressus* Horn, 1870
A *Gnathocerus cornutus* (Fabricius, 1798)
A *Gnathocerus maxillosus* (Fabricius, 1801)
A *Latheticus oryzae* Waterhouse, 1880
A *Lyphia tetraphylla* (Fairmaire, 1856)
A *Palorus ratzeburgi* (Wissmann, 1848)
A *Palorus subdepressus* (Wollaston, 1864)
E *Scaurus punctatus* Fabricius, 1798
E *Tenebrio obscurus* Fabricius, 1792
E *Trachyscelis aphodioides* Latreille, 1809
C *Tribolium castaneum* (Herbst, 1797)
A *Tribolium confusum* Jacquelin du Val, 1868
A *Tribolium destructor* Uyttenboogaart, 1933
A *Zophobas morio* Fabricius, 1776
Thorictidae
C *Thorictodes heydeni* Reitter, 1875
Throscidae
E *Throscus dermestoides* Linnaeus, 1766
Trogidae
A *Omorgus subcarinatus* (MacLeay, 1864)
A *Omorgus suberosus* (Fabricius, 1775)
E *Trox scaber* Linnaeus, 1767
Trogossitidae
A *Lophocateres pusillus* (Klug, 1832)
A *Tenebroides maroccanus* Reitter, 1884
A *Tenebroides mauritanicus* (Linnaeus, 1758)
Zopheridae
E *Aulonium ruficorne* (Olivier, 1790)
A *Microprius rufulus* (Motschulsky, 1863)
A *Pycnomerus fuliginosus* Erichson, 1842
C *Pycnomerus inexpectus* (Jaquelin Du Val, 1859)

Arthropoda, Insecta, Trichoptera
Polycentropodidae
A *Pseudoneureclipsis lusitanicus* Malicky, 1980

Arthropoda, Insecta, Lepidoptera
Agonoxenidae
A *Haplochrois theae* (Kuznetzov, 1916)
Arctiidae
A *Aloa lactinea* (Cramer, 1777)
A *Antichloris viridis* Druce, 1884
E *Eilema caniola* (Hübner, 1808)
A *Euchromia lethe* Fabricius, 1775
A *Hyphantria cunea* (Drury, 1773)
A *Pyrrharctia isabella* Smith, 1797
Bedelliidae
E *Bedellia somnulentella* (Zeller, 1847)
Blastobasidae
A *Blastobasis phycidella* (Zeller, 1839)
A *Blastobasis maroccanella* Amsel, 1952
A *Blastobasis vittata* (Wollaston, 1858)
A *Blastobasis decolorella* (Wollaston, 1858)
Bombycidae
A *Bombyx mori* Linnaeus, 1758
Brassolidae
A *Opsiphanes tamarindi* Felder, 1861
Bucculatricidae
A *Bucculatrix chrysanthemella* Rebel, 1896
Castniidae
A *Paysandisia archon* (Burmeister, 1880)
A *Riechia acraeoides* (Guérin-Méneville, 1832)
Choreutidae
E *Tebenna micalis* (Mann, 1857)
Coleophoridae
E *Coleophora spiraeella* Rebel, 1916
E *Coleophora laricella* (Hübner, 1817)
E *Coleophora versurella* Zeller, 1849
Cosmopterigidae
A *Anatrachyntis badia* (Hodges, 1962)
A *Anatrachyntis simple* (Walsingham, 1891)
A *Ascalenia acaciella* Chretien, 1915
A *Bifascioides leucomelanellus* (Rebel, 1917)
E *Cosmopterix pulchrimella* Chambers, 1875
A *Gisilia stereodoxa* (Meyrick, 1925)
E *Pyroderces argyrogrammos* (Zeller, 1847)
Crambidae
A *Chilo suppressalis* (Walker, 1863)
A *Diaphania indica* (Saunders, 1851)
A *Diplopseustis perieresalis* (Walker, 1859)

E *Duponchelia fovealis* Zeller, 1847
A *Eustixia pupula* Hübner, 1823
A *Herpetogramma licarsisalis* (Walker, 1859)
A *Leucinodes orbonalis* (Guenée, 1854)
A *Nymphula diffiualis* (Snellen, 1880)
A *Nymphula manilensis* Hampson, 1917
A *Oligostigma polydectalis* Walker, 1859
A *Parapoynx diminutalis* Snellen, 1880
A *Parapoynx bilinealis* Snellen, 1876
A *Parapoynx obscuralis* Grote, 1881
A *Parapoynx fluctuosalis* (Zeller, 1852)
A *Parapoynx crisonalis* (Walker, 1859)
A *Sclerocona acutella* (Eversmann, 1842)
A *Spoladea recurvalis* (Fabricus, 1775)
A *Synclita obliteralis* (Walker, 1859)

Epermeniidae
   E *Epermenia aequidentellus* (Hoffmann, 1867)

Gelechiidae
   E *Anarsia lineatella* Zeller, 1839
   E *Athrips rancidella* (Herrich-Schäffer, 1854)
   E *Aproaerema anthyllidella* (Hübner, 1813)
   E *Chrysoesthia sexguttella* (Thunberg, 1794)
   A *Coleotechnites piceaella* (Kearfott, 1903)
   E *Gelechia senticetella* (Staudinger, 1859)
   E *Gelechia sabinellus* (Zeller, 1839)
   A *Pectinophora gossypiella* (Saunders, 1844)
   E *Pexicopia malvella* (Hübner, 1805)
   A *Phthorimaea operculella* (Zeller, 1873)
   E *Platyedra subcinerea* (Haworth, 1828)
   E *Scrobipalpa ocellatella* (Boyd, 1858)
   A *Sitotroga cerealella* (Olivier, 1789)
   A *Tecia solanivora* (Povolny, 1973)

Geometridae
   E *Bupalus piniaria* (Linnaeus, 1758)
   A *Cabera leptographa* Wehrli, 1936
   E *Erannis defoliaria* (Clerck, 1759)
   E *Eupithecia pulchellata* Stephens, 1831
   E *Eupithecia phoeniceata* (Rambur, 1834)
   E *Eupithecia lariciata* (Freyer, 1841)
   E *Eupithecia carpophagata* Staudinger, 1871
   E *Eupithecia indigata* (Hübner, 1813)
   E *Eupithecia abietaria* (Goeze, 1781)
   E *Eupithecia sinuosaria* (Eversmann, 1848)
   E *Eupithecia intricata* (Zetterstedt, 1839)
   E *Eurranthis plummistaria* (De Villers, 1789)
   E *Idaea inquinata* (Scopoli, 1763)
   A *Idaea bonifata* Hulst, 1887
   E *Lithostege griseata* (Denis & Schiffermüller, 1775)
   E *Macaria liturata* (Clerck, 1759)
   E *Operophtera brumata* (Linnaeus, 1758)
   E *Peribatodes perversaria* (Boisduval, 1840)
   E *Peribatodes secundaria* (Denis & Schiffermüller, 1775)
   E *Scopula minorata* (Boisduval, 1833)
   E *Thera britannica* (Turner, 1925)
   E *Xanthorhoe biriviata* (Borkhausen, 1794)

Gracillariidae
   A *Caloptilia azaleella* (Brants, 1913)
   A *Caloptilia roscipennella* (Hübner, 1796)
   E *Caloptilia rufipennella* (Hübner, 1796)
   E **Cameraria ohridella** Deschka & Dimić, 1986
   A *Parectopa robiniella* Clemens, 1863
   A *Phyllocnistis vitegenella* Clemens, 1859
   A *Phyllocnistis citrella* (Stainton, 1856)
   A *Phyllonorycter robiniella* (Clemens, 1859)
   E *Phyllonorycter platani* (Staudinger, 1870)
   A *Phyllonorycter issikii* (Kumata, 1963)
   E *Phyllonorycter leucographella* (Zeller, 1850)
   E *Phyllonorycter messaniella* (Zeller, 1846)
   E *Phyllonorycter strigulatella* (Zeller, 1846)
   E *Phyllonorycter joannisi* (Le Marchand, 1936)
   E *Phyllonorycter geniculella* (Ragonot, 1874)
   E *Phyllonorycter comparella* (Duponchel, 1843)

Hesperiidae
   E *Heteropterus morpheus* (Pallas, 1771)

Limacodidae
   A *Phobetron hipparchia* (Cramer, 1777)
   A *Sibine stimulea* (Clemens, 1860)

Lycaenidae
   A *Cacyreus marshalli* Butler, 1898
   E *Celastrina argiolus* (Linnaeus, 1758)
   E *Zizeeria knysna* (Trimen, 1862)

Lymantriidae
   A *Clethrogyna turbata* (Butler, 1879)

11 List of Species Alien in Europe and to Europe

Lyonetiidae
  E *Leucoptera malifoliella* (O. Costa, 1836)
  E *Leucoptera aburnella* (Stainton, 1851)
Nepticulidae
  E *Acalyptris platani* (Müller-Rutz, 1934)
  E *Ectoedemia heringella* (Mariani, 1939)
  E *Stigmella aurella* (Fabricius, 1775)
  E *Stigmella pyri* (Glitz, 1865)
  E *Stigmella suberivora* (Stainton, 1869)
  E *Stigmella speciosa* (Frey, 1857)
  E *Stigmella atricapitella* (Haworth, 1828)
Noctuidae
  A *Acontia crocata* Guenée, 1852
  E *Acronicta aceris* (Linnaeus, 1758)
  A *Adris tyrannus* Guenée, 1852
  E *Apopestes spectrum* (Esper, 1787)
  A *Ascalapha odorata* (Linnaeus, 1758)
  E *Autographa gamma* (Linnaeus, 1758)
  A *Callopistria maillardi* (Guenée, 1862)
  A *Chrysodeixis acuta* (Walker, 1858)
  E *Chrysodeixis chalcites* (Esper, 1789)
  A *Chrysodeixis eriosoma* (Doubleday, 1843)
  E *Cryphia algae* (Fabricius, 1775)
  E *Euplexia lucipara* (Linnaeus, 1758)
  E *Eutelia adulatri* (Hübner, 1813)
  A *Feltia subgothica* (Haworth, 1809)
  E *Hadena compta* (Denis & Schiffermüller, 1775)
  A *Helicoverpa armigera* (Hübner, 1808)
  E *Lithophane leautieri* (Boisduval, 1829)
  A *Mythimna languida* (Walker, 1858)
  E *Panolis flammea* (Denis & Schiffermüller, 1775)
  E *Pardasena virgulana* (Mabille, 1880)
  E *Platyperigea ingrata* (Staudinger, 1897)
  A *Sedina buettneri* (E. Hering, 1858)
  E *Sesamia nonagrioides* (Lefèbvre, 1827)
  A *Spodoptera litura* (Fabricius, 1775)
  A **Spodoptera littoralis** (Boisduval, 1833)
  A *Spodoptera dolichos* (Fabricius, 1794)
  A *Spodoptera cilium* Guenée, 1852
  A *Tarachidia candefacta* (Hübner, 1827)
Nolidae
  E *Earias vernana* (Fabricius, 1787)
Notodontidae
  E *Spatalia argentina* (Denis & Schiffermüller, 1775)
  E *Thaumetopoea pityocampa* (Denis & Schiffermüller, 1775)
Nymphalidae
  A *Anartia fatima* (Fabricius, 1793)
  E *Kirinia roxelana* (Cramer, 1777)
  E *Neptis rivularis* Scopoli, 1763
  A *Vanessa carye* (Hübner, 1816)
Oecophoridae
  A *Borkhausenia nefra* Hodges, 1974
  E *Endrosis sarcitrella* (Linnaeus, 1758)
  E *Ethmia terminella* Fletcher, 1938
  E *Hofmannophila pseudospretella* (Stainton, 1849)
  A *Tachystola acrox antha* (Meyrick, 1885)
Pieridae
  E *Pieris rapae* (Linnaeus, 1758)
Plutellidae
  E *Plutella porrectella* (Linnaeus, 1758)
Pterophoridae
  E *Emmelina monodactyla* (Linnaeus, 1758)
  A *Megalorhipida leucodactylus* (Fabricius, 1794)
  E *Stenoptilia millieridactylus* (Bruand, 1861)
Pyralidae
  C *Achroia grisella* (Fabricius, 1794)
  A *Agassiziella angulipennis* (Hampson, 1891)
  C *Aglossa pinguinalis* (Linnaeus, 1758)
  E *Aglossa caprealis* (Hübner, 1809)
  E *Apomyelois ceratoniae* (Zeller, 1839)
  C *Cadra figulilella* (Gregson, 1871)
  A *Cadra cautella* (Walker, 1863)
  E *Cadra calidella* (Guenée, 1845)
  A *Corcyra cephalonica* (Stainton, 1866)
  E *Cryptoblabes gnidiella* (Millière, 1867)
  A *Cyrtogramme melagynalis* (Agassiz, 1978)
  E *Dioryctria schuetzeella* Fuchs, 1899
  E *Eccopisa effractella* Zeller, 1848
  A *Elophila nymphaeata* (Linnaeus, 1758)
  C *Ephestia kuehniella* Zeller, 1879
  C *Ephestia elutella* (Hübner, 1796)
  C *Etiella zinckenella* (Treitschke, 1832)
  E *Euclasta varii* (Popescu-Gorj & Constantinescu, 1973)
  E *Galleria mellonella* (Linnaeus, 1758)
  C *Hypsopygia costalis* (Fabricius, 1775)
  A *Maruca vitrata* (Fabricius, 1787)
  A *Paralipsa gularis* (Zeller, 1877)
  A *Paramyelois transitella* (Walker, 1863)
  E *Phycita diaphana* (Staudinger, 1870)
  C *Plodia interpunctella* (Hübner, 1813)
  A *Pseudarenipses insularum* Speidel & Schmitz, 1991
  C *Pyralis farinalis* Linnaeus, 1758
  E *Pyralis lienigialis* (Zeller, 1843)
  A *Vitula edmandsii* (Packard, 1865)
  E *Zophodia grossulariella* (Hübner, 1809)

Riodinidae
  A *Calephelis virginiensis* (Gray, 1832)
Roeslerstammiidae
  E *Roeslerstammia erxlebella* (Fabricus, 1787)
Saturniidae
  E *Actias isabellae* (Graells, 1849)
  A *Antheraea paphia* (Linnaeus, 1758)
  A *Antheraea pernyi* (Guérin-Méneville, 1855)
  A *Antheraea polyphemus* (Cramer, 1775)
  A *Antheraea yamamai* (Guérin-Méneville, 1861)
  A *Attacus atlas* (Linnaeus, 1758)
  A *Samia cynthia* (Drury, 1773)
  E *Saturnia pyri* (Denis & Schiffermüller, 1775)
Sesiidae
  E *Synanthedon andrenaeformis* (Laspeyres, 1801)
  E *Paranthrene tabaniformis* (Rottemburg, 1775)
Stathmopodidae
  A *Neomariania rebeli* (Walsingham, 1894)
Symmocidae
  A *Oegoconia novimundi* Busck, 1915
Tineidae
  A *Dasyses incrustata* (Meyrick, 1930)
  E *Haplotinea insectella* (Fabricius, 1794)
  E *Haplotinea ditella* (Pierce & Metcalfe, 1938)
  C *Monopis crocicapitella* (Clemens, 1859)
  A *Nemapogon variatella* (Clemens, 1859)
  E *Nemapogon granella* (Linnaeus, 1758)
  A *Nemapogon gerasimovi* Zagulajev, 1961
  E *Neurothaumasia ankerella* (Mann, 1867)
  E *Niditinea fuscella* (Linnaeus, 1758)
  E *Oinophila v-flava* (Haworth, 1828)
  A *Opogona sacchari* (Bojer, 1856)
  A *Opogona omoscopa* (Meyrick, 1893)
  A *Praeacedes atomosella* (Walker, 1863)
  A *Psychoides filicivora* (Meyrick, 1937)
  A *Tinea translucens* Meyrick, 1917
  E *Tinea pellionella* Linnaeus, 1758
  A *Tinea pallescentella* Stainton, 1851
  E *Tinea dubiella* Stainton, 1859
  A *Tinea murariella* Staudinger, 1859
  E *Tinea flavescentella* Haworth, 1828
  C *Tineola bisselliella* (Hummel, 1823)
  A *Trichophaga tapetzella* (Linnaeus, 1758)

Tortricidae
  A *Acleris undulana* (Walsingham, 1900)
  E *Acleris variegana* (Denis & Schiffermüller, 1775)
  E *Adoxophyes orana* (Fischer von Röslerstamm, 1834)
  E *Cacoecimorpha pronubana* (Hübner, 1799)
  E *Clavigesta sylvestrana* (Curtis, 1850)
  A *Clepsis peritana* (Clemens, 1860)
  E *Cnephasia pumicana* (Zeller, 1847)
  E *Cnephasia longana* (Haworth, 1811)
  E *Crocidosema plebejana* Zeller, 1847
  A *Cryptophlebia leucotreta* (Meyrick, 1927)
  A *Cydia medicaginis* (Kuznetzov, 1962)
  E *Cydia grunertiana* (Ratzeburg, 1868)
  A *Cydia deshaisiana* (Lucas, 1858)
  E *Cydia milleniana* Adamczewski, 1967
  E *Cydia strobilella* (Linnaeus, 1758)
  E *Cydia splendana* (Hübner, 1799)
  E *Cydia pomonella* (Linnaeus, 1758)
  E *Cydia pactolana* (Zeller, 1840)
  E *Cydia illutana* (Herrich-Schäffer, 1851)
  A *Dichelia cedricola* (Diakonoff, 1974)
  E *Ditula angustiorana* (Haworth, 1811)
  A *Epichoristodes acerbella* (Walker, 1864)
  A *Epinotia cedricida* Diakonoff, 1969
  A *Epinotia algeriensis* Chambon, 1990
  A *Epiphyas postvittana* (Walker, 1863)
  A *Grapholita molesta* (Busck, 1916)
  A *Grapholita delineana* Walker, 1863
  E *Gypsonoma minutana* (Hübner, 1799)
  E *Lobesia botrana* (Denis & Schiffermüller, 1775)
  A *Lozotaenia cedrivora* Chambon, 1990
  E *Notocelia rosaecolana* (Doubleday, 1850)
  E *Phtheochroa pulvillana* (Herrich-Schäffer, 1851)
  E *Rhopobota naevana* (Hübner, 1817)
  E *Rhyacionia buoliana* (Denis & Schiffermüller, 1775)
  E *Selania leplastriana* (Curtis, 1831)
Yponomeutidae
  A *Argyresthia thuiella* (Packard, 1871)
  E *Argyresthia laevigatella* (Heydenreich, 1851)
  E *Argyresthia curvella* (Linnaeus, 1761)
  E *Argyresthia trifasciata* Staudinger, 1871
  A *Argyresthia cupressella* Walsingham, 1890
  E *Ocnerostoma friesei* Svensson, 1966
  E *Ocnerostoma piniariella* Zeller, 1847
  A *Prays citri* (Millière, 1873)

E *Prays oleae* (Bernard, 1788)
E *Yponomeuta malinellus* Zeller, 1838
Zygaenidae
   E *Theresimima ampellophaga*
      (Bayle-Barelle, 1808)

Arthropoda, Insecta, Neuroptera
Coniopterygidae
   E *Aleuropteryx juniperi* Ohm, 1968
Hemerobiidae
   E *Wesmaelius ravus* (Withcombe, 1923)

Arthropoda, Insecta, Diptera
Agromyzidae
   A *Cerodontha unisetiorbita* Zlobin, 1993
   A *Liriomyza chinensis* Kato, 1949
   A **Liriomyza huidobrensis** (Blanchard, 1926)
   A *Liriomyza trifolii* (Burgess, 1880)
Anthomyiidae
   E *Strobilomyia infrequens* (Ackland, 1965)
   E *Strobilomyia laricicola* (Karl, 1928)
   E *Strobilomyia melania* (Ackland, 1965)
Braulidae
   C *Braula schmitzi* Orosi Pal, 1939
Calliphoridae
   C *Chrysomya albiceps* (Wiedemann, 1819)
Cecidomyiidae
   E *Aphidoletes abietis* (Kieffer, 1896)
   E *Asphondylia borzi* (Stefani, 1898)
   A *Asphondylia buddleia* Felt, 1935
   A *Clinodiplosis cattleyae* (Molliard, 1903)
   A *Contarinia citri* Barnes, 1944
   E *Contarinia lentis* Aczél, 1944
   C *Contarinia pisi* (Loew, 1850)
   E *Contarinia pyrivora* (Riley, 1886)
   A *Contarinia quinquenotata* (F. Loew, 1888)
   E *Dasineura abietiperda* (Henschel, 1880)
   A *Dasineura gibsoni* Felt, 1911
   A *Dasineura gleditchiae* (Osten Sacken, 1866)
   E *Dasineura kellneri* (Henschel, 1875)
   A *Dasineura oxycoccana* (Johnson, 1899)
   E *Dasineura pyri* (Bouché, 1847)
   E *Dasineura rhododendri* (Kieffer, 1909)
   A *Dicrodiplosis pseudococci* (Felt, 1914)
   A *Epidiplosis filifera* (Nijveldt, 1965)
   C *Feltiella acarisuga* (Vallot, 1827)
   A *Horidiplosis ficifolii* Harris & de Goffau, 2003

A *Janetiella siskiyou* Felt, 1917
E *Kaltenbachiola strobi* (Winnertz, 1853)
E *Monarthropalpus flavus* (Schrank, 1776)
A *Obolodiplosis robiniae* (Haldeman, 1847)
A *Orseolia cynodontis* Kieffer & Massalongo, 1902
E *Phyllodiplosis cocciferae* (Tavares, 1901)
A *Procontarinia matteiana* Kieffer & Cecconi, 1906
A *Prodiplosis vaccinii* (Felt, 1926)
A *Prodiplosis violicola* (Coquillett, 1900)
A *Resseliella conicola* (Foote, 1956)
E *Resseliella lavandulae* (Barnes, 1953)
E *Resseliella skuhravyorum* Skrzypczyńska, 1975
A *Rhopalomyia chrysanthemi* (Ahlberg, 1939)
A *Rhopalomyia grossulariae* Felt, 1911
A *Stenodiplosis panici* Plotnikov, 1926
A *Stenodiplosis sorghicola* (Coquillett, 1899)
Chironomidae
   A *Telmatogeton japonicus* Tokunaga, 1933
Culicidae
   A **Aedes albopictus** (Skuse, 1894)
   E *Aedes vexans* (Meigen, 1830)
   A *Culex deserticola* Kirkpatrick, 1925
   A *Culex tritaeniorhynchus* Giles, 1901
   A *Culex vishnui* (Theobald, 1901)
   A *Ochlerotatus atropalpus* (Coquillett, 1902)
   A *Ochlerotatus japonicus* (Theobald, 1901)
   A *Ochlerotatus subdiversus* (Martini, 1926)
Dolichopodidae
   A *Micropygus vagans* Parent, 1933
Drosophilidae
   A *Chymomyza amoena* (Loew, 1862)
   A *Chymomyza procnemis* (Williston, 1896)
   A *Chymomyza procnemoides* Wheeler, 1952
   A *Chymomyza wirthi* Wheeler, 1954
   A *Dettopsomyia nigrovittata* (Malloch, 1924)
   C *Drosophila busckii* Cocquillett, 1901
   A *Drosophila curvispina* Watabe & Toda, 1984
   C *Drosophila hydei* Sturtevant, 1921
   C *Drosophila immigrans* Sturtevant, 1921
   C *Drosophila melanogaster* Meigen, 1830
   C *Drosophila repleta* Wollaston, 1858
   C *Drosophila tsigana* Burla & Gloor, 1952

    A *Scaptomyza adusta* (Loew, 1862)
    A *Scaptomyza vittata* (Coquillett, 1895)
    A *Zaprionus ghesquieri* Collart, 1937
    A *Zaprionus indianus* Gupta, 1970
    A *Zaprionus tuberculatus* Malloch, 1932
Ephydridae
    A *Elephantinosoma chnumi* Becker, 1903
    A *Placopsidella phaenota* Mathis, 1986
    A *Psilopa fratella* (Becker, 1903)
Fanniidae
    A *Fannia pusio* (Wiedemann, 1830)
Heleomyzidae
    A *Prosopantrum flavifrons* Tonnoir & Malloch, 1927
Hippoboscidae
    C *Crataerina melbae* (Rondani, 1879)
Milichiidae
    C *Desmometopa microps* Lamb, 1914
    C *Desmometopa varipalpis* Malloch, 1927
Muscidae
    C *Athrerigona soccata* Rondani, 1871
    A *Hydrotaea aenescens* (Wiedemann, 1830)
Mycetophilidae
    A *Leia arsona* Hutson, 1978
Phoridae
    A *Chonocephalus depressus* Meijere, 1912
    A *Chonocephalus heymonsi* Stobbe, 1913
    A *Dohrniphora cornuta* (Bigot in de la Sagra, 1857)
    A *Dohrniphora papuana* Brues, 1905
    A *Hypocerides nearcticus* (Borgmeier, 1966)
    A *Megaselia gregaria* (Wood, 1910)
    A *Megaselia scalaris* (Loew, 1866)
    A *Megaselia tamilnaduensis* Disney, 1996
    A *Puliciphora borinquenensis* Wheeler, 1906
Sciaridae
    C *Bradysia difformis* Frey, 1948
Sphaeroceridae
    A *Coproica rufifrons* Hayashi, 1991
    A *Thoracochaeta johnsoni* (Spuler, 1925)
    A *Thoracochaeta seticosta* (Spuler, 1925)
    A *Trachyopella straminea* Rohacek & Marshall, 1986
Stratiomyidae
    A *Hermetia illucens* (Linnaeus, 1758)
Syrphidae
    E *Chamaesyrphus caledonicus* Collin, 1940
    E *Didea intermedia* Loew, 1854
    E *Eriozona erratica* (Linnaeus, 1758)
    E *Eriozona syrphoides* (Fallén, 1817)
    E *Merodon equestris* (Fabricius, 1794)
    E *Parasyrphus malinellus* (Collin, 1952)
    E *Xylota caeruleiventris* Zetterstedt, 1838
Tachinidae
    C *Blepharipa schineri* (Mesnil, 1939)
    C *Catharosia pygmaea* (Fallén, 1815)
    C *Clytiomya continua* (Panzer, 1798)
    C *Phasia barbifrons* (Girschner, 1887)
    C *Sturmia bella* (Meigen, 1824)
    A *Trichopoda pennipes* (Fabricius, 1794)
    A *Zeuxia zejana* Kolomiets, 1971
Tephritidae
    E *Bactrocera oleae* (Rossi, 1790)
    A **Ceratitis capitata** (Wiedemann, 1824)
    E *Rhagoletis cingulata* (Loew, 1862)
    A *Rhagoletis completa* Cresson, 1929
    A *Rhagoletis indifferens* Curran, 1932
    E *Tephritis praeco* (Loew, 1844)
Tethinidae
    A *Pelomyia occidentalis* Williston, 1893
Ulidiidae
    A *Euxesta pechumani* Curran, 1938

Arthropoda, Insecta, Siphonaptera
Ceratophyllidae
    E *Ceratophyllus columbae* (Gervais, 1844)
    C *Leptopsylla segnis* (Schönherr, 1811)
    A *Nosopsyllus fasciatus* (Bosc d'Antic, 1800)
    A *Nosopsyllus londinensis* (Rothschild, 1903)
    A *Orchopeas howardi* Baker, 1895
Pulicidae
    A *Euhoplopsyllus glacialis* (Baker, 1904)
    A *Xenopsylla brasiliensis* (Baker, 1904)
    A *Xenopsylla cheopis* (Rothschild, 1903)

Arthropoda, Insecta, Hymenoptera
Agaonidae
    A *Eupristina verticillata* Waterston, 1921
    A *Josephiella microcarpae* Beardsley & Rasplus, 2001
    A *Odontofroggatia galili* Wiebes, 1980
    A *Platyscapa quadraticeps* (Mayr, 1885)
Aphelinidae
    A *Ablerus perspeciosus* Girault, 1916
    A *Aphelinus mali* (Haldeman, 1851)
    A *Aphytis coheni* DeBach, 1960
    A *Aphytis holoxanthus* DeBach, 1960
    A *Aphytis lepidosaphes* Compere, 1955
    A *Aphytis lingnanensis* Compere, 1955
    A *Aphytis melinus* DeBach, 1959

- A *Aphytis yanonensis* DeBach &Rosen, 1982
- A *Cales noacki* Howard, 1907
- A *Coccophagoides murtfeldtae* (Howard, 1894)
- A *Coccophagoides utilis* Doutt, 1966
- A *Coccophagus ceroplastae* (Howard, 1895)
- A *Coccophagus gurneyi* Compere, 1929
- A *Coccophagus saissetiae* (Annecke & Mynhardt, 1979)
- A *Coccophagus scutellaris* (Dalman, 1825)
- A *Encarsia berlesei* (Howard, 1906)
- C *Encarsia citrina* (Craw, 1891)
- A *Encarsia diaspidicola* (Silvestri, 1909)
- A *Encarsia fasciata* (Malenotti, 1917)
- A *Encarsia formosa* (Gahan, 1924)
- A *Encarsia herndoni* (Girault, 1935)
- A *Encarsia lahorensis* (Howard, 1911)
- A *Encarsia lounsburyi* (Berlese & Paoli, 1916)
- A *Encarsia meritoria* Gahan, 1927
- A *Encarsia pergandiella* Howard, 1907
- A *Encarsia perniciosi* (Tower, 1913)
- A *Encarsia sophia* (Girault & Dodd, 1915)
- A *Eretmocerus californicus* Howard, 1895
- A *Eretmocerus debachi* Rose & Rosen, 1992
- A *Eretmocerus eremicus* Rose & Zolnerowich, 1997
- A *Eretmocerus haldemani* Howard, 1908
- E *Eretmocerus mundus* Mercet, 1931
- A *Eretmocerus paulistus* Hempel, 1904
- A *Pteroptrix chinensis* (Howard, 1907)
- A *Pteroptrix orientalis* (Silvestri, 1909)
- A *Pteroptrix smithi* (Compere, 1953)

Apidae
- E *Apis mellifera carnica* (Pollmann, 1879)
- E *Apis mellifera ligustica* (Spinola, 1806)
- E *Bombus hortorum* (Linnaeus, 1761)
- E *Bombus lucorum* (Linnaeus, 1761)
- A *Osmia cornifrons* (Radoszkowski, 1887)

Argidae
- E *Arge berberidis* Schrank, 1802

Bethylidae
- E *Sclerodermus domesticus* Klug, 1809

Blasticotomidae
- E *Blasticotoma filiceti* Klug, 1834

Braconidae
- A *Aphidius colemani* Viereck, 1912
- A *Aphidius smithi* Sharma & Subba Rao, 1959
- A *Cotesia marginiventris* (Cresson, 1865)
- A *Hymenochaonia delicata* (Cresson, 1872)
- C *Lysiphlebus testaceipes* (Cresson, 1880)
- A *Opius dimidiatus* Ashmead, 1889
- A *Pauesia cedrobii* Stary & Leclant, 1977

Chalcididae
- A *Dirhinus giffardii* Silvestri, 1913

Cynipidae
- E *Andricus corruptrix* (Schlechtendal, 1870)
- E *Andricus grossulariae* Giraud, 1859
- E *Andricus kollari* (Hartig, 1843)
- E *Andricus lignicola* (Hartig, 1840)
- E *Andricus quercuscalicis* (Burgesdorff, 1783)
- E *Aphelonyx cerricola* (Giraud, 1859)
- A *Dryocosmus kuriphilus* (Yasumatsu, 1951)

Diprionidae
- E *Diprion pini* (Linnaeus, 1758)
- E *Diprion similis* (Hartig, 1836)
- E *Gilpinia hercyniae* (Hartig, 1837)
- E *Gilpinia virens* (Klug, 1812)
- E *Neodiprion sertifer* (Geoffroy, 1785)

Dryinidae
- A *Neodryinus typhlocybae* (Ashmead, 1893)

Encyrtidae
- A *Ageniaspis citricola* Logvinovskaya, 1983
- E *Ageniaspis fuscicollis* (Dalman, 1920)
- A *Aloencyrtus saissetiae* (Compere, 1939)
- A *Anagyrus agraensis* Saraswat, 1975
- A *Anagyrus fusciventris* (Girault, 1915)
- E *Anagyrus pseudococci* (Girault, 1915)
- A *Anagyrus sawadai* Ishii, 1928
- A *Anicetus ceroplastis* Ishii, 1928
- A *Avetianella longoi* Siscaro, 1992
- A *Clausenia purpurea* Ishii, 1923
- A *Coccidoxenoides perminutus* Girault, 1915
- A *Comperiella bifasciata* Howard, 1906
- A *Comperiella lemniscata* Compere & Annecke, 1961
- A *Copidosoma koehleri* Blanchard, 1940
- A *Diversinervus cervantesi* (Girault, 1933)
- A *Diversinervus elegans* Silvestri, 1915
- A *Encyrtus fuscus* (Howard, 1881)
- A *Encyrtus infelix* (Embleton, 1902)
- A *Leptomastix dactylopii* Howard, 1885
- A *Metaphycus angustifrons* Compere, 1957
- A *Metaphycus anneckei* Guerrieri & Noyes, 2000

- A *Metaphycus helvolus* (Compere, 1926)
- A *Metaphycus inviscus* Compere, 1940
- A *Metaphycus lounsburyi* (Howard, 1898)
- A *Metaphycus luteolus* (Timberlake, 1916)
- A *Metaphycus stanleyi* Compere, 1940
- A *Metaphycus swirskii* Annecke & Mynhardt, 1979
- A *Microterys clauseni* Compere, 1926
- A *Microterys nietneri* (Motschulsky, 1859)
- A *Microterys speciosus* Ishii, 1923
- A *Neodusmetia sangwani* (Subba Rao, 1957)
- A *Ooencyrtus kuwanae* (Howard, 1910)
- A *Pseudaphycus angelicus* (Howard, 1898)
- A *Pseudaphycus malinus* Gahan, 1946
- A *Pseudectroma signatum* (Prinsloo, 1982)
- A *Psyllaephagus pilosus* Noyes, 1988
- A *Tachinaephagus zealandicus* Ashmead, 1904
- A *Tetracnemoidea brevicornis* (Girault, 1915)
- A *Tetranecmoidea peregrina* (Compere, 1939)
- A *Zarhopalus sheldoni* Ashmead, 1900

Eulophidae
- A *Aceratoneuromyia indica* (Silvestri, 1910)
- A *Aprostocetus diplosidis* Crawford, 1907
- A *Chrysocharis ainsliei* Crawford, 1912
- A *Cirrospilus ingenuus* Gahan, 1932
- A *Citrostichus phyllocnistoides* (Narayanan, 1960)
- A *Closterocerus cinctipennis* Ashmead, 1888
- A *Edovum puttleri* Grissell, 1981
- A *Elachertus cidariae* (Ashmead, 1898)
- A *Hyssopus thymus* Girault, 1916
- A *Leptocybe invasa* Fisher & LaSalle, 2004
- A *Oomyzus brevistigma* (Gahan, 1936)
- A *Ophelimus maskelli* (Ashmead, 1900)
- A *Quadrastichodella nova* Girault, 1922
- A *Semielacher petiolata* (Girault, 1915)
- A *Thripobius javae* (Girault, 1917)

Eupelmidae
- A *Anastatus japonicus* Ashmead, 1904
- A *Anastatus tenuipes* Bolivar & Pieltain, 1925

Eurytomidae
- A *Eurytoma orchidearum* (Westwood, 1869)
- A *Prodecatoma cooki* (Howard, 1896)
- A *Tetramesa maderae* (Walker, 1849)
- A *Tetramesa romana* (Walker, 1873)
- A *Bruchophagus sophorae* Crosby & Crosby, 1929

Formicidae
- E *Aphaenogaster senilis* Mayr, 1853
- A *Brachymyrmex heeri* Forel, 1874
- A *Cardiocondyla emeryi* Forel, 1881
- A *Cardiocondyla mauritanica* Forel, 1890
- A *Cardiocondyla nuda* (Mayr, 1866)
- E *Cremastogaster scutellaris* (Olivier, 1792)
- A *Crematogaster brevispinosa* Mayr, 1870
- A *Hypoponera ergatandria* (Forel, 1893)
- A *Hypoponera punctatissima* (Roger, 1859)
- E *Lasius alienus* (Foerster, 1850)
- E *Lasius flavus* (Fabricius, 1781)
- E *Lasius fuliginosus* (Latreille, 1798)
- A *Lasius neglectus* Van Loon et al., 1990
- A *Lasius turcicus* Sanctchi, 1921
- A *Leptothorax longispinosus* Roger, 1863
- A **Linepithema humile** (Mayr, 1868)
- A *Linepithema leucomelas* Emery, 1894
- A *Monomorium floricola* (Jerdon, 1851)
- A *Monomorium pharaonis* (Linnaeus, 1758)
- A *Monomorium salomonis* (Linnaeus, 1758)
- C *Pachycondyla darwinii* Forel, 1893
- A *Paratrechina bourbonica* (Forel, 1886)
- A *Paratrechina flavipes* (Smith, 1874)
- A *Paratrechina jaegerskioeldi* (Mayr, 1904)
- A *Paratrechina longicornis* (Latreille, 1802)
- C *Paratrechina vividula* (Nylander, 1846)
- A *Pheidole bilimeki* Mayr, 1870
- A *Pheidole guineensis* (Fabricius, 1793)
- A *Pheidole megacephala* (Fabricius, 1793)
- A *Pheidole noda* (Smith, 1874)
- A *Plagiolepis alluaudi* (Emery, 1894)
- A *Plagiolepis exigua* Forel, 1894
- A *Plagiolepis obscuriscapa* Santschi, 1923
- E *Ponera coarctata* (Latreille, 1802)
- A *Pyramica membranifera* (Emery, 1869)
- A *Strumigenys lewisi* Cameron, 1886
- A *Strumigenys rogeri* Emery, 1890
- A *Strumigenys silvestrii* Emery, 1906
- A *Tapinoma melanocephalum* (Fabricius, 1793)
- A *Technomyrmex albipes* (Smith, 1861)

A *Technomyrmex detorquens* (Walker, 1859)
A *Temnothorax longispinosus* Roger, 1863
A *Tetramorium bicarinatum* (Nylander, 1846)
E *Tetramorium caldarium* (Roger, 1857)
A *Tetramorium insolens* (F. Smith, 1861)
A *Tetramorium lanuginosum* Mayr, 1870
A *Tetramorium simillimum* (F. Smith, 1851)
Ichneumonidae
A *Itoplectis conquisitor* (Say, 1835)
Mymaridae
A *Anaphes nitens* (Girault, 1931)
A *Polynema striaticorne* Girault, 1911
Pamphiliidae
E *Acantholyda erythrocephala* Linnaeus, 1758
E *Acantholyda laricis* (Giraud, 1861)
E *Cephalcia abietis* (Linnaeus, 1758)
A *Cephalcia alashanica* (Gussakovskij, 1935)
E *Cephalcia alpina* (Klug, 1808)
E *Cephalcia erythrogaster* (Hartig, 1837)
E *Cephalcia lariciphila* (Wachtl, 1898)
Platygastridae
A *Amitus fuscipennis* MacGown & Nebeker, 1978
A *Amitus spiniferus* (Brethes, 1914)
Pteromalidae
A *Gugolzia harmolitae* Delucchi & Steffan, 1956
E *Lariophagus distinguendus* (Förster, 1841)
A *Mesopolobus pinus* Hussey, 1960
A *Mesopolobus spermotrophus* Husey, 1960
A *Monoksa dorsiplana* Boucek, 1991
A *Moranila californica* (Howard, 1881)
A *Muscidifurax raptor* Girault & Sanders, 1910
A *Spalangia cameroni* Perkins, 1910
Siricidae
A *Sirex areolatus* (Cresson, 1868)
A *Sirex cyaneus cyaneus* Fabricius, 1781
E *Sirex juvencus* (Linnaeus, 1758)
E *Sirex noctilio* Fabricius, 1773
A *Tremex columba* (Linnaeus, 1763)
A *Urocerus albicornis* (Fabricius, 1781)
A *Urocerus californicus* Norton, 1869
E *Urocerus gigas* (Linnaeus, 1758)
E *Xeris pectrum* (Linnaeus, 1758)

Sphecidae
A *Isodontia mexicana* (Saussure, 1867)
A *Sceliphron caementarium* (Drury, 1773)
A *Sceliphron curvatum* (Smith, 1870)
A *Sceliphron deforme* (Smith, 1856)
Tenthredinidae
E *Ametastegia pallipes* (Spinola, 1808)
E *Anoplonyx destructor* Benson, 1952
E *Athalia rosae* (Linnaeus, 1758)
E *Hoplocampa brevis* (Klug, 1816)
E *Nematus spiraeae* Zaddach, 1883
A *Nematus tibialis* Newman, 1837
E *Pachynematus imperfectus* (Zaddach, 1876)
A *Pachynematus itoi* Okutani, 1955
E *Pachynematus montanus* (Zaddach, 1883)
E *Pachynematus scutellatus* (Hartig, 1837)
E *Pachyprotasis variegata* (Fallén, 1808)
E *Phymatocera aterrima* (Klug, 1816)
E *Pristiphora abietina* (Christ, 1791)
E *Pristiphora amphibola* (Förster, 1854)
E *Pristiphora angulata* Lindqvist, 1974
E *Pristiphora compressa* (Hartig, 1837)
E *Pristiphora erichsonii* (Hartig, 1837)
E *Pristiphora glauca* Benson, 1954
E *Pristiphora laricis* (Hartig, 1837)
E *Pristiphora leucopus* (Hellén, 1948)
E *Pristiphora nigella* Förster, 1854
E *Pristiphora saxesenii* (Hartig, 1837)
E *Pristiphora subarctica* (Forsslund, 1936)
E *Pristiphora thalictri* (Kriechbaumer, 1884)
E *Pristiphora wesmaeli* (Tischbein, 1853)
Torymidae
A *Megastigmus aculeatus nigroflavus* Hoffmeyer, 1929
A *Megastigmus atedius* Walker, 1851
A *Megastigmus borriesi* Crosby, 1913
A *Megastigmus milleri* Milliron, 1949
A *Megastigmus nigrovariegatus* Ashmead, 1890
E *Megastigmus pictus* (Förster, 1841)
A *Megastigmus pinsapinis* Hoffmeyer, 1931
A *Megastigmus pinus* Parfitt, 1857
A *Megastigmus rafni* Hoffmeyer, 1929
A *Megastigmus schimitscheki* Novitzky, 1954
A *Megastigmus specularis* Walley, 1932
A *Megastigmus spermotrophus* (Wachtl, 1893)

E *Megastigmus suspectus* Borries, 1895
A *Megastigmus transvaalensis* (Hussey, 1956)
E *Megastigmus wachtli* Seitner, 1916
Trichogrammatidae
E *Trichogramma brassicae* Bezdenko, 1968
A *Trichogramma chilonis* Ishii, 1941
A *Trichogramma dendrolimi* Matsumura, 1926
C *Trichogramma minutum* Riley, 1871
C *Trichogramma pretiosum* Riley, 1879
Vespidae
A *Vespa velutina* Lepeletier, 1836
E *Vespula germanica* (Fabricius, 1793)
E *Vespula vulgaris* (Linnaeus, 1758)

Arthropoda, Araneae
Agelenidae
E *Tegenaria agrestis* (Walckenaer, 1802)
E *Tegenaria atrica* C.L. Koch, 1843
C *Tegenaria domestica* (Clerck, 1758)
E *Tegenaria duellica* Simon, 1875
E *Tegenaria saeva* Blackwall, 1844
Amaurobiidae
E *Amaurobius similis* (Blackwall, 1861)
Araneidae
E *Argiope bruennichi* (Scopoli, 1772)
Clubionidae
A *Clubiona facilis* O.P.-Cambridge, 1910
Dictynidae
A *Cicurina japonica* (Simon, 1886)
E *Dictyna civica* (Lucas, 1850)
E *Lathys lepida* O.P.-Cambridge, 1909
E *Nigma walckenaeri* (Roewer, 1951)
Dysderidae
A *Dysdera aculeata* Kroneberg, 1875
C *Dysdera crocata* C.L. Koch, 1838
E *Dysdera erythrina* (Walckenaer, 1802)
E *Harpactea rubicunda* (C.L. Koch, 1838)
Eresidae
A *Seothyra perelegans* Simon, 1906
Gnaphosidae
E *Leptodrassus pupa* De Dalmas, 1919
E *Sosticus loricatus* (L. Koch, 1866)
A *Zelotes puritanus* Chamberlin, 1922
Linyphiidae
E *Collinsia inerrans* (O.P.-Cambridge, 1885)
E *Diplocephalus graecus* (O.P.-Cambridge, 1872)
A *Eperigone eschatologica* (Crosby, 1924)
A *Eperigone trilobata* (Emerton, 1882)
A *Erigone autumnalis* Emerton, 1882
E *Lessertia dentichelis* (Simon, 1884)
E *Megalepthyphantes collinus* (L. Koch, 1872)
A *Ostearius melanopygius* (O.P.-Cambridge, 1879)
Lycosidae
E *Hogna hispanica* (Walckenaer, 1837)
E *Lycosa singoriensis* (Laxmann, 1770)
E *Pardosa plumipes* (Thorell, 1875)
Nesticidae
E *Nesticus eremita* Simon, 1879
Oecobiidae
E *Oecobius maculatus* Simon, 1870
E *Oecobius navus* Blackwall, 1859
Oonopidae
A *Diblemma donisthorpei* O.P.-Cambridge, 1908
A *Ischnothyreus lymphaseus* Simon, 1893
A *Ischnothyreus velox* Jackson, 1908
E *Oonops domesticus* De Dalmas, 1916
E *Oonops pulcher* Tempelton, 1835
E *Silhouettella loricatula* (Roewer, 1942)
E *Tapinesthis inermis* (Simon, 1882)
A *Triaeris stenaspis* Simon, 1891
Pholcidae
A *Artema atlanta* Walckenaer, 1837
E *Crossopriza lyoni* (Blackwall, 1867)
E *Holocnemus pluchei* (Scopoli, 1763)
A *Micropholcus fauroti* (Simon, 1887)
E *Pholcus opilionoides* (Schrank, 1781)
C *Pholcus phalangioides* (Fuesslin, 1775)
E *Psilochorus simoni* (Berland, 1911)
A *Smeringopus pallidus* (Blackwall, 1858)
A *Spermophora senoculata* (Dugs, 1836)
Prodidomidae
A *Zimiris doriai* Simon, 1882
Salticidae
E *Evarcha jucunda* (Lucas, 1846)
A *Hasarius adansoni* (Audouin, 1826)
A *Menemerus bivittatus* (Dufour, 1831)
A *Panysinus nicholsoni* (O.P.-Cambridge, 1899)
E *Pellenes geniculatus* (Simon, 1868)
A *Phidippus regius* C.L. Koch, 1846
A *Plexippus paykulli* (Audouin, 1826)
E *Pseudeuophrys lanigera* (Simon, 1871)
E *Saitis barbipes* (Simon, 1868)
E *Sitticus pubescens* (Fabricius, 1775)
Scytodidae
E *Scytodes thoracica* (Latreille, 1802)
A *Scytodes venusta* (Thorell, 1890)

Sicariidae
  A *Loxosceles laeta* (Nicolet, 1849)
  C *Loxosceles rufescens* (Dufour, 1820)
Sparassidae
  A *Barylestis scutatus* (Pocock, 1903)
  A *Barylestis variatus* (Pocock, 1899)
  A *Heteropoda venatoria* (Linnaeus, 1767)
  A *Olios sanctivincenti* (Simon, 1897)
  A *Tychicus longipes* (Walckenaer, 1837)
Tetragnathidae
  A *Tetragnatha shoshone* (Levi, 1981)
Theridiidae
  A *Achaearanea tabulata* Levi, 1980
  C *Achaearanea tepidariorum* (C.L. Koch, 1841)
  A *Achaearanea veruculata* (Urquhart, 1885)
  A *Chrysso spiniventris* (O.P.-Cambridge, 1869)
  A *Coleosoma floridanum* Banks, 1900
  E *Dipoena lugens* (O.P.-Cambridge, 1909)
  A *Latrodectus hasselti* Thorell, 1870
  A *Nesticodes rufipes* (Lucas, 1846)
  E *Steatoda castanea* (Clerck, 1758)
  C *Steatoda grossa* (C.L. Koch, 1838)
  C *Steatoda triangulosa* (Walckenaer, 1802)
Thomisidae
  A *Bassaniana versicolor* Keyserling, 1880
Uloboridae
  A *Uloborus plumipes* Lucas, 1846
Zodariidae
  E *Zodarion rubidum* Simon, 1914
Zoropsidae
  E *Zoropsis spinimana* (Dufour, 1820)

Arthropoda, Acari
Amblyommidae
  A *Amblyomma exornatum* Koch, 1844
  A *Amblyomma latum* Koch, 1844
  A *Hyalomma aegyptium* (Linnaeus, 1758)
  A *Hyalomma anatolicum* Koch, 1844
  A *Hyalomma excavatum* Koch, 1844
  E *Hyalomma scupense* Schulze, 1918
  A *Hyalomma dromedarii* Koch, 1844
  A *Hyalomma truncatum* Koch, 1844
  A *Rhipicephalus rossicus* Yakimov & Kol-Yakimova, 1911
Epidermoptidae
  A *Epidermoptes bilobatus* Rivolta, 1876

Eriophyiidae
  A *Acaphylla theae* (Watt & Mann, 1903)
  A *Aceria erinea* (Nalepa, 1891)
  A *Aceria sheldoni* (Ewing, 1937)
  A *Aceria tristriata* (Nalepa, 1890)
  A *Aculops allotrichus* (Nalepa, 1894)
  A *Aculops fuchsiae* Keifer, 1972
  C *Aculops lycopersici* (Tryon, 1917)
  A *Aculops pelekassi* (Keifer, 1959)
  A *Calacarus carinatus* (Green, 1890)
  C *Eriophyes pyri* (Pagenstecher, 1857)
  A *Phyllocoptes azaleae* Nalepa, 1904
  A *Tegolophus califraxini* (Keifer, 1938)
Glycyphagidae
  E *Glycyphagus domesticus* (De Geer, 1778)
Laelapidae
  A *Laelaps echidninus* Berlese, 1887
  A *Ondatralaelaps multispinosus* (Banks, 1909)
Listrophoridae
  A *Listrophorus americanus* Radford, 1944
  A *Listrophorus dozieri* Redford, 1994
  A *Listrophorus faini* Dubinina, 1972
  A *Listrophorus validus* Banks, 1910
Macronyssidae
  A *Ornithonyssus bacoti* (Hirst, 1913)
  A *Ornithonyssus bursa* (Berlese, 1888)
Myocoptidae
  A *Myocoptes ondatrae* Lukoschus & Rouwet, 1968
Phytoseiidae
  A *Amblyseius californicus* (McGregor, 1954)
  E *Phytoseiulus persimilis* Athias-Henriot, 1957
Pyroglyphidae
  A *Dermatophagoides evansi* Fain, Hughes & Johnston, 1967
Tarsonemidae
  C *Polyphagotarsonemus latus* (Banks, 1904)
Tenuipalpidae
  A *Brevipalpus californicus* (Banks, 1904)
  C *Brevipalpus obovatus* Donnadieu, 1875
Tetranychidae
  A *Eotetranychus lewisi* (McGregor, 1943)
  A *Eotetranychus weldoni* (Ewing, 1913)
  A *Eurytetranychus admes* Pritchard & Baker, 1955
  A *Eurytetranychus furcisetus* Wainstein, 1956

A *Eutetranychus banksi* (McGregor, 1914)
A *Eutetranychus orientalis* (Klein, 1936)
A *Oligonychus bicolor* (Banks, 1894)
A *Oligonychus ilicis* (McGregor, 1917)
A *Oligonychus laricis* Reeves, 1963
A *Oligonychus perditus* Pritchard & Baker, 1955
A *Oligonychus perseae* Tuttle, Baker & Abbatiello, 1976
A *Oligonychus pritchardi* (McGregor, 1950)
A *Oligonychus punicae* (Hirst, 1926)
A *Panonychus citri* (McGregor, 1916)
A *Petrobia lupini* (McGregor, 1950)
A *Schizotetranychus bambusae* Reck, 1941
A *Schizotetranychus parasemus* Pritchard & Baker, 1955
A *Stigmaeopsis celarius* Banks, 1917
A *Tetranychus canadensis* (McGregor, 1950)
A *Tetranychus evansi* Baker & Pritchard, 1960
A *Tetranychus kanzawai* Kishida, 1927
C *Tetranychus ludeni* Zacher, 1913
A *Tetranychus macfarlanei* Baker & Pritchard, 1960
A *Tetranychus mcdanieli* McGregor, 1931
A *Tetranychus neocaledonicus* André, 1933
A *Tetranychus sinhai* Baker, 1962
A *Tetranychus tumidellus* Pritchard & Baker, 1955
A *Tetranychus yusti* McGregor, 1955
Varroidae
A *Varroa destructor* Anderson, 2000

Arthropoda, Opiliones
Phalangiidae
E *Opilio parietinus* (De Geer, 1778)

Echinodermata
Asteriidae
E *Coscinasterias tenuispina* (Lamarck, 1816)
Asterinidae
A *Asterina burtoni* Gray, 1840
Cidaridae
A *Eucidaris tribuloides* (Lamarck, 1816)
Diadematidae
C *Diadema antillarum* (Philippi, 1845)
Ophiactidae
A *Ophiactis parva* Mortensen, 1926
A *Ophiactis savignyi* Müller & Troschel, 1842

Synaptidae
A *Synaptula reciprocans* Forskål, 1775

Chordata, Tunicata
Ascidiidae
A *Phallusia nigra* Savigny, 1816
Clavelinidae
E *Clavelina lepadiformis* (Müller, 1776)
A *Clavelina oblonga* Herdman, 1880
Corellidae
A *Corella eumyota* Traustedt, 1882
Didemnidae
A *Didemnum vexillum* Kott, 2002
E *Diplosoma listerianum* (Milne-Edwards, 1841)
Holozoidae
A *Distaplia corolla* Monniot, 1975
Molgulidae
A *Molgula manhattensis* (De Kay, 1843)
A *Molgula plana* Monniot, 1971
Perophoridae
A *Perophora japonica* Oka, 1927
Polycitoridae
C *Cystodytes dellechiajei* (Della Valle, 1877)
C *Eudistoma angolanum* (Michaelsen, 1915)
Polyclinidae
E *Aplidium nordmanni* (Milne-Edwards, 1841)
E *Polyclinum aurantium* Milne-Edwards, 1841
Pyuridae
A *Microcosmus exasperatus* Heller, 1878
A *Microcosmus squmiger* Michaelsen, 1927
E *Pyura tesselata* (Forbes, 1848)
Styelidae
A *Alloeocarpa loculosa* Monniot, 1975
A *Botrylloides violaceus* Oka, 1927
C *Botrylloides leachi* (Savigny, 1816)
A *Botryllus schlosseri* (Pallas, 1766)
C *Phallusia mamillata* (Cuvier, 1815)
A *Polyandrocarpa zorritensis* Van Name, 1931
A **Styela clava** Herdman, 1882

Chordata, Elasmobranchii
Dasyatidae
A *Himantura uarnak* (Forsskål, 1775)
Torpedinidae
A *Torpedo sinuspersici* Olfers, 1831

11 List of Species Alien in Europe and to Europe

Chordata, Osteichthyes
Achiridae
   Achirus fasciatus (Lacépède, 1803)
  A *Trinectes maculatus* (Bloch & Schneider, 1801)
Acipenseridae
  A *Acipenser baeri* Brandt, 1869
  E *Acipenser gueldenstaedtii* Brandt & Ratzeburg, 1833
  E *Acipenser nudiventris* Lovetsky, 1828
  E *Acipenser ruthenus* Linnaeus, 1758
  A *Acipenser transmontanus* Richardson, 1836
  E *Huso huso* Linnaeus, 1758
Acropomatidae
  A *Synagrops japonicus* (Doderlein, 1884)
Adrianichthyidae
  A *Oryzias sinensis* Chen, Uwa & Chu, 1989
Anguillidae
  E *Anguilla anguilla* (Linnaeus, 1758)
  C *Anguilla rostrata* (Lesueur, 1817)
Apogonidae
  A *Apogon pharaonis* Cuvier, 1828
Atherinidae
  E *Atherina boyeri* Risso, 1810
  A *Atherinomorus lacunosus* (Forster in Bloch & Schneider, 1801)
  A *Odontesthes bonariensis* (Valenciennes, 1835)
Belonidae
  A *Tylosurus choram* (Rüppell, 1837)
Blenniidae
  A *Omobranchus punctatus* Valenciennes, 1836
  A *Petroscirtes ancylodon* Rüppell, 1838
Callionymidae
  A *Callionymus filamentosus* Valenciennes, 1837
Carangidae
  A *Alepes djedaba* (Forsskål, 1775)
Catostomidae
  A *Cassiopeia andromeda* (Forsskål, 1775)
  A *Catostomus catostomus* (Forster, 1773)
  A *Ictiobus bubalus* (Rafinesque, 1818)
  A *Ictiobus cyprinellus* (Valenciennes, 1844)
  A *Ictiobus niger* (Rafinesque, 1819)
Centrarchidae
  A *Lepomis cyanellus* Rafinesque, 1819
  A *Lepomis gibbosus* Linnaeus, 1758
  A *Micropterus dolomieui* (Lacépède, 1802)
  A *Micropterus salmoides* (Lacepède, 1802)
Chaetodontidae
  A *Heniochus intermedius* Steindachner, 1893
Characidae
  A *Piaractus brachypomus* (Cuvier, 1818)
  A *Piaractus mesopotamicus* (Holmberg, 1887) 1891
  A *Serrasalmo nattereri* Kner, 1858
Cichlidae
  A *Cichlasoma nigrofasciatum* (Günther, 1867)
  A *Hemichromis fasciatus* Peters, 1857
  A *Hemichromis letourneauxi* Sauvage, 1880
  A *Oreochromis aurea* (Steindachner, 1864)
  A *Oreochromis mossambicus* (Peters, 1852)
  A *Oreochromis niloticus* (Linnaeus, 1758)
Clariidae
  A *Clarias gariepinus* (Burchell, 1822)
Clupeidae
  A *Alosa sapidissima* (Wilson, 1811)
  E *Clupeonella cultriventris* (Nordmann, 1840)
  A *Dussumieria elopsoides* Bleeker, 1849
  A *Etrumeus teres* (DeKay, 1848)
  A *Herklotsichthys punctatus* (Rüppell, 1837)
  A *Spratelloides delicatulus* (Bennett, 1831)
Cobitidae
  A *Misgurnus anguillicaudatus* (Cantor, 1842)
  A *Noemacheilus barbatulus* (Linnaeus, 1758)
Congridae
  A *Rhynchoconger trewavasae* Ben-Tuvia, 1993
Cottidae
  A *Beroe cucumis* (Fabricius, 1780)
Cynoglossidae
  A *Cynoglossus sinusarabici* (Chabunaud, 1913)
Cyprinidae
  E *Abramis brama* (Linnaeus, 1758)
  E *Abramis sapa* (Pallas, 1814)
  A *Aristichthys nobilis* (Richardson, 1845)
  C *Aspius aspius* Linnaeus, 1758
  A *Barbus barbus* (Linnaeus, 1758)
  E *Barbus brachycephalus* Kessler, 1872
  C *Barbus graellsii* Steindachner, 1866

E *Barbus plebejus* Bonaparte, 1839
C *Blicca bjoerkna* (Linnaeus, 1758)
A *Carassius auratus* Linnaeus, 1758
A *Carassius carassius* Linnaeus, 1758
A *Carassius gibelio* (Bloch, 1782)
A *Chondrostoma nasus* (Linnaeus, 1758)
A *Ctenopharyngodon idella* (Valenciennes, 1844)
A *Cyprinus carpio* Linnaeus, 1758
C *Gobio gobio* Linnaeus, 1758
A *Hypophthalmichthys molitrix* (Valenciennes, 1844)
E *Leucaspius delineatus* (Heckel, 1843)
E *Leuciscus cephalus* (Linnaeus, 1758)
E *Leuciscus leuciscus* (Linnaeus, 1758)
A *Megalobrama amblycephala* Yih, 1955
A *Megalobrama terminalis* (Richardson, 1846)
A *Mylopharyngodon piceus* (Richardson, 1846)
C *Pachychilon pictum* (Heckel & Kner, 1858)
A *Parabramis pekinensis* (Basilewsky, 1855)
E *Pelecus cultratus* (Linnaeus, 1758)
E *Phoxinus phoxinus* (Linnaeus, 1758)
A *Pimephales promelas* Rafinesque, 1820
A **Pseudorasbora parva** (Temminck & Schlegel, 1846)
E *Rhodeus amarus* (Bloch, 1782)
C *Rhodeus sericeus* (Pallas, 1776)
E *Rutilus rutilus* (Linnaeus, 1758)
E *Scardinius erythrophthalmus* Linnaeus, 1758
E *Tinca tinca* Linnaeus, 1758
E *Vimba vimba* Linnaeus, 1758
Cyprinodontidae
A *Aphanius dispar* (Rueppell, 1829)
Dactylopteridae
E *Dactylopterus volitans* Linnaeus, 1758
Diodontidae
A *Chilomycterus spilostylus* Leis & Randall, 1982
Esocidae
E *Esox lucius* Linnaeus, 1758
Exocoetidae
A *Cheilopogon cyanopterus* (Valenciennes, 1847)
A *Parexocoetus mento* (Valenciennes, 1846)
Fistularidae
A **Fistularia commersonii** Rüppell, 1835
Fundulidae
A *Fundulus heteroclinus* (Linnaeus, 1766)

Gasterosteidae
A *Culaea inconstans* (Kirtland, 1840)
E *Gasterosteus aculeatus* Linnaeus, 1758
E *Pungitius platygaster* (Kessler, 1859)
E *Pungitius pungitius* Linnaeus, 1758
Gobiidae
E *Benthophilus stellatus* (Sauvage, 1874)
A *Coryogalops ochetica* (Norman, 1927)
E *Gobius niger* Linnaeus, 1758
E *Neogobius fluviatilis* (Pallas, 1814)
E *Neogobius gymnotrachelus* (Kessler, 1857)
E *Neogobius iljini* Vasiljeva & Vasiljev, 1996
E *Neogobius kessleri* (Günther, 1861)
E **Neogobius melanostomus** (Pallas, 1814)
A *Oxyurichthys petersi* (Klunzinger, 1871)
E *Proterorhinus marmoratus* (Pallas, 1814)
A *Silhouetta aegyptia* (Chabanaud, 1933)
Haemulidae
A *Pomadasys stridens* (Forsskål, 1775)
Hemiramphidae
A *Hemiramphus far* (Forsskål, 1775)
A *Hyporhamphus affinis* (Gunther, 1866)
Holocentridae
A *Sargocentron rubrum* (Forsskål, 1775)
Ictaluridae
A *Ictalurus catus* (Linnaeus, 1758)
A *Ictalurus melas* (Rafinesque, 1820)
A *Ictalurus nebulosus* (Lesueur, 1819)
A *Ictalurus punctatus* (Rafinesque, 1818)
Labridae
A *Pteragogus pelycus* Randall, 1981
Leiognathidae
A *Leiognathus klunzingeri* (Steindachner, 1898)
Lutjanidae
A *Lutjanus argentimaculatus* (Forskål, 1775)
Monacanthidae
A *Stephanolepis diaspros* Fraser-Brunner, 1940
Moronidae
E *Dicentrarchus labrax* Linnaeus, 1758
A *Morone saxatilis* (Walbaum, 1792)
Mugilidae
A *Chelon haematocheilus* (Temminck & Schlegel, 1845)
E *Chelon labrosus* (Risso, 1827)
E *Liza aurata* (Risso, 1810)
A *Liza carinata* (Valenciennes, 1836)
A *Liza haematocheilus* (Temminck & Schlegel, 1845)

E *Liza ramada* (Risso, 1810)
E *Liza saliens* (Risso, 1810)
E *Mugil cephalus* Linnaeus, 1758
A *Mugil soiuy* Basilewsky, 1855
Mullidae
   A *Upeneus moluccensis* (Bleeker, 1855)
   A *Upeneus pori* Ben-Tuvia & Golani, 1989
Muraenesocidae
   A *Muraenesox cinereus* (Forsskål, 1775)
Odontobutidae
   A *Micropercops cinctus* (Dabry de Thiersant, 1872)
   A *Perccottus glenii* Dybowski, 1877
Osmeridae
   E *Osmerus eperlanus* (Linnaeus, 1758)
Ostraciidae
   A *Tetrosomus gibbosus* Linnaeus, 1758
Pempheridae
   A *Pempheris vanicolensis* Cuvier, 1821
Percidae
   A *Gymnocephalus cernuus* (Linnaeus, 1758)
   E *Perca fluviatilis* Linnaeus, 1758
   A *Percarina demidoffii* Nordmann, 1840
   E *Sander lucioperca* Linnaeus, 1758
   E *Sander volgensis* (Gmelin, 1789)
   A *Stizostedion lucioperca* (Linnaeus, 1758)
Pinguipedidae
   A *Pinguipes brasilianus* Cuvier & Valenciennes, 1829
Platycephalidae
   A *Papilloculiceps longiceps* (Ehrenberg & Valenciennes, 1829)
   A *Platycephalus indicus* (Linnaeus, 1758)
   A *Sorsogona prionota* (Sauvage, 1873)
Pleuronectidae
   E *Platichthys flesus* Linnaeus, 1758
Plotosidae
   A *Plotosus lineatus* (Thunberg, 1787)
Poeciliidae
   A *Gambusia affinis* (Baird & Girard, 1853)
   A *Gambusia holbrooki* Girard, 1859
   A *Lebistes reticulatus* (Peters, 1859)
   A *Poecilia reticulata* Peters, 1859
   A *Poecilia sphenops* Valenciennes, 1846
   A *Poecilia velifera* (Regan, 1914)
   A *Xiphophorus hellerii* Heckel, 1848
   A *Xiphophorus maculatus* (Günther, 1866)
Polyodontidae
   A *Polyodon spathula* (Walbaum, 1792)

Pomacentridae
   A *Abudefduf vaigiensis* (Quoy & Gaimard, 1825)
Rachycentridae
   A *Rachycentron canadum* (Linnaeus, 1766)
Salmonidae
   E *Coregonus albula* Linnaeus, 1758
   A *Coregonus autumnalis* (Pallas, 1776)
   E *Coregonus lavaretus* (Linnaeus, 1758)
   A *Coregonus lavaretus maraenoides* (Poljakow, 1874)
   A *Coregonus maraena* (Bloch, 1779)
   A *Coregonus muksun* (Pallas, 1814)
   A *Coregonus nasus* (Pallas, 1776)
   C *Coregonus oxyrhinchus* (L., 1758)
   E *Coregonus peled* (Gmelin, 1789)
   E *Hucho hucho* Linnaeus, 1758
   A *Oncorhynchus clarkii* (Richardson, 1836)
   A *Oncorhynchus gorbuscha* (Walbaum, 1792)
   A *Oncorhynchus keta* (Walbaum, 1792)
   A *Oncorhynchus kisutch* (Walbaum, 1792)
   A *Oncorhynchus mykiss* (Walbaum, 1792)
   A *Oncorhynchus nerka* (Walbaum, 1792)
   A *Oncorhynchus tshawytscha* (Walbaum, 1792)
   E *Salmo salar* Linnaeus, 1758
   E *Salmo trutta* (Linnaeus, 1758)
   E *Salvelinus alpinus* (Linnaeus, 1758)
   A **Salvelinus fontinalis** (Mitchill, 1814)
   E *Salvelinus namaycush* (Walbaum, 1792)
   E *Stenodus leucichthys nelma* (Pallas, 1776)
   C *Thymallus thymallus* (Linnaeus, 1758)
Scaridae
   A *Scarus ghobban* Forsskål, 1775
Sciaenidae
   A *Micropogonias undulatus* (Linnaeus, 1766)
   A *Sciaenops ocellatus* (Linnaeus, 1766)
Scombridae
   A *Rastrelliger kanagurta* (Cuvier, 1816)
   A *Scomberomorus commerson* (Lacepède, 1802)
Scorpaenidae
   A *Pterois miles* (Bennett, 1803)
Serranidae
   A *Cephalopholis taeniops* (Valenciennes, 1828)
   A *Epinephelus coioides* (Hamilton, 1822)
   A *Epinephelus malabaricus* (Bloch & Schneider, 1804)

Siganidae
  A *Siganus luridus* Rüppell, 1828
  A *Siganus rivulatus* Forsskål, 1775
Sillaginidae
  A *Sillago sihama* (Forsskål, 1775)
Siluridae
  E *Silurus glanis* Linnaeus, 1758
Sparidae
  A *Crenidens crenidens* (Forsskål, 1775)
  C *Lithognathus mormyrus* (Linnaeus, 1758)
  A *Rhabdosargus haffara* (Forsskål, 1775)
Sphyraenidae
  A *Sphyraena chrysotaenia* Klunzinger, 1884
  A *Sphyraena flavicauda* Rüppell, 1838
Stromateidae
  A *Pampus argenteus* Euphrasen, 1788
Syngnathidae
  A *Hippocampus fuscus* Rüppell, 1838
  E *Syngnathus abaster* Risso, 1827
Synodontidae
  A **Saurida undosquamis** (Richardson, 1848)
Teraponidae
  A *Pelates quadrilineatus* (Bloch, 1790)
  A *Terapon puta* (Cuvier, 1829)
Tetraodontidae
  A *Lagocephalus sceleratus* Gmelin, 1789
  A *Lagocephalus spadiceus* (Richardson, 1844)
  A *Lagocephalus suezensis* Clark & Gohar, 1953
  A *Tetraodon fluviatilis* Hamilton, 1822
  A *Torquigener flavimaculosus* Hardy & Randall, 1983
  A *Tylerius spinosissimus* Regan, 1908
Umbridae
  A *Umbra krameri* Walbaum, 1792
  A *Umbra pygmea* (DeKay, 1842)

Chordata, Amphibia
Ambystomatidae
  A *Ambystoma mexicanum* (Shaw & Nodder, 1798)
  A *Ambystoma tigrinum* (Green, 1825)
Bombinatoridae
  E *Bombina bombina* (Linnaeus, 1761)
  A *Bombina orientalis* (Boulenger, 1890)
Brachycephalidae
  A *Eleutherodactylus martinicensis* (Tschudi, 1838)
Bufonidae
  E *Bufo bufo* (Linnaeus, 1758)
  E *Bufo calamita* Laurenti, 1768
  A *Bufo mauritanicus* Schlegel, 1841
  A *Chaunus marinus* (Linnaeus, 1758)
  E *Pseudepidalea viridis* (Laurenti, 1768)
  A *Rhaebo blombergi* (Myers & Funkhouser, 1951)
Discoglossidae
  E *Alytes obstetricans* (Laurenti, 1768)
  E *Bombina variegata* (Linnaeus, 1758)
  E *Discoglossus pictus* Otth, 1873
Hylidae
  E *Hyla arborea* (Linnaeus, 1761)
  E *Hyla meridionalis* Boettger, 1874
  A *Pseudacris regilla* (Baird & Girard, 1852)
Pipidae
  A *Xenopus laevis* (Daudin, 1802)
Plethodontidae
  E *Speleomantes ambrosii* (Lanza, 1955)
Proteidae
  E *Proteus anguinus* Laurenti, 1768
Ranidae
  A **Lithobates catesbeianus** (Shaw, 1802)
  E *Pelophylax kl. esculentus* (Linnaeus, 1758)
  E *Pelophylax bedriagae* (Camerano, 1882)
  E *Pelophylax kurtmuelleri* (Gayda, 1940)
  E *Pelophylax lessonae* (Camerano, 1882)
  E *Pelophylax perezi* (Seoane, 1885)
  E *Pelophylax ridibundus* (Pallas, 1771)
  A *Pelophylax saharicus* (Boulenger In Hartert, 1913)
  E *Rana temporaria* Linnaeus, 1759
Salamandridae
  A *Cynops pyrrhogaster* (Boie, 1826)
  E *Lissotriton montandoni* (Boulenger, 1880)
  E *Mesotriton alpestris* (Laurenti, 1768)
  E *Pleurodeles waltl* Michahelles, 1830
  E *Triturus carnifex* (Laurenti, 1768)
  E *Triturus marmoratus* (Latreille, 1800)

Chordata, Reptilia
Agamidae
  A *Agama agama* Linnaeus, 1758
  E *Laudakia stellio* (Linnaeus, 1758)
  A *Uromastyx acanthinura* Bell, 1825
Alligatoridae
  A *Caiman crocodylus* Linnaeus, 1758
Anguidae
  E *Anguis fragilis* Linnaeus, 1758
  E *Pseudopus apodus* Pallas, 1775
Boidae
  A *Boa constrictor* Linnaeus, 1758
  A *Python molurus* Linnaeus, 1758

- A *Python regius* Shaw, 1802
- A *Python reticulates* Schneider, 1801
- A *Python sebae* Gmelin, 1789

Chamaeleonidae
- E *Chamaeleo chameleon* (Linnaeus, 1758)
- A *Chamaeleo africanus* Laurenti, 1768

Chelidae
- A *Chelus fimbriatus* Schneider, 1783

Chelydridae
- A *Chelydra serpentine* Linnaeus, 1758
- A *Macrochelys temminckii* Troost, 1835

Colubridae
- A *Hemorrhois algirus* (Jan, 1863)
- A *Elaphe guttata* (Linnaeus, 1766)
- E *Hierophis gemonensis* (Laurenti, 1768)
- E *Hierophis viridiflavus* (Lacèpéde, 1789)
- A *Lampropeltis getulus* (Linnaeus, 1766)
- A *Macroprotodon cucullatus* (Geoffroy Saint-Hilaire, 1809)
- E *Natrix maura* (Linnaeus, 1758)
- E *Natrix tessellata* (Laurenti, 1768)
- E *Rhinechis scalaris* (Schinz, 1822)
- E *Telescopus fallax* Fleischmann, 1831)
- E *Zamenis longissimus* (Laurenti, 1768)

Crocodylidae
- A *Alligator mississippiensis* Daudin, 1802
- A *Crocodylus niloticus* (Laurenti, 1768)
- A *Crocodylus rhombifer* Cuvier, 1807

Emydidae
- A *Chinemys reevesii* Gray, 1831
- A *Chrysemys picta* Schneider, 1783
- E *Emys orbicularis* (Linnaeus, 1758)
- A *Graptemys geographica* Le Sueur, 1817
- E *Mauremys caspica* Gmelin, 1774
- A *Pseudemys concinna* Le Conte, 1830
- A **Trachemys scripta** (Schoepff, 1792)

Gekkonidae
- A *Cyrtopodion scabrum* Conant & Collins, 1991
- E *Hemidactylus turcicus* (Linnaeus, 1758)
- E *Mediodactylus kotschyi* (Steindachner, 1870)
- E *Tarentola boettgeri* Steindachner, 1891
- E *Tarentola delalandii* Duméril & Bibron, 1836
- E *Tarentola mauritanica* (Linnaeus, 1758)

Geoemydidae
- E *Mauremys leprosa* Schweigger, 1812

Iguanidae
- A *Ctenosaura similis* Gray, 1831
- A *Iguana iguana* (Linnaeus, 1758)
- A *Plica plica* Linnaeus, 1758

Lacertidae
- E *Lacerta trilineata* Bedriaga, 1886
- A *Darevskia armeniaca* Méhely, 1909
- E *Gallotia galloti* Oudart, 1839
- E *Iberolacerta horvathi* (Méhely, 1904)
- E *Lacerta viridis* (Laurenti, 1768)
- E *Podarcis muralis* (Laurenti, 1768)
- E *Podarcis pityusensis* (Boscà, 1883)
- E *Podarcis siculus* (Rafinesque, 1810)
- E *Psammodromus algirus* (Linnaeus, 1758)
- E *Psammodromus hispanicus* Fitzinger, 1826
- A *Teira dugesii* (Milne-Edwards, 1829)
- A *Teira perspicillata* (Duméril & Bibron, 1839)
- E *Timon lepidus* (Daudin, 1802)

Pelomedusidae
- A *Pelomedusa subrufa* Bonnaterre, 1789

Polychrotidae
- A *Anolis caroliniensis* (Voigt, 1832)
- A *Anolis equestris* Merrem, 1820
- A *Anolis sagrei* Dumeril & Bibron, 1837

Testudinidae
- E *Testudo graeca* Linnaeus, 1758
- E *Testudo hermanni* Gmelin, 1789
- E *Testudo marginata* Schoepff, 1792
- A *Testudo horsfieldii* Gray, 1844

Trionychidae
- A *Pelodiscus sinensis* (Wiegmann, 1834)
- A *Trionyx triunguis* (Forskål, 1775)

Varanidae
- A *Varanus niloticus* Linnaeus, 1758

Viperidae
- E *Vipera aspis* (Linnaeus, 1758)

Chordata, Aves

Accipitridae
- C *Aegypius monachus* (Linnaeus, 1766)
- A *Elanoides forficatus* (Linnaeus, 1758)

Anatidae
- A *Aix galericulata* (Linnaeus, 1758)
- A *Aix sponsa* (Linnaeus, 1758)
- A *Alopochen aegyptiacus* (Linnaeus, 1766)
- E *Anas acuta* (Linnaeus, 1758)
- A *Anas bahamensis* (Linnaeus, 1758)
- A *Anas discors* (Linnaeus, 1766)
- A *Anas falcata* (Georgi, 1775)
- C *Anas penelope* (Linnaeus, 1758)
- C *Anas strepera* (Linnaeus, 1758)
- A *Anser albifrons* (Scopoli, 1769)
- E *Anser anser* (Linnaeus, 1758)

- A *Anser brachyrhynchus* (Baillon, 1834)
- A *Anser caerulescens* (Linnaeus, 1758)
- A *Anser canagica* (Sevastianov, 1802)
- A *Anser cygnoides* (Linnaeus, 1758)
- E *Anser fabalis* (Latham, 1787)
- A *Anser indicus* (Latham, 1790)
- A *Anser rossii* (Cassin, 1861)
- A *Aythya americana* (Eyton, 1838)
- C *Aythya ferina* (Linnaeus, 1758)
- A **Branta canadensis** (Linnaeus, 1758)
- A *Branta leucopsis* (Bechstein, 1803)
- A *Bucephala albeola* (Linnaeus, 1758)
- A *Cairina moschata* (Linnaeus, 1758)
- A *Callonetta leucophrys* (Vieillot, 1816)
- A *Chloephaga picta* (Gmelin, 1789)
- A *Cygnus atratus* (Latham, 1790)
- E *Cygnus cygnus* (Linnaeus, 1758)
- E *Cygnus olor* (Gmelin, 1789)
- A *Dendrocygna autumnalis* (Linnaeus, 1758)
- E *Netta rufina* (Pallas, 1773)
- A **Oxyura jamaicensis** (Gmelin, 1789)
- E *Tadorna ferruginea* (Pallas, 1764)

Ardeidae
- E *Egretta alba* (Linnaeus, 1758)
- E *Egretta gularis* (Bosh, 1792)
- E *Nycticorax nycticorax* (Linnaeus, 1758)

Cacatuidae
- A *Cacatua galerita* (Latham, 1790)

Cardinalidae
- A *Cardinalis cardinalis* (Linnaeus, 1758)

Columbidae
- A *Columba guinea* (Linnaeus, 1758)
- E *Columba livia* (Gmelin, 1789)
- A *Columbina passerina* (Linnaeus, 1758)
- A *Geophaps lophotes* (Temminck, 1822)
- E *Streptopelia decaocto* (Frivaldszky, 1838)
- A *Streptopelia roseogrisea* (Sundevall, 1857)
- A *Streptopelia senegalensis* (Linnaeus, 1766)

Corvidae
- A *Corvus splendens* (Vieillot, 1817)
- A *Pica pica* (Linnaeus, 1758)
- A *Urocissa erythrorhyncha* (Boddaert, 1783)

Cracidae
- A *Penelope superciliaris* (Temminck, 1815)
- A *Pipile cumanensis* (Jacquin, 1784)

Emberizidae
- E *Junco hyemalis* (Linnaeus, 1758)
- A *Paroaria coronata* (Miller, 1776)

Estrildidae
- A *Amadina fasciata* (Gmelin, 1789)
- A *Amandava amandava* (Linnaeus, 1758)
- A *Amandava subflava* (Vieillot, 1819)
- A *Estrilda melpoda* (Vieillot, 1817)
- A *Estrilda astrild* (Linnaeus, 1758)
- A *Estrilda troglodytes* (Lichtenstein, 1823)
- A *Lagonosticta senegala* (Linnaeus, 1766)
- A *Lonchura cantans* (Gmelin, 1789)
- A *Lonchura maja* (Linnaeus, 1766)
- A *Lonchura malabarica* (Linnaeus, 1758)
- A *Lonchura malacca* (Linnaeus, 1766)
- A *Padda oryzivora* (Linnaeus, 1758)
- A *Taeniopygia guttata* (Vieillot, 1817)
- A *Uraeginthus bengalus* (Linnaeus, 1766)

Fringillidae
- A *Carduelis ambigua* (Oustalet, 1896)
- E *Carduelis carduelis* (Linnaeus, 1758)
- E *Carduelis chloris* (Linnaeus, 1758)
- A *Serinus canaria* (Linnaeus, 1758)

Gruidae
- A *Balearica pavonina* (Linnaeus, 1758)

Meleagrididae
- A *Meleagris gallopavo* (Linnaeus, 1758)

Numididae
- A *Numida meleagris* (Linnaeus, 1758)

Odontophoridae
- A *Callipepla californica* (Shaw, 1798)
- A *Colinus virginianus* (Linnaeus, 1758)

Passeridae
- E *Passer domesticus* (Linnaeus, 1758)
- E *Passer hispaniolensis* (Temminck, 1820)

Pelecanidae
- C *Pelecanus rufescens* (Gmelin, 1789)

Phalacrocoracidae
- E *Phalacrocorax carbo* (Linnaeus, 1758)

Phasianidae
- A *Alectoris barbara* (Bonnaterre, 1792)
- E *Alectoris chukar* (Gray, 1830)
- E *Alectoris graeca* (Meisner, 1804)
- E *Alectoris rufa* (Linnaeus, 1758)
- A *Ammoperdix heyi* (Temminck, 1825)
- A *Bambusicola thoracica* (Temminck, 1815)
- A *Catreus wallichii* (Hardwicke, 1827)
- A *Chrysolophus amherstiae* (Leadbeater, 1829)
- A *Chrysolophus pictus* (Linnaeus, 1758)
- A *Coturnix coromandelica* (Gmelin, 1789)
- E *Coturnix japonica* (Temminck & Schlegel, 1849)

- A *Francolinus clappertoni* (Children & Vigors, 1826)
- A *Francolinus erckelii* (Ruppell, 1835)
- E *Francolinus francolinus* (Linnaeus, 1766)
- A *Gallus gallus* (Linnaeus, 1758)
- A *Lagopus mutus* (Montin, 1781)
- C *Lophophorus leucomelanos* (Latham, 1790)
- A *Lophura nycthemera* (Linnaeus, 1758)
- A *Pavo cristatus* (Linnaeus, 1758)
- A *Perdicula asiatica* (Latham, 1790)
- A *Perdix dauurica* (Pallas, 1811)
- E *Perdix perdix* (Linnaeus, 1758)
- A *Phasianus colchicus* (Linnaeus, 1758)
- A *Phasianus versicolor* (Vieillot, 1825)
- A *Syrmaticus reevesii* (Gray, 1829)

Phoenicopteridae
- A *Phoenicopterus chilensis* (Molina, 1782)
- A *Phoenicopterus minor* (Saint-Hilaire, 1798)

Picidae
- A *Melanerpes carolinus* (Linnaeus, 1758)

Ploceidae
- A *Euplectes afer* (Gmelin, 1789)
- A *Euplectes franciscanus* (Isert, 1789)
- A *Euplectes hordeaceus* (Linnaeus, 1758)
- A *Euplectes nigroventris* (Cassin, 1848)
- A *Euplectes orix* (Linnaeus, 1758)
- A *Ploceus cucullatus* (Statius Muller, 1776)
- A *Ploceus galbula* (Ruppell, 1840)
- A *Ploceus melanocephalus* (Linnaeus, 1758)
- A *Ploceus subaureus* (A.Smith, 1839)
- A *Ploceus velatus* (Vieillot, 1819)
- A *Quelea erythrops* (Hartlaub, 1848)
- A *Quelea quelea* (Linnaeus, 1758)

Psittacidae
- A *Agapornis fischeri* (Reichenow, 1887)
- A *Agapornis personatus* (Reichenow, 1887)
- A *Agapornis roseicollis* (Vieillot, 1818)
- A *Amazona aestiva* (Linnaeus, 1758)
- A *Amazona amazonica* (Linnaeus, 1766)
- A *Amazona leucocephala* (Linnaeus, 1758)
- A *Amazona ochrocephala* (Gmelin, 1788)
- A *Amazona oratrix* (Ridgway, 1887)
- A *Aratinga acuticaudata* (Vieillot, 1818)
- A *Aratinga erythrogenys* (Lesson, 1844)
- A *Aratinga mitrata* (Tschudi, 1844)
- A *Brotogeris pyrrhopterus* (Latham, 1802)
- A *Brotogeris tirica* (Gmelin, 1788)
- A *Cyanoliseus patagonus* (Vieillot, 1818)
- A *Melopsittacus undulatus* (Shaw, 1805)
- A *Myiopsitta monachus* (Boddaert, 1783)
- A *Nandayus nenday* (Vieillot, 1823)
- A *Nannopsittaca panychlora* (Salvin & Godman, 1883)
- A *Nymphicus hollandicus* (Kerr, 1792)
- A *Poicephalus senegalus* (Linnaeus, 1766)
- A *Psittacula eupatria* (Linnaeus, 1766)
- A **Psittacula krameri** (Scopoli, 1769)

Pteroclididae
- A *Pterocles exustus* (Temminck, 1825)

Pycnonotidae
- A *Pycnonotus cafer* (Linnaeus, 1766)
- A *Pycnonotus jocosus* (Linnaeus, 1758)

Rheidae
- A *Rhea americana* (Linnaeus, 1758)
- A *Rhea pennata* (d'Orbigny, 1834)

Spheniscidae
- A *Aptenodytes patagonica* (Miller, 1778)
- A *Spheniscus demersus* (Linnaeus, 1758)

Strigidae
- E *Athene noctua* (Scopoli, 1769)
- E *Bubo bubo* (Linnaeus, 1758)

Sturnidae
- A *Acridotheres cristatellus* (Linnaeus, 1766)
- A *Acridotheres ginginianus* (Latham, 1790)
- A *Acridotheres tristis* (Linnaeus, 1766)
- A *Gracula religiosa* (Linnaeus, 1758)
- A *Lamprotornis caudatus* (Statius Muller, 1776)
- A *Lamprotornis chalybaeus* (Hemprich & Ehrenberg, 1828)
- A *Lamprotornis purpureus* (Statius Muller, 1776)
- A *Lamprotornis splendidus* (Vieillot, 1822)
- A *Lamprotornis superbus* (Ruppell, 1845)
- A *Sturnus burmannicus* (Jerdon, 1862)
- A *Sturnus nigricollis* (Paykull, 1807)

Tetraonidae
- E *Bonasa bonasia* (Linnaeus, 1758)
- A *Tympanuchus cupido* (Linnaeus 1758)

Threskiornithidae
- A **Threskionis aethiopicus** (Latham, 1790)

Timaliidae
- A *Garrulax formosus* (Verreaux, 1869)
- A *Leiothrix lutea* (Scopoli, 1786)
- A *Paradoxornis alphonsianus* (Verreaux, 1870)
- A *Paradoxornis webbianus* (Gould, 1852)

Tinamidae
- A *Eudromia elegans* (Saint-Hilaire, 1832)
- A *Rynchotus rufescens* (Temminck, 1815)

Turdidae
- A *Sialia sialis* (Linnaeus, 1758)
- A *Turdus migratorius* (Linnaeus, 1766)

Chordata, Mammalia

Arvicolidae
- E *Clethrionomys glareolus* (Schreber, 1780)

Bovidae
- A *Ammotragus lervia* (Pallas, 1777)
- A *Bison bison* (Linnaeus, 1758)
- A *Boselaphus tragocamelus* Pallas, 1766
- E *Capra ibex* Linnaeus, 1758
- A *Connochaetes taurinus* (Burchell, 1823)
- A *Hemitragus jemlahicus* (H. Smith, 1826)
- A *Ovibos moschatus* (Zimmermann, 1780)
- E *Rupicapra rupicapra* (Linnaeus, 1758)

Camelidae
- A *Lama guanicoe* Müller, 1776

Canidae
- E *Alopex lagopus* (Linnaeus, 1758)
- A *Cuon alpinus* (Pallas, 1811)
- A **Nyctereutes procyonoides** (Gray, 1834)

Castoridae
- A *Castor canadensis* Kuhl, 1820

Caviidae
- A *Cavia porcellus* (Linnaeus, 1758)

Cercopithecidae
- A *Macaca mulatta* (Zimmermann, 1780)
- A *Macaca sylvanus* (Linnaeus, 1758)

Cervidae
- A *Axis axis* (Erxleben, 1777)
- E *Capreolus capreolus* (Linnaeus, 1758)
- E *Capreolus pygargus* (Pallas, 1771)
- C *Cervus canadensis* (Erxleben, 1777)
- A **Cervus nippon** Temminck, 1838
- A *Cervus porcinus* Zimmermann, 1780
- A *Dama dama* (Linnaeus, 1758)
- A *Hydropotes inermis* Swinhoe, 1870
- A *Muntiacus muntjak* Zimmermann, 1780
- A *Muntiacus reevesi* (Ogilby, 1839)
- A *Odocoilus hemionus* (Rafinesque, 1817)
- A *Odocoileus virginianus* (Zimmermann, 1780)
- E *Rangifer tarandus* (Linnaeus, 1758)

Chinchillidae
- A *Chinchilla lanigera* (Molina, 1782)

Cricetidae
- A *Mesocricetus auratus* (Waterhouse, 1839)

Dasypodidae
- A *Dasypus novemcinctus* Linnaeus, 1758

Didelphidae
- A *Marmosa cinerea* (Temminck, 1843)

Equidae
- A *Equus hemionus* Pallas, 1775

Erinaceidae
- A *Atelerix algirus* (Lereboullet, 1842)
- E *Erinaceus europaeus* Linnaeus, 1758

Galagonidae
- A *Galago senegalensis* (Geoffroy, 1796)

Gliridae
- E *Eliomys quercinus* (Linnaeus, 1766)
- E *Glis glis* Linnaeus, 1766

Herpestidae
- A *Herpestes auropunctatus* (Hodgson, 1836)
- A *Herpestes edwardsii* Geoffroy, 1818

Istricidae
- A *Hystrix brachyura* Linnaeus, 1758
- C *Hystrix cristata* Linnaeus, 1758

Leporidae
- A *Lepus californicus* Gray, 1837
- A *Lepus capensis* Linnaeus, 1758
- E *Lepus europaeus* Pallas, 1778
- E *Lepus granatensis* Rosenhauer, 1856
- A *Sylvilagus floridanus* (J.A. Allen, 1890)
- A *Sylvilagus transitionalis* (Bangs, 1895)

Macropodidae
- A *Macropus rufogriseus* (Desmarest, 1817)

Monodontidae
- A *Delphinapterus leucas* (Pallas, 1776)

Moschidae
- A *Moschus moschiferus* Linnaeus, 1758

Muridae
- E *Cricetus cricetus* (Linnaeus, 1758)
- A *Meriones unguiculatus* Milne-Edwards, 1867
- E *Micromys minutus* (Pallas, 1771)
- E *Microtus rossiaemeridionalis* Ognev, 1924
- A *Mus musculus* Linnaeus, 1758
- A **Ondatra zibethicus** (Linnaeus, 1766)
- A *Phodopus sungorus* Pallas, 1773
- A **Rattus norvegicus** (Berkenhout, 1769)
- A *Rattus rattus* (Linnaeus, 1758)
- A *Sigmodon hispidus* Say & Ord, 1825

Mustelidae
- A *Mephitis mephitis* (Schreber, 1776)
- E *Mustela lutreola* (Linnaeus, 1761)
- E *Mustela nivalis* Linnaeus, 1766
- A **Mustela vison** Schreber, 1777

Myocastoridae
- A **Myocastor coypus** (Molina, 1782)

Octodontidae
  A *Octodon degus* (Molina, 1782)
Procyonidae
  A *Nasua nasua* (Linnaeus, 1766)
  A *Potos flavus* (Schreber, 1774)
  A **Procyon lotor** (Linnaeus, 1758)
Pteropodidae
  A *Rousettus aegyptiacus* (Geoffroy, 1810)
Rhinolophidae
  E *Rhinolophus ferrumequinum* Schreber, 1774
Sciuridae
  A *Atlantoxerus getulus* (Linnaeus, 1758)
  A *Callosciurus erythraeus* (Pallas, 1779)
  A *Callosciurus finlaysonii* (Horsfield, 1824)
  A *Funambulus pennanti* Wroughton, 1905
  E *Marmota marmota* (Linnaeus, 1758)
  A *Sciurotamias davidianus* (Milne-Edwards, 1867)
  A *Sciurus anomalus* Schreber, 1785
  A **Sciurus carolinensis** Gmelin, 1788
  E **Tamias sibiricus** (Laxmann, 1769)
  A *Tamias striatus* (Linnaeus, 1758)
Soricidae
  E *Suncus etruscus* (Savi, 1822)
Talpidae
  C *Desmana moschata* Linnaeus, 1758
Tayassuidae
  A *Pecari tajacu* (Linnaeus, 1758)
Viverridae
  A *Civettictis civetta* (Schreber, 1776)
  C *Genetta genetta* (Linnaeus, 1758)

Chapter 12
One Hundred of the Most Invasive Alien Species in Europe

Montserrat Vilà, Corina Başnou, Stephan Gollasch, Melanie Josefsson, Jan Pergl, and Riccardo Scalera

One of the primary tools for raising awareness on biological invasions has been the publication of species accounts of the most prominent alien invaders. Until now such compilations have been available only for particular taxa, biomes and/or regions (Cronk and Fuller 2001; Weber 2003; Weidema 2000). In Europe, species accounts for selected invasive species have been published for a few countries or regions: the Czech Republic (Mlíkovský and Stýblo 2006), France (Pascal et al. 2006), Italy (Andreotti et al. 2001; Scalera 2001), Spain (Capdevila-Argüelles and Zilletti 2006); the Mediterranean Sea (CIESM 2007), and the North European and Baltic region (Gollasch et al. 1999; NOBANIS 2007). These accounts highlight invasive alien species which cause significant harm to biological diversity, socio-economic values and human health in these regions. The main purpose of these accounts is to provide guidance to environmental managers and raise public awareness of the biological, ecological and socio-economic impacts of the most harmful invaders, together with a description of the main management options to prevent their spread and reduce their impacts. The importance of the role of such tools has been clearly shown by the IUCN's 100 of the World's Worst Invasive Species list (Love et al. 2000) which has been very influential in raising awareness and supporting the development of policy conservation instruments relevant to biological invasions (Shine et al. 2000).

The European Environmental Agency has produced, within the SEBI 2010 project, a list of the worst invasive alien species threatening biological diversity in Europe (EEA 2007). This list contributes to the general indicator of changes in biological diversity caused by invasive alien species. The SEBI 2010 list is primarily a means to communicate the issue of invasive species to policymakers, stakeholders and the general public. The selection of the 168 species on the list was carried out in an open consultative process with an expert group, the scientific community and national environmental authorities. The main criterion used for selection was that the species have a serious impact on biological diversity at the regional level. Serious impact implies that the species has severe effects on ecosystem structure and function, it can replace native species throughout a significant proportion of its range, it can hybridise with native species or threaten biodiversity. In addition, the species can have negative consequences for human activities, health and/or economic interests.

Following the experience of SEBI 2010, the species accounts realised within the DAISIE project have been prepared with the purpose of delivering a synthesis of the most relevant up-to-date information on the ecology, distribution and impact of 100 of the most invasive species. These accounts are particularly designed for supporting actions dealing with biological invasions in Europe. For this reason, although they cannot be considered to be exhaustive, the DAISIE accounts may play a major role in raising public awareness and supporting the activities of a broad spectrum of professionals including land-use and wildlife managers, environmental policymakers, environmental educators, journalists, students and other stakeholders. On this regard, there are already examples on how DAISIE might offer a major contribution to the development of policy tools to face the threats of alien species. Indeed, it is a remarkable fact that the 100 of the most invasive species have been used as a basis for the realisation of a document for the Council of Europe titled "Towards a black list of invasive alien species entering Europe through trade, and proposed responses" (Genovesi and Scalera 2007), which led to Recommendation No. 125 (2007) of the Standing Committee, adopted on 29 November 2007, on trade in invasive and potentially invasive alien species in Europe.

The DAISIE research project has produced invasive species accounts for three terrestrial fungi species, 18 terrestrial plant species, 16 terrestrial invertebrate species, 15 terrestrial vertebrate species, 16 species found in inland waters and 32 species from coastal waters. These species invade European natural and semi-natural habitats and already cause or have the potential to cause severe environmental, economic and/or health problems. The species were nominated to the list by experts working within the DAISIE research project. They are perhaps not the 100 most invasive alien species in Europe, but rather representatives of all main taxonomic groups and all environments and were selected so as to represent diverse impacts on ecology, socio-economic values and human and animal health. The species accounts were written by experts in the specific taxon and are based on the most up to date information available, both published and unpublished. The names of these species are given in **bold** throught the book. Each DAISIE species account includes information on the following aspects:

Description: First we give a short description of the species aimed at helping the reader to identify the targeted species. Further information focuses on the characteristics related to the species' potential to reproduce, establish and successfully invade new areas as well as dispersal mechanisms for the species.

Habitat: The account also includes the habitat type where the species is found in its native and introduced range. In order to make habitat types comparable for a wide range of taxa across diverse biomes we adopted the EUNIS habitats classification, a standard classification of European habitats according to Davies and Moss (2003). We used habitats described at the hierarchical level 1 which indicates 10 broad habitat types (e.g., C: inland surface waters, F: heathland, scrub and tundra habitats, etc.). Where these habitats were too heterogeneous with respect to the level of invasion, we also used habitats at the finer hierarchical level 2 which accounts for geographic and topographic differences. (e.g., F2: arctic, alpine and sub-alpine scrub; F3: temperate and Mediterraneo-montane scrub). Where information was available, we included specific habitat requirements that might improve our understanding of environmental factors limiting species spread.

Distribution: The area of distribution of the species in its native and invaded range is described. For some species there is very precise information on its global area of distribution, but for other species only references to a continent or region could be made. The distribution trend of the species in its European range is also described, but the quality of this information varies greatly. For many species expertise could only indicate if the species is currently increasing, decreasing or remaining stable.

Maps of the most up-to-date known distribution were based on the available information on the species distribution in Europe and in some adjacent regions such as the Mediterranean Sea. Maps are based on detailed surveys of regional sources of information. However, the missing occurrence of species from some parts of Europe, especially in south-eastern countries, does not always mean that these areas are not invaded, rather this can be the result of a lack of data. It is important to highlight that maps do not include assumed distribution. If enough precise information on species distribution was available then the CGRS grid (Common European Chorological Grid Reference System; size of the grid 50×50 km) was used. Everywhere else the distribution is shown by using geographic/political regions or by the combination of CGRS grids and regions. For each geographic region the species distribution was also classified as native or alien. Whenever it was relevant and data were reliable enough, the maps also indicate eradication or extinction records.

Introduction pathway: This information is essential for early detection and management. As well as mentioning if the species has been intentionally or unintentionally introduced, we include information on whether the primary route of introduction to and within Europe has taken place through commodities, if the species has been transported by a vector, or if it has been dispersed unaided or through a man-made corridor (Hulme et al. 2008).

Impacts: Since actions against biological invasions are often only supported where there is evidence of some type of ecological and socio-economic impact, and particularly when affecting human health, special emphasis has been given to the description of all known impacts that have been reported or could potentially occur (Binimelis et al. 2007). As an example, 71% of the species listed reduce species diversity or alter the invaded community and 19% affect the viability of endangered species. Overall, the 100 invaders represent a broad spectrum of impacts to ecosystem services.

Management: This section is particularly dedicated to land and wildlife managers already concerned about the hazards of biological invasions. Experience regarding mechanical, chemical and biological control methods, either successful or unsuccessful are reported. Prevention strategies are also included. Unfortunately, for many species, especially in aquatic biomes, successful management options are unknown.

References: No references are included in this printed version of the DAISIE accounts. However, an extensive literature list for each species can be found at www.europe-aliens.org.

Overall, the DAISIE species accounts offer information on the main ecological aspects and impacts of 100 of the most invasive alien species in Europe, and contribute to the initiatives on raising awareness on this phenomenon of global change. We hope that the DAISIE accounts will be of benefit to many people with interest and responsibilities for preventing and managing biological invasions in Europe and beyond.

References

Andreotti A, Baccetti N, Perfetti A, Besa M, Genovesi P, Guberti V (2001) Mammiferi ed uccelli esotici in Italia: analisi del fenomeno, impatto sulla biodiversità e linee guida gestionali. Quad Cons Natura, 2, Min. Ambiente, Ist Naz Fauna Selvatica

Binimelis R, Born W, Monterroso I, Rodríguez-Labajos B (2007) Socio-economic impacts and assessment of biological invasions. In: Nentwig W (ed) Biological invasions. Ecological Studies 193, Springer, Berlin. 331–347

Capdevila-Argüelles L, Zilletti B (2006) Top 20. Las 20 especies exóticas invasoras más dañinas presentes en España. Edition GEIB, León

CIESM The Mediterranean Science Commission (2007) Atlas of exotic species in the Mediterranean. www.ciesm.org/online/atlas/index.htm. Cited Dec 2007

Cronk QCB, Fuller JL (2001) Plant invaders: the threat to natural ecosystems. Island Press, Washington DC

Davies CE, Moss D (2003) EUNIS habitat classification, August 2003. European Topic Centre on Nature Protection and Biodiversity, Paris

EEA (2007) Halting the loss of biodiversity by 2010: proposal for a first set of indicators to monitor progress in Europe. European Environment Agency, Technical Report 11/2007

Genovesi P, Scalera R (2007) Towards a black list of invasive alien species entering Europe through trade, and proposed responses. Convention on the conservation of European wildlife and natural habitats. Standing Committee 27th Meeting, Strasbourg, 26–29 November 2007. T-PVS/Inf 9

Gollasch S, Minchin D, Rosenthal H, Voigt M (eds) (1999) Exotics across the ocean. Case histories on introduced species: their general biology, distribution, range expansion and impact. Logos, Berlin

Hulme PE, Bacher S, Kenis M, Klotz, S, Kühn I, Minchin D, Nentwig W, Olenin S, Panov V, Pergl J, Pyšek P, Roques A, Sol D, Solarz W, Vilà M (2008) Grasping at the routes of biological invasions: a framework for integrating pathways into policy. J Appl Ecol 45:403–414

Love S, Browne M, Boudjelas S, De Poorter M (2000) 100 of the World's Worst Invasive Species. A Selection from the Global Invasive Species Database. www.issg.org. Cited December 2007

Mlíkovský J, Stýblo P (eds) (2006) Nepůvodní druhy fauny a flóry České republiky. ČSOP, Praha

NOBANIS (2007) North European and Baltic Network on Invasive Alien Species. www.nobanis.org. Cited 18 September 2007

Pascal M, Lorvelec O, Vigne JD (2006) Invasions biologiques et extinctions. 11,000 ans d'histoire des vertébrés en France. Quae-Belin Editions, Paris

Shine C, Williams N, Gündling L (2000) A guide to designing legal and institutional frameworks on alien invasive species. IUCN, Gland

Scalera R (2001) Invasioni biologiche. Le introduzioni di vertebrati in Italia: un problema tra conservazione e globalizzazione. Collana Verde, 103. Corpo Forestale dello Stato. Ministero delle Politiche Agricole e Forestali, Roma

Weber E (2003) Invasive plant species of the world: a reference guide to environmental weeds. CABI, Cambridge

Weidema IR (ed) (2000) Introduced species in the Nordic countries. Nord Environ 13:1–24

Chapter 13
Species Accounts of 100 of the Most Invasive Alien Species in Europe

Aquatic marine species

Alexandrium catenella
Balanus improvisus
Bonnemaisonia hamifera
Brachidontes pharaonis
Caulerpa racemosa cylindracea
Caulerpa taxifolia
Chattonella verruculosa
Codium fragile tomentosoides
Coscinodiscus wailesii
Crassostrea gigas
Crepidula fornicata
Ensis americanus
Ficopomatus enigmaticus
Fistularia commersoni
Halophila stipulacea
Marenzelleria neglecta
Marsupenaeus japonicus
Musculista senhousia
Odontella sinensis
Paralithodes camtschaticus
Percnon gibbesi
Pinctada radiata
Portunus pelagicus
Rapana venosa
Rhopilema nomadica
Saurida undosquamis
Siganus rivulatus
Spartina anglica
Styela clava
Teredo navalis
Tricellaria inopinata
Undaria pinnatifida

Alexandrium catenella (Whedon & Kofoid) Balech (Goniodomataceae, Pyrrophycophyta)

Irina Olenina and Sergej Olenin

It is an armoured, marine, planktonic dinoflagellate typically occurring in characteristic short chains of two, four or eight cells, swimming together in a snake-like fashion. Single cells are almost round, 20–48 µm in length and 18–32 µm in width. It is dispersed by water currents. Asexual reproduction by binary fission. This species also has a sexual cycle with opposite mating types (heterothallism). After gamete fusion, a planozygote forms, then it encysts into a characteristic resting cyst. The life cycle has several stages: motile vegetative cells, haploid gametes, diploid zygotes, resting cysts and temporary cysts. The colourless resting cyst is ellipsoidal with rounded ends, and is covered by a smooth wall and a mucilaginous substance, surviving long periods in darkness.

Native habitat (EUNIS code): A7: Pelagic water column, occupies the upper water layers in coastal and estuarine waters. The same habitats are occupied in the invaded range. A coldwater species that is seldom found at temperatures over 12°C but survives and even blooms in Japanese and Spanish Mediterranean waters at temperatures over 20°C. The salinity range is 20–37‰. The optimal experimental conditions for growth are pH 8.5, salinity 30–35‰, temperature 20–25°C. Native range: California, USA. Known introduced range: many Pacific coasts, Australia, South Africa, Europe. There is a rapid expansion and increasing abundance in the Mediterranean Sea.

A. catenella was probably introduced with ballast water discharges. Its resting cells were found in sediment samples from ballast tanks. It is responsible for creating "red tides", it is a known paralytic shellfish poisoning (PSP) toxins-producing species. The toxins can affect humans, other mammals, fish and birds. Recently, it has been shown that PSP toxins can also be found in crabs and lobsters. This species is responsible for numerous human illnesses and several deaths after consumption of infected shellfish. Its toxicity may cause considerable economic damage to aquaculture and the shellfish harvest. The PSP toxins affect living sea resources that feed by filtering plankton (e.g., gastropods, bivalves). Avoiding ballast water uptake during the red tides and exchange ballast water mid ocean. Chemical control: chemical treatment in ship ballast tanks can work.

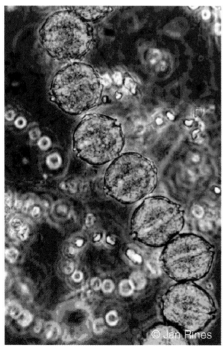

Balanus improvisus Darwin, bay barnacle (Balanidae, Crustacea)

Sergej Olenin and Irina Olenina

Sessile crustacean with a white, conical shell (up to 17 mm in diameter and 10 mm in height) and a diamond-shaped slightly toothed opening. It occurs in marine and brackish environments and filter feeds on detritus and phytoplankton. Planktonic larvae are dispersed with water currents. Acorn barnacles are hermaphrodites and are also able to be self-fertilisers. Fertilised eggs develop within the ovisac, present in the mantle cavity. Free swimming nauplial larvae hatch from the eggs. Other nauplial instars occur before the transformation to cyprid larvae. The cyprid larvae settle on hard substrate and transform into barnacles. Settlement is influenced by light (larvae are positively phototactic), flow velocity and quality of the substratum.

Native habitat (EUNIS code): A1: Littoral rock and other hard substrata, A3: Infralittoral rock and other hard substrata, A4: Circalittoral rock and other hard substrata, X1: Estuaries, X3: Brackish coastal lagoons, adult forms, A7: Pelagic water column, larvae. Inhabits littoral and sublittoral stony and rocky bottoms, often can be found on ship hulls, hydro-technical constructions, on sluices, sometimes attaches to other animals. The same habitats are occupied in the invaded range. It prefers brackish water bays, estuaries and various marine habitats with hard substrata (stones, rocky shores and man-made constructions such as breakwaters and ships) from boreal to tropical waters. Depth range 0–90 m, temperature range 0–30°C; optimum conditions for free swimming larvae at 14°C. Does not reproduce in fresh water, best activity at 6–30‰, maximal larval settlement in mid-salinities. Lives up to the splash zone, does not tolerate desiccation. Native range: Atlantic American coast. Introduced to the Atlantic coast of Europe, Baltic, Black and Caspian Seas, Africa, Japan, Australia, New Zealand and the Pacific coast of the Americas. Spreading.

Transported as a fouling organism on ship's hulls, or as planktonic larvae in ballast water; also common as an epibiont on imported oysters. It can dominate the community by competing for space and food. Changes the habitat, by fouling blue mussels and oysters. It causes fouling of water intake pipes and heat exchangers, underwater constructions and ships' hulls. Mid ocean exchange of ballast water is necessary to get rid of planktonic larvae. It is important to control mussels and oyster export, as well as boats and other movable equipment. Mechanical control: they can be physically removed from ship hulls, by high temperatures (tolerates 36°C for 30 h) and by oxygen deficiency (e.g., one of parallel pipelines closed for 3–4 weeks). Antifouling paints and chlorine treatment of water intake pipelines during the most intensive settling period (0.1–0.5 mg/L) can be efficient.

Bonnemaisonia hamifera Hariot (Bonnemaisoniaceae, Rhodophyta)

Stephan Gollasch

These red macroalgae occur in marine waters. Gametophytes disperse with water currents, adult plants with drift, clinging to floating objects. Its life-history involves an alternation between gametophytes and tetrasporophytes. Gametophytes occur in spring and are up to 350 mm in length, tetrasporophytes occur year-round, but are most common in October–March. Asexual reproduction occurs with stem fragmentation.

Native habitat (EUNIS code): A3: Sublittoral rock and other hard substrata. Adults are exclusively epiphytic. Tetrasporophytes are occasionally found on hard substrates including man-made structures up to 8 m water depth. The same habitats are occupied in the invaded range. Tetrasporophytes and gametophytes from Ireland survived and grew from -1°C to 29°C. The maximum growth of tetrasporophytes was between 15°C and 25°C. Gametophytes showed optimum growth at 15°C. For gametophyte production the tetrasporangia require short daylengths and water temperatures above 11°C. The production of male plants occurs at lower temperatures compared to temperatures at which females develop. In its native range the reproductive cycle is synchronised with water temperatures permitting timely overlap of sexes. The conditions for production of tetrasporophytes are within 15–20°C and long day-lengths (16:8 h) (L:D). Tolerates salinities 11–35‰. Native range: NW Pacific (Japan). Introduced to European seas. Actual trend is stable.

It is assumed that the alga was introduced unintentionally with shellfish or in the hull fouling of vessels. Secondary spread occurs by drift with water currents or attachment to floating objects (hooks enable entanglement). It may become the dominant alga in certain regions competing with other algae and seagrasses. In S Norway, along the Danish North Sea, the Kattegat coasts of Denmark and Sweden, Helgoland (Germany) and the UK the algae is abundant in some sites, and in certain regions of Norway it is the most commonly found red-algae. No prevention or control methods known.

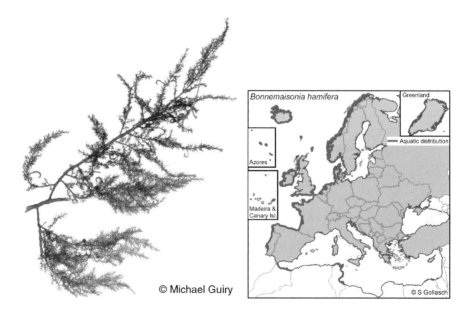

Brachidontes pharaonis (Fischer) (Mytilidae, Mollusca)

Bella S. Galil

A small gregarious intertidal bivalve with a 40 mm shell, externally dark brown-black and internally tinged violet-black. Equivalve, inequilateral, attached to substrate by stout byssus. Outline mussel-like with terminal umbones but variable in shape and in its height/length ratio; sometimes greatly expanded posteriorly, sometimes arcuate; occasionally subcylindrical with beaks not quite terminal. Shell sculpture consists of numerous fine radial bifurcating ribs, coarser posteriorly and margin crenulate. The hinge has dysodont teeth. Dispersal by planktonic larvae. Year-round reproduction.

Native habitat (EUNIS code): A1: Littoral rock and other hard substrata. Marine intertidal hard. Habitat occupied in invaded range: A1, and marine midlittoral on rocky platforms and man-made structures. In the Mediterranean, adult snails show large temperature tolerances (9–31°C), and occur at salinities of 35–53‰. However, lower winter temperatures limit their physiological activity. Native range: Indian Ocean, Red Sea. Known introduced range: East and Central Mediterranean areas. Constitutes large, stable populations.

The Levant Sea populations originated from propagules that entered the Mediterranean through the Suez Canal. Records in the central Mediterranean are likely to be due to ship transport. It locally displaces the native mytilid *Mytilaster minimus*. In the early 1970s it was much rarer than the native *Mytilaster*, that formed dense '*Mytilaster* beds' on intertidal rocky ledges along the Israeli coastline, with up to 26 specimens/cm^2. By the end of the 1980s, it was determined that *Brachidontes* interferes with recruitment of *Mytilaster*, and detrimentally affects its survival and growth. A survey conducted in some of the same sites in the late 1990s have shown a rapid shift in dominance, with some *Brachidontes* populations up to 300 specimens/100 cm^2, while *M. minimus* is only rarely encountered. Economic impact as a fouling organism. No prevention or control method known.

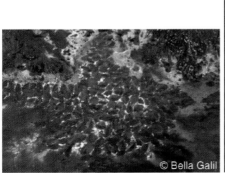

Caulerpa racemosa cylindracea (Sond.) Verlaque et al., grape alga (Caulerpaceae, Chlorophyta)

Bella S. Galil

A green macroalgae with slender thallus, lacking large rhizoidal pillars, basal part of the upright axes slightly inflated immediately above the attachment to the stolon, clavate branchlets, uncrowded and radially to distichously disposed. The rapid dissemination is due to vegetative propagation by random fragmentation, and by specialised propagules formed by detached ramuli. The propagules/fragments may be dispersed by currents or by anthropogenic means (vessels, nets, aquaculture products). Sexual reproduction with a low production of planozygotes. Vegetative reproduction after ramification and development of stolons, or after natural or anthropogenic fragmentation by man, hydrodynamic forces, or marine animals like sea urchins and crabs. The fragmentation may occur in any part of the alga. Another method is the specific fragmentation process involving detachment of ramuli which detach themselves from the fronds and form propagules consisting of fine coenocytic chlorophyllous filaments.

Native habitat (EUNIS code): A3: Sublittoral rock and other hard substrata, A4: Sublittoral sediments. Marine sublittoral hard and soft, polluted and unpolluted, intertidal to 70 m. The same habitats are occupied in the native range. Growth rate is closely correlated to seawater temperature and decreases rapidly in winter. The alga probably survives winter temperatures of the NW Mediterranean Sea (10°C) as zygotes or small fragments. In southern localities no significant winter regression has been observed. Native range: SW coast of W Australia. Introduced into the Mediterranean and Atlantic. Increasing.

Imported by aquarium trade and shipping. It is known to attain total coverage in certain areas within 6 months of entry, its fast growing stolons allowing it to overgrow other macroalgae, mainly turf and encrusting species, and to curtail species number, percentage of cover and diversity of the macroalgal community. This feature is achieved even in highly diverse, native macroalgal assemblages with dense coverage within a few years. The drastic change in the composition of the phytobenthos brought about a modification of the macrobenthos: a proliferation of polychaetes, bivalves and echinoderms and a reduction in the numbers of gastropods and crustaceans. It impacts fisheries by the obstruction of fishing nets by the uprooted alga.

Mechanical control: covering colonies with black PVC plastic; manual removal by SCUBA divers or by suction pump were ineffective. Chemical control: injecting liquid or solid chlorine or coarse sea salt to sealed off areas. Off the Montenegrin coast, copper-sulphate solution and lime were injected under the PVC foil – with no success.

Caulerpa taxifolia (Vahl) Agardh (Caulerpaceae, Chlorophyta)

Bella S. Galil

A green macroalga with upright leaf-like fronds arising from creeping stolons, fronds compressed laterally, small side branchlets constricted at the base, opposite in their attachment to the midrib. Frond diameter 6–8 mm, frond length 3–15 cm in the shallows, 40–60 cm in deeper waters. Fragments are transported by anchors or nets, or with natural currents. Sexual reproduction unknown (only male gametes formed), reproduces vegetatively via fragmentation. During summer (June to September) the thallus of the aquarium strain attains extreme growth rates of up to 32 mm of new stolon/day and a new frond every other day (August), resulting in frond densities of 5,000 fronds/m^2. The alga can survive out of water and under humid conditions for 10 days.

Native habitat (EUNIS code): A2 Littoral sediments, A4: Sublittoral sediments. Marine sublittoral soft. Habitat occupied in invaded range: A2, A3: Sublittoral rock and other hard substrata, A4: Sublittoral sediments. Marine sublittoral, dense coverage between depths of 1–35 m, small patches as far down as 100 m. On a wide variety of substrates, including sandy bottoms, rocky outcrops, mud, sheltered bays, seagrass meadows, and artificial substrates (concrete jetties, metal buoys, rubber bumpers, pipes, ship and nylon ropes). Able to withstand severe nutrient limitation, but also eutrophic or polluted conditions. Native range: tropical coastal areas in the Caribbean, Africa, Indian and Pacific Ocean. In 1984 a patch about 1 m^2 was discovered at the base of the Oceanographic Museum in Monaco, in 1989 it extended to 1 ha, in 1991 it was found near the French and Spanish border. By the end of the 1990s it was dominating large patches along the Mediterranean coastline.

Widely available through aquarium trade, it was unintentionally introduced into the Mediterranean, secondary spread by shipping and currents. Its rapid spread and formation of dense meadows (14,000 blades/m^2) leads to homogenised microhabitats and replacement of native algal species. The alga's dense clumps of rhizomes and stolons form an obstruction to fish feeding on benthic invertebrates. Caulerpenyne, the most potent of the endotoxins protecting this macroalgae, is toxic to molluscs, sea urchins, herbivorous fish, at least during summer and autumn.

Legislation on controlling practices of aquarium trade, shipping, and mariculture is necessary. Manual uprooting, different underwater suction devices, physical control with dry ice, hot water jets and underwater welding devices to boil the plant have been suggested. Except for a few failed eradication attempts made at the onset of the invasion, no control strategy has been established. Intervention utilizing chlorine or Cu and Al salts have been suggested. Studies were conducted on biocontrol using molluscs.

Chattonella cf. *verruculosa* Hara & Chihara (Chattonellaceae, Ochrophyta)

Stephan Gollasch

This small phytoplankton alga is found in brackish and marine waters. There is a controversial discussion on the species determination and its correct taxonomic group. Some experts believe that this species should belong to the dictyophyte group. Further, due to nomenclatural problems a new species name was suggested, i.e. *Verrucophora verruculosa*. After genetic studies it became clear that the Norwegian populations are different from the German and Japanese. Consequently it is much too early to assess whether or not the species was introduced or if it was previously overlooked or misidentified. Dispersal by water currents. The dominant reproductive mode is asexual fission. Cells, 12–30 µm in length with variable shape, are found during the North Sea blooms, and concentrations can range up to 10 million cells/L. Often towards the end of a bloom, this species is capable of producing a resting cyst, formed through a sexual process.

Native habitat (EUNIS code): A7: Pelagic water column. Upper water layers up to 15 m depth in coastal waters and also offshore. The same habitats are occupied in the invaded range. In laboratory experiments it tolerates temperatures of 5–30°C and salinities from 10–35‰. The maximum growth rate has been observed at 15°C and 25‰. During the 1998 bloom temperature ranged from 9–10°C, phosphate 0.3–0.6 µmol/L and nitrate 0.4–4 µmol/L. Native range: Japan. Known introduced range: It was never found forming large blooms in Europe before 1998. During this bloom, algae were found in S Norway, in the Skagerrak region with highest concentrations along the west coast of Sweden. Small amounts of the algae have been observed in the Kattegat. Other blooms occurred in 2000 and 2001. High cell densities were also observed NW of the German island Sylt and around Helgoland. It has been observed in lower concentrations in 2002 and 2003. Future trend unknown.

The species may have been introduced with ballast water. It is a potentially toxic raphidophyte killing fish. The toxin is a fatty acid which affects the gill tissue of fish resulting in the production of mucus which makes the fish suffocate. Economic Impact: harvest loss in fish cultures. In spring 1998 the species killed 350 t of farmed Norwegian salmon. Mechanical control: during harmful algae blooms commercial stocks may be saved by clay spraying, but it needs to be proven that this method is really effective at controlling the algae.

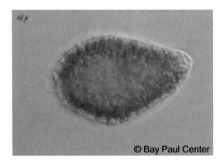

Codium fragile tomentosoides (v. Goor) Silva, green sea fingers (Codiaceae, Chlorophyta)

Bella S. Galil

A large marine green algae attaining 1 m in length, cylindrical branches branching dichotomously, anchored to the substrate in a spongy basal holdfast. Dispersed by advective transport of thallus fragments or entire plants. High growth rates in favourable conditions, sexual reproduction, parthenogenetic and vegetative reproductive capacity, low levels of genetic variation.

Native habitat (EUNIS code): A3: Sublittoral rock and other hard substrata. Marine intertidal and sublittoral hard. Habitat occupied in invaded range: The same as in the native range, natural and artificial substrata such as wharf pilings, jetties, ropes; on exposed and sheltered shores, polluted and unpolluted. Broad physiological tolerance, morphological and functional plasticity, capable of using various nitrogen sources in oligotrophic or eutrophic conditions. Though a warm temperate species with temperature optimum at 24°C, growth and reproduction are still possible at 12°C, adults can survive winter temperature of −2°C. Native range: Japan. Introduced to North American and European Atlantic coasts, the Mediterranean, and the Pacific Ocean from Australia to the Americas. In Europe, it was first reported in 1900 from the Netherlands. It is common but not dominant. Recent analysis of plastid microsatellite markers and DNA sequence data postulate that the Mediterranean and the NE Atlantic populations stem from separate introductions from its native range.

The vector to Europe is unknown; secondary dispersal was by movement of shellfish for mariculture, transport on ship hulls and net fouling. It alters benthic communities and habitats, its dense fronds hinder movement of large invertebrates and fish along the bottom, and increases sedimentation. It causes a nuisance to humans when it is swept ashore and rots. It fouls shellfish beds, smothering mussels and scallops, clogging scallop dredges, and interferes with harvesting. It fouls fishing nets, wharf pilings and jetties. Quarantine measures and public education may be the only way to prevent its spread. Mechanical control: impractical as it readily reproduces from fragments.

Coscinodiscus wailesii (Gran & Angst) (Coscinodiscaeae, Bacillariophyta)

Stephan Gollasch

This very large centric diatom, typically 175–500 μm in diameter, is a primary producer in brackish and marine waters. Dispersal via water currents. Most cells complete their first cell division within 48 h. They then continue to multiply by binary division. In the sea a doubling of biomass has been estimated in 70 h. In blooming conditions a biomass of 1.4 mg C L^{-1} may be reached. The seasonal cycle in the North Sea includes highest abundances in April and September to October. However, annually the abundance oscillates. Resting cells are known to survive dark conditions for long periods (at least 15 months) and may be found in sediment. These resting cells can rapidly rejuvenate under favourable light, temperature and nutrient conditions.

Native habitat (EUNIS code): A7: Pelagic water column. Occupies the upper water layers in coastal waters and also offshore. The same habitats are occupied in the invaded range. It shows a wide tolerance to temperature (0–32°C), salinity (10–35‰) and nutrients. The diatom is native to the N Pacific. Known introduced range: It was first detected in Europe near Plymouth in 1977. It reached the Atlantic coast of France and the Irish Sea by 1978 and Norway by 1979. The first record in the (western) Baltic occurred 1983. Today it is observed from the Atlantic coast of France to Norway.

It was probably introduced with ballast water discharges since resting cells were found in sediment samples from ballast tanks. Another possible introduction vector is shellfish movements. Cells may be carried within the gut/pseudofaeces of shellfish. This non-toxic species is considered as a nuisance as it forms dense blooms which produce copious amounts of mucilage and due to its large size it is inedible to most grazing zooplankton. Blooms may occur with highest abundances in the southern North Sea and Skagerrak area. During blooms the species may form up to 90% of the total algal biomass. Especially in blooming situations benthic organisms are threatened. The damage is caused by the copious mucilage, which can aggregate, sink and cover the seabed. The decay of a bloom is likely to cause anoxic conditions. It may also compete with phytoplankton and macroalgae species for space and nutrients. Impact on fisheries and aquaculture is known as the mucilage causes extensive clogging of fishing nets, cages and other equipment. No prevention or control method known.

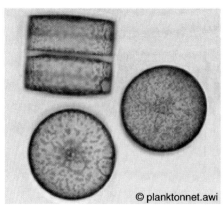
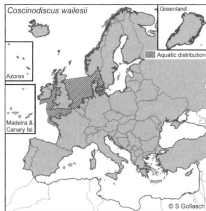

Crassostrea gigas (Thunberg), Pacific (giant) oyster (Ostreidae, Mollusca)

Stephan Gollasch and Dan Minchin

The Pacific oyster is a filter-feeder consuming phytoplankton and detritus in coastal brackish and marine waters. The two elongated valves are variable in shape and size with wild settling individuals cemented by one valve to a firm substrate. Shells with irregular radial folds, 8–31 cm in length. Pelagic larvae are dispersed by water currents. Oysters have separate sexes with external fertilisation. Spawning at temperatures of 18–26°C, salinity range 20–35‰. Each individual releases up to 100 million eggs. Larvae settle after 11–30 days, at a shell length of 290 μm, on hard surfaces to which they become cemented. They mature after 1 year and may live up to 10 years. This oyster is host to a wide range of pests and parasites including *Haplosporidium nelsoni* causing MSX disease. Adults can survive several days in air under damp and cool conditions.

Native habitat (EUNIS codes): A1: Littoral rock and other hard substrata, A3: Sublittoral rock and other hard substrata. Littoral zone, lower intertidal to subtidal. The same habitats are occupied in the invaded range. Littoral zone (3 m depth) on hard substrates in areas with low to moderate wave exposure, may occur to 40 m. Often found cemented to artificial hard substrates in ports and marinas. A euryhaline species (12–42‰, optimum range 20–30‰) and tolerant of a wide temperature (3–35°C) and pH range (6–9). May survive in water with oxygen concentrations down to 3 μg/L. Native range: NW Pacific. Introduced to the European Atlantic coasts, the Mediterranean and the Black sea.

Deliberate introductions of wild stock from Japan to France in the 1960s for cultivation, and from Canada to Britain through quarantine. It has been recorded attached to ships' hulls and in ballast water. It is a widely cultivated oyster in more than 40 countries. Pacific oysters directly introduced from the wild have been a source of several cryptic diseases, including oyster pests. Extensive settlements lead to competition with native biota for both food and space. There are some concerns for mussel cultivation due to heavy settlements and fast growth. In North America it has been found to hybridise with *C. virginica*, however, few survive to metamorphosis. Human impact: regular monitoring of cultivated oysters is necessary as toxic algal blooms can render oysters unmarketable. Uncontrolled harvests of oysters contaminated by microbiota can lead to diseases in humans. Economic impact: in tourist areas wild settlements fouling ladders can lacerate bathers' feet. It is cultivated and fished. This species is responsible for the main biomass of mollusc production in Europe. Biological control: Although specific pests, parasites and diseases have had impacts on production, an effective biological control agent has not been identified.

Crepidula fornicata (Linnaeus), slipper limpet (Calyptraeidae, Mollusca)

Dan Minchin

The slipper limpet is a snail with an asymmetrical shell with an inner shelf. It can attain 5 cm. It is a filter-feeder occurring within sheltered coastal bays and estuaries and sometimes in deeper water. It attaches firmly to objects with its muscular foot. Individuals may attach to each other to form 'chains'. It is dispersed locally as free-swimming larva. Settled stages can be dispersed attached to flotsam or attached to errant crustaceans and molluscs. Females annually produce 200,000 eggs in N Europe. Capsules containing eggs are laid in early summer, larvae hatch after 3 weeks and they have a similar planktotrophic larval duration. Settled individuals crawl seeking out a female to attach to and then act as a male for about 2 years. Further males attach to form chains of up to 12 individuals with the oldest limpet at the base. Solitary individuals may become self-fertile. It can survive aerial exposure under cool damp conditions for several days.

Native habitat (EUNIS code): A3: Sublittoral rock and other hard substrata, A4 Sublittoral sediments. The same habitats are occupied in the invaded range. It occurs mainly in shallow sheltered estuaries, bays and channels from low water to 30 m depth, on muddy and sandy sediments with shells, stones and rocks. It can survive light frosts, temperatures up to 30°C, and in turbid and brackish water. Native range: N American Atlantic coast. Introduced: Europe, N American Pacific coast, Japan, Uruguay. Expanding its range.

Introduced for aquaculture and spread with oyster movements, with ships as hull fouling and on moved floating structures and also by natural dispersal. May occur at densities >1,700/m^2 = 10 kg/m^2 resulting in trophic competition, causing reduced growth of commercial bivalves. Their abundance changes sediments to mud deposits of feces, pseudofeces and shell drifts, thus reducing diversity and abundance of living plants. May also reduce recruitment of some benthic commercial fishes. It has not successfully been developed as a food. Often associated with oyster layings and on scallop beds, slipper limpets need to be removed before marketing. Fouls artificial structures in port regions. Prevention by regulation and regular monitoring of transfers of oysters and mussels, in particular, used for stocking uninfested areas. Mechanical control: cultivation of oysters in bags laid on trestles reduces impacts as small slipper limpets often become crushed. Slipper limpets on open grounds compete with oysters for food unless dredged. They should be removed from bivalves used for stocking bays outside of their current range, but it is best not to make any such transfers. If removed from ships or other floating structures in dry-dock, all removed fouling biota should be destroyed and not returned to the water.

Ensis americanus (Gould), American Jack knife clam (Solenidae, Mollusca)

Irina Ovcharenko, Sergej Olenin, and Stephan Gollasch

The red-brown bivalves (16–17 cm in length) occur in muddy or fine sand, marine and brackish waters. It prefers the lower zone of the intertidal areas, where they burrow into the sediment and filter-feed on algae. The free-swimming larvae are distributed by currents, secondary dispersal of post-larval stages occurs in summer. It is also able to swim or drift by use of byssus threads. Juveniles settle on clean sands in the lower zone of the intertidal areas, where they burrow in the sediment. The survival of recruits is limited to areas below the level of mean low tides. Migrating juveniles are mostly 1–3 mm long. They reach about 6 cm in length after the first winter. The life-span is up to 5 years.

Native habitat (EUNIS code): A2: Littoral sediment, A5: Sublittoral sediment, brackish littoral soft sediment, brackish sublittoral soft sediment (adults); A7: Marine pelagic water column and brackish pelagic water column (pelagic larvae). The same habitats are occupied in the invaded range, also littoral and sublittoral soft sediment bottoms in fully marine and brackish environments. It is found in marine and estuarine areas and tolerates relatively low salinities but low winter temperatures limits the development. It prefers unstable clean fine sand with small amounts of silt and burrows to 3–18 cm depth. Native to the Atlantic coast of North America. Introduced to the North Sea and surrounding waters. Increasing.

The spread is associated with dispersal of long-lasting pelagic larvae, transported in ballast waters of ships and by water currents. It is also able to swim or use byssus threads for drifting and rapidly extends its distribution in Europe. Although dense populations may change the community structure of the benthic fauna or compete for space and food, there were no significant interactions with resident species along the Island of Sylt (North Sea). Dense populations may have an impact on the sediment structure by their burying activities. In dense beds of razor clams, fine sediment particles accumulated which may have altered abundances of polychaetes. In spite of high annual variability, *E. americanus* has become a prominent component of the macrobenthos in shallow subtidal sands of the North Sea. The sharp shells can cause deep cuts on bathers feet. Such injuries can also occur when stepping on native species, but *E. americanus* lives at much shallower depths than native species and consequently injuries are more likely. The species can damage trawls and other fishing nets on the seabed, causing economic losses to fisheries. Mechanical control: mid-ocean exchange or filtration and/or preventive disinfection of ballast water should be used to reduce transfer of planktonic larvae. Chemical control: chemical treatment in ship ballast tanks.

Ficopomatus enigmaticus (Fauvel), tube worm (Serpulidae, Annelida)

Dan Minchin

This is a small worm that forms concretions with their calcareous interlacing tubes. It feeds on seston in brackish to hypersaline sheltered environments, estuaries and lagoons. It is dispersed as larvae or attached to floating substrata. Populations are made up of dioecious individuals and some hermaphrodites. As it requires 18°C to reproduce, warm areas will have prolonged periods of reproduction. Spawning mainly takes place July–September. Larvae are brooded before release to a planktonic stage for some months. In N Europe it produces normally one generation. It can survive in limey tube cases for some hours out of water.

Native habitat (EUNIS code): A1: Littoral rock and other hard substrata, A3: Sublittoral rock and other hard substrata. Normally low velocity sheltered environments with varying salinity. The same habitats are occupied in the invaded range, also lagoons, estuaries and docks from mean water neap tides to 10 m depth, normally abundant near to the surface when permanently covered. Tolerates 2–56‰ in temperate to sub-tropical low current, turbid water with high nutrients. Native range unknown. Once thought of as Indonesia/India; however specimens there are recognised as *F. ushakovi*. May be from E South America but considered to be an introduction there. Introduced to America, S Africa, Australasia, E Asia, Europe, Mediterranean, Black and Caspian Seas. May extend further with warming seas.

Probably introduced as hull fouling or as larvae in ballast water. Can form extensive reefs of up to 7 m in diameter, but usually 3–20 cm in temperate areas and 1.5 m under mixohaline and hyperhaline conditions in warm climates. Individual worms can grow at 1.5–2 cm per month and collectively produce up to 13 kg of limey tubes in 3 months. Established reefs may provide refuge for invertebrates including snails and crabs that may have an impact on native species communities. Their dense tube colonies attach to abstraction pipes, reducing water flow and causing blockages. They also foul surfaces in aquaculture ponds, ports and docks, which requiring cleaning and maintenance. It fouls the hulls of leisure craft and floating structures in lagoons and docks. Areas with thermal effluents may develop large colonies. Controls on the movement of aquaculture equipment and of hull-fouled craft may reduce the rate of spread. Mechanical control: this species is difficult to manage. By removing tubes large numbers of embryos can be released that may subsequently colonise. In dry-docks, and areas where boats are serviced, fouling biota should be removed and destroyed. Flushing of water in docks during peak larval abundance may reduce settlements. Chemical control: antifouling paints reduce fouling on ship and boat hulls.

Fistularia commersonii Rüppell, blue-spotted cornetfish (Fistularidae, Osteichthyes)

Bella S. Galil

A marine, mainly piscivorous fish, with a grey to olive-green body, commonly 20–100 cm long (max. 150 cm). The body is extremely elongated, the head is more than one third of the body length, snout tubular, ending in small mouth. Dorsal and anal fins are posterior in position, opposite to each other. The caudal fin is forked, with two very elongated and filamented middle rays. The skin is smooth, without bony plates along the midline of the back. Dispersal as planktonic eggs and larvae, adults by swimming. Off California spawning occurs in June–August.

Native habitat (EUNIS code): A4: Sublittoral sediments. Marine sublittoral. Adults inhabit reef habitats to a depth of at least 128 m, but also found in sandy bottoms adjacent to reef areas, and seagrass beds. Habitat occupied in invaded range: A4, soft, sandy bottoms and seagrass meadows. Temperature tolerance 15–30°C. Native range: Indo-Pacific: Red Sea, E Africa to Easter Island, Japan, Australia, New Zealand. Eastern Central Pacific: Mexico to Panama. Known introduced range: Mediterranean Sea. Spreading to the West and meanwhile already established in the Tyrrhenian and Ligurian Seas.

Entered the Mediterranean through the Suez Canal. Possible impact from competition for food with native piscivorous fish and it preys on native commercially important fish. It is of minor commercial importance. No specific prevention or control method known.

Halophila stipulacea (Forssk.) Ascherson, Halophila seagrass (Hydrocharitaceae, Magnoliophyta)

Bella S. Galil

A euryhaline marine angiosperm (seagrass). Plants are dioecious with male and female flowers produced at each leaf node. Rhizomes are creeping, branched and fleshy, and roots appear solitary at each node of the rhizome, unbranched and thick with dense soft root hairs. Pairs of leaves are distributed on petioles along a rhizome, rooted in the sand. Leaves 3–8 mm wide, obovate, not narrowing at base, thin and hairy; margin spinulose, petiole 3–15 mm long. Dispersal via current, vessel-borne plant fragments, fruits. Flowers are solitary, axillary covered by spathes. It disperses strings of four reniform trinucleate pollen grains contained in a mucilaginous moniliform tube. In the Mediterranean the main flowering season is July–August, fruits ripening in September.

Native habitat (EUNIS code): A4: Sublittoral sediments. Marine sublittoral soft, grows in sheltered localities as isolated patches, on muddy bottom and coral rubble. Habitat occupied in invaded range: A4, marine, sandy and muddy bottoms, intertidal to 65 m, but mainly at depth of 30–45 m, mostly in harbours or in their vicinity. Native range: W Indian Ocean, Red Sea and Persian Gulf. Introduced into the E Mediterranean Sea to Italy. It forms extensive and stable meadows, characterised by a high density of 20,000 shoots/m^2 and an abundant and diversified fauna.

It entered the Mediterranean through the Suez Canal. The Aegean populations may have originated from fragments carried by Greek fishing boats, and secondary spread is likely due to ship transport. It out-competes the native Mediterranean seagrasses and can induce changes in the sublittoral communities. A comparison between the associated algal assemblages of an invaded meadow and two contiguous meadows dominated by *Posidonia oceanica* and *Cymodicea nodosa* revealed significant differences in species composition. No prevention or control method known.

Marenzelleria neglecta Mesnil, red-gilled mud worm (Spionidae, Annelida)

Sergej Olenin

The red-gilled mud worms (up to 157 mm in length) occur in marine and brackish waters. They inhabit vertical mucus lined burrows (up to 40 cm in depth), feeding on sediment particles, meiobenthic and planktonic organisms. Planktonic larvae disperse with water currents. Nocturnal swimming of adult worms with gametes (most probably associated with reproduction) may also facilitate dispersal. Fecundity of animals depends on salinity, temperature, age and body size. Development of gametes starts in May, individuals reach maturity in September after 20 weeks. Animals spawn in autumn and the pelagic larvae can be found September to November, but may also occur up to March. Larval development largely depends on water temperature and lasts about 4–12 weeks.

Native habitat (EUNIS code): A2: Littoral sediment, A5: Sublittoral sediment, X1: Estuaries, X3: Brackish coastal lagoons, for adults. A7: Pelagic water column, for larvae. *M. neglecta* is an estuarine species which inhabits sandy and muddy sediments. The same habitats are occupied in the invaded range. Occurs on gravel, sandy and muddy bottoms 1–90 m depth. It is highly tolerant of very low salinities (<1‰) and temporal fluctuations. However, successful larval development from egg to juvenile is not possible below 5‰, but colonisation of oligohaline regions can be accomplished by larvae with more than four setigers or by swimming juveniles. It is tolerant to short term oxygen deficiency and may survive hydrogen sulphide exposure. Native in the Atlantic coast of North America. Introduced to the Baltic Sea were it gradually invaded nearly all NW European intertidal and estuarine areas. Expanding its range.

The spread is mostly associated with dispersal and development of planktonic larvae, which can be transported by currents and in ballast water of ships. Also, adult worms with gametes may enter ballast water during their nocturnal swimming. The red-gilled mud worm competes with native benthic macrofauna for food and space. Being numerically dominant, it can change the structure of a native benthic community. Burrowing activity of this worm has a high impact on fluid-exchange rates between bottom water and sediments, especially in muddy sediments. The burrow walls make good substrates for aerobic degradation of organic matter. May compete for food with aquaculture organisms. Mechanical control: to remove planktonic larvae, exchange of ballast water should take place in the mid ocean. Due to presence of bottom sediments in ballast tanks, removal of adult worms is complicated. Chemical control: chemical treatment in ship ballast tanks has been shown to be effective.

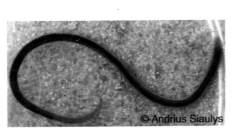

Marsupenaeus japonicus (Bate), Kuruma prawn (Penaeidae, Crustacea)

Bella S. Galil

A nocturnal prawn (body length males 17 cm, females 27 cm), adults buried in substrate during daytime. The body is pale bearing uninterrupted brown transverse bands; walking legs and pleopods are pale yellow proximally and blue distally; uropods are distally striped with yellow and blue, setal fringe red. The rostrum has 7–11 teeth on upper margin and a single tooth on the lower margin. The carapace bears gastrofrontal, hepatic crests; the post-rostral crest is medially grooved. Abdominal segments are 4–6 keeled; sixth abdominal segments bear three lateral scars. The telson has a pointed tip and three pairs of movable distal spines. First three pairs of walking legs are chelate. Dispersal by long-living planktonic larvae. Minimum size at maturity is 13–14 mm. However, after a female reaches maturity it is able to produce mature ova regardless of size. Repeated copulations follow moulting in females. Copulation does not directly accelerate maturation. The proportion of inseminated females increased with increasing body size up to 17 cm. Multiple spawnings occur between April–November. Recruitment takes place between August and March.

Native habitat (EUNIS code): A2: Littoral sediments, A4: Sublittoral sediments. Marine sublittoral soft. Habitat occupied in invaded range: The same as in the native range, on sandy bottoms, up to 90 m, usually less than 50 m. The larvae require water temperature above 24°C; the rate of larval growth increases with temperature to 32°C, the optimal range being 28–30°C. Salinities below 27‰ and above 35‰ inhibit hatching and induce high mortalities. Adults are poor osmoregulators compared with the young, preferring salinity >35‰. Native range: Indo-Pacific: Red Sea and E Africa to Fiji, though recent molecular studies suggest that the western Indian Ocean population differs from the rest. Introduced into the Mediterranean Sea where it is very abundant along the Levantine coast.

It entered the Levant through the Suez Canal. It has been released by mariculture in Italy, France, Greece, and Marmara Sea. Appears to have outcompeted the native penaeid prawn *Melicertus kerathurus* which has almost disappeared. Of major commercial importance, a highly prized species considered a boon to the Levantine fisheries. It composes much of the prawn catch off the Mediterranean coast of Egypt and in the Nile delta lagoons. Off the Israeli coast a small fleet of coastal "mini" trawlers has specialised in shrimping, bringing in a quarter of the total trawl catch volume and a third of the trawl gross income. It is also of major commercial importance in the Bay of Iskenderun, Turkey. No specific prevention or control method known.

Musculista senhousia (Benson in Cantor), Asian date mussel (Mytilidae, Mollusca)

Bella S. Galil

A small (10–25 x 12 mm) intertidal bivalve with thin shell, equivalve, oval, elongate with sculpture of radiating lines posteriorly. Outline modioliform; umbones subterminal; ligament and dorsal margins not continuous, slightly angled; anterior end rounded. Ventral margin slightly concave. Shell pale olive-green with irregular brownish-purple markings. Larvae spend 3–6 weeks as plankton, dispersed by water currents. A broadcast spawner, with fertilisation occurring in the water column, spawning in the Mediterranean in September–November. Larvae settle after 14–55 days on hard surfaces to which they become cemented. They mature in about 9 months and can live for 2 years. Adults can survive several days out of water.

Native habitat (EUNIS codes): A1: Littoral rock and other hard substrata, A3: Sublittoral rock and other hard substrata. Littoral zone, intertidal to subtidal. The same habitats are occupied in the invaded range, also on hard and soft substrates in the intertidal and shallow subtidal zones to 20 m depth. They may settle on hard surfaces, but mostly settle gregariously on soft substrates, burrowing until only the hind part of their shell protrudes, and then secrete fibrous threads that attach to sediment particles to form a kind of nest or bag around them. Tolerant of low salinity and low oxygen levels. Euryhaline (17–37‰, optimum range 20–25‰) and tolerant of a wide range of temperatures (5–30°C). Native range: W Pacific, from Siberia and Japan to Singapore. Introduced into the Mediterranean Sea, American Pacific coast, Australia and New Zealand. Spreading with shellfish culture and shipping.

It was probably introduced with Japanese oysters *Crassostrea gigas*, since it has been collected in lagoons that are used for shellfish cultivation. Other possible mechanisms include transport in ships' seawater systems or in ballast water, or as hull fouling. The presence of this species at high densities increases the abundance of detritovorous amphipods, tanaids, small snails and polychaete worms, but the abundance of suspension-feeding and filter-feeding organisms may decline. When in large enough densities, it shifts the community from suspension-feeding to primarily deposit-feeding. It has been blamed for smothering and killing commercially important bivalves, including the manila clam *Ruditapes philippinarum*. Mechanical control: dredging as an eradication or control method is not practical as it will cause the mat to fragment, and individuals may be swept away and settle to form new mats. Scraping *M. senhousia* foulings from man-made substrates is similarly not efficacious.

Odontella sinensis (Greville) Grunow, Chinese diatom (Eupoodiscaeae, Bacillariophyta)

Stephan Gollasch

This phytoplankton species is known to form mass developments (plankton blooms). Dispersed by water currents. This dioecious diatom occurs solitarily or in pairs, reproduces year-round and shows a tendency to bloom as late as November/December. In European waters, the species has the highest abundance from fall to spring. The life cycle of the diatom includes asexual and occasionally sexual reproduction. Asexual reproduction occurs as cell division and sexual reproduction results from a fusion of the gamete nucleii. Culture growth results in approximately 1–2 cell divisions per day. Auxospore formation may take place.

Native habitat (EUNIS code): A7: Pelagic water column. Upper water layers in coastal waters and offshore. The same habitats are occupied in the invaded range. It is common at water temperatures of 2–12°C and salinities from 27–35‰. However, temperatures of 1–27°C and salinities of 2–35‰ may be tolerated. Native range: Red Sea, Indian Ocean to the East Chinese Sea. Introduced range: North Sea and adjacent waters. The population is stable.

Living cells are frequently found in ballast water samples. Ecosystem Impact: Even during blooms, other native species occur in higher numbers. However, long-term investigations have shown that the population growth of other phytoplankton species was depressed by high populations of this species. No specific prevention or control method known.

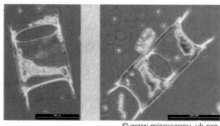

© www.microscopy-uk.org

Paralithodes camtschaticus (Tilesius), red king crab (Lithodidae, Crustaceae)

Stephan Gollasch

This very large crab is an omnivorous predator, it grows up to >220 mm carapace length (CL) and has a leg span up to 1.4 m. Weight > 10 kg. Adult crabs are fast migrating (3–13 km daily, 426 km during a year). Larval settlement occurs in shallower waters (<20 m). Once spawning is completed the crabs undertake a feeding migration to deeper waters (300 m). Individuals <20 mm CL remain solitary and may be found on hard substrates. In the second year (20–25 mm CL) grouping behaviour is seen. Sexually immature crabs (<120 mm CL) generally remain in shallow water at 20–50 m depth. Two-year-old crabs migrate to deeper waters. Their life span may cover 20 years.

Native habitat (EUNIS codes): A3: Sublittoral rock and other hard substrata, A4: Sublittoral sediments. Hard and soft bottom habitats. The same habitats are occupied in the invaded range. The crab tolerates water temperatures of −1.7–11°C. As soon as the temperature decreases the crabs migrate to deeper waters where they over-winter. In Alaska crabs are found at salinities from 22–34‰. It is native to the Okhotsk, Japan and Bering Sea and N Pacific. Its eastern distribution limits in the Barents Sea is near Cape Kanin, east of the Kola Peninsula. From 1992 the crab became abundant in NE Norwegian waters and is spreading southwards.

Russian scientists intentionally introduced king crabs to the southern Russian Barents Sea 1961–1969. Ten years later a reproductive population became established. It continues to spread in Russia and N Norway. Especially during mass developments, the crab has an enormous predatory impact on local species. The pelagic larvae consume both phytoplankton and zooplankton and once settled feed on hydroids. Adult crabs prey upon a large variety of species. Crabs in the by-catch of local fishermen cause concern as they damage fishing gear in gillnet, longline and trap fisheries.

Avoid moving the species to areas where it does not occur to create a fishing resource. Mechanical control: today, selective crab fishing has been established in Norway. The population stock is managed jointly by Russian and Norwegian authorities, thereby creating a valuable fishing business and a total annual catch limit was set to avoid over-fishing. In 2002 newly negotiated fishery regulations were introduced and in Norway the king crab fishery became an ordinary commercial fishery. East of the 26° E line the fishery is regulated. and it is hoped that this valuable resource can be maintained and at the same time avoid further spread of the crab southwards. However, no decision has yet been made on how to prevent the species from migrating south and westwards.

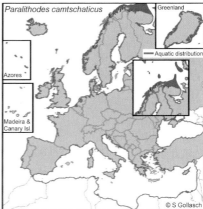

Percnon gibbesi (Milne-Edwards), sally lightfoot crab (Grapsidae, Crustacea)

Bella S. Galil

An agile, aposomatic mimetic crab. The carapace is brownish, with golden yellow rings on joints of walking legs. Carapace length to 3 cm, disc-like, with tridentate front and antero-lateral margins with four acute teeth. All walking legs have a row of spines on anterior margin of merus. Chelipeds have spinose merus and carpus. The palm has a small pilose area proximally on the inner surface and a pilose groove proximally on the upper surface. The crab is active during daytime with feeding peaks at sunset. It is strictly herbivorous, using its small claws to pick off the fine film of algal growth. Dispersal by long-living planktonic larvae.

Native habitat (EUNIS code): A1: Littoral rock and other hard substrata, A3: Sublittoral rock and other hard substrata. Rocky infralittoral to 30 m depth. The same habitats are occupied in the invaded range, intertidal and subtidal rocky bottoms to 8 m. It lives under boulders, or in narrow crevices, where it seeks shelter when threatened. Native range: West coast of America (California to Chile), East coast of America (Florida to Brazil), E Atlantic (Madeira to Gulf of Guinea). Known introduced range: Mediterranean Sea from the Balearic Islands to Turkey. It forms thriving high-density populations in an amazingly short space of time.

Vessel-mediated transport (in ballast or fouling), and/or larval transport by currents. Ecosystem impact: its habitat overlaps with the native grapsid *Pachygrapsus marmoratus* and the pebble crab *Eriphia verrucosa*. Exclusion of native crabs may occur in some areas. Prevention by controlling vessel fouling.

© Baki Yokes

© B Galil

Pinctada radiata (Leach), gulf pearl oyster (Pteriidae, Mollusca)

Bella S. Galil

A large bivalve, length commonly 50–65 mm, up to 106 mm. The shell is compressed, inequivalve with the outline almost quadrate. The shell is brownish with shades of red, inner surface pearly. The dorsal margin is longer than the body of shell and the posterior margin slightly concave. The beaks are pointing anteriorly. Shell sculpture consists of concentric lamellae often with rows of flattened spines, margins spinous. Hinge line straight. Pelagic larvae are dispersed by water currents. Protandric hermaphrodite species with sex inversion occurring in shells 32–57 mm. Gonad maturity is controlled by temperature. In the Mediterranean, gonad activity is nearly year-round, spawning occurs mainly in summer and early autumn.

Native (EUNIS code): A1: Littoral rock and other hard substrata, A3: Sublittoral rock and other hard substrata. Littoral zone, subtidal. The same habitats are occupied in the invaded range. It may occur in 0.5–150 m depth. Known to foul artificial surfaces. Tolerant of a wide temperature range (13–30°C). Native range: Indo Pacific. Introduced into the Mediterranean Sea. Stable in the southern Mediterranean, increasing northwards.

It entered the Mediterranean through the Suez Canal. The populations in France and N Adriatic Sea are due to ship transport. It was intentionally introduced to Greece for mariculture. It was recorded as an epibiont on a loggerhead turtle off Lampedusa Island, Italy. Ecosystem Impact: It is considered a habitat-modifying, gregarious bivalve capable of impacting native fauna by forming oyster banks. It fouls mussel lines and commercial shellfish collectors. No specific prevention or control method known.

Portunus pelagicus (Linnaeus), blue swimming crab (Portunidae, Crustacea)

Bella S. Galil

A medium-sized marine crab (carapace length of males 7 cm, females 6.5 cm), carapace greenish-brown with irregular pale mottling with dark edges, chelipeds are purplish, mottled and fingers blue. Carapace broad, with transverse granulate lines. The front has four acute lobes and the antero-lateral margin bears nine triangular teeth, the last tooth is the largest, projecting laterally. Chelipeds are long, massive, spinous and ridged. An active swimmer, but during inactive periods buried in sediment. Nocturnal.

Dispersal through planktonic larvae. Copulation occurs year-round, sperm is stored in the spermathecae for a year or until eggs mature and are fertilised internally. Females migrate to shallow waters to spawn and return to deeper waters to hatch. The larvae undertake diel vertical migrations and travel inshore for settlement. The number of egg batches (1–3) and the batch fecundity are correlated with size. The estimated number of eggs produced during a spawning season ranges from 78,000 in small crabs (carapace width 80 mm) to 1 million in large crabs (carapace width 180 mm). In tropical regions ovigerous females occur year round, in subtropical regions spawning occurs during summer months.

Native habitat (EUNIS code): A4: Sublittoral sediments. Adults occur on sand, sandy-mud, from intertidal to 65 m, near reefs, mangroves, seagrass and algal beds. Juveniles commonly occur in the intertidal and in estuaries. Habitat occupied in invaded range: A4, sandy or muddy substrate, to depth of 55 m. Euryhaline, 30–40‰; temperature tolerance range from 15–25°C. Native range: Indo-Pacific Ocean, Red Sea to Tahiti. Introduced into the Mediterranean Sea.

Entered the Mediterranean through the Suez Canal. Ecosystem impact: competes for food with the native biota. Commercially important in nearshore fisheries in the Levantine Basin. No specific prevention or control method known.

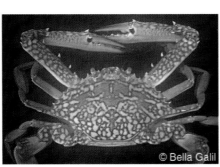

Rapana venosa (Valenciennes), veined rapa whelk (Muricidae, Mollusca)

Stephan Gollasch

A large marine and brackish water Asian gastropod with voracious predatory behaviour. Shell lengths from 67–137 mm, wet weight up to 554 g. Dispersal as larvae with water currents. It is a dioecious gastropod with separate sexes. Mating occurs during winter and spring. Masses of egg cases are laid April to July. The egg cases are attached to hard substrates and may contain 1,000 developing embryos. One female can lay multiple egg cases throughout summer. Upon hatching the larvae are planktotrophic.

Native habitat (EUNIS codes): A3: Sublittoral rock and other hard substrata, A4: Sublittoral sediments. Hard and soft bottom habitats. The same habitats are occupied in the invaded range. In its native range, adult snails show large temperature tolerances (4–27°C). Surface freezing in winter is tolerated by migration into deeper waters. In the Black Sea it occurs at salinities of 25–32‰ and at lower salinities in the Sea of Azov. In the Black Sea the species is tolerant to water pollution and low oxygen conditions. Native range: Sea of Japan, Yellow Sea, Bohai Sea, and the E China Sea to Taiwan. Introduced to the Black Sea, the Mediterranean Sea and the French Atlantic coast. Spreading.

Although ballast water and hull fouling is a possibility, the most likely vector is oyster shipments. They were used to "ballast" clam culture bags of *Tapes philippinarum*, which were transferred from the Adriatic, where the whelk was established, to France. The large population in the N Adriatic Sea is generally considered to have had no major detrimental effect. However, areas with substantial oyster cultures may be at risk once the gastropod becomes established and occurs in high densities. In the North Sea the whelk may become a competitor of the native whelk *Buccinum undatum*. The ecological impacts in the Black Sea have been severe. *R. venosa* predation was identified as the key reason for the decline of the commercially fished *Mytilus galloprovincialis* population in Bulgarian waters, the Kerch Strait and the Caucasian shelf.

Avoid transfers and release of living organisms. The species is suspected to be introduced with oyster shipments and therefore oysters should be cleaned properly and inspected before laying. Substantial fishery for the species exists in the Black Sea, which may contribute to its control. Unfortunately, all control strategies show weaknesses. Egg cases, although visible and often concentrated, may spread over large areas making them difficult to be manually collected. Adults may be collected with dredges or pots/traps. It is recommended to catch as many as possible to minimise the risk of the species to become established.

Rhopilema nomadica Galil, nomad jellyfish (Rhizostomatidae, Cnidaria)

Bella S. Galil

A neritic epipelagic, swarming jellyfish, planktotroph. The bell is up to 90 cm in diameter, commonly 40–60 cm. The body is light blue, bell rounded, with blunt tuberculation on the exumbrella. The mouth arms end in vermicular filaments. Dispersal as planktonic larvae (planulae). It has a two-stage life cycle consisting of the conspicuous, large sexually reproducing swimming medusa stage and a benthic polyp stage (scyphistoma) that, because of its small size (usually <2 mm), remains cryptic. These polyps reproduce asexually by budding. The medusa stage begins when polyps undergo budding of transverse fissions that form medusae; each polyp may produce several medusae. Depending upon food availability and other environmental variables, the scyphistomas form large numbers of pelagic medusae. Spawning occurs June to August.

Native habitat (EUNIS code): A7: Pelagic water column, marine neritic epipelagic. The same habitats are occupied in the invaded range. The sexually reproducing swimming scyphomedusae appear when the water temperature exceeds 24°C. Native range: E Africa, Red Sea. Introduced to the Mediterranean Sea in the late 1970s and since the mid-1980s huge swarms have appeared along the Levantine coast.

Entered the Mediterranean through the Suez Canal, may spread autochthonically as current-borne adults. Massive swarms stretching more than 100 km appear along the Levant coast every summer. These swarms of voracious planktotrophs must play havoc with the limited resources of the oligotrophic Mediterranean Sea. Local municipalities in Israel have reported a decrease in holiday-makers frequenting the beaches because of the public's concern over the painful stings inflicted by the jellyfish. When the shoals draw nearer shore, they adversely affect tourism, fisheries and coastal installations. Coastal trawling and purse-seine fishing are disrupted for the duration of the swarming due to net clogging and inability to sort yield, due to the overwhelming presence of these venomous medusas in the nets. Jellyfish-blocked water intake pipes pose a threat to cooling systems of port-bound vessels and coastal power plants. Juveniles of a commercially important Red Sea carangid, *Alepes djedaba*, find shelter among the jellyfish's tentacles. No specific prevention or control method known.

Saurida undosquamis (Richardsn), brushtooth lizardfish (Synodontidae, Osteichthyes)

Bella S. Galil

A marine, mainly piscivorous fish, commonly 15–35 cm long (max. 40 cm). The body is brown-beige on the back with a silvery white belly, and a series of 7–10 dark spots along the lateral line. The body is slender and cylindrical; the head is slightly depressed with a large mouth and long jaws terminating behind the eye. Numerous needle-like teeth are visible when the mouth is closed. An adipose fin is present above the anal fin. Dispersal as planktonic eggs and larvae. Fertilisation is external, scatters eggs. Spawning off the Turkish coast occurs in May–July and September–November. A mature female can produce up to 400,000 eggs.

Native habitat (EUNIS code): A2: Littoral sediments, A4: Sublittoral sediments, marine sublittoral soft, generally to 100 m depth, records to 350 m. The same habitats are occupied in the invaded range. The abrupt rise in catch of the lizardfish in the 1950s was attributed to a rise of 1–1.5°C in sea temperature during the winter months of 1955. Native range: Indo-West Pacific from the Red Sea to Japan and the Great Barrier Reef, Australia. Introduced into the E Mediterranean Sea. Everywhere it formed thriving populations in an amazingly short space of time.

Entered the Mediterranean through the Suez Canal, and spread with the prevailing currents. The sudden increase of the lizardfish came at the expense of the native hake *Merluccius merluccius*, which was displaced into deeper waters. Economic impact: In 1955–1956 the lizardfish became commercially important, constituting for a few years up to one fifth of the total annual trawl catch along the Mediterranean coast of Israel, and an important staple of the coastal fishery in the area stretching from Damietta eastward to Port Said. By the mid-1960s it formed the main catch of trawlers off Mersin, Turkey, and accounted for two thirds of the fish landing biomass in the autumn months of the 1980s. Since the mid-1980s the lizardfish's share in catches has declined. No specific prevention or control method known.

Siganus rivulatus Forsskål, marbled spinefoot, rabbitfish (Siganidae, Osteichthyes)

Bella S. Galil

A demersal, gregarious, herbivorous fish commonly 5–25 cm long (max. 27 cm). The oval body, laterally compressed, has a grey-green to brown back, a light-brown to yellow abdomen and fine yellow lines on each side. It has a small first dorsal spine directed forward and usually covered with skin. The caudal fin is forked. It has small embedded scales. Dispersal as planktonic eggs and larvae. Spawning occurs between May and August.

Native habitat (EUNIS code): A3: Sublittoral rock and other hard substrata, A4: Sublittoral sediments, marine infralittoral and sublittoral soft and hard bottoms to 60 m depth. Habitat occupied in invaded range: The same as in the native range, subtidal soft and hard bottoms, often covered by algae. Young specimens feed at high tide on algae-covered shallow rocky platforms. Larvae live near the surface, metamorphosis occurs after about 20 days when they form schools of thousands in very shallow water and then migrate into deeper waters (4–10 m) upon reaching 2–3 g. Euryhaline, temperature tolerance 15–28°C. Native range: W Indian Ocean: Red Sea, Gulf of Aden, E Africa. Known introduced range: eastern and central Mediterranean. Disperses autochthonically as adults, or as current-borne planktonic eggs and larvae.

Entered the Mediterranean through the Suez Canal, forming early stable populations off the Levantine coast. The siganids are the most abundant herbivorous fish in shallow coastal sites in the Levant, and comprise much of the fish biomass along its rocky habitats. They have replaced native herbivorous fish. Their grazing pressure on the intertidal rocky algae has a significant impact on the structure of the local algal community, locally eradicating algae. The spines of the dorsal and pelvic fins are venomous, causing painful injuries to the unwary. The siganids are of moderate commercial importance for inshore fisheries, caught by gill-nets, trammel-nets and beach seines, and cultured on an experimental scale in Egypt, Israel and Cyprus. No specific prevention or control method known.

Spartina anglica Hubbard, common cord grass (Poaceae, Magnoliophyta)

Dan Minchin

A coarse-leaved grass of 1 m in height. It colonises the upper tidal level of estuaries and lagoons. They form radiating clonal tussocks that can increase in diameter by 30 cm/year and knit into each other to form meadows. Its seeds may float to new localities or may be transmitted by wading birds. It also spreads by fragments. Growth is mainly vegetative. Seeds produced have a limited opportunity to germinate. Some populations have endured a die-back which may relate to self-inflicted environmental conditions. Seeds are viable for short periods.

Native habitat (EUNIS code): A2: On littoral sediments. Habitat occupied in invaded range: A2, occurs from mean high water neap tides to mean high water spring tides on muddy estuarine, sheltered marine shores and wetlands influenced by brackish water, normally in areas of salt marsh preferring sheltered muddy tidal flats. It is a hybrid that developed in Britain. Known introduced range: Europe, North America, South Africa, New Zealand, China. Spreading.

It has been widely used as a plant for stabilising mud and for land accretion to protect coastlines and prevent undue erosion. Its seeds may float to new localities and may also be dispersed by birds. It is recognised as an important environmental modifier. It has resulted in replacement of *S. maritima* and excluded native flora, such as *Salicornia* spp., *Zostera* sp. and benthic infauna, with consequences for some wildfowl and waders. Economic impact: it may compete with areas used for oyster farming and impede recreational activities.

Plants should not be transplanted, for example when used unnecessarily for land reclamation projects. In some world regions *S. anglica* is prohibited for sale or distribution. Mechanical control: smothering with plastic has resulted in success over small areas, and removal by digging out plants at an early stage has also been successful. Chemical control: chemical herbicides applied to large areas have been more successful with two time-separated applications. Biological control: attempts to control it as a fodder for cattle have failed. Biological control using a planthopper *Prokelisia marginata* is being implemented on the Pacific coast of North America for the parental species *S. alternifolia*.

Styela clava Herdman, Asian sea-squirt (Styelidae, Ascidiacea)

Dan Minchin

This Asian sea-squirt has a club-shaped body and a narrow base attaching by means of a membranous plate. The outer surface is leathery, often wrinkled and fouled. It is a filter-feeder occurring mainly in sheltered estuaries, docks and inlets. As larvae, attached to crabs, with drifting plants or as fouling on the hulls of ships or other floating structures. This tunicate is a hermaphrodite surviving up to 2 years. It may spawn twice in its lifetime. Larvae hatch from released eggs in late summer to early autumn and settle after about a day. They are poor swimmers and normally settle near to parent populations.

Native habitat (EUNIS code): A1: Littoral rock and other hard substrata, A3: Sublittoral rock and other hard substrata, A4: Sublittoral sediments. Shallow sheltered environments to 20 m on firm surfaces. The same habitats are occupied in the invaded range, also estuaries, channels and bays from mid-tide (on shaded shores) to 25 m attaching to shell, stones and rock and each other. Tolerates −2°C to 23°C and salinities >26‰ and lower salinities for short periods. Native range: The Sea of Othotsk, Korea and Siberia. Introduced to the European Atlantic coasts from Portugal to Denmark, the east and west coasts of North America, Australia and New Zealand. Spreading.

Probably introduced to Europe as fouling on warships arriving during the Korean War. Known to occur on ship and leisure craft hulls and may be spread with oyster stock movements. Local transmissions in ships' ballast water is possible. Movement of floating port structures may also result in spread. It can attain densities >1,000/m^2 in sheltered areas, creating a high biomass that results in competition with other filter-feeders. Young individuals often attach to larger specimens (up to 200 mm) to form clusters. Human impact: sprays produced from damaged tissues when removing them from oysters are known to result in a respiratory condition in humans. Economic impact: it can foul artificial structures in port regions, ranched oysters and shellfish held in hanging culture and attach to fish cages. It may also impede fishing activities. In the St Lawrence Estuary, Canada, their abundance has caused declines in cultured mussel production. Stock movements of oysters or mussels from infested areas should be carefully monitored. Cleaning of equipment and boat hulls before transfers reduces risk. Mechanical control: apart from scraping, no other physical method is known. Chemical control: brine dips kill tunicates associated with oysters. Tunicates are sensitive to copper salts.

Teredo navalis Linnaeus, common shipworm (Teredinidae, Mollusca)

Stephan Gollasch

This mollusc has a wormlike, elongated body and two small shells at the anterior body end acting as wood-boring organ. Larvae are planktotrophic and adults are herbivorous suspension feeders. Individuals are found completely embedded in wood, which is digested with the help of endosymbiotic bacteria. A tiny opening allows water ventilation with two retractable siphons. The tunnels are covered with calcareous substances excreted by the species and may be closed with two calcareous plates. Dispersal as larvae with water currents. It is a protandric hermaphrodite. During metamorphosis male gametes develop after 6 weeks. They can also self-fertilise. After reproduction it changes back to the male phase and another cycle is started. Fertilisation is internal. Up to 5 million larvae are developed per cycle. The larval phase is usually 28 days.

Native habitat (EUNIS code): A3: Sublittoral rock and other hard substrata. Wood in the littoral zone. The same habitats are occupied in the invaded range. The key habitats are permanently submerged wooden structures, such as pilings and marina piers. Inside the wood, in sealed tunnels, it survives low anoxic conditions for more than 5 weeks and also tolerates air or freshwater exposure. In wood, ice cover is tolerated. It survives water temperatures up to 30°C. Larvae tolerate salinities down to 9‰ and adults even lower salinity levels. The maximum tolerable salinity is above 40‰. Native range: The native range is unclear, this is a truly cryptogenic species. In Europe it is found along the British Channel and North Sea coasts. The easternmost border of shipworm settlement in the Baltic is the Island of Rügen, Germany. Stable.

Adults naturally spread with floating wooden objects. Larvae are pelagic for approximately 2 weeks and are dispersed by water currents. Teredenid larvae were found in ballast water samples indicating the likeliness of this dispersal vector. It was first recorded in Europe in 1731 when it destroyed wooden dyke gates in the Netherlands causing a terrible flood. Recently caused damages along the German coast in the western Baltic were estimated to cost approximately €25–€50 million. Canadian logging companies used explosives near to floating timber to kill the shipworms with the shock waves. Wooden pilings may also be covered with polyethylene or polyvinyl or replaced by metal or concrete. Replacement with tropical hard-wood is also effective but any wooden structure in the sea has to be protected from marine borers. The earliest strategies to reduce the shipworm impact were extended ship hull exposure to air or fresh water. Today, chemical impregnation is probably the most effective shipworm deterrent. However, unwanted toxic side effects may occur.

Tricellaria inopinata d'Hondt & Occhipinti-Ambrogi (Scrupocellariidae, Ectoprocta)

Bella S. Galil and Anna Occhipinti-Ambrogi

A robust opportunistic bryozoan, capable of enduring a wide spectrum of temperatures and salinities, as well as high organic content. Settles on a wide range of anthropogenic and natural substrata. Dispersal with natural rafting such as current-dispersed algal fragments or pumice; or as fouling organisms on vessels and marine flotsam. Hermaphroditic. Broods its lecitotrophic eggs in a special external chamber (ovicell). Larvae are planktonic, non-feeding and last only for a few hours. Metamorphosis occurs within minutes of settlement to substrate. The first zooid (ancestrula), is rapidly fixed with two short rhizomes. Mature colonies with ovicells and embryos are found over much of the year, allowing a continuous recruiting.

Native habitat (EUNIS code): A1: Littoral rock and other hard substrata, marine infralittoral. Habitat occupied in invaded range: A1, marine benthic intertidal and subtidal, preferably in marinas, harbours and brackish water bodies. Has a wide temperature (2–35°C) and salinity (20–35‰) tolerance, also tolerant to polluted and turbid waters. It grows on a variety of artificial hard substrata: vessels' hulls, buoys, ropes, and epiphytic on macrophytes *Sargassum muticum*, *Undaria pinnatifida* and *Codium fragile*, mussels, sponges, ascidians and other bryozoans. Native range: temperate Pacific from North America, Japan to Australia. Introduced to the Mediterranean Sea and the European Atlantic coast. It seems to form rapidly expanding populations.

The undetermined Pacific origin of the species (belonging to the species complex *T. porteri-occidentalis-inopinata*) suggests that the first Mediterranean introduction was very likely due to oyster import. Secondary introductions are possibly due to small commercial vessels, yachts, fishing boats in coastal routes or dispersed with currents on floating fragments of *Sargassum*. The invasion of the Venice Lagoon appears to have caused a drastic reduction in frequency and abundance of native bryozoan species. It is a fast growing fouling organism, settling on buoys, vessels and ropes. Prevention: Aquaculture imports should be strictly quarantined, and only 2nd generation spat used; special care should be taken when mariculture facilities adjoin ports and marinas.

Undaria pinnatifida (Harvey) Suringar, Wakame (Alariaceae, Ochrophyta)

Stephan Gollasch

Marine brown alga, growing on hard substrates, reaching 3 m in length (sporophyte stage). Dispersal as gametophytes with water currents. It has a heteromorphic, diplohaplontic annual life cycle with a large sporophyte and microscopic female and male gametophytes. In the Mediterranean Sea the sporophyte is usually <1 m, some introduced populations with more than 1 generation/year. Vegetative reproduction by fragmentation unknown, gametophytes may undergo a dormancy period. Especially at low light, capable of surviving adverse conditions as thick-walled resting stages.

Native habitat (EUNIS code): A3: Sublittoral rock and other hard substrata. It grows on hard bottom habitats. 1–15 m depth. In the invaded range it occupies the same habitat, also other hard substrata. Hard bottom habitats from 1–18 m depth. Seems to prefer artificial substrates. Growth of sporophytes was documented in Japan from 4–25°C where plants appear in winter and disappear in summer. Optimum growth for young sporophytes is 20°C. Microscopic gametophytes may survive −1–30°C. Gametophytes are able to survive in darkness >6 months. Salinities >27‰ are necessary for growth of sporophytes and gametophyte development. Zoospores attach above 19‰. Native range: NW Pacific shores from Japan to Korea, China and Russia. Known introduced range: European Atlantic coast from Portugal to the Netherlands and the UK, sporadic records from the Mediterranean Sea. Spreading.

Unintentional introduction with oyster imports from Japan, secondary dispersal by vessels. In some regions it is the dominant seaweed and several co-occurring species decrease when it becomes abundant. It prefers artificial substrates and due to its fouling behaviour, frequent cleaning of aquaculture equipment and boats is required. In the Netherlands it grows mainly on *Crassostrea gigas*, but also on mussels. Being slippery, it causes problems for fishermen harvesting oysters. During mass developments further detrimental impacts may occur, which may impair aquaculture harvests.

Hulls of ships should only be cleaned out of the water and detached organisms should be disposed out of the reach of the sea. It should not be used as a display organism in public aquaria. Even if the water is treated, a risk remains that fertile parts may reach the sea. Mechanical control: one problem with eradication efforts is that the microscopic gametophytes are very tolerant and not visible to the naked eye. In several cases attempts at manual eradications were unsuccessful or showed a limited efficacy. Chemical control: trials with herbicides and antifouling paints showed that some toxins are efficient at preventing zoospore germination or at causing gametophyte mortality.

Aquatic inland species

Anguillicola crassus
Aphanomyces astaci
Cercopagis pengoi
Corbicula fluminea
Cordylophora caspia
Crassula helmsii
Dikerogammarus villosus
Dreissena polymorpha
Elodea canadensis
Eriocheir sinensis
Gyrodactylus salaris
Mnemiopsis leidyi
Neogobius melanostomus
Procambarus clarkii
Pseudorasbora parva
Salvelinus fontinalis

Anguillicola crassus Kuwahara et al., eel swim-bladder nematode (Anguillicolidae, Nematoda)

Dan Minchin

This is a parasitic nematode often 3 cm in length only noticed by opening up the body cavity of the freshwater eel. It has a large girth and a transparent outer skin that allows the inner organs to be seen. It is dispersed by movements of its various hosts and water currents. It occurs as an adult in the freshwater eel *Anguilla anguilla* and matures after 8–10 months, but may do so in 2 months at 20°C. Present in eels throughout the year. Adult eels may contain >70 nematodes. Each nematode can contain 500,000 eggs laid in the swim bladder. These and newly hatched larvae pass through the pneumatic duct to the gut and are expelled with feces. Larvae are consumed by planktonic crustaceans, to develop in their haemocoel and in turn are consumed by the glass eel stage. Other fishes, amphibians and larval insects can act as transfer hosts should they feed on infected crustaceans. If eaten by an eel the nematode burrows through the stomach wall to lodge on the air bladder and moults to the adult stage.

Native habitat: endoparasite of the Japanese eel *Anguilla japonicus*. In invaded range, as an endoparasite in intermediate hosts, usually arthropods, with its final stage in the European eel *A. anguilla* and the American eel *A. rostrata*. Larval stages normally occur in freshwater but can tolerate 8‰ salinity. May occur in hosts in marine conditions. Native range: E Asia, from China to Vietnam and Japan. Known introduced range: Europe and eastern North America. Spreading.

Dispersed by aquaculture with infected eels used for culture, by stocking glass eels upstream or spread by transfer hosts or moved water containing eggs and larval stages. Larvae may be distributed in crustaceans released by ships' ballast water. Adult nematodes feed on blood supplied to the swimbladder wall and can result in eel mortality. Nematodes may contribute to the decline of the North Atlantic eel stock because eels may not reach spawning areas due to damaged swim-bladders. Infected eels may be more susceptible to stresses following damage to the air bladder, spleen and liver. No human impacts. May compromise eel-culture production and may contribute to a decline in fishery landings. The International Council for the Exploration of the Sea presently advises a discontinuation of the fishery.

Stock transfers of fish from infested areas should be controlled. No live imports of infested eels should be permitted to islands. Mechanical control: filtration and treatment of water used for culturing eels.

Aphanomyces astaci Schikora, crayfish plague (Saprolegniaceae, Chromista)

David Alderman

This oomycete pseudofungus is the aetiologic agent for the disease which is known as crayfish plague. Crayfish plague is a disease which, as an acute disease, has only created problems in Europe, not in the native range of North America where crayfish act only as carrier vectors. It presents an extreme example of a pathogen that rarely kills its established hosts in its normal geographical range. Native European crayfish populations were totally destroyed by the very aggressive pathogen. Over the 150 years that the disease has been present in European rivers, no resistant European crayfish have appeared. This oomycete disperses by biflagellate zoospores, it has only asexual stages and produces gemmae as resistant stages.

Native habitat (EUNIS code): C1: Surface standing waters, C2: Surface running waters, C3: Littoral zone of inland surface waterbodies). The same habitats are occupied in the invaded range and it affects all non North American fresh water crayfish. Native range: North America. Introduced to Europe and Asia Minor. Increasing trend.

It was accidentally introduced from North America in the 19th century with the crayfish *Orconectes limosus*, *Pacifastacus leniusculus* and *Procambarus clarkii*. This oomycete destroys European crayfish species in all infected watersheds. Relict populations survive, but when populations recover a further mass mortality will occur. Social impact through destruction of native crayfish populations, highly affecting crayfish trappers and dealers 1870–1930. The lack of native crayfish stocks led to the introduction of replacement North American species from 1960s onward, for farming and repopulation, with the introduction of new *A. astaci* strains. Economic impact was very significant 100 years ago. To prevent further damage, the movements of crayfish should be stopped. Fish movements can also transmit the disease.

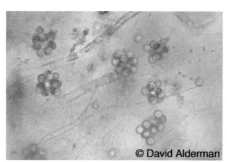

Cercopagis pengoi (Ostroumov), fish-hook waterflea (Cercopagididae, Crustacea)

Vadim E. Panov

A water flea with body size up to 2 mm and caudal process with length up to 10 mm. Parthenogenic females of the first generation that hatch from resting eggs are anatomically distinct from parthenogenic females of following generations. They have a short straight caudal spine unlike the characteristically looped caudal spine of parthenogenically produced individuals. They prey mainly on small plankton crustaceans. Long-distance dispersal take place from resting eggs in ballast tanks of ships and local dispersal by fishing boats and on fishing lines. Parthenogenesis prevails during periods of rapid population growth. In the Caspian Sea, sexual reproduction is more typical at the last stages of population growth, and results in the production of resting eggs. In invaded habitats, it may switch to prolonged sexual reproduction during summer. Important prey item for planktivorous fish. Most gamogenetic females (94%) carry two resting eggs. As with other invasive cladocerans, may possess adaptive life cycles, switching to the early gamogenetic reproduction which facilitates establishment in the recipient ecosystems and further dispersal.

Native habitat (EUNIS code): A7: Pelagic water column, C2: Surface running waters. The same habitats are occupied in the invaded range, also large freshwater reservoirs, lakes and coastal waters. It is a brackish water euryhaline species, found from freshwater to brackish water up to 13‰. It is eurythermic, first appears in summer plankton at water temperatures 15–17°C, during autumn it appears in zooplankton at relatively low temperatures of 8°C. Native range: Caspian endemic species, which spread during different geological periods to the Ponto-Azov and Aral Sea basins. Introduced to reservoirs of Don and Dnieper rivers, the Baltic Sea, Great Lakes of North America. Generally increasing in Europe and North America.

Introduced by ships' ballast water. It is a potential competitor with young stages of planktivorous fish for herbivorous zooplankton. It affects resident zooplankton communities by selective predation. May cause allergy in humans during cleaning of fishing nets. May attach to fishing gear, clog nets and trawls, causing problems and substantial economic losses for fishermen. Preventive measures for long-distance dispersal may include ballast water management. On the local level, fishing boats and gear should be properly cleaned.

Corbicula fluminea (Müller), Asian clam (Corbiculidae, Mollusca)

Dan Minchin

Inland water filter feeding bivalve with a globular shell. Has the capability of collecting food in sediments with its extendable foot. Tan to brown, ridged solid shells, it occurs in large numbers in sediments, usually <3 cm long. In Europe there are two distinct morphotypes, currently treated at species level, with varied length/shell height ratios and shell colour with *C. fluminalis* occurring in river mouths with small variations in salinity. Pediveliger stage produces a byssus causing it to be dragged by water currents, juveniles and adults may produce tacky mucus strings that can also result in dispersal. May also be spread by birds and mammals. It can rapidly recolonise areas following purges. It is a hermaphrodite (cross- and self-fertilizing) releasing a brooded non-swimming pediveliger stage of 200 μm length. Reproduces at 15°C from about 6–10 mm from 3 months of age with more than one brood a year with releases from late spring to autumn. Tolerant of aerial exposure for weeks but intolerant of low oxygen levels.

Native habitat (EUNIS code): In river and lake sediments. Habitat occupied in invaded range: Oligotrophic to eutrophic flowing streams, rivers and lakes on oxygenated muddy to sandy sediments, but also occurring among gravel and cobbles. Also can occur in irrigation and drainage cuts. Tolerates 2–34°C and salinities to 5‰ with short periods of up to 14‰. Intolerant of areas with high nutrient loads. Native range: SE Asia, Australia and Africa. Introduced to North and South America and Europe. Arrived in Europe in the 1970s in Portugal, still spreading eastwards.

Might have been carried in ships' freshwater ballast to Portugal. It is used in ornamental ponds and aquaria and as angling bait. Could be moved entangled in macrophytes and with overland boat transmissions. Natural dispersal by birds is suspected on account of the tacky mucus threads that may adhere to wading birds or fishes. Since it is a self fertile species a single individual might be sufficient to develop a new population. Competes with other filter feeding bivalves (unionids) and with snails feeding on organics in sediments. High density occurrences in gravels make some building materials worthless. Capable of reducing flows in drainage and abstraction pipes in low-flow areas and during periods of low peak usage. Shells can clog the narrow gauge piping of condensers and heat exchangers of power plants.

Its use as an ornamental should be discouraged and legislation and inspections are needed to prevent its arrival on islands and other geographically separated areas. May be removed from piping by passing wads under pressure. Screens and shell traps can reduce impacts. May be controlled by flushing oxidising molluscicides with water at 40°C and by oxygen depletion.

Cordylophora caspia (Pallas), freshwater hydroid (Clavidae, Cnidaria)

Sergej Olenin

This colonial hydroid (up to 10 cm high) light brown in colour occurs in brackish waters, branching occasionally from alternate sides. Branches are ringed at the base and have terminal polyps with colourless, extensile tentacles that are 12–16 mm long. Feeds on small planktonic organisms. Pelagic larvae are dispersed with water currents. Each upright branch may bear 1–3 gonophores with 6–10 eggs each. The larvae are released as planulae and no medusoid stage occurs. In some cases the larvae may develop directly into juvenile polyps in the gonophore before release. Hydroids may reproduce asexually by budding to from another colony. A common form of asexual reproduction in hydroids is the formation of vertical stolons, which then adhere to adjacent substratum, detach and form another colony. It has the ability to produce dormant resting stages (menonts) that are far more resistant to environmental change than the colony itself.

Native habitat (EUNIS code): X1: Estuaries, X3: Brackish coastal lagoons. The same habitats are occupied in the invaded range. It occurs on hard substrates including rocks, shells and artificial substrates (pilings, harbour installations, bridge supports), floating debris and occasionally from the leaves of reeds *Phragmites* or stalks of water lilies. Colonies tolerate 5–35°C, and reproduce between 10–28°C. It can also survive from 0–35‰ as resistant stages, grows between 0.2–30‰, and reproduce between 0.2–2‰. In nature, well-developed colonies are usually found in water of 1–2‰ where tidal influence is considerable, or between 2–6‰ where conditions are constant. It may also occur at full salinities, and fast flowing, well oxygenated freshwater containing Ca, Mg, Na, Cl and K ions. Native range: Ponto-Caspian region. Introduced into the Baltic Sea in early 1800s; it has spread rapidly to inland waters and estuaries of Europe, reaching Ireland by 1842, Australia by 1885 and the Panama Canal by 1944; now it is globally known in temperate and tropical coastal regions and in many fresh waters.

Transported as a fouling organism on ships' hulls, or as planktonic larvae in ballast water. Competes with native species for space and food. The large dense colonies essentially modify benthic habitats causing structural changes in pelagic and benthic communities. It is an important fouling animal in industrial cooling water systems. To get rid of planktonic larvae exchange of ballast water should take place in the mid ocean. Appropriate control of boats and pontoons would minimise the risk of inoculation. Mechanical control: physical removal from ship hulls, increased temperature, oxygen deficiency (e.g., one of parallel pipelines closed for 3–4 weeks). Chemical control: chlorination.

Crassula helmsii (Kirk) Cockayne, New Zealand pigmyweed (Crassulaceae, Magnolopsida)

Dan Minchin

A small succulent flowering perennial that grows rapidly to form an extensive lush-green 'carpet' that floats on freshwater or may be submerged. Growth can extend from the margins of sheltered waterbodies to completely cover the water surface with tangles of stems and shoots. Plants 10–130 cm in length, flowers <4 mm, white to pale pink. Dispersed by plant fragments or buoyant shoots (turions) carried by birds and mammals, downstream movements and flooding. May also be moved with mud. Reproduction is mainly vegetative, propagating from fragments <5 mm containing a node. It has a high growth rate growing most of the year without serious winter die-back. May overwinter as turions.

Native habitat (EUNIS code): C1: Surface standing waters, C3: Littoral zone of inland surface waterbodies. Slow-flowing water, ponds, ditches, canals, reservoirs, lakes and wetlands. The same habitats are occupied in the invaded range. On wet ground, as a marginal plant or submerged to ~3 m. Prefers acidic to alkaline nutrient rich water-bodies, tolerates temperatures –6 to 30°C. Native range: Australia, Tasmania and New Zealand. Introduced to Europe, Russia, and in SE USA. Introduced to Europe in 1911 from Tasmania. Spreading.

Ornamental plant, sold in garden centres and for aquaria. May be carried overseas with ferried leisure craft or with fishing gear. Forms dense marginal and floating mats that can shade-out other water plants and result in oxygen depletion of the underlying water causing a decline in invertebrates, frogs, newts and fishes. Health impact: floating mats can be mistaken for dry land. Economic impact: reduced opportunities for angling, obstructing boat movements.

Legislation and inspections of ornamental plants are needed to prevent its sale in vulnerable regions. Mechanical control: physical removal results in many small viable fragments being left in the water, which may spread the plant downstream or elsewhere within lakes and reservoirs. Removal may be practical for small water-bodies. Shading out with dark-plastic sheeting has been successful. Chemical control: it has been shown to be resistant to many available herbicides. Efficacy depends of the density of the plants to be managed. Mechanical removal followed by chemical treatments using diquat have been shown to be effective. Application of diquat is most effective in autumn and winter with water temperatures >12°C, below 8°C absorption is poor. For emergent plants glycophosphate may be used but could be a hazard for grazing animals. More than one application may be needed spaced about 3 weeks apart. Complete eradication is often difficult. There are regulations on the use of chemicals in many countries.

Dikerogammarus villosus (Sowinsky), killer shrimp (Gammaridae, Crustacea)

Simon Devin and Jean-Nicolas Beisel

The killer shrimp is an omnivorous predator. It can feed on a variety of macroinvertebrates, including other gammarid species. Dispersal probably through shipping activity. The females are sexually mature at 6 mm in length, when they are 4–8 weeks old. They can reproduce when water temperature is above 13°C, with a mean fecundity of 27 eggs per female. Hatching length is about 1.8 mm. Several fish species feed on this shrimp, but no invertebrate species is known to predate it.

Native habitat (EUNIS codes): A1: Littoral rock and other hard substrata, A2: Littoral sediments, A3: Sublittoral rock and other hard substrata, A4: Sublittoral sediments, B1: Coastal dune and sand habitats, B3: Rock cliffs, ledges and shores, including the supralittoral, C1: Surface standing waters, C2: Surface running waters, C3: Littoral zone of inland surface waterbodies. Hard substratum, macrophytes. The same habitats are occupied in the invaded range. The species exhibits a wide range of environmental tolerance. It can live in a broad spectrum of temperature (0–30°C) and salinity (up to 12‰), and can occupy every substratum except sand. The species is only present in areas with low current velocity. Native range: Ponto-Caspian basin. Introduced to almost all W Europe large rivers and the Baltic Sea basin. Spreading.

The most likely introduction vector is shipping (ballast water and hull fouling of vessels). The colonisation of W European water systems probably occurred through the southern corridor, via the Danube and Rhine rivers. It locally eliminates other gammarid species through competition and predation. It eats fish eggs and attacks small fish. Prevention includes ballast water treatment for transcontinental dispersion. No effective solution has been proposed for intracontinental dispersion.

Dreissena polymorpha (Pallas), zebra mussel (Dreissenidae, Mollusca)

Anastasija Zaiko and Sergej Olenin

Sessile bivalve mollusc, forming dense colonies on hard substrates in fresh and slightly brackish waters. Brownish-yellowish triangular shells, up to 50 mm wide, with dark and light coloured ("zebra") zigzag banding. Filter feeder on plankton organisms and organic particles. Pelagic veliger larvae and post-veligers are transported by currents, secondary dispersal by drifting of post-larvae and young adults using byssal or mucous threads. It has separate sexes, ratio 1:1, external fertilisation. Synchronised spawning occurs once >8 mm. A mature female may produce 1 million eggs per year. Spawning begins at 12–15°C, profuse at 18–20°C, over a period of 3–5 months.

Native habitat (EUNIS code): C1: Surface standing waters, C2: Surface running waters, C3: Littoral zone of inland surface waterbodies, X1: Estuaries, X3: Brackish coastal lagoons. Large estuaries and inland waters, hard and soft bottom habitats. The same habitats are occupied in the invaded range. Typical habitats are estuaries, rivers and lakes with firm surfaces suitable for attachment. Tolerates temperatures from -2°C to 40°C, best growth at 18–20°C. Tolerates brackish water up to 7‰ and prefers mesotrophic water bodies, up to depths of 12 m in brackish water and to 60 m in lakes. Tolerates low oxygen content in water for several days and survives out of water under cool damp conditions for 3 weeks. Native to the drainage basins of the Black, Caspian and Aral Seas. Introduced to neighboring parts of Russia and most parts of Europe. In 1988 it reached North America, future expansion is possible.

Introduced by ballast water and hull fouling of vessels. It competes for space and food with native mussels and other filter-feeding organisms. Bioaccumulates pollutants. Shell deposits cause severe habitat alteration. Its high consumption of phytoplankton results in increased water clarity. Multiple economic impacts, including: fisheries (interference with fishing gear, alteration of fish communities), aquaculture (fouling of cages); water abstractions (clogging of water intake pipes); aquatic transport (fouling of ship hulls and navigational constructions). Sharp shells cause injuries in recreational areas. Invasion to North America is causing annual multimillion losses to the economy.

Preventing overseas transfer can only be achieved by mid-ocean exchange or by disinfection of ballast water. Appropriate control measures (e.g., inspection, removal of attached mussels, drying) should be taken to minimise risk of inoculation by transfer of boats and fishing gears. Mechanical control: removal (scraping, mechanical scrubbers in pipes), thermal, UV light and electric currents have been used. Chemical control: using anti-fouling paints and surfaces, metal-organic chemicals, chlorine, etc.

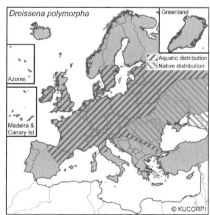

Elodea canadensis Michaux, Canadian waterweed (Hydrocharitaceae, Magnoliophyta)

Stephan Gollasch

It is an aquatic herb with branching stems 20–30 cm long, tending to form dense monospecific stands, covering hundreds of hectares. Leaves oblong-linear in groups of three, flowers white or pale purple at the water surface, fruits are capsules <1 cm in length. Dispersed by seeds and fragments via water currents. It is a dioecious plant flowering June–August. Pollination occurs near water surface and pollen is distributed by wind and water currents. Vegetative reproduction by fragments is very common. Mass development has been reported.

Native habitat (EUNIS code): C1: Surface standing waters, C2: Surface running waters. Shallow lakes, ponds, pools, ditches and streams with slow moving water. The same habitats are occupied in the invaded range, up to 3 m water depth. Tolerates pH values from 6.0–7.5 and temperatures from 1–25°C. Native range: North American inland waters. Introduced to Europe where it became widespread in north and central Europe. After a rapid colonisation the populations declined due to the introduction of *Elodea nuttallii*. Today the population is stable.

Accidental release after intentional import for ornamental purposes in aquaria and ponds. As it can be very dominant, it competes for nutrients and space with other plants. It can bioaccumulate nutrients and modify the habitat by reducing water movement. The species is known to outcompete other plants. During dense blooms, impairs boating, fishing, swimming, and water skiing. Clogging of water intake pipes of power plants and other industries were reported.

Should not be released in the wild. Mechanical control: covering the plants to block light may result in eradication. In reservoirs and lake systems the water level may be lowered in winter with the aim of controlling the population. As the plant spreads through fragmentation, mechanical controls should only be undertaken during mass developments and when the risk of spread to other water systems is minimal. Using mechanical controls during an ongoing invasion may promote the spread due to fragmentation. Chemical control: trials were undertaken by using various chemical formulations, such as complexed copper, dipotassium salt and fluridone. Biological control: recently a fungus (*Fusarium* sp.) was identified which damaged *Elodea* in laboratory tests. The enhancement of native or introduced herbivorous fish may pose another biocontrol option. Carp prey upon *E. canadensis*. However, a risk remains, as most biocontrol species do not selectively prey upon the invader only.

Eriocheir sinensis Milne-Edwards, Chinese mitten crab (Varunidae, Crustacea)

Stephan Gollasch

Crab with a carapax up to 5 cm, brownish in colour, with characteristic mitten like "fur" on the claws. It is an omnivorous predator, feeding on a wide range of plants, invertebrates, fishes and detritus. Gastropods and bivalves are the dominant food component. Larvae disperse with water currents, juveniles and adults show active migration, even across dikes and streets. Larval stages in marine and higher saline estuarine waters, upstream larvae migration supported by currents in estuaries. Juveniles actively migrate upstream up to 1,500 km inland in China. Adults migrate downstream to the marine environment in summer. This migration may take several months, during which they become reproductively mature. Most crabs live for 2 years.

Native habitat (EUNIS codes): A1: Littoral rock and other hard substrata, A2: Littoral sediments, A3: Sublittoral rock and other hard substrata, A4: Sublittoral sediments, B1: Coastal dune and sand habitats, B3: Rock cliffs, ledges and shores, including the supralittoral, C1: Surface standing waters, C2: Surface running waters, C3: Littoral zone of inland surface waterbodies. Larger estuaries and inland waters, hard and soft bottom habitats. The same habitats are occupied in the invaded range. Due to its inland migration it colonises lakes and streams several 100 km from the sea. Tolerance to temperature down to 0°C, high salinity, low oxygen conditions and air exposure for several hours. Native range: between E Russia, China and Japan. Introduced to most coastal parts and river systems of NW Europe, still spreading.

Most likely introduced by shipping (ballast water and hull fouling of vessels) or imports of living species for aquaria and for human consumption. The crabs are the second intermediate host for the human lung fluke parasite in Asia (not yet recorded in crabs in Europe). Crabs damage nets and prey upon fishes caught in traps and nets. In freshwater ponds they feed on cultured fish and their food as well. The burrowing activities of crabs result in increased erosion of dikes, river and lake embankments. They can also clog up industrial water intake filters. In some European countries crabs are caught as by-catch in inland fisheries and sold to Asian restaurants. Also used as fishing bait, for fish meal production, and for cosmetic products.

Release in the wild should be avoided. Mechanical control: attempts to catch as many juvenile crabs as possible during their upstream migration have been undertaken, especially during mass developments. However, trapping of crabs has not been found to be effective in controlling crab populations.

Gyrodactylus salaris Malmberg, salmon fluke (Gyrodactylidae, Platyhelminthes)

Dan Minchin

This is a very small worm parasite (<1 mm) attached to the outer body and gills of fish in the salmon family, damaging the skin, thus leading to infections. Unless heavily infested, it is unlikely that its presence will be noticed. There are many related *Gyrodactylus* species that are difficult to distinguish. This genus contains specialist and generalist parasites. It has been dispersed by movements of Atlantic salmon and rainbow trout in fresh and brackish water. Might be spread to different catchments by piscivorous birds. Females have up to four viviparous broods of 2–7 individuals, the first may be asexually produced. Populations rapidly develop at 6–13°C, but can also grow at 2.5°C (surviving 33 days) to 19°C (4.5 days). It may reproduce on different salmonids.

It is an ectoparasite on the skin, gills and fins of Atlantic salmon *Salmo salar*, heavy infections normally occur on the body and caudal fin. It is a cold water species surviving at temperatures 0–20°C and salinities up to 7‰. At higher salinities reproductive capability and longevity declines. Native range: Adapted to an isolated evolutionary form of Atlantic salmon in Baltic river catchments in Russia, Sweden, Finland, Latvia and Lithuania. Introduced to Norway, Finland and the White Sea region of Russia. Found on farmed rainbow trout in Denmark. It presently has a restricted range that may expand.

Imported by aquaculture and stocking and spread with movements of infected salmon or rainbow trout fingerlings and perhaps with fish-farm, fishing and water sports equipment. It can survive in air under damp conditions for some days. It causes a significant mortality of young salmon in river catchments outside of the N Baltic Sea, in Norway, N Finland and the White Sea area. Reports from Portugal, Spain, France and Germany require confirmation. It may cause reduced unionid recruitment caused by declines in abundance of salmonids. No effects on humans known. Economic impact: high mortalities of salmon outside of its native range. In Norway, estimates of annual losses in production in the late 1990s from salmon farms was 250–500 t of production.

Stock transfers of fish from risk areas to uninfested areas should not take place. Chemical control: eradication using rotenone has been effective in eliminating its presence in 16 Norwegian waterbodies. Aqueous aluminium may be effective for treating fish for stocking. Detached trematodes can infect free-swimming salmon although they have no swimming ability themselves. The use of brackish treatments needs to be reviewed. Populations have declined arising from control measures in Norway.

Mnemiopsis leidyi Agassiz, sea walnut, comb jelly (Bolinopsidae, Ctenophora)

Tamara A. Shiganova and Vadim E. Panov

A comb jelly, length up to 100 mm, body laterally compressed with four rows of small ciliated combs, iridescent by day, glowing green by night. Usually transparent or slightly milky, feeds on plankton fish eggs and larvae. Dispersal with ballast water of ships. This species is hermaphrodite (capable of self-fertilisation) and also possesses paedogenesis (sexual maturity of larvae and juveniles) and dissogony (sexual maturity of larvae followed by regression of gonads and subsequent rematuring of adults). Large specimens produce 2,000–8,000 eggs during spawning.

Native habitat (EUNIS code): A7: Pelagic water column. Marine and brackish water, coastal waters (temperate and subtropical estuaries and coastal areas). The same habitats are occupied in the invaded range when salinity is >3–4‰, maximum is 39‰. Temperature is a most important factor (from 6°C in winter to 31°C in summer), followed by food availability and mortality by predation. Native range: Atlantic coastal waters of North and South America. Known introduced range: Azov, Black, Caspian, North, Baltic and NE Mediterranean seas.

Introduced by ballast waters of ships. It was first found in Europe in the Black Sea in 1982, where it caused a dramatic reduction in zooplankton, ichthyoplankton, and zooplanktivorous fish populations. In the Caspian Sea density and biomass of zooplankton has been decreasing with increasing size of the comb jelly population. Catches of three species of kilka greatly decreased in all Caspian countries. Since the decrease of kilka stocks, rations and share of kilka in diet composition of beluga has been reduced. This invasion causes cascading effects. The bottom-up effects include the collapse of planktivorous fish, vanishing dolphins in the Black Sea and seals in the Caspian Sea. Top-down effects include an increase in phytoplankton, free from grazing pressure, and increasing bacterioplankton populations, triggering increases in zooflagellates and infusoria populations. Significant economic losses for the Black Sea and Caspian Sea coastal countries due to drastic decline in pelagic fish catch, estimated to several hundred million dollars for the Black Sea.

Prevention by ballast water management. Biological control: the ctenophore *Beroe ovata* is considered as a biological control agent for *Mnemiopsis leidyi*. Accidental introduction of *Beroe ovata* in the Black Sea resulted in significant decline in *Mnemiopsis* in late summer periods and subsequent increase in zooplankton in some areas. Currently intentional introduction of *Beroe ovata* is suggested for biological control of *Mnemiopsis* in the Caspian.

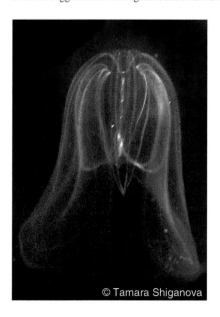

Neogobius melanostomus (Pallas), round goby (Gobiidae, Osteichthyes)

Vadim E. Panov

It is a small, soft-bodied fish, with a fused pelvic fin forming a suction disk on the ventral surface. Body brownish grey with dark brown lateral spots, mature males completely black during spawning and nest guarding, with yellowish spots on body and median fins fringed yellow or white. A large black spot is usually present at the end of the first dorsal fin, beginning at the fifth ray. It is a benthic feeder. Its diet composes of crustaceans and molluscs, including zebra mussels *Dreissena polymorpha*, polychaetes, small fish, goby eggs and chironomid larvae. Females spawn approximately every 20 days April–September. 500–3,000 eggs are deposited on hard substrate and then guarded by the male until hatching. Females mature at the age of 2 years, males at the age of 3 years.

Native habitat (EUNIS code): C3: Littoral zone of inland surface waterbodies. It is a bottom dweller in the nearshore region of lakes and in rivers, and prefers rocky habitats providing many hiding opportunities. Although juvenile and adult round gobies prefer rocky substrates, the fish also is found in fine gravel and sandy substrates in which they may burrow. It is found in slowly flowing rivers, lagoons, and brackish coastal water to 20 m, but it migrates to deeper water (50–60 m) in winter. The same habitats are occupied in the invaded range. It is a euryhaline and eurythermic species, thus tolerating a salinity range of 20–37‰ and water temperature between −1 to +30°C. Native range: Basins of the Caspian, Black and Azov Seas. Introduced into the Baltic and North Sea basins and the Great Lakes of North America. It is increasing in the Baltic Sea area.

It has been introduced by ballast water of ships. Species richness of native fish has declined in areas where the round goby has become abundant. It preys on darters, other small fish, and lake trout eggs and fry in laboratory experiments. They also may feed on eggs and fry of sculpins, darters and logperch. Adults aggressively defend spawning sites and may occupy prime spawning areas, keeping natives out. Health impact: It often eats bivalves that filter the water, thus accumulating many contaminants which are passed on to larger game fish and possibly on to humans. Prevention by management of ballast water. Mechanical control: the use of electrical barriers can be successful.

Procambarus clarkii (Girard), red swamp crayfish/crawfish (Cambaridae, Crustacea)

Francesca Gherardi and Vadim E. Panov

Crayfish up to 15 cm body length, adults usually dark red, orange, or reddish brown, chelae red, typically S-shaped and covered in spines and tubercles. Adult individuals mainly feed on plants and detritus; juveniles consume a higher proportion of animal food; sometimes cannibalistic. Long-distance dispersal is facilitated by intentional introductions. It can migrate long distances on land, exceeding 3 km per day. It has a short life cycle and a high fecundity with an extremely plastic reproductive cycle. Two generations per year are possible at low latitudes.

Native habitat (EUNIS code): C2: Surface running waters, C3: Littoral zone of inland surface waterbodies. It is able to occupy a wide variety of aquatic habitats. The same habitats are occupied in the invaded range. Found in natural and agricultural areas throughout south central Europe, most often occurring in small but permanent waterbodies. Tolerant to saline waters and dry periods up to 4 months. It is native to NE Mexico and south-central USA. Introduced to several states of the USA and S America, E Asia, parts of Africa, and Europe. Legally introduced into Spain in 1973, later illegally introduced throughout Europe, increasing in many areas.

Intentionally introduced for aquaculture and used as bait. Contributed to the decline of native European crayfish by competition and as a vector of the fungus-like *Aphanomyces astaci*, agent of the crayfish plague. It reduces the value of habitats by consuming invertebrates and macrophytes, and by degrading riverbanks because of its burrowing activity. It bioaccumulates heavy metals and toxins from Cyanobacteria, thus transferring them to its consumers, including humans. Intermediate host of *Paragonimus* trematodes, a potential pathogen of humans if undercooked crayfish are consumed. It may cause significant economic losses in irrigation structures, such as reservoirs, channels or rice fields, due to its burrowing activity and feeding on rice plants.

Import of live crayfish from abroad is banned by customs legislation in some European countries, transfer to areas not yet invaded is banned in the UK. Identification of new populations in the wild is necessary for quick eradication. Mechanical methods include the use of traps, fyke and seine nets, and electro-fishing. The use of sexual pheromones to attract males is under investigation. Physical methods of control include the drainage of ponds, the diversion of rivers and the construction of barriers, either physical or electrical. Chemical control: biocides have been used. Biological control: possible methods include the use of fish predators, disease-causing organisms and the use of microbes that produce toxins. The use of sterile male release technique is under investigation.

Pseudorasbora parva (Temminck & Schlegel), stone moroko (Cyprinidae, Osteichthyes)

Vadim E. Panov

Zooplanktivorous fish with an elongated body, slightly flattened on sides, resembling *Gobio* species. Maximum size up to 110 mm, 17–19 g body mass. Colouration similar in both sexes, with grey back, light sides and belly passing from yellowish-green to silver. Spawning takes place when 1 year old, in Europe usually April–June. It can produce some 100 to some 1,000 eggs. Spawning is multi-litter, taking place in the littoral zone. Eggs are laid on plants, sand, stones, mollusc shells and other substrates. The male guards the eggs until hatching, and aggressively drives away other, often larger fishes.

Native habitat (EUNIS code): C1: Surface standing waters, C2: Surface running waters, C3: Littoral zone of inland surface waterbodies. The same habitats are occupied in the invaded range, shallow lakes, carp ponds, irrigation canals, ditches and slow sections of lowland rivers. Minimum temperature for reproduction is 15–19°C. Native range: E Asian region from China and Japan to Korea. Introduced to Europe and still increasing.

In Europe it has been intentionally introduced through aquaculture with stocking material of herbivorous fishes imported from China. It competes for food with farmed fish species and feeds on juvenile stages of many locally valuable native fish species. Being a vector of infectious diseases (including *Spherotecum destruens*), it constitutes a serious threat to both native and farmed fish in Europe. In the open waters of S Europe it has probably contributed to a decrease in abundance or even disappearance of other cyprinids such as *Scardinius erythrophthalmus*, *Carassius carassius*, *Rhodeus sericeus*, *Gobio gobio* and *Leucaspius delineatus*. In ponds, during a mass occurrence, it depletes the food supplies of farmed species like carps, decreasing their productivity.

Stocking material imported for fish farms or in order to stock open waters should be checked carefully to ensure the absence of this invader. Additionally, using the stone moroko as live bait for predatory fish should be stopped.

Salvelinus fontinalis (Mitchill), brook trout (Salmonidae, Osteichthyes)

Melanie Josefsson

Salmonid predatory fish with long streamlined body, adipose fin close to the tail and a large mouth extending past the eye. Olive back, blue-grey to dark brown, very similar to the native brown trout *Salmo trutta*, but with a distinct marbled pattern of lighter colour (vermiculation). Sides lighter than back with pale and red spots, surrounded by blue "haloes". Lower fins with white edges and a contrasting black stripe. During the breeding season in the autumn, males can become very bright orange-red along the sides. Brook trout reaches maturity after 2–4 years, spawns in running water occurs in late summer or autumn. Eggs hatch in spring. Eggs are deposited in a redd constructed by the female on a gravely substrate.

Native habitat (EUNIS code): C1: Surface standing waters, C2: Surface running waters, A7: Pelagic water column. The same habitats are occupied in the invaded range. Occurs in freshwater, brackish water, and marine environments, growing between 7–20°C. Native to the eastern parts of North America, it has been introduced to temperate areas all over the world. Although stocking with the brook trout is decreasing, the fish is expanding its distribution through reproduction and secondary spread.

Intentionally introduced for aquaculture, sport fisheries and for food production. It has escaped from fish farms and has been intentionally released into the wild for sport fishing and as a replacement for depleted native salmonids in acidified waters. Competes with and predates on native fish such as other salmonids for food and cover. It may replace native brown trout *Salmo trutta* in high altitude lakes and streams. It predates on amphibians, zooplankton and other invertebrates. When stocked in previously fishless oligotrophic mountain lakes, brook trout alter nutrient cycles and stimulate primary production by accessing benthic sources of phosphorus. The replacement of native salmonids with brook trout negatively affects the freshwater pearl mussel *Margaritifera margaritifera*, as the brook trout cannot serve as host to the glochidiae larvae. The brook trout hybridises with the native brown trout, of which some hybrids are fertile. Recreational fishing for brook trout may be of social and economic importance for local communities. Sport fishing for brook trout has a positive economic effect for local communities, but it is negative if native salmonids disappear.

Intentional introduction should be forbidden. Established populations are difficult and costly to control, further introductions or stocking with brook trout should be avoided. Electrofishing and gill netting can be effective in small, contained environments such as mountain lakes. The use of piscicides such as rotenone can be effective, but pose serious risks to other species.

Terrestrial fungi

Ophiostoma novo-ulmi
Phytophthora cinnamomi
Seiridium cardinale

Ophiostoma novo-ulmi (Brasier), Dutch elm disease (Ophiostomataceae, Ascomycetes)

Marie-Laure Desprez-Loustau

Ophiostoma novo-ulmi is responsible for the second pandemic of Dutch elm disease. It has progressively replaced the closely related species *O. ulmi*, less aggressive and competitive, which caused the first pandemic. In Europe, *O. novo-ulmi* is spread mainly by the bark beetles *Scolytus scolytus* and *S. multistriatus*. The fungus can also spread via root grafts between trees. Native habitat (EUNIS code): G: Woodland, forest and other wooded land. In the invaded area it occupies the same range of habitats and I2: Cultivated areas of gardens and parks. Host trees include all Euro-American native elms, which are much more susceptible than Asian elms. Disease expression is strongly correlated with temperature, levels of defoliation being greatest above 17°C. Higher temperatures also increase the size and number of annual bark beetle generations. Native range: both *O. ulmi* and *O. novo-ulmi* are believed to come from Asia. Known introduced range: both species have affected the North American and Eurasian continents. Most European countries are in a post epidemic situation. However, the disease is still spreading in N Europe, especially Scotland. Climate warming might favour its northwards expansion.

O. novo-ulmi first appeared in Europe in the 1940s in the Moldova-Ukraine region and also in the southern Great Lakes area in North America. From there it was introduced to Great Britain ca 1960, probably with a shipment of elm logs. Subsequent spread throughout Europe was mainly caused by natural migrations, involving bark beetle vectors. Trees infected by beetles first show wilting, curling and yellowing of leaves on one or several branches. Once the fungus is established within a tree, it spreads rapidly via the water-conducting vessels. Gums and tyloses are produced by the tree in response to infection, causing the occlusion of vessels and eventually wilting and death of the tree. Few mature native elm trees are left in much of Europe. In the UK alone, it has been estimated that more than 25 million trees have died. The elms mostly survive as shrubs, especially in hedgerows, as the roots are not killed and produce root sprouts ("suckers"). These suckers rarely reach more than 5 m tall before succumbing to a new attack of the fungus. Health and social impact: elms have a high landscape and ornamental value.

Although some surviving native elms have shown some resistance, none reach the resistance level of cultivars for ornamental planting obtained by hybridisation with Asian elms. Mechanical control: sanitation programs, involving the removal of infected trees are the key to slowing disease spread, Chemical control: in urban situations, insecticide spraying of high value trees can be effective in preventing bark beetles from attacking trees.

Phytophthora cinnamomi (Rands), ink disease (Pythiaceae, Chromista)

Marie-Laure Desprez-Loustau

This fungus-like organism is a soil-borne pathogen causing disease on many woody hosts. Primary infections in the root system result in two main types of (visible) symptoms: bleeding cankers on trunks and/or crown symptoms (little leaves, yellowing, wilting) eventually leading to chronic decline or death. Native habitat (EUNIS code): G: Woodland, forest and other wooded land. In the invaded area it occupies the same range of habitats as well as I2: Cultivated areas of gardens and parks. It attacks nearly 1,000 species, mainly woody ornamental, fruit and forest species. *P. cinnamomi* is highly sensitive to frost and is generally thermophilic, with the minimum temperature for growth 5–6°C, optimum 24–28°C and maximum 32–34°C. Native probably from Indonesia/Papua New Guinea, it occurs now in most temperate and subtropical areas of the world (Americas, Europe, Australia, New Zealand, Asia). In Europe, *P. cinnamomi* has been observed mostly in natural habitats of France, Italy, Spain and Portugal, in other countries it occurs mainly in nurseries. Climatic models predict increasing range and damage with climate warming.

P. cinnamomi was likely to have been introduced into Europe in the 18th century. Subsequent spread occurred through transport of contaminated soil and infected nursery stock. In Europe, *P. cinnamomi* has been associated with the widespread mortality of chestnut and evergreen oak trees, especially cork oak and holm oak. *P. cinnamomi* could also be a crucial factor in their natural regeneration. Deciduous oaks, especially pedunculate oak, are hosts to the pathogen, but are generally tolerant. Other native species, mainly from Ericaceae, Cistaceae and Leguminosae plant families, have been known to be infected. It is a serious problem in chestnut and avocado orchards, as well as in ornamental and forest nurseries (on *Chamaecyparis*, heathers, rhododendrons, many conifers and Ericaceae, in addition to chestnuts and oaks). Red oak is highly susceptible to ink disease and is no longer planted in high hazard areas.

Prevention of disease spread involves clean nursery stock but infected plants can remain asymptomatic for a long time, and therefore be distributed since no systematic detection of the pathogen is prescribed by current regulation. Physical control: the potential of soil solarisation to control the pathogen has been demonstrated in infested orchards. Chemical control: phosphite is effective but application of fungicides is mostly not applicable for environmental and economic reasons except in a few situations. Biological control: some soils prevent disease expression probably due to higher levels of antagonisitic microflora. The "Ashburner system", based on amendments to soil and cover crops has been used with success in infested avocado groves.

Seiridium cardinale (Wag.) Sutton & Gibson, cypress canker (Amphisphaeriaceae, Ascomycetes)

Marie-Laure Desprez-Loustau

Seiridium cardinale is a micro-fungus that causes a lethal canker disease on cypress and related conifers. The first evidence of the disease is a browning or reddening of the live bark around the point of entry of the pathogen, followed by dieback and eventually whole-tree death. Conidia are dispersed by wind, rain and insects but the most effective long-distance dispersal has probably been through trade of infected nursery stock. Native habitat (EUNIS code): G: Woodland, forest and other wooded land. In the invaded area it occupies the same range of habitats as well as I2: Cultivated areas of gardens and parks. The pathogen affects several species of *Cupressus*, *Chamaecyparis*, *Cryptomeria*, *Cupressocyparis*, *Juniperus*, *Thuja* and related species. Infection is optimal at 25°C and high relative humidity (autumn to spring) and it is favoured by frost or strong winds producing wounds on trees. The origin of *S. cardinale* remains uncertain but it has been introduced to all continents. In Europe, most records are from Mediterranean countries, but also from Germany, Ireland and the UK.

The outbreak of a destructive cypress blight was first reported from N California on planted Monterey cypress in 1927. The pathogen was then transported into Europe, where it was first reported in the middle of the 20th century. The most li-kely explanation for both epidemics is an accidental introduction of the pathogen on imported nursery stocks of ornamental trees. The disease has caused the loss of millions of cypress trees in S Europe but it is much more prevalent and severe in areas where cypress has been introduced than in regions where it is considered to be native. The incidence and severity has been especially high in some areas of Greece, Italy and S France, reaching 25–75% mortality. An after-effect of the disease is soil erosion in devastated hills. Health and social impact: Italian Cypress is a major feature of the Mediterranean landscape. In ancient times, cypresses were considered as the emblem of wisdom and immortality. Cypress trees today still embellish historical sites and gardens. Economic impact: cypress plantations, grown for highly valued timber and oils for use as pharmaceuticals, have been decimated. Cypresses have also been widely used as efficient windbreaks, so agricultural losses occurred after they were destroyed. Serious economic losses have also affected ornamental cypress trade.

Several resistant clones of Italian cypress are now available for planting. Sanitation measures, including the removal of infected trees and prevention of contact with healthy plants, are key to controlling disease spread. Some fungicides can provide a high degree of control of cypress canker, for use in specific conditions (nurseries, ornamental plantings).

Terrestrial invertebrates

Aedes albopictus
Anoplophora chinensis
Anoplophora glabripennis
Aphis gossypi
Arion vulgaris
Bemisia tabaci
Bursaphelenchus xylophilus
Cameraria ohridella
Ceratitis capitata
Diabrotica virgifera
Frankliniella occidentalis
Harmonia axyridis
Leptinotarsa decemlineata
Linepithema humile
Liriomyza huidobrensis
Spodoptera littoralis

Aedes albopictus (Skuse), Asian tiger mosquito (Culicidae, Diptera)

Alain Roques

Mosquito with a black adult body with conspicuous white stripes. Females are active during the day and are blood-feeders on vertebrates including humans. Adult flight range is limited (200–400 m). Long-distance dispersal (eggs, larvae) mediated by human activity. Average fecundity of 150–250 eggs, up to 5 generations per year. Eggs are laid in the water in tree holes and domestic containers. Breeding populations are present from March to November; overwintering at egg stage. Eggs are resistant to desiccation and cold. Larvae require only 6 mm of water depth to complete the life cycle.

Native habitat (EUNIS code): G: Woodland and forest habitats and other wooded land, J6: Waste deposits. Typically breeds in tree holes and others small water collections surrounded by vegetation but also in peridomestic containers filled with water. Habitat occupied in invaded range: J6: Waste deposits. Mostly opportunistic container breeder capable of using any type of artificial water container, especially discarded tyres, but also saucers under flower pots, bird baths, tin cans and plastic buckets. It can establish in non-urbanised areas lacking artificial containers. Areas at risk have mean winter temperatures higher than 0°C, at least 500 mm precipitation and a warm-month mean temperature higher than 20°C. Native in SE Asia and introduced to SW Europe, Middle East, Africa, the Caribbean and North and South America. Continuous spread all over the world since the late 1970s; accelerated expansion in SW Europe since 2000.

Passive transport by aircrafts, boats and terrestrial vehicles as dormant eggs in goods (used tyres) and as larvae in lucky bamboo *Dracaena* spp. and other phytotelmata shipped with standing water. Interspecific larval competition causes displacement of native mosquito species. Considerable health risk and economic costs result from the biting nuisance and the potential as vector for at least 22 arboviruses (including dengue, chikungunya, Ross River, West Nile virus, Japanese encephalitis, eastern equine encephalitis), avian plasmodia and dog heartworm filariasis *Dirofilaria*.

For monitoring, ovitraps are used: artificial breeding containers (e.g., tyres) baited with frozen CO_2 from dry ice. Mechanical control: removal of discarded tyres. All sources of standing water should be emptied every 3 days in areas at risk; water reserves that cannot be dumped can be treated with a spoonful of vegetable oil to suffocate mosquito larvae. To control larvae, spray water with derivates of *Bacillus thuringiensis israelensis* or larval growth inhibitors (diflubenzuron). To control adults, spray with deltamethrine. Cyclopoid copepod predators (e.g., *Macrocyclops*, *Mesocyclops*) could be used for container-breeding larvae, and fishes and dragonflies in other situations.

Anoplophora chinensis (Förster), citrus longhorned beetle (Cerambycidae, Coleoptera)

Daniel Sauvard

Large, 21–37 mm long, stout beetle with shiny black elytra marked with 10–12 white round spots; long antennae basally marked with white or light blue bands; polyphagous insect attacking over 100 species of broadleaved trees and shrubs; of major concern on *Citrus* spp. Adults can fly up to 1.5 km from their emergence place. Man-mediated long-distance dispersal is possible by infested wood movement or adults hitch-hiking on vehicles. the Netherlands, Females lay eggs throughout their life from spring to late summer. Fecundity varies from tens to more than a hundred eggs per female. Full development is achieved in 1 or 2 years depending on climate and egg laying date. Larvae and pupae overwinter inside their tunnels in wood.

Native habitat (EUNIS code): G1: Broadleaved deciduous woodland, G5: Lines of trees, small anthropogenic woodlands. Habitat occupied in invaded range: G5: Lines of trees, small anthropogenic woodlands. Prefers subtropical to temperate climate; can survive in a large part of Europe. Native range: E Asia. Introduced range: USA, The Netherlands, France, Italy. Increasing frequency of interceptions during the last 10 years in Europe. At the places of introduction, all in urban areas, the beetles are presumed to have been eradicated, except in Lombardy (Italy) where populations are in expansion and their eradication is beginning to prove problematic.

Introduced with infested woody materials, especially bonsai plants. It may disturb broadleaved forest ecosystems by selective tree killing or direct/indirect competition with native xylophagous insects, including protected ones. Causing social impact because in urban areas (streets, private and public gardens) the species is killing trees and *Rosa* shrubs. It is one of the most destructive cerambycid pests of fruit orchards in its native range, especially on *Citrus* trees. Larval tunnels also depreciate harvested wood.

It is difficult to trap; surveys are generally based on visual detection of damage. Mechanical control: destruction of infested trees by chipping or burning; trees could also be protected with fine wire meshes to prevent oviposition. Chemical control: limited because the insects are deep within the tree; possible use of systemic insecticides. Biological control: Natural enemies (parasitoid insects, entomopathogenic nematodes, fungi or bacteria) are under investigation but not yet being used.

Anoplophora glabripennis (Motschulsky), Asian longhorned beetle (Cerambycidae, Coleoptera)

Daniel Sauvard

Large, stout beetle, 20–35 mm long with a jet-black body with white spots on the elytra; the antennae are longer than the body, black with blue rings at segment base. It is a xylophagous species, feeding on a wide range of deciduous trees, mostly species with soft wood such as *Acer* or *Populus* where the larvae live inside the wood, in tree bole or large branches, adults eat bark on small branches. Adults fly up to 1.5 km from the emergence place. Possible man-mediated long-distance dispersal by infested wood movement or adults hitchhiking on vehicles. Eggs are laid throughout female life from spring to late summer; fecundity is variable from tens to more than a 100 eggs per female. Full development is achieved in 1 or 2 years depending on climate and egg laying date. Larvae and pupae overwinter inside wood tunnels.

Native habitat (EUNIS code): G1: Broadleaved deciduous woodland, G5: Lines of trees, small anthropogenic woodlands. Habitat occupied in invaded range: G5: Lines of trees, small anthropogenic woodlands. Prefers subtropical to temperate climate; can survive in a large part of Europe up to S Sweden. Native range: E Asia. Known introduced range: USA, Canada, Austria, France, Germany, Italy. Increasing frequency of interceptions and introductions in Europe during the last 10 years: where the species has been introduced, all in urban areas, eradication attempts were undertaken.

Introduced repeatedly with infested woody materials, especially wood packaging, pallets and waste materials. May disturb European broadleaved ecosystems by selective tree killing or direct/indirect competition with native xylophagous insects, including protected ones. Causing social impact because primary introduction is always in urban areas where it weakens or kills trees in streets, private and public gardens. It is one of the most destructive cerambycid forest pests in its native range, inducing heavy damage in broadleaved stands, including poplar plantations. Larval tunnels also depreciate harvested wood.

Difficult to trap; survey generally based on visual detection of damage. Mechanical control: destruction of infested trees by chipping or burning; trees could also be protected with fine wire mesh to prevent oviposition. Chemical control: limited because the insects live deep within the tree; possible use of systemic insecticides. Biological control: natural enemies (parasitoid insects, entomopathogenic nematodes, fungi or bacteria) under investigation but not yet being used.

Aphis gossypii Glover, cotton aphid, melon aphid (Aphididae, Hemiptera)

Alain Roques

Small aphid, about 2 mm long, phloem-feeding with two viviparous forms. Winged and wingless, highly variable in colour from yellowish green to partly black; immature stages pale yellow to pale green. It is a highly polyphagous species, major pest of cultivated plants in the families Cucurbitaceae, Rutaceae, Malvaceae and of Citrus trees. Flight range of winged adults is limited. Long-range dispersal of eggs, immature stages and adults is man-mediated with the transport of infested plant material. In Europe, it reproduces by apomictic parthenogenesis, and can produce nearly 60 generations a year. The optimal temperature is 21–27°C. Viviparous females produce 70–80 offspring at a rate of 4.3 per day. Developmental periods of immature stages vary from 21 days at 10°C to 4 days at 30°C.

Native habitat (EUNIS code): Unknown. Habitat occupied in invaded range: I1: Arable land and market gardens, I2: Cultivated areas of gardens and parks; glasshouses. Good resistance to summer heat. Dry weather conditions are favourable and heavy rainfall decreases population sizes. Native range unknown. Introduced range: Found in tropical and temperate regions throughout the world except northern areas. It is common in Africa, Australia, Brazil, East Indies, Mexico and Hawaii, Present in most of Europe. Can develop outdoors in S Europe but survives only in glasshouses in N Europe. Increasing introductions all over Europe.

Passive transport with plant trade including vegetables, fruits, cut flowers, ornamental plants, bonsais, and nursery stock. Economically important because nymphs and adults feed on the underside of leaves, or on growing tip of vines, sucking nutrients from the plant. The foliage may become chlorotic and die prematurely. Their feeding also causes distortion and leaf curling, hindering photosynthetic capacity of the plant. In addition, they produce honeydew which allows growth of sooty moulds, resulting in a decrease of fruit/vegetable quantity and quality. It is a vector of crinkle, mosaic, rosette, Tristeza citrus fruit (CTV) and other virus diseases. Its impact is especially important on courgette, melon, cucumber, aubergine, strawberry, cotton, mallow and citrus.

It has become resistant to many pesticides. Insecticides should be used sparingly and in conjunction with other non-chemical control methods. Parasitoid aphidiid wasps (e.g., *Aphidius manii*, *Lysiphlebus testaceipes*), aphelinid wasps (e.g., *Aphelinus gossypii*), predatory midges (e.g., *Aphidoletes aphidimyza*), predatory anthocorid bugs (e.g., *Anthocoris* spp.), predatory coccinelids, and entomopathogenic fungi (e.g., *Neozygites fresenii*) are efficient and available for biocontrol in glasshouse crops.

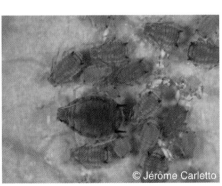

Arion vulgaris (Moquin-Tandon), Spanish slug (Arionidae, Mollusca)

Wolfgang Rabitsch

Large, 7–15 cm long, polyphagous slug feeding on a wide range of green plants, on decayed organic matter and animal carcasses; colour variable, usually orange to dark brown. Although this species is highly mobile for a slug (5–9 m/h) natural dispersal is low. Large anthropogenic distance dispersal of eggs, immature and adult slugs is predominantly via plant material. It is a hermaphroditic species. Mating usually takes place in spring. A slug can produce up to 400 eggs in autumn. Maturity of young slugs is reached within 1 year. It has 1 (sometimes 2) generations per year. Eggs may tolerate adverse conditions.

Native habitat (EUNIS code): G1: Broadleaved deciduous woodland (in the lowlands). Habitat occupied in invaded range: I2: Cultivated areas of gardens and parks, I: Regularly or recently cultivated agricultural, horticultural and domestic habitats, G1: Broadleaved deciduous woodland (in the lowlands). Prefers moist habitats. Native range: SW Europe (parts of Spain, France, the UK). Known introduced range: Large parts of central and the southern part of N Europe, USA (since 1998). In Europe, increasing in abundance, distributional and altitudinal range.

The slug was unintentionally introduced with plant material, package and waste materials. Because of scattered first records across Europe, several independent introduction events are presumed. Has considerable ecosystem impact as important plant defoliator. Outcompetes native slug species due to its large size and its high population densities. It hybridises with the native *A. ater*. If combated with toxic baits, the toxicants can accumulate in predators. Health and social impact as intermediate host of nematode parasites affecting pets. The use of toxic baits could have adverse effects on children and pets in private gardens. It is the most important slug pest in Europe causing severe damage to horticultural plants in private and public gardens and cultivated crops in agriculture. It is also known for transmission of plant pathogens.

Prevention by screening of introduced plant material and packaging. No intentional releases from private gardens to natural sites. Mechanical control: traps; slug fences; collecting by hand and killing slugs with boiling water. Chemical control: several toxicants (e.g., metaldehyds, carbamates) are available. Biological control: providing near-natural habitats so that natural predators are supported; use of nematodes (e.g., *Phasmarhabditis hermaphrodita*) as biocontrol agents.

Bemisia tabaci (Gennadius), cotton whitefly (Aleyrodidae, Hemiptera)

Alain Roques

Small, about 1 mm long, sap-sucking insect with two pairs of white wings (whitefly) with a white to light yellow body, covered with waxy powdery material. Larvae also sap-sucking, feeding on >900 plant species. This taxon corresponds to a species complex with 19 identified biotypes and 2 described cryptic species. Directional adult flight is limited but winds may carry flying adults over long distances due to their small size. Intercontinental dispersal of eggs, nymphs and adults occurs with plant trade. One female produces 80–300 eggs per lifetime. Unmated females produce parthenogenetically only male progeny. Development needs 15–70 days from egg to adult depending on temperature (10–32°C, 27°C is optimal), 11–15 generations per year are possible.

Native habitat (EUNIS code): Unknown. Habitat occupied in invaded range: I: Regularly or recently cultivated agricultural, horticultural and domestic habitats, I1: Arable land and market gardens; glasshouses. Native range: Not known, possibly India. Introduced range: Reported from all continents; present in the field in most of S Europe, restricted to glasshouses in W, central and N Europe. Apparently eradicated in Finland, Ireland and the UK. Widely spread in the last 15 years.

Introduced via trade of ornamental plants. Heavy infestations cause important yield losses, ranging from 20% to 100% depending on the crop and season, to both field and glasshouse agricultural crops and ornamental plants. Three types of damage are observed. Direct feeding damage by adults and larvae may reduce host vigour and growth, cause chlorosis and uneven ripening, and induce physiological disorders. Indirect damage results from the accumulation of honeydew produced by the nymphs, which serves as a substrate for the growth of black sooty mould on leaves and fruit. The mould reduces photosynthesis and lessens the market value of the plant or yields it unmarketable. Finally, it is the most important vector of plant viruses worldwide. As vector of over 100 plant viruses a small population of whiteflies is sufficient to cause considerable damage.

Avoid importations from infested areas. Sequential plantings, avoiding the establishment of affected crops near infested fields, can be used. Adult activity and abundance can be monitored using yellow sticky traps. Chemical control: a number of insecticides provided effective control in the past but resistance has developed rapidly. Biological control: the use of natural enemies such as chalcids (e.g., *Encarsia formosa*, *Eretmocerus* spp.) and entomopathogenic fungus *Verticillium lecanii* is moderately efficient but cannot sufficiently decrease infestations to stop virus transmission.

Bursaphelenchus xylophilus (Steiner & Bührer), pine wood nematode (Parasitaphelenchidae, Nematoda)

Alain Roques

Small, 0.5–1.3 mm long, slender nematode, infesting wood and causing pine wilt disease. A known vector for the nematode is insects, especially longhorned beetles of the genus *Monochamus* (Cerambycidae). Feeds on fungal hyphae (usually *Ceratocystis* spp.) within the wood. Can move actively from one piece of wood to another but incapable of moving from one tree to another. Transcontinental dispersal is man-mediated via infested wood transported with the insect vector, but introduction is also possible without this vector. There are two modes of life cycle. In the propagative mode, nematode larvae penetrate into fresh logs or dying trees through the oviposition scars cut in the bark by the insect vector. They leave the insect, enter the tree, then moult to adult and begin to lay eggs. In the dispersal mode, 3rd instar larvae present in the wood gather in insect pupal chambers, then moult in a special larval stage, which enters the callow adult insect through thoracic spiracles to settle, usually in the tracheae. As many as 100,000 larvae may enter an adult insect vector. The non-feeding third larval dispersal stage is adapted to survive unfavourable conditions.

Native habitat (EUNIS code): G3: Coniferous woodland. Habitat occupied in invaded range: G3: Coniferous woodland, G4: Mixed deciduous and coniferous woodland. The temperature threshold for development is 9.5°C. Native range: North America. Known introduced range: Asia; Europe: observed only in a small area of maritime pine in Portugal (Setubal) since 1999. The Portuguese population has not expanded yet but the nematode has been intercepted in a number of occasions by quarantine services in Europe during recent years.

Often transported by trade of wood and derivates (sawn wood, round wood, wood chips, wood packing material). Most species of conifers endemic to North America are resistant to the wilt disease, but many species are highly susceptible. *Pinus* spp. are the most susceptible species, but the nematode host list includes species of *Abies*, *Chamaecyparis*, *Cedrus*, *Larix*, *Picea* and *Pseudotsuga* as well. The introduction of the nematode into Japan had devastating effects on the native pines in that country. Therefore, it constitutes a real threat for European coniferous forests. The risk associated with introductions of the pinewood nematode from infested countries has led to embargoes on untreated wood chips and timber. An annual loss of US$100 million has been estimated for green lumber exports from the United States to Europe during the 1990s.

Any kind of wood material for exportation must have been heat-treated to a core temperature of 56°C for 30 min in order to kill the nematodes. Cultural practices consist of removing dead or dying trees from the forest to prevent their use as a source of further infection.

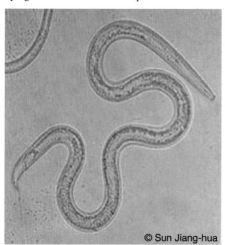

Cameraria ohridella Deschka & Dimić, horse chestnut leaf-miner (Gracillariidae, Lepidoptera)

Sylvie Augustin

Tiny moth, 3–5 mm long with phytophagous larvae, mining leaves of white-flowered horse chestnut *Aesculus hippocastanum*; can also develop on other *Aesculus* species and occasionally on maple *Acer pseudoplatanus*. Local dispersal through adult flight and infested leaves blown by the wind. Long-distance dispersal is human-mediated (vehicles, infested nursery stock). From April onwards; an average of 75 eggs are laid per female on the upper epidermis of horse-chestnut leaves. Produces four (rarely five) mining and two spinning larval instars; usually 3 generations per year in W Europe, but up to 5 overlapping generations depending on weather conditions and climate. Pupae diapausing in leaves.

Native habitat (EUNIS code): Unknown. Habitat occupied in invaded range: G1: Broadleaved deciduous woodland, I2: Cultivated areas of gardens and parks, X13: Land sparsely wooded with broadleaved deciduous trees, X11: Large parks, J: Constructed, industrial and other artificial habitats. Native range: Unknown. Known introduced range: Most of Europe except a part of N Europe and W Russia. It is increasing in distributional range and in abundance in newly colonised areas.

Severely defoliated trees produce smaller seeds with a lower fitness that affects tree regeneration and seriously impairs recruitment of horse chestnut in the last endemic forests in the Balkans. A single tree can host up to 10^6 leaf-miners. Parasitism rates are low as most of the parasitoids emerge when larvae or pupae are not yet available, this may have an important impact on native leaf-miners. There is significant public concern because of aesthetical impact. Main costs are caused by severely damaged horse chestnut trees planted in cities and villages due to replacement of trees.

Mechanical control: complete removal of leaf litter, in which pupae hibernate, is the only effective measure available to lessen the damage. The majority of adults can be prevented from emerging when the leaves are properly composted (e.g., coverage of horse chestnut leaves with a layer of soil or uninfested plant material). Chemical control: aerial spraying with dimilin is efficient. Other "biological pesticides" with fewer non-target effects, such as neem, are also possible but their efficiency is considered to be lower. Stem injection is also efficient, but is not widely registered. It injures trees through necrosis and infections, and the systemic insecticide may cause side effects on non-target species.

Ceratitis capitata (Wiedemann), Mediterranean fruit fly (Tephritidae, Diptera)

Alain Roques

Small fly, 4–5 mm long. Adults with yellowish body, brown abdomen and legs and yellow bands on the wings. Larva, 6–8 mm long at maturity, elongated, cream coloured, and of cylindrical maggot shape. Phytophagous on a wide range of temperate and subtropical fruits. Adult flight range up to 20 km but winds can carry flying adults over longer distances; intercontinental dispersal (eggs, larvae) via infested fruits transported by humans. Before reaching sexual maturation, adults feed 6–8 days on fruit juices. Females lay up to 22 eggs per day, 300–800 eggs during lifetime, under the skin of a fruit just beginning to ripen. Under tropical conditions, overall life cycle is completed in 21–30 days. Adults may survive for up to 6 months.

Native habitat (EUNIS code): G: Woodland and forest habitats and other wooded land. Habitat occupied in invaded range: I: Regularly or recently cultivated agricultural, horticultural and domestic habitats, I1: Arable land and market gardens. Native range: tropical E Africa. Known introduced range: S Europe; regularly observed but not established in other parts of Europe; Africa, Middle East, Central and South America, the Caribbean, Hawaii, Australia. Eradicated in USA except Hawaii. Continuously introduced to European countries with infested fruits; global warming may allow populations to establish at higher latitudes than at present.

Imported with the fruit trade but also with passengers transporting infested fruits during trips. It is probably the most important fruit fly pest, inducing large damage in fruit crops, especially citrus fruits and peach. Fly damage results from both oviposition in fruit, feeding by the larvae, and decomposition of plant tissue by invading secondary microorganisms (bacteria, fungi) that cause the fruit to rot. Their presence often requires host crops to undergo quarantine treatments, other disinfestation procedures or certification of fly-free areas. The costs of such activities and phyto-sanitary regulatory compliance can be significant and definitely affect global trade.

To ensure early detection, traps baited with chemical attractants (especially trimedlure) can be used. Larvae can be killed by soaking, freezing, cooking or pureeing infested fruits. Fruits can be bagged to prevent egg laying. Field sanitation needs to destroy all unmarketable and infested fruits; harvesting fruit weekly also reduces food sources by keeping the quantity of ripe fruit on the trees to a minimum. Chemical sprays are not completely effective. It is better to use foliage baits combining a source of protein with an insecticide to attract both males and females. Biological control: use of sterile insects; release of parasitoids.

Diabrotica virgifera virgifera LeConte, western corn rootworm (Chrysomelidae, Coleoptera)

Wolfgang Rabitsch

Small beetle, 5–6 mm long, with a basic pale greenish-yellow body colouration. Larvae are wrinkled, yellowish-white, with a brown head capsule. It is a major crop pest on maize (*Zea mays*), repeatedly introduced to Europe from North America in the early 1990s and it spreads rapidly. Flight dispersal of adults 20–100 km per year; intercontinental dispersal via the transfer of goods. Up to 1,000 eggs per female during lifetime, preferably in the soil at the base of maize plants. Larvae develop in and on roots of the food plant; adults move upwards and feed on the plant. It develops 1 generation per year, eggs overwinter in diapause.

Native habitat (EUNIS code): E: Grassland. Habitat occupied in invaded range: I: Regularly or recently cultivated agricultural, horticultural and domestic habitats, I1: Arable land and market gardens. Temperature not only influences the development of the larvae, but also triggers flight activity which governs the rate of dispersal. Increased habitat diversity slows the rate of spread. Native range: Probably in the tropics and subtropics of Mexico and Central America. Known introduced range: North America, Europe (Serbia 1992; Croatia, Hungary 1995; Romania 1996; Bosnia and Herzegovina 1997; Bulgaria, Italy, Montenegro 1998; Slovakia, Switzerland 2000; Ukraine 2001; Austria, the Czech Republic, France 2002; Belgium, Netherlands Slovenia, the UK 2003; Poland 2005). Genetic data provides evidence for repeated introductions from America. Spread in Europe continues and the species is expected to colonise all maize producing countries in Eurasia.

Repeatedly transported via vehicles (airplanes, railways, ships). Ecosystem impact include side-effects on non-target species as a consequence of insecticide treatment or biological control are possible but not demonstrated. This species is regarded as one of the most serious pest species of corn in the USA and its damage to crops and chemical control amounts to US$1 billion per year. Current economic damage in Europe is restricted to some countries, but there is clearly a time lag of several years between first record and economic damage. Predictive models forecast an economic impact of about €500 million/year in Europe.

Crop rotation is the most feasible preventative measure, although crop rotation resistant rootworm variants are already known in the USA. Monitoring the spread of adults via pheromone-traps is used as a predictor of damage and further treatment in the following season. Chemical control: several toxicants are applied as granular soil insecticides against the larvae and as aerial spraying against adults (the latter not permitted in most European countries). Biological control: the use of natural enemies (particularly tachinid flies) is currently under investigation.

© Pierre Zagatti

© W Rabitsch

Frankliniella occidentalis (Pergande), western flower thrips (Thripidae, Thysanoptera)

Alain Roques

Tiny, slender insect with narrow fringed wings. Males, 1.2–1.3 mm long, are pale yellow, females, 1.6–1.7 mm long, are yellow to brown, larvae are yellowish-white. Adults and larvae suck plant fluids from flowers and leaves from at least 244 species from 62 families. It is an outdoor pest as well as a glasshouse pest. Adults can be easily carried by winds, but also by clothes, equipment and containers not properly cleaned. Intercontinental dispersal of eggs, larvae and adults is taking place with the plant trade. This species reproduces in glasshouses with 12–15 generations/year. The overall life cycle lasts from 44 days at 15°C to 15 days at 30°C. A female can lay 20–40 eggs; unmated females produce males. Different developmental stages are typically found in different parts of plants: eggs in leaves, flower tissue and fruits; nymphs on leaves, in buds and flowers; pupae in soil or in hiding places on host plants such as the bases of leaves; adults on leaves, in buds and flowers.

Native habitat (EUNIS code): I: Regularly or recently cultivated agricultural, horticultural and domestic habitats. In the invaded area it occupies the same range of habitats, I1: Arable land and market gardens; glasshouses. Native range: North America. Introduced range: Reported from all continents; present in glasshouses in N and central Europe but already in the field in S Europe. Continuous and rapid spread since 1980.

Imported with the ornamental plants (e.g., cut flowers, potted plants). Flowers and foliage of a great number of economically important crops are affected, in glasshouses as well as outdoors. On ornamental flower crops, feeding induces discolouration, indentation, distortion and silvering of the upper leaf surface as well as scarring and discolouration of petals and deformation of flower heads, largely reducing their economic value. In orchids, eggs laid in petal tissues cause a 'pimpling' effect on flowers. The thrips also kills or weakens terminal buds and blossoms in fruit trees (e.g., apricot, peach) and roses, and on most fruiting vegetables, especially cucumbers. In addition, nymphs are vectors of tobacco streak ilarvirus (TSV) and tomato spotted wilt virus (TSWV), which is inducing severe diseases on ornamental and vegetable crops in Europe.

Blue sticky traps can be used to detect initial infestation and monitor adult population levels. Chemical control: it is difficult because the thrips are resistant to most pesticides and feed deep within the flower or on developing leaves. Predatory mites (e.g., *Neoseiulus cucumeris*, *Amblyseius* spp. and *Hypoaspis* spp.) and minute pirate bugs (e.g., *Orius laevigatus*, *O. insidiosus*) provide effective biological control, in glasshouses.

Harmonia axyridis (Pallas), harlequin ladybird (Coccinellidae, Coleoptera)

Helen Roy and David B. Roy

Polyphagous predatory ladybird, 5–8 mm long, variable in colour pattern (yellow to orange to black) with a variable number of spots (0–21). Highly dispersive, flying readily between host plants during breeding periods, migrates over long distances in Asia and America. 20–50 eggs produced per day, 1,000–4,000 in their lifetime; adults typically live for a year, reproducing for 3 months; generally bivoltine but can produce 4 generations per year in favourable conditions.

Native habitat (EUNIS code): G: Woodland, forest habitats and other wooded land. In the invaded area it occupies the same range of habitats as well as G3: Coniferous woodland, G5: Lines of trees, small anthropogenic woodlands, recently felled woodland, early-stage woodland and coppice, I: Regularly or recently cultivated agricultural, horticultural and domestic habitats, I1: Arable land and market gardens, I2: Cultivated areas of gardens and parks, J1: Buildings of cities, towns and villages. The wide native range in Asia shows that it can reproduce in both warm and cool climates and it is well adapted to temperature extremes. Native range: Central and E Asia. Known introduced range: America, South Africa, Egypt, Europe. Increasing trend.

Introduced intentionally as a biocontrol agent for aphids and unintentionally in horticultural/ornamental material. Causes reduction in biodiversity of other aphidophages and non-pest insects by resource competition, intraguild predation and direct intra-specific competition. They are also a pest of orchard crops (apples and pears) because as aphids become scarce in autumn the beetles feed on soft fruit causing blemishing and an associated reduction in the market value. Their tendency to aggregate in clusters of grapes prior to harvest makes them difficult to separate from the fruit and so are sometimes processed during wine making. The alkaloids contained within these beetles adversely affect the taste of the vintage. The beetle's propensity to swarm and its large aggregations formed in buildings during the winter are regarded as a nuisance. Economic impact derives from the wine industry, reduction in fruit quality and management measures required in domestic dwellings.

Stopping its use as a biocontrol agent and ensuring that fruit and cut flower imports are free from the ladybird will reduce introduction events. Invasion into households can be limited by covering entrances with fine mesh. Adults and late instar larvae can be removed from unwanted locations manually, e.g., using a vacuum cleaner. Light traps can attract adults but the efficiency of these is not yet quantified. Chemical control in field situations such as orchards and vineyards is not applicable because of the impact of insecticides on other aphidophages and beneficial insects.

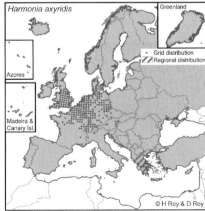

Leptinotarsa decemlineata Say, Colorado beetle (Chrysomelidae, Coleoptera)

Carlos Lopez-Vaamonde

Adult beetles up to 11 mm long, yellow elytra with 10 characteristic black longitudinal bands. Main natural spread of beetle over large areas is by wind-borne migration. Females usually deposit eggs on the underside surface of the host plant leaves. An egg mass may contain 10–40 eggs. Most adult females deposit over 300 eggs during 4–5 weeks, but they can lay up to 800 eggs. Potatoes are the preferred host, but it may feed and survive on a number of other Solanaceae: eggplant, tomato, pepper, tobacco, ground cherry, horse-nettle, common nightshade, belladonna, thorn apple, henbane, and its first recorded host plant: buffalo-bur *Solanum rostratum*. Larvae are hardy and resistant to unfavourable weather.

Native habitat (EUNIS code): G1: Broadleaved deciduous woodland. Habitat occupied in invaded range: I1: Arable land and market gardens, I2: Cultivated areas of gardens and parks. Beetles are sensitive to cold temperatures. They need at least 60 days of temperature over 15°C in summer and winter temperatures not falling below 8°C. Native range: Mexico, where beetles are still present and feed on wild Solanaceae such as *Solanum rostratum*. Known introduced range: beetles were accidentally introduced into USA. In 1922 it was introduced to France from where it expanded almost throughout the European continent and parts of Asia in about 30 years. Capable of adapting to different climatic conditions and different host plants this beetle is constantly moving to new areas. Its distribution is limited by temperature and therefore climate warming could further expand its distribution range.

International trade appears to be the most likely pathway for introduction on imported commodities such as fresh vegetables from infested areas. Beetles can also be spread through wind and attachment to all forms of packaging and transport. It is a serious pest of potatoes. Both adults and larvae feed on potato leaves and the damage can greatly reduce potato yields. Beetles can also be a pest of other solanaceous plants such as tomato, aubergine, tobacco and peppers.

This beetle may be managed culturally by crop rotation. Mechanical control: destruction of crop debris is very effective at reducing population levels. Chemical control: insecticides are commonly used to control populations of Colorado potato beetle, but resistance to insecticides develops rapidly. Biological control: there is a long list of natural enemies. *Bacillus thuringiensis* and some species of nematodes have been used as control agents.

Linepithema humile (Mayr), Argentine ant (Formicidae, Hymenoptera)

Wolfgang Rabitsch

Ant of light brown colour; females are 4.5–4.9 mm and workers 2.1–3.0 mm long. They are omnivorous, feeding on honeydew, nectar, insects and carrion. Local dispersal by budding of large unicolonial nests (up to 150 m/year); long-distance dispersal within the introduced ranges human-mediated. Haplodiploid system with sterile workers; polygynous (multi-queened) nests; social organisation variable in its native range (from multicolonial to unicolonial), but entirely unicolonial in introduced range, with surface area covered by single supercolonies ranging from 2,500 m² to many square kilometers. In the absence of queens, workers can lay unfertilised eggs, which develop into fully functional males.

Native habitat (EUNIS code): G: tropical and subtropical natural forests. Habitat occupied in invaded range: I: Regularly or recently cultivated agricultural, horticultural and domestic habitats, G4: Mixed deciduous and coniferous woodland; preferably associated with disturbed, human-modified habitats in its introduced range, but may also invade natural habitats (e.g., oak and pine woodland in the Mediterranean). Prefers moderate temperature and moisture level. Native range: South America. Known introduced range: The species occurs throughout the world on all continents, especially in mediterranean-type climates, and many oceanic islands. Ecological niche models predict that with changing climate the species will expand at higher latitudes.

Transported with vehicles (airplanes, ships) together with goods and materials, soil, plants, etc. The supercolonies, by reducing costs associated with territoriality, allow high worker densities and interspecific dominance in invaded habitats. It has displaced, even leading to species extinction in some cases, native ant species in many parts of the world. It also competes with other arthropod species for resources (e.g., for nectar with bees) and reduces local arthropod diversity; taxa other than arthropods are also affected (e.g., nest failure of birds). Ecosystem level impacts such as reduction of seed dispersal capacity and disruption of mutualistic associations with other species are documented. Regarded as a nuisance for tourism at some places on the Mediterranean coast. Homoptera-tending may increase Homoptera populations causing some crop loss. However, costs are considered to be low.

Several toxicants applied via ant baits, including insect growth regulators. Application needs supervision to optimise results and to minimise side-effects on non-target species. Biological control: since Argentine ants prefer disturbed sites, any extensification of land use or reduction in monoculture may help prevent high densities of this species.

Liriomyza huidobrensis (Blanchard), serpentine leaf miner (Agromyzidae, Diptera)

Alain Roques

Adult small, 1.3–2.3 mm long, compact-bodied fly, of greyish-black colour; larvae headless maggot up to 3.3 mm in length, yellow-orange at maturity. Larvae are leaf miners on a wide range of hosts, especially economically important vegetables and ornamental plants in both glasshouses and outdoors. Adult flight range is limited. Long-range dispersal (eggs, larvae) with human-transported infested plant material, including cut flowers. The vase life of chrysanthemums is sufficient to allow completion of the life-cycle. Under laboratory conditions, a female lays about 100–130 eggs but as much as 250 eggs have been observed. Eggs are laid into the leaf tissue. Larvae tunnel within the leaf tissue forming characteristic mines, then cut a semi-circular opening in the tissue and drop to the soil to pupate. The life cycle could be as short as 14 days at 30°C or as long as 64 days at 14°C. Generations follow in quick succession as long as the growing conditions of the host-plant provide suitable food.

Native habitat (EUNIS code): F5: semi-arid and subtropical habitats in pre-saharian Africa. Habitat occupied in invaded range: I1: Arable land and market gardens, I2: Cultivated areas of gardens and parks; glasshouses. Optimal temperatures for feeding and egg laying range between 21°C and 32°C. Egg-laying is reduced at temperatures below 10°C. All stages are killed within a few weeks by cold storage at 0°C and above 40°C. Native range: South America. Introduced range: Central America, most of Asia, Kenya; present outdoors in S Europe, but mainly a glasshouse pest in N Europe. Spreading tendency.

Passive transport with plant trade including vegetables, cut flowers and nursery stock. A serious pest for the floriculture industry where leaf-miner damage directly affects the marketable portion or in vegetable crops where the leaves are sold as edible part.

Sticky traps can be used to monitor adult flies. Crop rotation is an effective pest management tool as is avoiding varieties which are highly susceptible to leaf-miner infestations in glasshouses. There is little information about leaf-miner tolerance of vegetables in the field. In field vegetables, cultivation of crop debris or removal of infected plant material is recommended. *L. huidobrensis* adults are resistant to conventional insecticides. At present, the only effective insecticides are translaminar insecticides (abamectin, cyromazine, neem and spinosad), which penetrate the leaves to affect the leaf-miner larvae. Parasitoid wasps, e.g., *Diglyphus isaea* and *Dacnusa sibirica* are available for control in glasshouse crops. These parasites will not be effective for vegetables growing in the field. However there may be natural parasites present that can reduce the population.

Spodoptera littoralis (Boisduval), African cotton leaf worm (Noctuidae, Lepidoptera)

Carlos Lopez-Vaamonde

It is a polyphagous moth, up to 2 cm long with a wingspan of 4 cm; larvae 35–45 mm long, its colour varying from grey to reddish or yellowish; eggs laid in batches covered with orange-brown hairs. Flight range of moths can be 1.5 km during a period of 4 h overnight. Long distance dispersal occurs through eggs and larvae present on plant material, cut flowers and vegetables. 1,000–2,000 eggs laid per female 2–5 days after emergence; egg masses of 100–300 on the lower leaf surface of host plants. Life cycle lasts 19–144 days.

Native habitat (EUNIS code): F5: semi-arid and subtropical habitats in pre-Saharan Africa. Habitat occupied in invaded range: F5: Maquis, matorral and thermo-Mediterranean brushes, F6: Garrigue, F8: Thermo-Atlantic xerophytic habitats, H5: Miscellaneous inland habitats with very sparse or no vegetation, I1: Arable land and market gardens, I2: Cultivated areas of gardens and parks. Larvae are extremely sensitive to climatic conditions, especially to combinations of high temperature and low humidity; temperatures above 40°C or below 13°C increase mortality. Native range: tropical and subtropical Africa. Introduced range: Africa, S Europe and Asia Minor. It is one of the most commonly intercepted species in Europe, for example on imported ornamentals. Not yet established in NW Europe. It is a potentially serious pest of glasshouse crops in N Europe.

Trade appears to be the most likely pathway for introduction, on imported commodities such as glasshouse crops, both ornamentals and vegetables from infested areas. Adult moths can also be spread through wind, attached to or transport by another organism or through other natural means. It is one of the most destructive agricultural lepidopteran pests within its subtropical and tropical range. It is attacking plants from 44 families including grasses, legumes, crucifers and deciduous fruit trees. In N Africa it is damaging vegetables, in Egypt cotton and in S Europe plant and flower production in glasshouses or vegetables and fodder crops.

It is important to seek assurance from suppliers that plants are free from this pest as part of any commercial contract. Avoid importing plant material from infested areas. Carefully inspect new plants on arrival, including any packaging material, to check for eggs and caterpillars and for signs of damage. As the adults are nocturnal, light or pheromone traps should be used for monitoring purposes. Mechanical control: physical destruction of insects and any plant material infested by this pest is recommended. Egg masses can be hand collected. Chemical control: there are many cases of resistance to insecticides. Biological control: includes the use of microbial pesticides, insect growth regulators and slow-release pheromone formulations for mating disruption.

Terrestrial plants

Acacia dealbata
Ailanthus altissima
Ambrosia artemisifolia
Campylopus introflexus
Carpobrotus edulis
Cortaderia selloana
Echinocystis lobata
Fallopia japonica
Hedychium gardnerianum
Heracleum mantegazzianum
Impatiens glandulifera
Opuntia maxima
Oxalis pes-caprae
Paspalum paspalodes
Prunus serotina
Rhododendron ponticum
Robinia pseudoacacia
Rosa rugosa

Acacia dealbata Link, silver wattle, blue wattle (Fabaceae, Magnoliophyta)

Hélia Marchante

This fast growing tree can reach up to 30 m in height. Seeds are dispersed by animals, namely birds and ants, and by strong winds. However, the majority of the seeds accumulate under the tree. Seeds are triggered to germinate en masse following fires. It also has vegetative reproduction, forming new shoots from lateral roots. In the native habitat several Coleoptera, Lepidoptera and Hemiptera attack this species. Large amounts of long-lived seeds can accumulate in the soil seed banks persisting for 50 years.

Native habitat (EUNIS code): C3: Littoral zone of inland surface waterbodies, G: Woodland and forest habitats and other wooded land. In the invaded area it occupies the same range of habitats, mainly G1: Broadleaved deciduous woodland, G2: Broadleaved evergreen woodland, G3: Coniferous woodland, G5: Lines of trees, small anthropogenic woodlands, recently felled woodland, early-stage woodland and coppice), B1: Coastal dune and sand habitats, I1: Arable land and market gardens (specifically abandoned fields), J1: Buildings of cities, towns and villages, J4: Transport networks and other constructed hard-surfaced areas. It prefers moist but not waterlogged soils, especially stream-sides, tolerates drier soils, strong wind, and frosts down to −7°C. Native range: Australasia. Introduced to S Europe, South Africa, New Zealand, California, India, Chile, Madagascar. Increasing invasion after fire and in disturbed areas.

In Europe known since 1824, it is still planted for forestry, as an ornamental plant and for soil stabilisation. It can form dense, impenetrable stands that prevent the development of other species. It has allelopathic properties and is an N fixing species that increases N soil content. Dense thickets disrupt water flow and increase erosion along stream banks. Pollen allergies are frequently reported. Invasion of forests decreases its productivity and control actions involve enormous economic costs, mainly due to the necessity of several follow up control actions.

Preserving the commercialisation of *A. dealbata* reduces the risk of invasion. Seedlings and small trees can be pulled or dug out, large trees must be cut. It readily re-sprouts both from the cut stump and from lateral roots, implying that steps must be taken to kill the stump and its root system. Cutting the stump level, covering it with 10 mm black plastic and follow up control must be assured to prevent resprouts from lateral roots or stumps. Bark removal from 1 m high to ground level is also effective and decreases resprouting. Glyphosate application at the stump immediately after cutting is quite efficient. Foliar sprays can also be applied. *Melanterius maculatus* (Curculionidae) is a seed-feeding weevil, released in South Africa, but its damage has not been quantified yet.

Ailanthus altissima (Mill.) Swingle, tree of heaven (Simaroubaceae, Magnoliophyta)

Corina Başnou and Montserrat Vilà

This fast growing deciduous tree, 8–10 m high, has fruits that are very distinctive for their long samaras forming large bunches, turning reddish in summer. The winged seeds are dispersed by wind, water, some birds and machinery. It is a dioecious plant (either male or female flowers). Flowering and pollination occur in late spring. Although disagreeable to humans, the strong odour of flowers attracts honeybees and other insects. One tree can produce up to 325,000 seeds per year. It also reproduces vegetatively, spreading profusely from suckers, sprouts emerging up to 15 m from the nearest stem. It has high resistance to herbivory and seed predation. Most seeds are viable, even those that overwinter on the tree.

Native habitat (EUNIS code): Very little is known about the native habitat. In the invaded range it mostly invades interior forests via roads or windthrow. F5: Maquis, matorral and thermo-Mediterranean brushes, G: Woodland and forest habitats and other wooded land, I: Regularly or recently cultivated agricultural, horticultural and domestic habitats, J1: Cities, towns and villages, J4: Transport networks and other constructed hard-surfaced areas, J6: Waste deposits, X24: Domestic gardens of city and town centres. Markedly resistant to disturbed or stressed habitats, tolerates varying temperatures, humidity, light and moisture levels. It is highly tolerant of poor air quality and poor soils, but intolerant of shade and sensitive to ozone. Native in China and introduced into Europe, Macaronesian Islands, Africa, America, Asia, and Australia. In Europe it is increasing close to urban areas, probably due to climate warming.

It was introduced in Europe in the 1700s as an ornamental. Its dense thickets competes with the native vegetation, there are also allelopathic effects due to ailanthone. Contact with plant sap can produce dermatitis. Long exposure to the sap can produce myocarditis due to plant quassinoid proteins. The root system can damage pavements, archaeological remains, walls, etc.

Avoid planting it as an ornamental. Treated areas should be rechecked several times a year, in order to avoid new root suckers. Repeated cutting, mowing or hand pulling could be effective for young infestations. Prolific sucker resprouting needs a chemical application. Spraying the foliage with glyphosate products is very effective. Foliar applications should be followed 3 weeks later by basal bark applications. Another possibility is the stump treatment using Garlon, which must be conducted immediately after cutting. Although suckering from the roots is inevitable after cutting, this method will prevent vigorous stump sprouts. It may be attacked by several fungal pathogens, such as *Verticillium dahliae* and *Fusarium oxysporum*.

Ambrosia artemisiifolia Linnaeus, common ragweed (Asteraceae, Magnoliophyta)

François Bretagnolle and Bruno Chauvel

Summer monoecious annual plant 0.2–2.5 m tall. The male flowers (2–4 mm) are grouped at the end of branches, while female flowers are located at the bases of upper leaves. Pollinated by wind, produces woody reddish-brown indehiscent fruits (akenes) with one seed per fruit, 3–4 mm long. Akenes fall down but are also dispersed via birds or by water. Anthropogenic activities are suspected to be a strong source of dispersal along roadsides and in fields. Seeds germinate after rainy periods and cold temperatures. Rodents, birds and ants are probably efficient seed predators. Seeds may remain viable for at least 40 years in the soil seed banks. It has a strong ability for re-growth after mowing.

Native habitat (EUNIS code): Streams and temporary watercourses, arable land and market gardens, cultivated areas of gardens and parks, transport networks. Habitat occupied in invaded range: C2: Running waters, including springs, streams and temporary water courses. I1: Arable land and market gardens, I2: Cultivated areas of gardens and parks, J1: Building of cities, towns and villages, J2: Low density building, J4: Transport networks and other constructed hard-surface area, J6: Waste deposits. This plant establishes itself in freshly moved soil and disturbed areas. The species prefers nutrient rich bare soils with neutral or acid pH. Plants are resistant to high summer temperatures, drought and moderate soil salinity. Soil fertilisation increases fruit production. Native in North America. Introduced in many European countries, South America, Australia, China and Japan. The species is actually increasing in all European countries, in particular in agricultural fields, along roadsides and in river banks.

It was introduced in the mid 19th century in France and Germany as contaminants of agricultural products from North America. The species is mainly spread as contaminant of agricultural products, machinery or construction materials. The species is highly allergenic and is the prime cause of hay fever. During the pollen release period, causes rhino-conjunctivitis, asthma and more rarely contact dermatitis and urticaria. In colonised areas, ragweed rapidly becomes the main allergenic species. Medical costs of hay fever are important in highly infested regions.

Avoid transport of contaminated soil. Food for birds can also be highly contaminated. Agricultural machinery may also contaminate non-invaded fields. Plants can sprout vigorously when cut. The sensitivity of common ragweed to herbicides is high. A single treatment, carried out toward the end of the season, would stop the cycle of the plant and help to control its propagation. Although tried in Russia, no successful biological controls have been developed to date.

Campylopus introflexus (Hedw.) Brid., heath star moss (Dicranaceae, Bryophyta)

Philip W. Lambdon

A moss with leaves tapering to a long, whitish hair-point. Spores disperse with wind and vegetative fragments are important in short-distance colonisation since plants regenerate from fragments of leaf. May be grazed by typical moss herbivores such as slugs or pill beetles (Bhyrridae). The leaf fragments can persist for up to a year. Native habitat (EUNIS code): probably similar to European habitats which include B1: Coastal dune and sand habitats, D1: Raised and blanket bogs, F4: Temperate shrub heathland, H3: Inland cliffs, rock pavements and outcrops, J1: Buildings of cities, towns and villages, J2: Low density buildings, J6: Waste deposits. It occurs also sporadically on decaying logs. It is a very successful pioneer species, often amongst the first to colonise recently burned or disturbed areas, especially in forests and on heathland. It thrives best in acid conditions with moderately high nutrient levels (e.g., areas which suffer from atmospheric pollution). It almost certainly originated in the southern hemisphere, possibly in South America where it is abundant on NW savannahs. The first European observations were from England and W France in 1941. It was found throughout the UK and Ireland by the 1960s, and had become problematic in the Netherlands and Germany by the 1970s. Most common in NW oceanic parts of Europe, including islands as remote as Iceland, and especially abundant on North Sea and German Baltic coasts. In Europe it is spreading rapidly eastwards, where it has reached Lithuania and Russia, and southwards, to N Spain and Menorca. It has recently penetrated to Switzerland, N Italy and the Czech Republic. Similar expansion is occurring in SW North America.

The arrival to Europe was possibly by ship. Local dispersal may be facilitated by the hooves of grazing animals or vehicles. Studies from dune heath have shown that it can replace much of the ephemeral cover of specialist lichens, which are relatively common growing on trees elsewhere, but localised to dunes in exposed places. In north-central Europe, it has become problematic on sandy heaths dominated by *Cladonia* lichens and grey hair grass *Corynephorus canescens*. However, invasions do not slow the rate of succession, and although heather seedlings germinate less well than on bare ground, they grow better once established.

Burning or mechanical removal may sometimes be practical. Trials using livestock to trample and fragment the mats have been largely unsuccessful because they merely disperse vegetative propagules. Heavy liming to reduce the acidity of the soil is another option, but the effectiveness of this measure is unproven and the risk of environmental damage may often be prohibitive.

Carpobrotus edulis (L.) N.E.Br., freeway iceplant (Aizoaceae, Magnoliophyta)

Pinelopi Delipetrou

Succulent, trailing perennial, rooting at nodes and forming large, dense mats. Flowers are large, yellow, pinkish or purple. In the Mediterranean basin, *C. edulis* hybridises with *C. acinaciformis* forming a hybrid complex known as *C. affine acinaciformis*; therefore flower size and colour can be variable. The fleshy, indehiscent fruits provide food for deer, rats, rabbits etc. and seed germination is enhanced by the ingestion of fruits. Vegetative propagation by runners (rooting at nodes). Herbivory by mammals may cause significant seedling mortality, but once established it is not affected by herbivory or competition. Rodents or insects may be important seed predators. Ungerminated seeds remain viable for at least 2 years.

Native habitat (EUNIS code): B1: Coastal dune and sand habitats, B2: Coastal shingle habitats, B3: Rock cliffs, ledges and shores, including the supralittoral, H3: Inland cliffs, rock pavements and outcrops, H5: Miscellaneous inland habitats with very sparse or no vegetation, H6: Recent volcanic features. In the invaded range the plant occupies the same habitats as in the native range and C3: Littoral zone of inland surface waterbodies, F: Heathland, scrub and tundra habitats, F6: Garrigue, G1: Broadleaved deciduous woodland, G3: Coniferous woodland, I: Agricultural, horticultural and domestic habitats, J: Constructed, industrial and other artificial habitats. Warm temperate to dry climate, sensitive to frost, resistant to drought. Grows on well-drained acid to alkaline and also saline soils. Can grow on nutrient poor soils. It prefers to grow in the sun but can develop well in the shade. Native in the Cape region of South Africa, invasive from W Europe to the Mediterranean, America and Australasia. Apparently increasing due to increasing landscape use.

Intentionally introduced, mainly as an ornamental or landscaping plant, also as a medicinal plant. Widely used for erosion control on sandy habitats and planted along highways, in open sites and gardens. It forms impenetrable mats and competes aggressively with native species, threatening rare and endangered species. Can modify soil properties by increasing soil N and organic C and by reducing soil pH. In dune habitats it hinders the disturbance regime. Hybrids are very vigorous and may lead to intensified invasion. Avoid planting as ornamentals and dumping plant debris in the wild. Prescribed burnings can reduce the seedbank. Manual eradication is a most effective and cost-efficient method. Plant remains should be removed because they regenerate. Broad spectrum herbicides such as glyphosate kill the plant. Chlorflurenol has been used along roadways. Biological control: intense herbivory may suppress seedling establishment and slow invasion.

Cortaderia selloana (Schult. & Schult.f.) Asch. & Graebn., pampas grass (Poaceae, Magnoliophyta)

Corina Başnou

An erect perennial, gynodioecious tussock grass, up to 2–4 m tall. It has 1–3 m long, deep green, attenuate leaves with serrulate margins. Inflorescences consist of several large nodding plumose panicles and flower from August to December. They are wind-pollinated and seeds wind dispersed. Female plants produce a high number of seeds (up to 10^6/individual), they have long hairs and are capable of reaching distances up to 30 km. It is a good fodder for cattle. Plants resist water stress by developing an extensive root system and by minimizing above ground growth.

Native habitat (EUNIS code): The grass grows on moist, sandy soils along river margins. Habitat occupied in invaded range: B1: Coastal dune and sand habitats, C: Inland surface water habitats, E: Grassland and tall forb habitats, G5: Lines of trees, small anthropogenic woodlands, recently felled woodland, early-stage woodland and coppice, J4: Transport networks and other constructed hard-surfaced areas, J6: Waste deposits, X1: Estuaries. The plant tolerates high light intensity, water stress, harsh winters and high summer temperatures but prefers low pH sandy soils. Native in southern South America. Introduced into S Europe, South Africa, Australia, New Zealand, Hawaiian islands and the Pacific coast of the USA. Increasing in old fields and marshes close to urbanised areas.

Planted as an ornamental. In Europe, it was first introduced into cultivation in Ireland and France; it was also introduced for erosion control. It has some value as fodder and it is ideal for barrier or windbreak plantings. It is a very invasive plant, forming dense, often impenetrable, stands. It can impact shorebird nesting sites on dunelands. Pampas grass rapidly increases its density and colonises semi-natural areas, being a threat to native diversity. Risks to human health are low. The leaves can produce superficial cuttings and flowers may provoke allergies in summer. *C. selloana* increases fire hazard, damages grazing lands and affects visibility on roads.

Avoid planting it as an ornamental, especially female plants. Early detection of invaded habitats. Small plants and seedlings can be hand-pulled with protective gloves. Larger plants can also be removed mechanically. For large scale infestations, other mechanical methods could involve the use of bulldozers. Removals should be conducted before flowering to avoid seed dispersal. Herbicide application of glyphosate can improve the mechanical removal of plants. Biological control: grazing by cattle at an early stage of its invasion proved successful in New Zealand.

Echinocystis lobata (Michx.) Torr. & Gray., wild cucumber (Cucurbitaceae, Magnoliophyta)

Stefan Klotz

This annual growing vine can climb up to 12 m. It germinates in May, flowers from July to September and leaves die in October. The flowers are pollinated by insects but are also self-fertile. They are greenish to white and both sexes can be found on the same plant. The plant is often damaged by late and early frosts. The relatively heavy seeds fall down out of the fruits which open when mature. Additionally, the seeds are often transported by water during flooding along river margins. Relatively high soil temperatures are necessary for seeds to germinate in spring. In North America, the striped cucumber beetle *Acalymma vittata* is known as herbivore and transmitter of cucumber-wilt bacteria. *Anasa armigera* (Heteroptera) was recorded as another herbivore in the native range. Seeds in the soil may remain viable for more than a year.

Native habitat (EUNIS code): G: Woodland, forest and other wooded land, C3: Littoral zone of inland surface water bodies. Habitat occupied in invaded range: E5: Woodland fringes and clearings, tall form stands, F9: Riverine and fen scrubs, G1: Broad leaved deciduous woodland, I2: Cultivated areas of gardens and parks. It usually grows in floodplains and forest fringes, and is therefore associated with high light levels. Native in North America, invasive in temperate and continental Europe. There has been increasing invasion within the last 20 years along the main rivers and in floodplains from western to eastern Europe up to the Asian border in Russia.

The plant was introduced in late 19th and early 20th century as an ornamental and medical plant and planted in several botanical gardens. The first information on escaped plants is from Slovakia (1906). This vine branches very fast, covering large areas and overgrowing native vegetation. The plant contains toxic substances (cucurbitacines).

Planting as an ornamental plant in and near floodplains should be avoided. Mechanical control: seedlings can be removed easily. Chemical control: herbicide use is impossible in floodplain areas.

Fallopia japonica (Houtt.) Ronse Decr., Japanese knotweed (Polygonaceae, Magnoliophyta)

Petr Pyšek

This fact sheet also concerns *Fallopia sachalinensis* (F. Schmidt) Ronse Decr. and *Fallopia* × *bohemica* (Chrtek & Chrtková) Bailey. Herbaceous perennials with robust erect stems up to 4 m tall and extensive system of rhizomes, 15–20 m long, penetrating 2–3 m deep in soil. Human-transported soil contaminated with rhizomes is the major dispersal mode. Rhizome fragments of 7 g are able to regenerate. Functionally dioecious, flowers are exclusively entomophilous. Only one *Fallopia japonica* var. *japonica* female clone has been introduced to Europe. The winged achenes are dispersed by wind and water.

In its native range, *F. japonica* is a pioneer on volcanic slopes. *F. sachalinensis* occurs in tall-forb communities at forest edges, avalanche clearings, riverbanks and coastal cliffs, but also colonises lava flows. Habitat occupied in invaded range (EUNIS code): E2: Mesic grasslands, E3: Seasonally wet and wet grasslands, F9: Riverine and fen scrubs, FA: Hedgerows, J6: Garrigue. Knotweeds invade disturbed habitats and thrive on a wide range of soils, with pH ranging from 3 to 8. *F. japonica* is native to Japan, Korea, Taiwan, and N China, *F. sachalinensis* to the Sakhalin Island through Hokkaido to Honshu. Introduced range: *F. japonica* is invasive in Europe, Canada, USA, Australia and New Zealand. The distribution of *F. sachalinensis* in Europe is similar. It is also reported from USA, Canada, New Zealand and S Japan. All three taxa are increasing, the hybrid is spreading at a faster rate than its parents in Central Europe.

F. japonica and *F. sachalinensis* were introduced into Europe as garden ornamentals in the 19th century and soon escaped from cultivation. The hybrid resulted from hybridisation in the invaded range. *F. japonica* damages native riparian communities by reducing light availability, through the alteration of the soil environment and through the release of allelochemicals. Soil K and Mn is greater under *F. japonica* than under native vegetation. *F. japonica* decreases soil bulk density and increases organic matter content, water content and nutrient levels. Prolific rhizome and shoot growth can damage foundations, walls, pavements, and drainage works, and causes flood hazards by increasing resistance to water flow and damaging flood prevention structures.

Plant debris should not be released in the wild. The combination of digging the soil surface and spraying with glyphosate is the most efficient. *F. sachalinensis* is the easiest to control of the three taxa, the hybrid is the most resistant. Biological control: in the native range of Japan, the leaf-feeding chrysomelid beetle *Gallerucida nigromaculata* regulates *F. japonica* population growth, and is under consideration as a biocontrol agent.

Hedychium gardnerianum Shepard ex Ker-Gawl, Kahili ginger (Zingiberaceae, Magnoliophyta)

Philip E. Hulme

A large leafy herb of 1–3 m in height. Lance-shaped leaves arise of stems or basal stock. Fragrant yellow flowers are borne in long 25–30 cm spikes. Capsules contain seeds that are initially red then grey. This species is found in open habitats in warm moist climates. Dispersed by stolons where already established and by root fragments. Conspicuous, fleshy, red seeds are dispersed by frugivorous birds as well as humans. Dumping of ginger rhizomes on roadsides or in bushland has also been a major source of spread.

Native habitat (EUNIS code): Semi-evergreen rainforest. Habitat occupied in invaded range: F9: Riverine and fen scrubs, H3: Inland cliffs, rock pavements and outcrops, G2: Broadleaved evergreen woodland, G3: Coniferous woodland. Invaded habitat also includes Macaronesian laurisilvas. The plant grows in wet habitats up to 1,500 m in Macaronesia. Prefers to grow in open, light-filled environments which are warm and moist, but will readily grow in semi and full shade beneath forest canopy, such as in regenerating forest, streamside and alluvial forests, forest gaps and gullies. Native from India, Bhutan, Nepal. Known introduced range: Widely cultivated in the tropics, invasive in New Zealand, Réunion, Hawaii and Macaronesia (especially Azores). Probably increasing where introduced.

Ornamental gingers are spread via the horticulture industry. Its spatial occupation competes with native species. Permanent smothering of stream ecosystems and forest ground flora, almost entirely preventing regeneration. Once fully established it is extremely difficult for native seedlings to regenerate. May permanently displace rare plants, or cause serious losses to populations of uncommon plants or specialised communities.

Avoid planting as an ornamental and prevent dumping of plant debris in the wild. Mechanical control: only small plants and seedlings can be effectively removed manually. Removal requires slashing stems and digging out all rhizomes and tubers. Dry and burn rhizomes to prevent resprouting. Chemical control: the herbicide metsulfuron-methyl can be effective when applied following slashing of established plants. Biological control: the pathogenic bacterium *Ralstonia solanacearum* was found in Hawaii and tested for biological control but its effectiveness is unclear.

Heracleum mantegazzianum Sommier & Levier, giant hogweed (Apiaceae, Magnoliophyta)

Jan Pergl, Irena Perglová, and Petr Pyšek

It is monocarpic perennial, which persists usually 3–5 years in rosette stage, with leaves up to 2.5 m long. After reaching the mature stage it flowers and dies. The flowering stem can be up to 5 m high and bears large umbels with small white flowers. The species reproduces only by seeds, which are dispersed by wind, water and humans. The species is monocarpic, thus, it reproduces only once in its lifetime. A single plant produces about 20,000 seeds which have to be stratified in the soil in cold and wet conditions during winter and then are highly germinable. The majority of seeds germinate the following year after release and only about 1% of seeds are able to survive more than 3 years in the soil.

Native habitat (EUNIS code): Native to mountain meadows below tree line. E4: Alpine and subalpine grasslands, E5: Woodland fringes and clearings and tall forb habitats. Habitat occupied in invaded range: E2: Mesic grasslands, E5: Woodland fringes and clearings and tall forb habitats, I: Regularly or recently cultivated agricultural, horticultural and domestic habitats, J: Constructed, industrial and other artificial habitats. Occupies man-made or semi-natural habitats along roads and water corridors, abandoned meadows, forest clearings and areas near parks. Native from W Greater Caucasus. Introduced range: covers Europe N of the Alps and parts of North America. Other invasive relatives, *H. sosnowskyi* and *H. persicum*, occur in NE European countries. Introduced as ornamental plant (first record from the UK in 1817). The species may form dense stands reducing species diversity. The plant produces phytotoxic sap containing photosensitizing furanocoumarins, which in contact with human skin and UV radiation cause skin burnings. The danger to human health complicates eradication efforts and it clearly lowers the recreational value of the landscape.

In large infested areas, repeated grazing with consecutive cutting of flowering stems is recommended. It is also necessary to control plants in neighbouring areas to prevent seed input. As the species has high regeneration potential, the success of eradication depends on the removal of all regenerating inflorescences. Grazing can significantly decrease the reproductive output but also prolongs the lifespan before flowering: long-term control program necessary. In small populations, the plants can be effectively killed by root cutting 10 cm below the ground level. As the plant flowers only once in its lifetime and dies after setting seeds, root cutting can be applied only to vegetative plants, together with destruction of all flowers/seeds from flowering plants. Insect or pathogens have little effect. The species is sensitive to herbicides based on glyphosate and triclopyr.

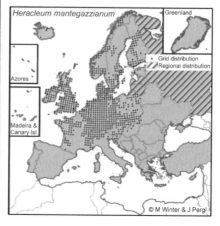

Impatiens glandulifera Royle, Himalayan balsam (Balsaminaceae, Magnoliophyta)

Martin Hejda

Annual plant up to 2.5 m tall with pink to purple flowers and green fruits around 5 cm long. Usually grows in riparian habitats and in other disturbed places with good water and nutrient supply. The seeds are ejected from the fruits via ballochory. Seeds are not able to float for a long time but do not loose the ability to germinate and are therefore effectively transported with other material. The fruits germinate later than the surrounding vegetation, so the frost sensitive seedlings are protected. Most seeds germinate at the beginning of the next spring, however, some seeds remain fertile over another winter and germinate the next spring.

Native habitat (EUNIS code): Wet, open places in forests, shrubs and hedges, 1,800–3,200 m elevation. F9: Riverine and fen scrubs, FA: Hedgerows, G5: Lines of trees, small anthropogenic woodlands, recently felled woodland, early-stage woodland and coppice. Habitat occupied in invaded range: riparian vegetation, wet disturbed places, forest edges, wet roadsides. F9: Riverine and fen scrubs, J4: Transport networks and other constructed hard-surfaced areas, J5: Highly artificial man-made waters and associated structures. The plants are shade tolerant and the seedlings are frost sensitive. Native in the Himalayas. Introduced to almost all temperate European countries and W and NE states of the USA. Due to climatic changes, *I. glandulifera* is expected to move its boundaries northwards and to higher elevations as well.

As a tall impressive species with large colourful flowers, it used to be planted for ornamental purposes. Besides this, it used to be favoured by beekeepers because of high nectar production. *I. glandulifera* reduces the diversity of invaded communities but this reduction concerns mostly widespread weed and even other non–native species. It competes successfully for pollinators, e.g., with *Stachys palustris*. The species is capable of changing the appearance of riverbanks completely, especially when in bloom. When *I. glandulifera* usurps the dominance in riparian vegetation it can promote erosion due to its modest root system, especially compared to the clonal native dominants of these communities.

Prevention by reducing its use as an ornamental, especially in wet areas. Due to the modest root system, the whole plant can be removed easily. However, the effect of such attempts is rather questionable due to the effective transportation of seeds through the river corridor, which usually results in a quick reinvasion. Chemical control: Juvenile plants respond to spraying by herbicides, however, when the flowering plants are sprayed, they are still able to produce viable seeds.

Opuntia ficus-indica (L.) Miller, prickly-pear cactus (Cactaceae, Magnoliophyta)

Montserrat Vilà

This cactus reaches up to 2 m height. Succulent stems are flat, oval and segmented. Flowers 5–6 cm in diameter are orange and the fig shaped fruits are purple. It has a CAM physiology with very high water-use efficiency. Flowers are pollinated by insects. Seeds are dispersed by birds, feral pigs and lizards that feed upon fruits. Seeds from fruits and scats germinate after long rainfall and warm temperatures (21°C). Vegetative reproduction from succulent plant parts falling close to parental plants is important. Seeds in the soil can remain viable for several years.

Native habitat (EUNIS code): Semi-arid habitats. Habitat occupied in invaded range: B1: Coastal dune and sand habitats, B3: Rock cliffs, ledges and shores, including the supralittoral, F7: Spiny Mediterranean heaths (phrygana), F8: Thermo-Atlantic xerophytic habitats, G3: Coniferous woodland, J6: Waste deposits. Planted as a garden plant, it usually occurs close to buildings. It is mostly found in sunny, rocky and well-drained slopes. Sensitive to freezing temperatures below -6°C and tolerant to high temperatures up to 65°C. Tolerant to moderate soil salinity. Native range: tropical America from Mexico to Colombia. Known introduced range: Mediterranean Basin, Macaronesian Islands, Australia, tropical and S Africa, western USA, Caribbean, temperate Asia. There is increasing invasion in abandoned Mediterranean agricultural fields. Climatic models predict increasing productivity with climate warming.

It was introduced by the Spanish conquerors between 1548 and 1570 for mass-rearing of the cochineal insect *Dactylopius coccus* (Homoptera) for the production of a red dye. In Italy and Israel it is planted for fruit consumption. It is also planted as an ornamental, for wind protection fencing, land reclamation and rehabilitation, and erosion control. Its spatial occupation competes with native species recruitment. Famous for the injury its spines can cause to animals and humans. Invaded woodlands are misperceived as typical Mediterranean landscapes. Invaded old-fields interfere with sheep and cattle grazing. Organic acids accumulated during night on young stems might cause diarrhoea to livestock.

Avoid planting as an ornamental and dumping of plant debris in the wild. Plants sprout vigorously after removal (e.g., fire, grazing, clipping). Successful regrowth can occur from any plant fragment. Summer glyphosate injection into cladodes or areas were it had been cut has been proved to be effective. Biological control: in Hawaii, Australia and South Africa there has been successful biological control by the scale insect *Dactylopius opuntiae* and by the moth *Cactoblastis cactorum*.

Oxalis pes-caprae L., Bermuda buttercup (Oxalidaceae, Magnoliophyta)

Philip W. Lambdon

A short perennial herb forming large clonal colonies reproducing by annual bulbs. Flowering takes place between January and March and plants die back by late spring. Bulbs are dispersed by agricultural activity and also washed along gullies in rain or stream water. It is a heterostylous species, with at least three forms known globally. The forms are not easily cross-pollinated and are not self-compatible. Only the short-styled form is common in Europe, which means that plants do not produce seeds and must rely on vegetative reproduction. Mammals may graze the leaves and rodents consume bulbs.

It occupies the same range of habitats in both its native and invaded part of its area (EUNIS code). Often extremely abundant in cultivated areas, especially in the shade of olive and citrus groves or vineyards. Bulbs are often spread into secondary disturbed habitats such as riverbeds, ditches, dunes, ruderal areas, screes and degraded Mediterranean shrublands. C3: Littoral zone of inland surface waterbodies, F6: Garrigue, FA: Hedgerows, FB: Shrub plantations, H2: Screes, I1: Arable land and market gardens, I2: Cultivated areas of gardens and parks, J2: Low density buildings, J5: Highly artificial man-made waters and associated structures. Prefers warm, light soils and remains sensitive to frost, which limits the distribution to S Europe and below 700 m. Abundance often declines following the cessation of cultivation, although it may persist for some years in abandoned fields. Native in the Cape region of South Africa, it has spread widely across S Europe, N Africa, Asia, Australia, New Zealand and America. May expand northwards in Europe due to global warming.

Oxalis pes-caprae L was brought to Europe in 1757, reached habitable Mediterranean areas via Sicily in 1796 as ornamental plant, and subsequently spread through soil movement. It can suppress other ruderal weedy species. This is a serious problem where the arable flora is of conservation value, as over much of the Mediterranean old fields. The leaves contain large quantities of toxic oxalates, and in grassy areas the species can therefore be a danger to livestock. Significant losses of cattle and sheep have been recorded from Sardinia and Menorca. In annual crops, it can be a significant pest, reducing yields and becoming a nuisance during harvesting.

Prevention of soil contamination is essential. Grazing by pigs or turkeys, which consume the bulbs, may sometimes be practical. Soil ploughing in January can reduce plant growth and bulb production. Pre-emergence treatments with glyphosate or sulphonyl urea-based herbicides are generally effective, although these must be used in moderately doses in order to not affect crop performance or the native flora. It is resistant to certain classes of chemical such as dinitroanilines.

Paspalum paspalodes (Michx) Scribner, knotgrass (Poaceae, Magnoliophyta)

Paulina Anastasiu

It is a creeping stoloniferous perennial herb, up to 50 cm tall; leaves are stiff, narrow, about 10 cm in length, with sheaths ciliate on margin. Usually every inflorescence has two racemes 1.5–7 cm, with ovate, pale-green spikelets. The upper glume is appressed and puberulent. It is adapted to marshy, brackish conditions and saline soils, which are moist in summer. The productivity of seeds is low. Fragments of rhizomes or creeping stems could be water dispersed. Mainly vegetative reproduction from rhizomes. Creeping stems also root at the nodes and give rise to flowering stems. It flowers freely in summer, but some clones are markedly self-sterile so that little seed is produced. Some clones are reasonably self-fertile and cross-pollination between clones may result in seed set. Good fodder for cattle and horses. Stolons and rhizomes survive over the winter season.

Native habitat (EUNIS code): A littoral species occurring in sands and muds near the seashore, and in saline soils and swamps. Habitat occupied in invaded range: Fresh or brackish marshes, coastal salt marshes, ponds, ditches, shorelines, beaches, and dunes. A2: Littoral sediments, C1: Surface standing waters, C3: Surface running waters. It needs moist areas and summer rains, but persists during the dry season; it tolerates environmental extremes and grows well in shade. Native range: Tropical Africa and America. Known introduced range: Europe, Australia, New Zealand. Even though it is a tropical weed it is increasing its distribution in Europe.

In some regions it has been introduced intentionally for erosion control and it is a common commodity contaminant. Its dense populations can cover large surfaces in a short time, competing very successfully with other weeds. It is a harmful weed in rice fields. Sometimes it can be troublesome by blocking irrigation ditches.

Prevention by monitoring and early detection in susceptible habitats. Mechanical control: the effective removal of knotgrass from invaded habitats is very difficult. Burning is not recommended because rhizomes survive a fire. Chemical control: the best results have been obtained with fluazifop, quizalofop, glyphosate, and glufosinate.

Prunus serotina Ehrh., black cherry (Rosaceae, Magnoliophyta)

Stefan Klotz

This tree can reach up to 35 m in height. Leaves are entire and shiny. Flowers are white and organised in an oblong-cylindrical raceme 10–15 cm long. Flowers are pollinated by insects. Fruits are blackberries, about 8 mm in diameter. Wood and fruits are used commercially. Seeds are dispersed by vertebrates such as birds, foxes and other mammals that feed upon fruits. Many seedlings are found beneath trees. Germination rates are high and vegetative reproduction from extensive lateral roots is also common. It forms dense, highly competitive thickets. Seeds in the soil can remain viable up to 5 years. Plants respond to cutting by resprouting.

Native habitat (EUNIS code): G: woodland, forest and other wooded land. Habitat occupied in invaded range: G1: Broadleaved deciduous woodland, G3: Coniferous woodland, G4: Mixed deciduous and coniferous woodland, G5: Lines of trees, small anthropogenic woodlands, recently felled woodland, early-stage woodland and coppice, J6: Waste deposits. Planted as a forest and park tree, it usually occurs in managed forests and clearings. It is mostly found in pine or mixed pine and oak forests on sandy soils. The tree can invade open ground and grasslands and is tolerant to atmospheric pollution. Native range: eastern North America from Nova Scotia to SW Guatemala. Known introduced range: Temperate and continental Europe. There is increasing invasion in forests and abandoned fields, and a clear range expansion.

It was introduced in 1623 to Europe as an ornamental tree and used as a forest tree in the late 19th century. It was planted mainly on poor sandy soils to increase forest production. Additionally, the tree was used for restoration of mining land. It competes for resources with native plant species, especially during natural forest regeneration. The litter changes humus quality. Bark and seeds are toxic (cyanogenic glycoside). Maintaining natural forest regeneration causes increasing costs for forest management.

Avoid planting as a forest and ornamental tree near forested areas. Mechanical control: plants sprout fast after tree and shrub cutting. Successful regrowth can occur from any root system. Chemical control: herbicide (Round-Up) used in combination with mechanical management can be successful.

Rhododendron ponticum L., rhododendron (Ericaceae, Magnoliophyta)

Philip E. Hulme

This evergreen shrub is densely branched growing to 5 m. Flowers are violet to purple and pollination by insects results in the production of 3,000–7,000 small seeds per woody capsule, several million seeds per bush. Seeds are dispersed up to 100 m by wind and water under favourable open conditions, but less far in closed canopy forest. Seeds require light for germination. Vegetative reproduction is limited to branches rooting in contact with soil, usually only at forest edges. Potentially toxic chemicals, particularly free phenols and diterpenes, occur in the tissues of rhododendron, such that foliage is unpalatable to vertebrates and few insects feed on the plant. Seeds in the soil can remain viable for several years.

Native habitat (EUNIS code): G1: Mixed deciduous forest. Habitat occupied in invaded range: G1: Mixed deciduous forest, F4: Temperate heaths, D1: Raised and blanket bogs. Tolerant of a wide range of temperatures but intolerant to drought. It grows best in uniformly damp climates. Seedlings have difficulty becoming established in areas where there is already continuous ground cover from native plants. Establishment is best in disturbed areas where the native vegetation has been in some way disrupted, providing an opening in the plant cover. Native range is disjunct with *R. ponticum* ssp. *baeticum* in WS Spain and S Portugal, whereas ssp. *ponticum* is found in Turkey, Lebanon, Bulgaria and the Caucasus. Naturalised in the UK, Ireland, Belgium, France, Netherlands and Austria. There is increasing invasion in continental Europe.

Introduced as an ornamental, it is still available from nurseries. Once rhododendron has invaded an area, few native plants survive. In woodlands, only those trees which manage to grow above the level of the rhododendron canopy will persist. When such trees die, they cannot be replaced because seedlings cannot become established under the lightless canopy. At this point, the rhododendron completely dominates the area. Stands accumulate thick litter layers. Anecdotal information suggests that honey from rhododendron is toxic to humans. It is considered a problem in commercial forests and moorlands managed for gamebirds.

Avoid planting as an ornamental. Plants sprout vigorously after cutting. Mechanical clearance involves a tracked swing shovel with a rotary flail mounted on a moving hydraulic arm. In sensitive conservation areas, such techniques may well not be appropriate. Such mechanical devices often leave a thick layer of smashed rhododendron on the ground that may have to be removed using manual labour. The leaves are waxy and herbicide treatment must include a chemical additive to help break this surface down. Usually spraying is ineffective. Glyphosate injection into root stumps should be combined with mechanical clearing.

Robinia pseudoacacia L., black locust (Fabaceae, Magnoliophyta)

Corina Başnou

This deciduous N-fixing tree can reach up to 30 m in height. The bark is thick, deeply furrowed and has stipular spines on the twigs. The inflorescence is a pendant large raceme of white highly scented flowers. The fruit is a legume, 5–10 cm long, which remains attached until splitting open in winter. Fruits can be wind dispersed but germination is low. Seedlings established on sites free of competition show rapid growth. The tree also reproduces by root suckering and stump sprouting to form a connected root system.

Native habitat (EUNIS code): The tree grows on moist, limestone-derived soils, in upland oak-hickory forests. Habitat occupied in invaded range: C2: Surface running waters, E: Grassland and tall forb habitats, G5: Lines of trees, small anthropogenic woodlands, recently felled woodland, early-stage woodland and coppice, J4: Transport networks and other constructed hard-surfaced areas. It cannot grow in the shade but grows best in full sun and well-drained soils. It is drought tolerant and a pioneer on old fields. As a nitrogen fixing species, it can rapidly colonise acidic or polluted soils. Native range: Appalachian Mountains (USA). Known introduced range: Europe, Asia, Africa, Australia, New Zealand, N America. The trend in Europe is increasing, as it is still the most widely planted American tree.

Black locust was introduced from N America and planted for the first time in France as an ornamental in 1601. It is also planted for reforestation, erosion control and nectar production. It produces excellent firewood. The inner bark and roots have tonic and purgative properties. Once introduced in an area, the tree expands rapidly, creating dense clones of shaded islands with little ground vegetation. The large blossoms of black locust compete with native plants for pollinating bees. As a nitrogen fixing species, black locust can achieve early dominance on open sites where nitrogen is limiting to other species. The robinine contained in flowers and seeds are toxic to humans and provoke gastroenteritis. The tree makes large roots near the surface, sometimes buckling sidewalks or interfering with mowing.

Avoid planting it for reforestation. Cutting and burning works only temporarily because the species spreads vegetatively. Bulldozing may be used on disturbed lands. This tree can be controlled using the following herbicides: dicamba, fosamine, glyphosate, imazapyr, picloram, tricopyr. The management of this suckering species is very difficult and follow-up treatments are required. Biological control: in the United States, the major pest of this tree is the locust borer *Megacyllene robiniae*. Other insect pests include leafminers and twig borers.

Rosa rugosa Thunb. ex Murray, rugosa rose (Rosaceae, Magnoliophyta)

Franz Essl

It is a small sprouting shrub that forms dense thickets, mainly in coastal habitats. The twigs are stout and covered with thin, straight sharp spines of various sizes. The flowers are big (8–10 cm across) and can be white or light to dark pink. It has hermaphroditic flowers which are insect pollinated. The fruits (rose hips) are large and slightly flattened, shiny, deep red and ripen in late summer. Rose hips are very tasty to animals and humans and so seeds are dispersed by birds and small mammals. The plants also reproduce by rhizomes. The formation of dense thickets occurs via vegetative reproduction by root suckers. Seeds in the soil can remain viable for several years.

Native habitat (EUNIS code): colonises mainly old, stabilised coastal sand dunes, forming shrubs with other woody species. If succession continues, the shrubs are finally replaced by dune forests. Habitat occupied in invaded range: B1: Coastal dunes and sandy shores, B3: Rock cliffs, ledges and shores, including the supralittoral, E5: Woodland fringes and clearings and tall forb stands, F4: Temperate shrub heathland, FA: Hedgerows. It shows a preference to open, fresh to dry habitats. It can colonise acidic and basic soil alike, and is able to invade nutrient-poor habitats. Native to E Asia, its range encompasses the Islands of Hokkaido, Sakhalin, the Kuriles and the coasts of Kamtchatka to NE China. Known introduced range: It is widespread at the coasts of the North and Baltic Sea, as well as the NW European Atlantic coasts. Distribution area and population sizes have been increasing in the last few decades in the British Isles and Germany.

It was introduced to Europe as an ornamental plant in 1796 and the first records of naturalised populations are from Germany in 1845 and Denmark in 1875. Once invaded, dune plant communities are altered to monospecific stands, with greatly reduced light availability and decreased number of native species. Invaded dunes are becoming impenetrable to humans due to the spiny thickets. It is also a common plant for landscaping, e.g., along highways and in cities. Invaded coastal areas are misinterpreted, as in Denmark, and are displayed in tourist brochures and on postcards. It controls erosion on shores and riverbanks.

It is important to stop plantings in the countryside, particularly in coastal areas. Hand grubbing of smaller populations is effective, but roots and rhizomes must be removed as far as possible to prevent recolonisation. Grazing by goats or sheep is effective at destroying seedlings and older plants alike. Digging up the plants can be combined with herbicide application (e.g., glyphosate).

Terrestrial vertebrates

Branta canadensis
Cervus nippon
Lithobates catesbeianus
Mustela vison
Myocastor coypus
Nyctereutes procyonoides
Ondatra zibethicus
Oxyura jamaicensis
Procyon lotor
Psittakula krameri
Rattus norvegicus
Sciurus carolinensis
Tamia sibiricus
Threskiornis aethiopicus
Trachemys scripta

Branta canadensis (Linnaeus), Canada goose (Anatidae, Aves)

Susan Shirley

A large grey-brown goose inhabiting terrestrial and freshwater habitats. Omnivorous, feeding mostly on plant materials, rhizomes, stems, leaves, seeds and fruit. Distinguished by a black head, neck and tail with a large white chin strap. North American populations are migratory; many European populations are sedentary though some N populations move south to central Europe. Breeds March–June, semi-colonial, produces 2–6 eggs, normally single brood, incubation takes 28–30 days and fledging 40–48 days. Widely hunted.

Native habitat (EUNIS code): B: Coastal habitats, C: Inland surface water habitats, E: Grassland and tall forb habitats, F: Heathland, scrub and tundra habitats, I: Regularly or recently cultivated agricultural, horticultural and domestic habitats. The same habitats are occupied in the invaded range. It avoids large, deep lakes with oligotrophic waters and rocky banks; also avoids rivers. Native range: North America. Introduced range: Established in N Europe, and across north-central Europe from Belgium east to Russia. Introduced, but not yet established, in an additional seven central and S Europe countries such as Austria, Italy, Poland, Czech Republic and Switzerland. Increasing in many European countries.

It has escaped from aviaries, has been released for hunting and naturally spread to neighbouring European countries. It hybridises with 16 Anatidae species in captivity. There is concern for the potential for hybridisation with other goose species such as *Anser anser* (greylag goose), particularly in introduced areas. Competition with greylag goose has been documented. Very aggressive to small waterfowl, displacing territory and killing young and adults. Some benefit to dabbling ducks, which steal floating vegetation from submerged vegetation dislodged by Canada geese during feeding. There is some concern about human health hazard from soil and water contamination caused by excess droppings. Some threat to air safety from collisions with aircraft have been noted. It is a pest species causing habitat modification by trampling and algal blooms from eutrophication caused by nutrients from roosting geese. It is also a minor feeder on crops.

Major steps for control are to establish baseline information and monitor existing wild and captive populations, improve legislation to prevent deliberate introductions and to limit or remove populations, and institute measures such as licensing to prevent escapes from captive collections. Despite growing awareness of negative impacts and large populations, to date there has been no organised international effort to control populations. Unregulated introductions continue to support hunting in several N, central and E European countries.

13 Species Accounts of 100 of the Most Invasive Alien Species in Europe

Cervus nippon Temminck, sika deer (Cervidae, Mammalia)

Piero Genovesi and Rory Putman

Small to medium-sized deer, weight of females 25–45 kg, males 40–110 kg, height at shoulder 95–140 cm. The most distinctive characteristic is a white caudal patch outlined in black. Mainly feeding on foliage, forbs, twigs, mast and many grass species. In a continuous optimal habitat, sika show a steady expansion in range, estimated in the UK at 3–5 km per year. Rutting seasons in Europe is in September–November. Pregnancy rates 85–100%, gestation 210–246 days, usually one calf in May–June. Most hinds breed successfully for the first time as yearlings.

Native habitat (EUNIS code): E: Grassland and tall forb habitats, G: Woodland and forest habitats and other wooded land. Habitat occupied in invaded range is similar to the native range, F3: Temperate and mediterraneo-montane scrub habitats, F4: Temperate shrub heathland, F8: Thermo-Atlantic xerophytic habitats, I: Regularly or recently cultivated agricultural, horticultural and domestic habitats. Favoured by warm climates (12° to 46° N), selects areas where snowfall does not exceed 10–20 cm and snow-free sites are also available. Prefers early seral stages over mature forests. Native range: E Asia. Introduced range: New Zealand, South Africa, Morocco, Australia, Papua New Guinea, North America, Europe. Increasing in many European countries (5% per year in the UK 1972–2002).

Introduced in many areas of the world as ornamental or game species. Commonly bred in farms. Feral populations originated from deliberate releases or escapees. Damage is due to ring barking, browsing, trampling, antler rubbing, erosion due to creation of trails. Mature trees may also suffer damage through bole-scoring (characteristic of this species). In open heathland and wetland areas, sika can cause significant change in vegetational structure and species composition of both plants and animals. Hybrids with the native congeneric red deer *Cervus elaphus* are fertile, and further hybridisation or back-crossing to either parental type is rapid threatening the genetic integrity of the native species. In E Europe, sika play a role in the epidemiology of *Asworthius sidemi*, a nematode affecting bison, roe deer, red deer and potentially livestock. Both bovine and avian TB recorded in captive and wild populations. Sika are a serious forest pest, causing significant damage to broadleaved and conifer plantations.

Sika deer can be excluded from vulnerable croplands by erecting fences. Deterrents can in some cases be temporarily effective, ultrasounds are largely ineffective. Control usually by shooting. Capture through individual traps or corral devices can be effective management alternatives. Chemical repellents are ineffective. Methods to control fertility through chemicals or immunocontraceptives are being explored, but are not yet available.

Lithobates catesbeianus (Shaw), American bullfrog (Ranidae, Amphibia)

Olivier Lorvelec and Mathieu Détaint

It is the largest North American frog. In its natural range, adults can reach at most a 184 mm snout-vent length, while in Europe, adults may exceed 195 mm, with an average weight of 430 g. It is likely to feed on a wide range of prey: amphibians, fishes, small mammals, ducklings and small bird species, molluscs, crustaceans and insects, with amphibians and insects most prominent. Natural spread of juveniles and adults is both via rivers and on land. In SW France, the breeding period is May to September, tadpole development takes 1–2 years. Larval and young frog stages undergo significant predation. Known predators include carnivorous fish, in particular *Micropterus salmoides* introduced from North America, and probably some Mustelidae and Ardeidae. The main source of predation, in high-density areas, appears to be cannibalism.

Native habitat (EUNIS code): C: Inland surface water habitats; D: Mire, bog and fen habitats. In its native range, it usually lives in lakes or large ponds. Where introduced, it occupies any type of habitat that is lentic or with slowly moving water, especially if aquatic and with abundant bank vegetation. Native range: eastern part of North America. Known introduced range: It is currently established in the western part of North America, in several countries of South America, Greater Antilles, Asia, Hawaii and Europe. Although it is presently illegal to import this species to the EU, it can still be ordered on the Internet and shipped worldwide. Farming could increase its distribution area by escapees.

It has been introduced in many parts of the world, often for farming purposes but also as a predator of unwanted species like insect pests in Hawaii, and sometimes as a pet. Where introduced, it has the ability to occupy a wide range of habitats and to feed on many species. A negative impact on native ranids has been stressed. American bullfrogs carry the pathogenic fungus *Batrachochytrium dendrobatidis* which has been implicated in global amphibian decline. The cost of eradication has been GB£29,000 in the UK.

Except for a limited area in the UK and France, no eradication strategy has been formulated in Europe. Only few control operations have been attempted (i.e., in Germany). In SW France, a long-term regional program was started in May 2003. It aimed at the development of strategies and management tools (in particular egg collection, tadpoles trapping, and juvenile and adult shooting) to limit American bullfrog populations.

Mustela vison (Schreber), American mink (Mustelidae, Mammalia)

Laura Bonesi

Small, semi-aquatic carnivore living in freshwater and marine habitats. It is a generalist and opportunist predator with a variable diet that includes aquatic, semi-aquatic and terrestrial prey. Males disperse further than females, up to 50 km, typically along water bodies. Mink are sexually dimorphic, males weighing 0.9–1.6 kg, females 0.6–1.1 kg. Mating February to April, implantation is delayed, gestation lasts 39 days. On average six young are born between April and May, disperse in August and reach sexual maturity at 10 months. Life expectancy is 3–4 years in the wild.

Native habitat (EUNIS code): B1: Coastal dune and sand habitat, B2: Coastal shingle habitats, B3: Rock cliffs, ledges and shores, including the supralittoral, C1: surface standing waters, C2: Surface running waters, C3: Littoral zone of inland surface waterbodies, F9: Riverine and fen shrub, G: Woodland and forest habitats and other wooded land, I: Regularly or recently cultivated agricultural, horticultural and domestic habitats. Habitat occupied in invaded range same as above plus J: Constructed, industrial and other artificial habitats. Associated with water, habitat requirements are determined mainly by food availability, and secondarily by the availability of dens. Mink are sensitive to pollution by PCBs. Native range: North America. Introduced range: Europe, the former Soviet Union, and southern countries of South America. Increasing worldwide but apparently decreasing in some European countries (the UK, Sweden).

Introduced for fur farming industry, feral populations due to intentional or accidental releases. Intentional releases from farms are often carried out by animal rights activists. The impact on native species can occur through predation, competition, and as a vector of Aleutian disease. Significant population declines of ground nesting birds (e.g., *Larus ridibundus*, *Sterna hirundo*) and small mammals (e.g., *Arvicola terrestris*) due to mink predation. The European mink *Mustela lutreola*, whose range is now restricted to a few fragmented populations, is threatened by the American mink through direct aggression. Can inflict damage to free ranging chickens, reared game birds, fisheries (salmon farming) and the eco-tourist industry through predation on ground nesting birds. Germany estimates the costs of impacts to be €4,200,000 per year.

Prevention by regulating licenses to fur farms and improving fencing around the farms. At the moment lethal trapping is the only feasible method for containing or eradicating mink. In most areas live-traps are recommended to avoid non-target impacts. Exclusion devices for otters should be used when appropriate. Research is currently being carried out to investigate effectiveness and best strategies for control trapping.

Myocastor coypus (Molina), coypu, nutria (Myocastoridae, Mammalia)

Sandro Bertolino

Large, brown semi-aquatic rodent along rivers, lakes, and marshes, weight 2–4 kg, adult males up to 8 kg. Superficially rat-like, with short legs and a long cylindrical tail, the first four digits of the hind feet webbed. It is herbivorous except for occasional feeding on mussels. Coypus are good swimmers and fast colonisers, able to rapidly occupy suitable vacant habitats using freshwater as a pathway. Breeding throughout the year, first parturition with 3–8 months. Productivity influenced by 50–60% prenatal embryo losses and abortion of litters. Litter size 4–6, up to 3 litters/year. Limited predation by foxes, dogs, and marsh harriers.

Native habitat (EUNIS code): Aquatic habitats. Habitat occupied in invaded range: C1: Inland surface water habitats, D1: Mire, bog and fen habitats. It can adapt to a wide variety of aquatic habitats, from freshwaters and lakes to drainage canals, usually in the lowlands. Cold winter reduces breeding success and influences population dynamics. Native range: the Patagonian subregion of South America. Naturalised populations occur in North America, Europe, central and N Asia, Japan, E Africa and the Middle East. In the UK eradicated after an 11-year removal campaign. Distribution range and population density increasing in many countries.

Often naturalised after escapes or releases from fur farms. The impact on wetlands through feeding on aquatic vegetation could be severe. Selective feeding by coypu caused massive reduction in reedswamp areas, eliminated *Rumex* spp. and *Nuphar lutea* over large areas. It destroys nests and preys on eggs of several aquatic birds, including some endangered species. It has been hypothesised that the species has (after rats) also a role in the epidemiology of leptospirosis. It is considered a pest for its feeding on crops, such as sugar beets and maize, and for its burrowing activity that disrupts riverbanks and dikes. In Italy during 1995–2000, despite control activities involving the removal of 220,688 coypus with a cost of €2,614,408, damage to the riverbanks exceeded €10 million and impact on agriculture reached €935,138.

Where farming is still active, fences and security should be verified and improved. In some small areas, buried fences have been used to avoid burrowing by animals to protect crops. Mechanical control: shooting is effective for population control when environmental conditions force the animals into the open, while cage trapping has also been used in the English eradication program. In some countries like France and the USA, baits with toxicants are used, e.g., zinc phosphide on carrots or sweet potatoes.

Nyctereutes procyonoides (Gray), raccoon dog (Canidae, Mammalia)

Kaarina Kauhala and Marten Winter

A fox-sized mammal with short legs and tail and a typical black face with long hair on cheeks. This omnivorous carnivore is the only canid with winter lethargy. Juveniles disperse at 4–5 months of age (August–October). Average dispersal distance 20 km, sizes of home range 150–700 ha. Sexual maturity at 9–11 months, mating season in March. The gestation period is 9 weeks, and cubs are born April–June. Mean litter size is 7–9, higher than in the native range. Proportion of breeding females in the population averages 80%. The wolf *Canis lupus*, red fox *Vulpes vulpes*, dog *Canis familiaris*, lynx *Lynx lynx* and large predatory birds are predators mainly of young animals.

Native habitat (EUNIS code): B: Coastal habitats, G: Woodland and forest habitats and other wooded land, I: Regularly or recently cultivated agricultural, horticultural and domestic habitats, X8: Rural mosaics, consisting of woods, hedges, pastures and crops, J: Constructed, industrial and other artificial habitats. Occupy in invaded range the same habitats as in the native range, C3: Littoral zone of inland surface waterbodies, D: Mire, bog and fen habitats; E: Grassland and tall forb habitats, F4: Temperate shrub heathland. Raccoon dogs often live near water and prefer moist forests with abundant undergrowth. N range is limited by annual mean temperature below 0°C, a snow cover of 800 mm or longer than 175 days and length of growing season of 135 days. Native range: NE Asia, introduced to Europe. Trend of fast range expansion towards SW Europe.

Several times intentionally introduced to E Europe, from these starting points the raccoon dog migrated westwards into new areas, also escaped from fur farms. Suspected impacts, especially on islands, are predation on birds and amphibians with resulting decreased nesting success and decreased population size. There may be competition for food and den sites with badger *Meles meles* and red fox *Vulpes vules*. The raccoon dog causes economic costs being one of the main vectors of rabies in Europe and an important vector of sarcoptic mange, the fox tapeworm *Echinococcus multilocularis* and trichinellosis. In Finland, raccoon dogs are still hunted for fur. Hunting has been increasing in recent years.

Management plans should focus on communication programs and diminishing conflict, especially in protected and isolated areas (islands). It is important to avoid providing freely available food on compost piles or from pet feeding places. Mechanical control: there is no possibility of eradicating the raccoon dog from the wild. Like other canids, the raccoon dog tends to increase litter size with increased hunting pressure.

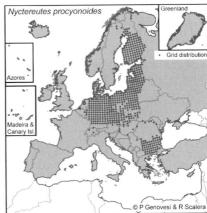

Ondatra zibethicus (Linnaeus), muskrat (Muridae, Mammalia)

Piero Genovesi

Large, dark brown stocky aquatic rodent, 41–62 cm length, weight 0.7–1.8 kg, with large head, small eyes, short rounded ears, tail as long as head and body (18–30 cm), hairless and flattened laterally. Common name refers to the musky odour. They eat almost any aquatic vegetation as well as crops. Also feed on crayfish, mussels, turtles, frogs and fish in ponds where vegetation is scarce. Young usually disperse after their first winter along streams and water bodies. Average expansion speed in central Europe in the first half of last century was around 11 km per year. They have 2–3 litters per year after a gestation period of 25–30 days. Able to swim by 2 weeks of age, weaned at 3–4 weeks when they begin to feed independently. Young become sexually mature the spring following birth.

Native habitat (EUNIS code): C: Inland surface water habitats, D: Mire, bog and fen habitats, F9: Riverine and fen scrubs. The same habitats are occupied in the invaded range. Lives in brackish and freshwater lakes, ponds, rivers and marshes, well adapted to cold climates. Higher reproductive rates in southern latitudes. Tidal fluctuations, periodic flooding or droughts limit species distribution. Native range: North America. Introduced into most of the Palearctic, Argentina and Chile. Successfully eradicated from the UK and Ireland in the 1930s. Increasing and expanding in many countries.

Imported for fur farming in many countries, escaped into the wild or intentionally released. Strongly affects vegetation dynamics through grazing. Threatens endemic species such as the desman *Desmana moschata*, impacts shellfishes, fishes and ground nesting birds; endangered mussel populations are particularly impacted. In some areas they are vector of leptospirosis and intermediate host for the cestode *Echinococcus multilocularis*. Burrowing damages riverbanks, railroads, dams and fences. It also causes extensive damage to crops, irrigation structures and aquaculture industry. In Germany annual damage costs are estimated at €12.4 million.

Fencing is effective to prevent damage to valuable crops or gardens. Damage to pond dams can be prevented through stone rip-rapping of dams, or by constructing dams with proper slope and size. Also drawing of ponds in winter is used to remove muskrats. Frightening devices are seldom effective. Mechanical control: control usually by trapping, less frequently by shooting. Most effective capturing devices are stove-pipe traps. Control usually through live traps. Also snares and iron bow-nets placed in front of entrances of holes are used. Chemical control: zinc phosphide and anticoagulants, however, undesired impacts on non-target species have been reported. Poison baits are usually placed on floating platforms to minimise risks to non-targets.

Oxyura jamaicensis (Gmelin), ruddy duck (Anatidae, Aves)

Susan Shirley

A small diving duck inhabiting freshwater habitats. Omnivorous, feeding on molluscs, insects and their larvae, but also on seeds and parts of water plants. Males have a bright blue bill, black crown and nape, white cheeks and reddish body. Only seasonal dispersal in resident UK populations. Breeds singly or colonially from April–August, usually once per season. It lays 4–12 eggs. Adults are hunted.

Native habitat (EUNIS code): B: Coastal habitats, C: Inland surface water habitats, D: Mire, bog and fen habitats. Habitat occupied in invaded range: B: Coastal habitats, C: Inland surface water habitats, D: Mire, bog and fen habitats. Prefers pools with fairly shallow bottoms and rich in aquatic plants; avoids flowing fresh water. Native range: North and Central America, Caribbean. Introduced range: Europe. The UK has the largest population with around 5,000 individuals followed by France with 50 breeding pairs while other countries currently have very low numbers. Increasing in several European countries.

Escapes and accidental releases from waterfowl collections and breeding farms, natural spread to continental Europe from the UK populations. Hybridisation with two species including the vulnerable white-headed duck *Oxyura leucocephala*. It is dominant over this species in the wild. First and second generation hybrid back-crosses with the ruddy duck are fertile and dominant also. Costs of eradication are considerable. There is an ongoing eradication program in the UK since 1992 with the goal of reducing the population to less than 175 birds or 5% of the 1999 population at an estimated cost of €4.4 million over a 4–6 year period. By 2004, at least 15 countries were taking actions to control populations.

Major steps are to establish baseline information and monitor existing wild and captive populations. Legislation should be improved to prevent deliberate introductions and to limit or remove populations. Strict controls, such as licensing, should be put in place to prevent escapes or ban their inclusion in captive collections. Mechanical control: culling (shooting) has been applied in France since 1998, Spain since 1993 and Portugal. The largest population and assumed source of some introductions is the UK where culling has been controversial.

Procyon lotor (Linnaeus), raccoon (Procyonidae, Mammalia)

Marten Winter

This cat-sized omnivore, mostly nocturnal carnivore, with its black and white face, can live in almost all terrestrial habitats. It has a distinct sense of touch and an excellent climbing and swimming ability. For females the adult home range is often the same as the birth area, mean home range 40–400 ha, mean migration distances 5–10 km. Mostly 1 litter per year with 2–4 young in April after a mean gestation period of 50–70 days. The mean weaning time is 16 weeks, up to 75% of females reproduce within their first year.

Native habitat (EUNIS code): B: Coastal habitats, D: Mire, bog and fen habitats, E: Grassland and tall forb habitats, F: Heathland, scrub and tundra habitats, G: Woodland and forest habitats and other wooded land, H: Inland unvegetated or sparsely vegetated habitats, I: Regularly or recently cultivated agricultural, horticultural and domestic habitats, J: Constructed, industrial and other artificial habitats, X11: Large parks, X22: Small city centre non-domestic gardens, X23: Large non-domestic gardens, X24: Domestic gardens of city and town centres, X25: Domestic gardens of villages and urban peripheries. The same habitats are occupied in the invaded range. They prefer woody habitats adjacent to fresh water or urbanised areas. Its climatic range is very large, surviving harsh winters and desert like conditions. Native range: S Canada to Panama. Introduced to Europe, Caucasus and Japan. Range expansion towards SE Europe.

The first time in Germany, it was intentionally released for hunting and because of its fur. Escapes from fur farms, zoological gardens and animal husbandries in several other countries of Europe. Occasionally predation on birds (nests) and amphibians with resulting decreased nesting success and decreased population sizes. Due to the raccoon roundworm *Baylisascaris procyonis* there is a high potential of zoonosis for humans and its animal vectors. It is considered a pest in some urban areas.

Management plans should focus on communication programs and diminishing conflicts. In urban areas it is possible to control the denning behaviour by reducing climbing opportunities at houses and on roofs. It is important to avoid providing freely available food on compost piles or from pet feeding places. In its native range eradications from islands are documented. But population control by hunting is only possible with enormous efforts. An effective control program was done after fencing and trapping near Berlin for protection of great bustards *Otis tarda*.

Psittacula krameri (Scopoli), rose-ringed parakeet (Psittacidae, Aves)

Assaf Shwartz and Susan Shirley

This pale yellow-green parakeet with a distinguishing long tail lives in tropical and subtropical lightly wooded habitats feeding mainly on seeds, fruit, flowers and nectar. Males have a black and rose-red ring encircling their throat. Sedentary, with some local movements for food and seasonal changes in roost sites. Breeds singly or in small loose groups in the same tree from January to July. Clutch size 2–6, normally single brooded, but known to occasionally have second broods.

Native habitat (EUNIS code): G: Woodland and forest habitats and other wooded land (Mainly lowland, most abundant in moist and dry deciduous lightly wooded areas, secondary jungle), I: Regularly or recently cultivated agricultural, horticultural and domestic habitats, J: Constructed, industrial and other artificial habitats (gardens, orchard, cultivated areas and city suburbs). Habitat occupied in invaded range: I and J, mainly parks, local gardens and suburban areas, but also cultivated areas with fields and orchards. Avoids mountainous and arid areas, in a large range of temperatures, precipitation and light regimes in anthropogenic-influenced habitats. In Europe, they depend on bird feeders during winter. Native range: From W to E Africa, Afghanistan, W Pakistan, Indian Subcontinent, Myanmar. Introduced range: Europe, Turkey and Israel. Increasing in population size and distribution.

Rose-ringed parakeets were highly traded during the late 1960s and 1970s and have escaped from aviaries. However, spreading has occurred naturally in human-dominated habitats. It would be useful to establish baseline information and monitor existing wild and captive populations, improve legislation to prevent deliberate introductions and to remove populations, for example by trapping, and institute strict controls such as licensing to prevent escapes or ban their inclusion in captive collections. It has been suggested that the rose-ringed parakeets may have detrimental effects on other cavity-nesters. In many habitats, the number of cavities is a major factor regulating population densities of cavity-nesters. Parakeets, which begin breeding prior to most other secondary cavity-nester species, may limit resources available for species such as house sparrow *Passer domesticus*, stock dove *Columba oenas*, European nuthatch *Sitta europaea* and European starling *Sturnus vulgaris*. Noise disturbance from loud squawking and screeching at large roost sites may be considered as social impact. Rose-ringed parakeet is considered by some to be the worst avian pest. It is a major crop pest in India, damaging grain products and fruits. It is also a pest of sunflower, dates and other fruit orchard crops. Trapping has been conducted in Australia to remove individuals from the wild.

Rattus norvegicus (Berkenhout), Norway rat, brown rat (Muridae, Mammalia)

Michel Pascal and Olivier Lorvelec

An omnivorous and opportunistic terrestrial rodent, mostly grey or brown, adults weigh 230–550 g, body length 190–265 mm, tail length 160–205 mm. Its albino form is used in laboratories and bred as a pet. When introduced, natural spread occurs, following paths and roads, rivers, lake banks and the seashore. In Europe, reproduction stops in natural habitats during winter, but is observed all year round in human dwellings. Sexual maturity is reached within 50–60 days, pregnancy 21 days, 7–8 embryos per litter. Post-partum fertilisation is frequent.

Native habitat (EUNIS code): I: Regularly or recently cultivated agricultural, horticultural and domestic habitats, J: Constructed, industrial and other artificial habitats, B: Coastal habitats, C: Inland surface water habitats, D: Mire, bog and fen habitats, E: Grassland and tall forb habitats, F: Heathland, scrub and tundra habitats, G: Woodland and forest habitats and other wooded land, H: Inland unvegetated or sparsely vegetated habitats. Habitat occupied in invaded range: D, I, J. Where introduced, the Norway rat occupies many types of habitats except high mountains, but seems to need some fresh water. Native range: The Norway rat is likely to be native to SE Siberia, N China and Hondo (Japan). Introduced to all continents, still increasing its distribution area especially on islands.

Imported mainly via maritime and terrestrial traffic as a stowaway. On islands, it leads to declines of shrews, bird and reptile populations. It has contributed to the disappearance of several bird populations on islands. Norway rat serves as a reservoir and vector of *Leptospira interrogans* and as a reservoir of the hepatitis E virus. The costs of Norway rat population control in towns and warehouses are very high.

Since it is difficult to eradicate rats, it is better to prevent their colonisation, particularly on islands. Rats can be captured effectively through live-traps and snap-traps. In Europe, Norway rat populations in human dwellings, towns and warehouses are controlled with toxic baits. Recent attempts to eradicate several European island populations using successively trapping and toxic baiting were successful and had positive effects on populations of native species. Over the last 15 years, eradication of rats have been carried out successfully using chemical poisons on increasingly larger offshore islands. Rats, as with many rodents, are susceptible to anticoagulant poisons such as chlorofacinone, bromadialone, difetialone and brodifacoum. Contraceptive methods of control are currently experimental, but the potential for effective control using such methods is promising and National Wildlife Research Center (USA) scientists are working on formulations for an effective oral immunisation.

Sciurus carolinensis, Gmelin, grey squirrel (Sciuridae, Mammalia)

Piero Genovesi and Sandro Bertolino

Medium sized tree squirrel, without sexual dimorphism. Body-head length 380–525 mm, tail 150–250 mm, adults 480–650 g. Dorsal grey to pale grey, with cinnamon tones, ears and tail pale grey to white. Feeds mostly on seeds, flowers, buds, fruits, fungi, some insects and occasionally bird eggs. During low food periods, strips bark to get to inner bast and cambian layers. Dispersal facilitated by wooded corridors. Mean rate of colonisation in the UK and Italy is about 18 km^2/year. Most females reproduce in the second year, usually 2 litters per year (December–February, May–June), gestation 44 days, litter size 2–4.

Native habitat (EUNIS code): G: Woodland and forest habitats and other wooded land. The same habitats are occupied in the invaded range. Well adapted to live in broadleaved woods, can colonise conifer and mixed forests. Native range: Eastern part of North America. Introduced range: North America, South Africa, Australia (extinct) and Europe. Constantly expanding in the UK, Ireland and Italy, it is expected to colonise France and Switzerland in the next few decades. Can potentially expand to a large portion of Eurasia. Density is usually below 3 individuals/ha, but higher in optimal habitats. On several occasions populations originated from the release of very few individuals.

Imported as a pet in many countries, accidentally escaped into the wild or intentionally released for ornamental purposes. Still traded in Europe as a pet. In overlap areas, grey squirrel causes the extinction of the red squirrel *Sciurus vulgaris* through competitive exclusion. Moreover, host pox-virus, fatal to red squirrels but benign to grey squirrels, appear to increase the rate of replacement. Potential impact on nesting birds. Squirrel poxvirus potentially transmissible to humans. Severe damage to trees by bark stripping. They cause local damage to fruit orchards and nut growers.

Prevention by import ban through inclusion of the grey squirrel in Council Regulation 338/97/EC. Regulation of pet trade and information campaign to traders and owners is necessary. Mechanical control: in the UK, intensive control is carried out through nest destruction, shooting and trapping. In Italy, control is exclusively carried out through live-trapping and anaesthesia in order to maximise selectivity and animal welfare. Attempted eradication in Italy failed because of opposition by animal rights movements. Chemical control: warfarin (anti-coagulant) widely used in the UK. Biological control: research to develop species-specific immunocontraceptive agents is ongoing in the UK but not yet available.

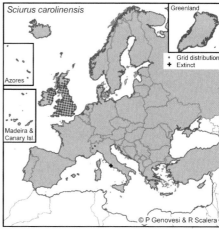

Tamias sibiricus (Laxmann), Siberian chipmunk (Sciuridae, Mammalia)

Jean Louis Chapuis

A small, diurnal and omnivorous terrestrial squirrel, 100 g body weight, living exclusively in forests. Fur characterised by five longitudinal brown dorsal stripes. Reproduction in a burrow where it hibernates from October to March. Young individuals disperse 2–3 weeks after emerging from their nest. They set up their nest, up to 100–200 m away. Adults are extremely sedentary with a home range of about 1 ha. Sexual maturity is reached at 8–11 months. 1–2 litters/year with 4–5 young each. Weaned young (6–8 weeks old, 40–50 g) leave the nest and disperse. Hibernation 5–6 months.

Native habitat (EUNIS code): F: Heathland, scrub and tundra habitats, G: Woodland and forest habitats and other wooded land. The same habitats are occupied in the invaded range and I2: Cultivated areas of gardens and parks. The native habitats of the Siberian chipmunk contain a wide range of geographical areas with different climatic conditions, including continental and oceanic forests. A deep soil and the presence of stumps are important for the establishment of a burrow. Native range: From N European Russia to China, Korea, and Japan. Introduced range: Europe. Populations are stable or increasing.

Ten populations introduced to France originated from the release of pets, and one population from individuals escaping breeding. In Belgium and the Netherlands individuals have been deliberately introduced in parks. May compete directly (trophic availability) with native forest rodents, mainly *Sciurus vulgaris*, *Apodemus sylvaticus* and *Clethrionomys glareolus* or indirectly (pathogen transmission) with *Sciurus vulgaris*. Potential impact to ground and burrow nesting birds. Chipmunks harboured the spirochete *Borrelia burgdorferi s.l.* and three species of ticks: *Ixodes acuminatus*, *I. ricinus* and *Dermacentor reticulatus*, thus chipmunks could contribute to an increased lyme disease transmission risk. Economic impact is unknown in Europe but in its native area, impacts to grain crops are recorded. Prevention by prohibiting the sale of this species in pet shops in order to limit the risk of introduction to other sites.

Threskiornis aethiopicus (Latham), sacred ibis (Threskiornithidae, Aves)

Philippe Clergeau and Pierre Yésou

Large white bird, black head and neck, 1.4 kg body weight. Mostly found in wet meadows, inland wetlands to coastal areas, gregarious, often forming large groups. It is carnivorous with a tendency to omnivory. The diet is based on terrestrial and aquatic insects, fish, amphibians, molluscs and crustaceans, also feeding upon small mammals, bird eggs and young, and also animal or vegetable refuse. Nomadic species able to change its breeding sites to suit environmental conditions. In the introduced range, there is regular exchange between colonies, with nomadic individuals moving up to several 100 km from colonies. Colonies up to several 1,000 pairs, sometimes with other ciconiiformes and herons. Nests often closely aggregated in trees, bushes and on the ground near water. 2–4 eggs give 1–2 chicks. Breeding success in France appears to be higher than in Africa.

Native habitat (EUNIS code): D: Mire, bog and fen habitats (large wetland), E: Grassland and tall forb habitats, B: Coastal habitats, I: Regularly or recently cultivated agricultural, horticultural and domestic habitats. The same habitats are occupied in the invaded range. Large tolerance to various landscapes but presence of water essential. Native range: Africa south of Sahara, Iraq, Madagascar and Aldabra Island. Introduced to Europe, Arabian Peninsula, Taiwan, Florida. In W France, feral population raised over 5,000 individuals in 30 years from a single source, in S France 200 individuals in 6 years, in Italian Piedmont 2 increasing populations.

Escapes from zoos where ibises were breeding and allowed to fly freely. Predation on several threatened species (e.g., insects, amphibians) and especially on protected colonies of terns and herons. Vegetation rapidly affected at breeding sites. Epidemiological role suspected since foraging ibises frequently visit rubbish dumps and slurry pits. An economic impact has not been documented, but destruction of salt pans structure has been observed.

In most cases, no action has been undertaken against ibises. However, the feral population in Barcelona, Spain, has been culled in 2001. Mechanical: in France, decision-making is in progress, aiming at removing the majority of ibises in western and southern areas by shooting them at roost, feeding places or colonies.

Trachemys scripta (Schoepff), common slider (Emydidae, Reptilia)

Riccardo Scalera

A medium to large freshwater turtle, prominent yellow to red patches on each side of the head, 20–60 cm body length. Carapace and skin olive to brown with yellow stripes or spots. The diet of this opportunistic predator changes from highly carnivorous as juveniles to omnivorous as adults. Sliders have the potential to spread throughout waterways. They can live for 40 years. Courtship and mating may occur in spring and fall. Nesting in temperate zones occurs from April–July. Nests are usually excavated on the shore of a fresh waterbody. Up to six clutches with 2–30 eggs, incubation 59–112 days.

Native habitat (EUNIS code): C: Inland surface waters. The same habitats are occupied in the invaded range. Sliders occur in most freshwater habitats, but prefer quiet waters with soft bottoms, abundance of aquatic vegetation, and suitable basking sites. For a sound hibernation in winter, clean waters with sufficient amounts of oxygen are needed. Native range: E USA, Mexico. Introduced to Europe, Israel, Asia, the Caribbean and South Africa. In Europe, breeding only in Spain, Italy and France. Notwithstanding the EU import suspension of this species since 1997, every year new records are reported in most European countries, due to continuous dumping in the wild of the animals still kept as a pet.

Sliders are among the world's most commonly traded pet reptile, also marketed for human consumption, particularly in Asia. In total, 1989–1997, the USA exported 52 million individuals. Introductions are therefore a side-effect of trade. Sliders feed on plants and animals, from invertebrates to all vertebrates, including amphibians and reptiles, small mammals and birds. Competition dynamics with indigenous turtles, particularly with the endangered European pond turtle *Emys orbicularis* are known for food, basking sites and nesting sites. May contribute to the spread of diseases and parasites that could affect native turtles and other aquatic wildlife. Considered a potential vector of salmonella, in the USA such epidemiological risk has resulted in a national ban of sales of sliders since 1975.

The import of this species has been suspended within the EU since 1997, however, other replacement taxa have been found being traded since then. An information campaign to raise public awareness of the risk posed by dumping pets in the wild is considered a priority. In some countries, unwanted pets are disposed of in rescue centres and zoological gardens. Sliders can be captured by hand or through various trapping devices. Floating boards used by sliders as basking sites seem very effective when equipped with baited cages. Sniffer dogs can be used to detect and remove turtles and their eggs.

Chapter 14
Glossary of the Main Technical Terms Used in the Handbook

Petr Pyšek, Philip E Hulme, and Wolfgang Nentwig

Throughout the *Handbook* a variety of terms have been used to describe the origin and status of alien species, their residency, the invasibility of ecosystems and the pathways of introduction. We have attempted to use these terms consistently in the *Handbook* and provide a glossary of definitions. The meaning of these technical terms is based on previously published terminology and reflects how particular categories were understood during the production of the *Handbook*. It should be made clear that we do not propose a new set of definitions; rather we hope to achieve a broad consensus among different subdisciplines of invasion biologists. Further details of terminology, including additional terms, can be found in the reference list at the end of the glossary.

14.1 Origin and Invasion Status

Acclimatised/casual taxa (synonymous not established, adventive) are aliens that may reproduce occasionally outside cultivation or captivity in a region, but eventually die out because they do not form self-sustaining populations without human intervention, and rely on repeated introductions for their persistence (Richardson et al. 2000; Pyšek et al. 2004; Copp et al. 2005). The latter terms is usually used for plants, the former for animals.

Adventive: see Acclimatised taxa

Alien taxa (synonymous exotic, non-native, non-indigenous, allochthonous) are species, subspecies or lower taxa introduced outside of their natural range (past or present) and outside of their natural dispersal potential. Their presence in the given region is due to intentional or unintentional introduction or care by humans, or they have arrived there without the help of people from an area in which they are alien. This includes any part, gamete or propagule of such species that might survive and subsequently reproduce (IUCN 2000, 2002; Pyšek et al. 2004).

Allochthonous: see Alien taxa

Casual: see Acclimatised taxa

Cryptogenic taxa are those of unknown origin which can not be ascribed as being native or alien (Carlton 1996).

Established: see Naturalised taxa

Feral animals/crops are those that have reverted/escaped to the wild from domesticated/cultivated stock (e.g., has undergone some change in phenotype, genotype and/or behaviour as a result of artificial selection in captivity) (IUCN 2000, 2002; Elvira 2001).

Invasive taxa are a subset of naturalised/established alien taxa, that produce reproductive offspring, often in very large numbers and have potential to spread exponentially over a large area, thus rapidly extending their range (Richardson et al. 2000; Occhipinti-Ambrogi and Galil 2004; Pyšek et al. 2004). This is usually associated, although not necessarily for an organism to qualify as invasive (Richardson et al. 2000; Elvira 2001), with causing significant harm to biological diversity, ecosystem functioning, socio-economic values and human health in invaded regions. From an ecological point of view, invasiveness is not bound to a type of habitat, hence a species may be invasive in natural/semi-natural or human-made habitats (Richardson et al. 2000). For conservation purposes, the term invasive usually relates to natural or semi-natural ecosystems or habitats (IUCN 2000, 2002).

Native taxa (synonymous indigenous) are those that have originated in a given area without human involvement or that have arrived there from an area in which they are native without intentional or unintentional intervention of humans. The definition excludes products of hybridisation involving alien taxa since human involvement in this case includes the introduction of an alien parent (Pyšek et al. 2004).

Naturalised/established taxa are aliens that form free-living, self-sustaining (reproducing) and durable populations persisting in the wild in a region unsupported by and independent of humans (IUCN 2000, 2002; Richardson et al. 2000; Occhipinti-Ambrogi and Galil 2004; Pyšek et al. 2004). The former terms is usually used for plants, the latter for animals.

Non-indigenous: see Alien taxa

Non-native: see Alien taxa

Pests are animals (not necessarily alien) that live in places where they are not wanted and which have detectable economic or environmental impact or both.

Reintroduced taxa are those deliberately released by humans into a geographic area, in which they were native in historical times but where they subsequently became extinct (Elvira 2001).

Weeds are plants (not necessarily alien) that grow in sites where they are not wanted and which have detectable economic or environmental impact or both (Pyšek et al. 2004).

14.2 Residence Time Status

Archaeophytes/archaeomycetes/archaeozoans are alien plants/fungi/animals introduced to a region during the period since the beginning of Neolithic agriculture and before the discovery of America by Columbus in 1492 (Kowarik and Starfinger 2003; Pyšek et al. 2004).

Neophytes/neomycetes/neozoans are alien plants/fungi/animals introduced to a region after 1492, together referred to as neobiota (Kowarik and Starfinger 2003; Pyšek et al. 2004).

Residence time is the time since the introduction of a taxon to a region; as it is usually not known exactly when a taxon was introduced, the term 'minimum residence time' (MRT) has been suggested and used in the literature (Rejmánek 2000).

14.3 Invasibility of Habitats, Ecosystems and Regions

Invasibility: an inherent property of habitats/ecosystems/regions, resulting from the habitat/region/ecosystem's resistance to invasion and manifested in the rate of mortality of alien taxa (Lonsdale 1999). Technically, it can be expressed as the number or proportion of alien taxa in a habitat/region/ecosystem when the effects of propagule pressure and confounding variables are held constant (Chytrý et al. 2008).

Level of invasion: the actual number of alien taxa in a habitat/region/ecosystem (Hierro et al. 2005)

Propagule: a structure with the capacity to give rise to a new individual. For plants this can be a seed, a spore, a bulb, or a part of the vegetative body capable of independent growth if detached from the parent. For animals, this includes eggs, larvae, neonates or individual organisms.

Propagule pressure: the number of propagules arriving to a site, habitat, ecosystem or region (Williamson 1996; Lonsdale 1999). Propagule is any part of a plant or animal which can be dispersed and give rise to an individual.

14.4 Pathways of Introduction

Contaminant: unintentional introduction with a specific commodity, e.g., parasites, pests and commensals of traded plants and animals (Hulme et al. 2008).

Dispersal by corridor: unintentional introduction via human infrastructures linking previously unconnected regions, e.g., species migrating from the Red to the Mediterranean Sea through the Suez canal (Hulme et al. 2008).

Escape: intentional introduction as a commodity but escapes unintentionally, e.g., feral crops and livestock, pets, garden plants, live bait (Hulme et al. 2008).

Intentional introduction: deliberate movement and/or releases by humans, past or present, of an alien species outside its natural distribution range (Occhipinti-Ambrogi and Galil 2004; Hulme et al. 2008).

Introduction: the movement, by human agency, of a species, subspecies, or lower taxon (including any part, gamete or propagule that might survive and subsequently reproduce) outside its past or present natural range (IUCN 2000, 2002). This movement can be either within a country or between countries.

Release: intentional introduction as a commodity for release in the wild, e.g., as biocontrol agents, game animals, plants for erosion control, landscaping or enrichment of native flora (Hulme et al. 2008).

Stowaway: unintentional introduction attached to or within a transport vector, e.g., hull fouling, ballast/water/soil and sediment organism, on car tyres; refers to species that have been introduced accidentally but are not known to be associated with any particular commodity (Hulme et al. 2008).

Unaided dispersal: unintentional introduction through natural dispersal of alien species across political borders; refers to species that have spread via spontaneous means from an introduced population elsewhere in non-native distribution range (Hulme et al. 2008).

Unintentional introduction: all other introductions which are not intentional (Hulme et al. 2008).

References

Carlton JT (1996) Biological invasions and cryptogenic species. Ecology 77:1653–1655

Chytrý M, Jarošík V, Pyšek P, Hájek O, Knollová I, Tichý L, Danihelka J (2008) Separating habitat invasibility by alien plants from the actual level of invasion. Ecology 89:1541–1553

Copp GH, Bianco PG, Bogutskaya NG, Ers T, Falka I, Ferreira MT, Fox MG, Freyhof J, Gozlan RE, Grabowska J, Kovář V, Moreno-Amich R, Naseka AM, Peáz M, Pov M, Przybylski M, Robillard M, Russell IC, Staknas S, Šumer S, Vila-Gispert A, Wiesner C (2005) To be, or not to be, a non-native freshwater fish? J Appl Ichthyol 21:242–262

Elvira B (2001) Identification of non-native freshwater fishes established in Europe and assessment of their potential threats to the biological diversity. Convention on the conservation of European wildlife and natural habitats. Standing Committee 21st Meeting, Strasbourg, 26–30 November 2001

Hierro JL, Maron JL, Callaway RM (2005) A biogeographical approach to plant invasions: the importance of studying exotics in their introduced and native range. J Ecol 93:5–15

Hulme PE, Bacher S, Kenis M, Klotz S, Kühn I, Minchin D, Nentwig W, Olenin S, Panov V, Pergl J, Pyšek P, Roque A, Sol D, Solarz W, Vilà M (2008) Grasping at the routes of biological invasions: a framework for integrating pathways into policy. J Appl Ecol 45:403–414

IUCN (2000) Guidelines for the prevention of biodiversity loss caused by alien invasive species prepared by the Species Survival Commission (SSC) invasive species specialist group. Approved by the 51st Meeting of the IUCN Council, Gland, www.iucn.org/themes/ssc/publications/policy/invasivesEng.htm. Cited Dec 2007

IUCN (2002) Policy recommendations papers for sixth meeting of the Conference of the Parties to the Convention on Biological Diversity (COP6). The Hague, The Netherlands, 7–19 April 2002. www.iucn.org/themes/pbia/wl/docs/biodiversity/cop6/invasives.doc. Cited Sept 2007

Kowarik I, Starfinger U (2003) Introduction. In: Kowarik I, Starfinger U (eds), Biological invasions in Central Europe: a challenge to act? Biol Inv 5:279

Lonsdale M (1999) Global patterns of plant invasions and the concept of invasibility. Ecology 80:1522–1536

Occhipinti-Ambrogi A, Galil BS (2004) A uniform terminology on bioinvasions: a chimera or an operative tool? Mar Poll Bull 49:688–694

Pyšek P, Richardson DM, Rejmánek M, Webster G, Williamson M, Kirschner J (2004) Alien plants in checklists and floras: towards better communication between taxonomists and ecologists. Taxon 53:131–143

Rejmánek M (2000) Invasive plants: approaches and predictions. Austral Ecol 25:497–506

Richardson DM, Pyšek P, Rejmánek M, Barbour MG, Panetta FD, West CJ (2000) Naturalization and invasion of alien plants: concepts and definitions. Diversity Distrib 6:93–107

Williamson M (1996) Biological invasions. Chapman & Hall, London

Index

Species print in **bold** belong to the 100 of the most invasive alien species in Europe (Chapter 13).

A
Acacia dealbata 341
Acacia spp. 2
Acanthaceae 177
Acanthocephala 221
Acanthocnemidae 234
Acanthodrilidae 217
Acari 65, 253
Acartiidae 221
Accipitridae 259
acclimatisation society 84
acclimatised 375
Acheta domestica 66
Achiridae 255
Acipenseridae 255
Acoraceae 198
Acrididae 226
Acridotheres tristis 112
Acrochaetiaceae 146
Acropomatidae 255
Actinidiaceae 145, 170
Aculops fuchsiae 65
Adelgidae 229
Adeonidae 219
Adiantaceae 147
Adoxaceae 169
Adrianichthyidae 255
adventive 375
Aedes albopictus 69, 72, 324
Aelosomatidae 218
Aeolidiidae 212
Aeolothripidae 228
Aerococcaceae 133
Aeteidae 219
Aethina tumida 75
African cotton leaf worm 339

Agamidae 115, 258
Agaonidae 248
Agapanthaceae 198
Agaricales 135
Agaricus bisporus 20
Agavaceae 198
Agelenidae 252
Aglajidae 212
Agonoxenidae 243
Agriolimacidae 212
Agromyzidae 247
Ailanthus altissima 50, 56, 69, 342
Aiptasiidae 210
Aizoaceae 162
Alariaceae 144
Albuginales 133
Alburnus alburnus 896
Alburnus arborella 82
Aleurotuba jelineki 74
Alexandrium catenella 270
Aleyrodidae 229
alien 375
Alismataceae 198
Alliaceae 198
Alligatoridae 115, 258
allochthonous 375
Alpheidae 221
Alstroemeriaceae 199
Amanitales 135
Amaranthaceae 48ff, 163
Amaranthus retroflexus 49
Amaryllidaceae 199
Amathinidae 212
Amaurobiidae 252
Amblyommidae 65, 72, 253
Amblyseius californicus 69

Ambrosia artemisifolia 2, 50, 343
Ambystomatidae 258
Ameiridae 221
Ameiurus melas 84
Ameiurus nebulosus 84
American bullfrog 21, 362
American Jack knife clam 281
American mink 363
Ametropodidae 226
Ammotheidae 221
Ampelodesmos mauritanica 57
Ampharetidae 218
Amphibia 105, 258
Amphinomidae 218
Amphipoda 86
Ampithoidae 221
Ampullariidae 212
Anacardiaceae 194
Anatidae 259
Ancylidae 212
Ancylostomatidae 220
Anguidae 258
Anguilla anguilla 99
Anguillicola crassus 2, 83, 88, 99, 303
Anguillicolidae 220
Anguillidae 255
Anisolabididae 227
Anisomeridium nyssaegenum 32
Annelida 217
Anobiidae 234
Anomiidae 212
Anoplocephalidae 211
Anoplophora chinensis 70, 75, 325
Anoplophora glabripennis 66, 71, 74, 326
Anthicidae 235
Anthocoridae 229
Anthomyiidae 247
Anthribidae 235
Aoridae 221
Aphanomyces astaci 21, 304
Aphelinidae 248
Aphididae 65, 229
Aphis gossypii 67, 327
Aphodiidae 235
Aphrophoridae 231
Apiaceae 48, 150
Apiales 150
Apidae 249
Apionidae 235
Apis mellifera 73
Aplysiidae 212
Apocynaceae 175
Apodemus sylvaticus 119
Apogonidae 255

Aponogetonaceae 199
Apterygota 225
Aquifoliaceae 152
Aquifoliales 152
Araceae 199
Araliaceae 152
Araneae 65, 252
Araneidae 252
Araucariaceae 149
Archaeobalanidae 221
archaeomycete 377
archaeophyte 377
archaeozoan 377
Arcidae 212
Arctiidae 243
Ardeidae 260
Arecaceae 199
Areschougiaceae 146
Argentine ant 337
Argidae 249
Arion vulgaris 65, 328
Arionidae 212
Aristichthys nobilis 87
Aristolochiaceae 188
Armadillidiidae 221
Artemiidae 221
Arthropoda 12, 64, 67f, 82, 85, 129f, 221, 303
Arthurdendyus triangulata 64
Arvicolidae 262
Aryaephyra desmaresti 88
Ascaridia dissimilis 72
Ascarididae 220
Ascidiidae 254
Asellidae 221
Ashworthius sidemi 72
Asian clam 306
Asian date mussel 287
Asian longhorned beetle 326
Asian sea-squirt 298
Asian tiger mosquito 324
Asparagaceae 200
Asparagales 47f
Asphodelaceae 200
Aspleniaceae 147
Astacidae 221
Asteraceae 47ff, 152
Asterales 152
Asteriidae 254
Asterinidae 254
Asterolampraceae 144
Asterolecaniidae 231
Atelerix algirus 127
Atheliales 135
Atherinidae 255

Index 383

Athyriaceae 148
Atlantoxerus getulus 127
Atrichum crispum 32
Aves 259
Azollaceae 148

B
Bacillariaceae 143
Bacillariophyta 143
Bacillidae 226
Bacteria 133
Baetidae 226
Balanidae 221
Balanus improvisus 271
ballast water 97ff, 270f, 276ff, 285ff, 293, 298f, 303, 305, 309ff, 378
Balsaminaceae 170
Bangiaceae 146
Barentsiidae 219
Basellaceae 165
Batrachochytrium dendrobatidis 21, 115
bay barnacle 271
Baylisascaris procyonis 64
Beaniidae 219
Bedelliidae 243
Begoniaceae 169
Belonidae 255
Bemisia tabaci 329
Berberidaceae 188
Bermuda buttercup 353
Bern Convention 3f
Beroidae 210
Bethylidae 249
Betulaceae 171
Bignoniaceae 177
Biodiversity Strategy and Biodiversity Action Plans 11
biological control 20, 24, 68f, 84, 110, 113, 267, 270ff
Bipaliidae 211
bird 10-12, 105
Bithyniidae 212
Blaberidae 226
black cherry 355
black locust 357
Blasticotomidae 249
Blastobasidae 243
Blastocladiales 135
Blattella germanica 66
Blattellidae 226
Blattidae 226
Blattodea 65, 226
Blechnaceae 148

Blenniidae 255
blue swimming crab 292
blue wattle 341
blue-spotted cornetfish 283
Boettgerilla pallens 64
Boettgerillidae 212
Boidae 115, 258
Boletales 135
Bolinopsidae 210
Bombinatoridae 258
Bombus terrestris 73
Bombycidae 243
Bonnemaisonia hamifera 272
Bonnemaisoniaceae 146
Boraginaceae 48, 177
Bosminidae 222
Bostrichidae 235
Bothriocephalidae 211
Botryosphaeriales 136
Bougainvilliidae 210
Bovidae 262
Brachidontes pharaonis 99, 273
Brachionidae 221
Brachycephalidae 258
Brachytheciaceae 148
Braconidae 249
Bradyporidae 226
Branchiobdellidae 218
Branta canadensis 107, 110, 112, 360
Brassicaceae 47ff, 159
Brassicales 159
Brassolidae 243
Braulidae 247
Bromeliaceae 200
Bromus catharticus 2
brook trout 318
brown rat 370
Bruchidae 72
brushtooth lizardfish 295
Bryaceae 148
Bryophytoa 29, 148
Bryozoa 219
Bryum gemmiferum 31
Buccinidae 212
Bucculatricidae 243
Buddelundiellidae 222
Bufonidae 258
Bugulidae 219
Bullidae 212
Buprestidae 235
Bursaphelenchus xylophilus 64, 330
Butomaceae 200
Buxaceae 162

Buxales 162
Byrrhidae 235

C

Cabombaceae 188
Cacatuidae 260
Cactaceae 165
Caeciliidae 227
Caeciliusidae 227
Caesalpinioideae 171
Calanidae 222
Calappidae 222
Calceolariaceae 178
Cales noaki 73
Caligidae 222
Callionymidae 255
Callioplanidae 211
Calliphoridae 247
Callosciurus erythraeus 120
Callosciurus fynlaisonii 127
Callosobruchus chinensis 67
Calomniaceae 148
Calycanthaceae 183
Calyptraeidae 212
Cambaridae 222
Cambarincolidae 217
Camelidae 262
Cameraria ohridella 67, 331
Campanulaceae 159
Campanulariidae 210
Campodeidae 225
Campylopus introflexus 30, 33, 36, 344
Canada goose 360
Canadian waterweed 311
Canellales 162
Canidae 262
Cannabaceae 190
Cannaceae 200
Capillariidae 220
Capitellidae 218
Capnodiales 136
Capparaceae 162
Caprellidae 222
Caprifoliaceae 169
Carabidae 235
Carangidae 255
Carassius auratus 88
Carcinophoridae 227
Cardiidae 212
Cardinalidae 260
Carditidae 212
Caricaceae 162

Carnivora 121
Carophyllaceae 48f
Carpobrotus edulis 2, 50, 56. 345
Carpobrotus spp. 57
Caryophyllaceae 165
Caryophyllales 47, 162
Cassiopeidae 210
Castniidae 65, 243
Castor canadensis 123
Castoridae 262
casual 34, 44f, 50, 52, 375
Casuarinaceae 171
Catostomidae 255
Caulacanthaceae 146
Caulerpa racemosa cylindracea 274
Caulerpa taxifolia 2, 99, 275
Caulerpaceae 145
Caviidae 262
Cecidomyiidae 247
Celastraceae 168
Celestrales 168
Centrarchidae 255
Centropagidae 222
Cephalotaxaceae 149
Cerambycidae 236
Ceramiaceae 146
Ceratiaceae 133
Ceratitis capitata 332
Ceratocystis fagacearum 23
Ceratocystis platani 21ff
Ceratophyllaceae 169
Ceratophyllales 169
Ceratophyllidae 248
Cercideae 172
Cercopagididae 222
Cercopagis pengoi 87, 305
Cercopithecidae 262
Cerithiidae 212
Cerithiopsidae 213
Cervidae 262
Cervus elaphus 2
Cervus nippon 2, 124, 361
Cerylonidae 236
CGRS grid 267
Chabertiidae 220
Chaetocerotaceae 144
Chaetodontidae 255
Chaetognatha 219
Chalcididae 249
Chalinidae 210
Chamaeleo chamaeleon 112
Chamaeleonidae 259
Chamidae 213

Index

Characeae 145
Characidae 255
Charybdis hellerii 98
Chatonellaceae 133
Chattonella cf. verruculosa 276
Chelicorophium curvispinum 87
Chelidae 259
Chelydridae 259
chemical control 267, 270ff
Chenopodium ambrosioides 2
Chinchillidae 262
Chinese diatom 288
Chinese mitten crab 312
Chironomidae 247
Chitonidae 213
Chlorophyta 145
Chondrinidae 213
Chondrostoma genei 82
Chordaceae 144
Chordariaceae 144
Chordata 254
Choreutidae 243
Chorizoporidae 219
Chromista 133
Chromodorididae 213
Chrysomelidae 236
Chthamalida 222
Chydoridae 222
Chytridiales 136
Cicadellidae 231
Cichlidae 255
Cidaridae 254
Ciidae 237
Cimex lectularis 66
Cirratulidae 218
Cistaceae 185
citrus longhorned beetle 325
Cladophoraceae 145
Clambidae 237
Clariidae 255
Clathrus archeri 25
Clausiliidae 213
Clavelinidae 254
Clavidae 210
Cleridae 237
Clethraceae 170
Clubionidae 252
Clupeidae 255
Cnidaria 210
Cobitidae 255
Cobitis taenia 82
Coccidae 232
Coccinellidae 237

Cochlicellidae 213
Cochlostomatidae 213
Codiaceae 145
Codium fragile tomentosoides 277
Colchicaceae 200
Coleophoridae 243
Coleoptera 65, 67, 234
Collembola 65
Collybia luxurians 22
Colorado beetle 336
Colubridae 115
Colubridae 259
Columbellidae 213
Columbidae 260
Colydiidae 237
comb jelly 314
Combretaceae 186
Commelinaceae 200
common cord grass 297
common ragweed 343
common shipworm 299
common slider 374
Congridae 255
Conidae 213
Coniopterygidae 247
contaminant 377
Convention on Biological Diversity 4, 11
Convolvulaceae 196
Conyza canadensis 2, 49
Corallinaceae 146
Corambidae 213
Corbicula fluminea 2, 306
Corbiculidae 213
Corbulidae 213
Cordylophora caspia 307
Coreidae 232
Corellidae 254
Corethraceae 143
Coriariaceae 169
Corixidae 232
Cornaceae 169
Cornales 169
Cornigerius maeoticus 87
Corophiidae 222
corridor 377
Cortaderia selloana 56f, 346
Cortinariales 136
Corvidae 260
Corylophidae 237
Corynidae 210
Corynocarpaceae 169
Corythucha arcuata 72
Corythucha ciliata 72

Coscinodiscaceae 144
Coscinodiscus wailesii 278
Cosmopterigidae 243
Cossuridae 218
Costellariidae 213
Cottidae 255
cotton aphid 327
cotton whitefly 329
coypu 364
Cracidae 260
Crambidae 243
Crangonyctidae 222
Craspedacusta sowerbyi 83
Crassostrea gigas 95, 99, 279
Crassula helmsii 308
Crassulaceae 49, 195
crayfish plague 304
Crepidula fornicata 280
Cribrilinidae 219
Cricetidae 262
Cricetus cricetus 120
Crocidura suaveolens 119
Crocodylidae 259
Cronartium ribicola 21
Crossosmatales 169
Crustacea 221
Cryphonectria parasitica 21, 24
cryptogenic 132, 376
Cryptomophalus aspersus 65
Cryptophagidae 237
Ctenocephalides felis 66
Ctenopharyngodon idella 83
Ctenophora 210
Cucurbitaceae 169
Cucurbitales 169
Culicidae 72, 247
Cumacea 86
Cupressaceae 149
Curculionidae 237
Cutleriaceae 144
Cyanobacteria 143
Cybocephalidae 238
Cycadaceae 150
Cyclopidae 222
Cyclopinidae 222
Cygnus olor 107
Cylichnidae 213
Cynipidae 249
Cynoglossidae 255
Cyperaceae 49, 200
Cypraeidae 213
cypress canker 322
Cyprididae 222, 255
Cyprinodontidae 256

Cytheaceae 148

D
Dacrymycetales 136
Dactylogyridae 211
Dactylopiidae 232
Dactylopteridae 256
Daltoniaceae 148
Daphniidae 222
Dasyaceae 146
Dasyatidae 254
Dasycladaceae 145
Dasypodidae 262
Datiscaceae 169
Datura stramonium 49
Davaineidae 211
Davalliaceae 148
Delesseriaceae 146
Delphacidae 232
Delphinapterus leucas 120f
Dendrocoelidae 211
Dendrodorididae 213
Dennstaedtiaceae 148
Dentaliidae 213
Derbesiaceae 145
Dermaptera 65, 227
Dermestidae 72, 238
Derodontidae 239
Desmana moschatta 120
Desmarestiaceae 144
Diabrotica virgifera 1, 74, 333
Diadematidae 254
Diadumenidae 210
Dialidae 213
Diaporthales 136
Diaptomidae 222
Diaspididae 65, 232
Diaspidiotus perniciosus 72
Dicksoniaceae 148
Dicotyledonae 150
Dicranaceae 148
Dictynidae 252
Dictyocaulidae 220
Dictyotaceae 144
Didelphidae 262
Didemnidae 254
Didymodon australasiae 39
Didymozoidae 211
Dikerogammarus villosus 87, 309
Dinophysiaceae 133
Dinophyta 145
Diodontidae 256
Diogenidae 222

Dioscoreaceae 201
Diphyidae 210
Diphyllobothriidae 211
Diplostomatidae 211
Diplozoidae 211
Diprionidae 249
Dipsacaceae 170
Dipsacales 169
Diptera 65, 67, 247
Discodorididae 213
Discoglossidae 258
distribution 5f, 8f, 267
Ditrichaceae 148
Dolichopodidae 247
Dorippidae 222
Dorvilleidae 218
Dothideales 137
Dreissena polymorpha 83, 87, 310
Dreissenidae 213
Dromiidae 222
Droseraceae 167
Drosophilidae 247
Dryinidae 249
Dryocosmus kuriphilus 72
Dryophthoridae 239
Dryopteridaceae 148
Dumontiaceae 146
Dutch elm disease 320
Dysderidae 252
Dytiscidae 239

E
Ebalidae 213
Ebenaceae 170
Echinocystis lobata 53, 347
Echinodermata 254
Echinogammarus ischnus 86
Echinorhynchidae 221
ecological impact 22, 38, 56, 72, 87, 99, 110, 114, 125
economic impact 22, 38, 56, 72, 87, 99, 110, 114, 125
Ectinosomatidae 222
Ectocarpaceae 144
Ectoprocta 219
Ectopsocidae 227
eel swim-bladder nematode 303
Eisenia japonica 64
Elaeagnaceae 190
Elasmobranchii 254
Elateridae 239
Elatinaceae 183
Electridae 219

Elodea canadensis 311
Elymus athericus 57
Elysiidae 213
Emberizidae 260
Emydidae 115, 259
Encarsia formosa 68
Encyrtidae 249
Endomychidae 239
Ensis americanus 281
Entomobryidae 225
Entylomatales 137
Epermeniidae 244
Ephedraceae 150
Ephemeroptera 226
Ephydridae 248
Epialtidae 222
Epidermoptidae 253
Epilampridae 226
Epitoniidae 213
Equidae 262
Equisetaceae 148
Eresidae 252
Ergasilidae 222
Ericaceae 170
Ericales 170
Erinaceidae 262
Erinaceomorpha 121
Eriocheir sinensis 312
Eriococcidae 233
Eriophyiidae 253
Erirhinidae 239
Erotylidae 239
Erpobdellidae 217
Erysiphales 137
escape 20, 51, 68f, 86, 101, 109f, 113ff, 120-124, 378
Esocidae 256
Esox lucius 86
established 376
Estrildidae 260
EU Biodiversity Strategy 4
EU Habitats Directive 4
Eudrilidae 218
Eudrilus eugeniae 64
Eulophidae 250
Eunicidae 218
EUNIS habitats classification 12, 266
Eupelmidae 250
Euphorbiaceae 49, 183
European Alien Species Database 7
European Alien Species Expertise Registry 6
European Biodiversity Strategy 1
European Community Biodiversity Clearing-House Mechanism 12

European Environment Agency 11
European Invasive Alien Species Information
  System 8
European Strategy on Invasive Alien
  Species 3, 5
European Topic Centre on Nature Protection
  and Biodiversity 12
Euryplacidae 222
Eurytomidae 250
Exobasidiales 138
Exocoetidae 256

F
Fabaceae 47ff, 171
Fabales 171
Faboideae 47f, 172
Facelinidae 213
Fagaceae 175
Fallopia japonica 1, 50, 348
Fallopia sachalinensis 348
Fallopia x bohemica 348
Fanniidae 248
Fasciolariidae 213
Fasciolidae 211
Fascioloides magna 64, 73
Faustulidae 211
feral 119, 376
Ficopomatus enigmaticus 100, 282
fish 12, 81f
fish-hook waterflea 305
Fissurellidae 213
Fistularia commersonii 283
Fistularidae 256
Flabellinidae 213
Flatidae 233
Flavobacteriaceae 133
Formicidae 250
Fossombroniaceae 148
Frankeniaceae 167
Frankliniella occidentalis 334
freeway iceplant 345
freshwater hydroid 307
Fringillidae 260
Fucaceae 144
Fumariaceae 188
Fundulidae 256
Fungi 15, 135

G
Galagonidae 262
Galaxauraceae 146
Galinsoga parviflora 49

Galinsoga quadriradiata 49
Gambusia holbrooki 84
Gambusia spp. 84
Gammaridae 222
Gammarus pulex 86
Gammarus roeseli 88
Gammarus tigrinus 86, 88
Gasterosteidae 256
Gastrochaenidae 213
Gastrodontidae 213
Gastropoda 64
Geastrales 138
Gekkonidae 259
Gelechiidae 244
Gentianaceae 176
Gentianales 175
Geocalycaceae 148
Geoemydidae 259
Geometridae 244
Geoplanidae 211
Gephyrocapsaceae 143
Geraniaceae 48, 176
Geraniales 176
Gesneriaceae 178
giant hogweed 350
Gigartinaceae 146
Gigaspermaceae 149
Ginkgoaceae 150
Gliridae 262
Globodera pallida 72
Globodera restochiensis 72
Glossoscolecidae 218
Glycymerididae 214
Glycyphagidae 253
Gnaphosidae 71, 252
Gnathotrichus materarius 72
Gobiidae 256
Gongylonematidae 220
Goniadidae 218
Goniodidae 227
Goniodomataceae 145
Gonyaulacaceae 145
Gracilariaceae 146
Gracillariidae 244
grape alga 274
Grapsidae 223
glasshouse 16, 30, 37, 39, 65, 68, 71, 73
greenhouse 22, 37, 64, 71
green sea fingers 277
grey squirrel 371
grid system 8, 267
Grossulariaceae 196
Gruidae 260
Gryllidae 226

Gryphaeidae 214
gulf pearl oyster 291
Gunnera tinctoria 50
Gunneraceae 177
Gunnerales 177
Gymnodiniaceae 145
Gymnospermae 149
Gyrodactylidae 211
Gyrodactylus salaris 2, 86, 313
Gyropidae 227
Gyropus ovalis 72

H
habitat 270ff
Haematopinidae 227
Haemulidae 256
Halgerdidae 214
Halichondriidae 211
Halimococcidae 233
Haliotidae 214
Haliplanellidae 210
Halophila stipulacea 284
halophilia seagrass 284
Haloragaceae 196
Halymeniaceae 147
Hamamelidaceae 196
Haminoeidae 214
Haplosporidiidae 133
Haptophyta 143
harlequin ladybird 335
Harmonia axyridis 68, 72, 335
heath star moss 344
Hedychium gardnerianum 50, 349
Heleomyzidae 248
Helianthus tuberosus 49
Helicidae 214
Helicodiscidae 214
Helotiales 138
Hemerobiidae 247
Hemerocallidaceae 201
Hemiaulaceae 144
Hemidiscaceae 144
Hemiptera 65, 67, 70, 229
Hemiramphidae 256
Hemiuridae 211
Henicopidae 225
Hennediella macrophylla 37
Heracleum mantegazzianum 2, 50, 56, 350
Heracleum sosnowskyi 53
Herpestidae 262
Hesionidae 218
Hesperiidae 244
Heteroderidae 220

Heterophyidae 211
Heteroptera 347
Heterosigmataceae 144
Hiatellidae 214
Himalayan balsam 351
Hippoboscidae 248
Hippolytidae 223
Hipponicidae 214
Hippothoidae 219
Hirudinea 217
Hirudinidae 217
Histeridae 239
Holocentridae 256
Holozoidae 254
homogenisation 88
Hoplopleuridae 227
horse chestnut leafminer 331
Hyacinthaceae 201
hybridisation 2, 17, 24, 52, 56f, 73, 88, 111, 126, 320, 348, 360f, 367, 376
Hydrangeaceae 169
Hydrobiidae 214
Hydrocharitaceae 50, 201
Hydrophilidae 239
Hydrophyllaceae 178, 197
Hygromiidae 214
Hyla meridionalis 112
Hylidae 258
Hymenochaetales 139
Hymenolepididae 211
Hymenoptera 65, 67, 248
Hypania invalida 87
Hypericaceae 184
Hypneaceae 147
Hypocreales 139
Hypolepidaceae 148
Hypopterygiaceae 149
Hysterangiales 139
Hystrix cristata 120

I
Ichneumonidae 251
Ictaluridae 256
Ictalurus punctatus 84
Idoteidae 223
Iguanidae 259
impact 22, 38, 56, 72, 87, 99, 110, 114, 125, 267
Impatiens glandulifera 50, 56f, 351
ink disease 321
Insecta 12, 63ff, 225
intentional introduction 378

introduction 3ff, 17ff, 29ff, 46ff, 65ff, 83ff, 93ff, 105ff, 119ff, 129f, 267, 378
invasibility 52, 55, 375, 377
invasive 376
invertebrate 12, 63ff, 81ff, 98, 130f, 266
Iridaceae 201
Isaeidae 223
Isognomonidae 214
Isoptera 65, 226
Isotomidae 225
Issidae 233
Istricidae 262

J
Janiridae 223
Japanese knotweed 348
Juglandaceae 175
Juncaceae 50, 202
Juncaginaceae 203

K
Kahili ginger 349
Kalotermitidae 226
Kateretidae 239
Katiannidae 226
killer shrimp 309
Kirchenpaueriidae 210
knotgrass 354
Kuruma prawn 286

L
Labia minor 66
Labiduridae 227
Labiidae 227
Labridae 256
Labyrinthulales 139
Lacertidae 115, 259
Lachesillidae 227
Laelapidae 253
Laemophloeidae 239
Lagomorpha 121
Lamiaceae 48, 178
Lamiales 47, 48, 177
Laminariaceae 144
Lamyctes coeculus 65
Lamyctes emarginatus 65
Languriidae 239
Lardizabalaceae 189
Laternulidae 214
latitude 8, 18, 68, 113, 126
Latridiidae 239

Latrodectus hasselti 73
Lauraceae 183
Laurales 183
Lauriidae 214
Leathesiaceae 144
Lecanora conizaeoides 32, 36, 38
Lecanoraceae 149
Leiodidae 240
Leiognathidae 256
Lemnaceae 203
Lentibulariaceae 180
Lentinula edodes 20
Leotiales 139
Lepadidae 223
Lepidium virginicum 49
Lepidopsocidae 227
Lepidoptera 65, 67, 243
Lepidoziaceae 149
Lepisma saccharina 66
Lepismatidae 226
Lepomis gibbosus 86
Leporidae 262
Lepraliellidae 219
Leptinotarsa decemlineata 74, 336
Leptochitonidae 214
Leptocreadiidae 211
Leptodontium gemmascens 38
Leptoglossus occidentalis 72
Leptophascum leptophyllum 36
Leptoplanidae 211
Lessoniaceae 144
Leucocoprinus birnbaumii 22
Leucosiidae 223
Lichenes 29, 149
Lichomolgidae 223
Ligiidae 223
Liliaceae 203
Limacidae 214
Limacodidae 244
Limnomysis benedeni 87
Limnoriidae 223
Limopsidae 214
Limulidae 223
Linaceae 184
Linderniaceae 180
Linephitema humile 70, 73, 337
Linyphiidae 65, 71, 252
Liposcelididae 227
Liriomyza huidobrensis 338
Listrophoridae 253
Lithobates catesbeianus 114, 362
Lithodidae 223
Litiopidae 214
Loasaceae 169

Index 391

Loliginidae 214
Lomentariaceae 147
Longidoridae 220
Lophocolea heterophylla 38
Lophocolea semiteres 38
Lophopodidae 219
Lotus cytisoides 57
Lovenellidae 210
Loxosomatidae 220
Luciferidae 223
Lucinidae 214
Lumbricidae 218
Lumbrineridae 218
Lunularia cruciata 30, 32, 36
Lunulariaceae 149
Lutjanidae 256
Lycaenidae 244
Lycopodiaceae 148
Lycosidae 252
Lyctidae 240
Lygaeidae 233
Lymantriidae 244
Lymnaeidae 215
Lyonetiidae 245
Lythraceae 186
Lytocestidae 211f

M
Macaca sylvanus 120
Macronyssidae 253
Macropodidae 262
Macrothricidae 223
Mactridae 215
Magnoliaceae 183
Magnoliales 183
Magnoliophytina 150
Maldanidae 218
Malleidae 215
Malpighiales 183
Malvaceae 185
Malvales 185
Mambraniporidae 220
Mamiellaceae 145
Mammalia 10, 12, 119, 262
management 4ff, 23, 38f, 55ff, 74, 97, 100, 115, 126, 267, 270ff
map 6ff, 267
marbled spinefoot 296
Marchantia planiloba 30
Marenzelleria neglecta 285
Margarodidae 233
marine biota 93
Marsileaceae 148

Marsupenaeus japonicus 99, 286
Martyniaceae 180
Mastigiidae 210
Matricaria discoidea 49
Matutidae 223
mechanical control 267, 270ff
Meconematidae 226
Mediterranean fruit fly 332
Megascolecidae 218
Melanoides tuberculata 87
Melanthiaceae 203
Melastomataceae 186
Meleagrididae 260
Meliaceae 194
Melithaeidae 210
Melitidae 223
Meloidae 240
Meloidogynidae 220
melon aphid 327
Melyridae 240
Membracidae 233
Membraniporidae 220
Menippidae 223
Menoponidae 227
Menyanthaceae 159
Merothripidae 228
Mesodesmatidae 215
Microascales 139
Microbotryales 140
Microporellidae 220
Micropterus dolomieu 84
Microstromatales 140
Microthyriales 140
Milacidae 215
Milichiidae 248
Mimosoideae 175
Miridae 233
Mnemiopsis leidyi 98, 314
Moerisiidae 210
Molgulidae 254
Molineidae 220
Molluginaceae 167
Mollusca 64, 82, 85, 130, 212
Monacanthidae 256
Monoblastiaceae 149
Monocelididae 212
Monocotyledonae 198
Monodontidae 262
Monostromataceae 145
Monotomidae 240
Moraceae 190
Mordellidae 240
Moronidae 256
Mugilidae 256

Mullidae 257
Muraenesocidae 257
Muricidae 215
Muridae 262
Mus musculus 119
Musaceae 203
Muscidae 248
Musculista senhousia 287
muskrat 366
Mustela lutreola 120
Mustela vison 2, 124, 363
Mustelidae 262
Mycalidae 211
Mycetophagidae 240
Mycetophilidae 248
Mycosphaerellales 140
Myicola ostreae 98
Myicolidae 223
Myidae 215
Mymaridae 251
Myocastor coypus 2, 124, 364
Myocastoridae 262
Myocoptidae 253
Myriangiales 140
Myriapoda 225
Myricaceae 175
Myrsinaceae 171
Myrtaceae 186
Myrtales 186
Mysidacea 86
Mysidae 223
Mytilicola orientalis 98
Mytilicolidae 223
Mytilidae 215

N
Nacellidae 215
Naididae 218
Nassariidae 215
Nasua nasua 127
Naticidae 215
native 376
naturalised 376
Nauphoetidae 226
Neanuridae 226
Nelumbonaceae 188
Nemastomataceae 147
Nematoda 64, 220
Nemertea 221
Nemonychidae 240
Neogobius melanostomus 315
neomycete 377
neophyte 377

neozoan 377
Nephropidae 223
Nepticulidae 245
Nereididae 218
Neritidae 215
Nesopsyllus fasciatus 72
Nesticidae 252
Neuroptera 247
New Zealand pigmyweed 308
Nitidulidae 72, 240
Noctilucaceae 145
Noctuidae 245
Nolidae 245
nomad jellyfish 294
non-indigenous 375
non-native 375
Norway rat 370
Nostocaceae 143
not established 375
Nothofagaceae 175
Notocotylidae 212
Notodontidae 245
Nucellicolidae 223
Numididae 260
nutria 364
Nyctaginaceae 167
Nyctereutes procyonoides 124, 365
Nymphaeaceae 188
Nymphaeales 188
Nymphalidae 245

O
Obesogammarus crassus 86
Obtortionidae 215
Ochnaceae 184
Ochrophyta 144
Ocinebrellus inornatus 98
Ocnerodrilidae 218
Octochaetidae 218
Octodontidae 263
Octopodidae 215
Oculinidae 210
Ocypodidae 223
Odontella sinensis 288
Odontobutidae 257
Odontophoridae 260
Oecobiidae 71, 252
Oecophoridae 245
Oedemeridae 240
Oenothera biennis 49
Ogyrididae 223
Oithonidae 223
Oleaceae 180

Index   393

Oligochaeta 217
Olindiidae 210
Onagraceae 48ff, 187
Oncorhynchus mykiss 83
Ondatra zibethicus 2, 124, 366
Oodiniaceae 133
Oonopidae 65, 71, 252
Ophiactidae 254
Ophioglossaceae 148
Ophiostoma novo-ulmi 21, 320
Ophiostoma ulmi 21, 25
Ophiostomatales 140
Opiliones 254
Opisthorchiidae 212
Opuntia ficus-indica 53, 352
Opuntia maxima 56
Orbiliales 140
Orchidaceae 203
Orconectes limosus 88
Oregoniidae 223
Ornithonyssus bursa 72
Orobanchaceae 180
Ortheziidae 233
Orthodontiaceae 149
Orthodontium lineare 30, 33, 36f,
Orthoptera 226
Osmeridae 257
Osmundaceae 148
Osteichthyes 255
Ostraciidae 257
Ostrea edulis 99
Ostreidae 215
Othoptera 65
Oxalidaceae 188
Oxalis pes-caprae 2, 56, 353
Oxidales 188
Oxychilidae 215
Oxynoidae 215
Oxytoxaceae 144
Oxyura jamaicensis 2, 111, 367
Oxyura leucephala 2
Oxyuridae 220

P

Pachytroctidae 228
Pacifastacus leniusculus 88
Pacific oyster 279
Paeoniaceae 196
Palaemonidae 223
Palinuridae 224
Palmariaceae 147
pampas grass 346
Pamphiliidae 251

Panicum miliaceum 49
Panopeidae 224
Papaveraceae 50, 189
Paradoxosomatidae 225
Paralithodes camtschaticus 100, 289
Parastacidae 224
Parmelia elegantula 32
Parmelia exasperatula 32
Parmelia laciniatula 32
Parmelia submontana 32, 36
Parmeliaceae 149
Partulidae 215
Pasiphaeidae 224
Paspalum paspalodes 2, 56, 354
Passandridae 240
Passeridae 260
Passifloraceae 184
pathway 20, 36, 51, 68, 84, 96, 109, 113, 123,
   267, 270ff, 377
Patulidae 215
Paysandisia archon 76
Pectinatellidae 220
Pectinidae 216
Pedaliaceae 181
Pelecanidae 260
Pelliaceae 149
Pelomedusidae 259
Pempheridae 257
Penaeidae 224
Pentatomidae 233
Percidae 257
Percnon gibbesi 290
Peridiniaceae 133
Peripsocidae 228
Perkinsidae 133
Peronosporales 133
Perophoridae 254
pest 1f, 23, 57, 64ff, 98f, 116, 376
Petricolidae 216
Pezizales 140
Phaeocystaceae 143
Phaeophyscia rubropulchra 32
Phalacridae 240
Phalacrocoracidae 260
Phalangiidae 254
Phallales 140
Phaneropteridae 226
Pharidae 216
Phasianidae 260
Phasianus colchicus 107
Phasmatidae 226
Phasmatodea 65, 226
Philopteridae 227
Philosciidae 224

Philotarsidae 228
Phlaeothripidae 228
Phoenicococcidae 233
Phoenicopteridae 261
Pholcidae 71, 252
Phoridae 248
Phoronida 221
Phoronidae 221
Phoxichilidiidae 224
Phragmonemataceae 147
Phrymaceae 181
Phtiraptera 65, 227
Phyllachorales 140
Phyllodicolidae 224
Phyllophoraceae 147
Phylloxeridae 233
Physciaceae 149
Physidae 216
Phytolacca americana 2
Phytolaccaceae 167
Phytophthora alni 23
Phytophthora cambivora 20
Phytophthora cinnamomi 17, 20, 321
Phytophthora infestans 23
Phytophthora ramorum 17, 21, 24
Phytoseiidae 253
Picidae 261
Pieridae 245
Pilayellaceae 144
Pilumnidae 224
Pimephales promelas 87
Pinaceae 47f, 150
Pinctada radiata 291
pine wood nematode 330
Pinguipedidae 257
Piperales 188
Pipidae 114, 258
Piscicolidae 217
Pisidae 224
Pisidiidae 216
Pistiaceae 203
Pittosporaceae 152
Plagusiidae 224
Planariidae 212
Planaxidae 216
Planorbidae 216
plant 2, 7ff, 16ff, 31ff, 43ff, 65ff, 130f
Plantaginaceae 48ff, 181
Platanaceae 188
Platyarthridae 224
Platycephalidae 257
Platygastridae 251
Platyhelminthes 211
Platypodidae 240

Plectonemertidae 221
Pleosporales 140
Plethodontidae 258
Pleurobranchidae 216
Pleurodiscidae 216
Pleuronectidae 257
Pleurosigmataceae 143
Plicatulidae 216
Plocamiaceae 147
Ploceidae 261
Plotosidae 257
Plumbaginaceae 167
Plumulariidae 210
Plutellidae 245
Poaceae 47ff, 203
Podarcis muralis 112
Podarcis sicula 112
Podonidae 224
Poeciliidae 257
Polemoniaceae 171
Pollicipedidae 224
Polycentropodidae 243
Polyceridae 216
Polychaeta 218
Polychrotidae 115, 259
Polycitoridae 254
Polyclinidae 254
Polygalaceae 175
Polygonaceae 48f, 167
Polynoidae 218
Polyodontidae 257
Polyphemidae 224
Polyphysaceae 145
Polypodiaceae 148
Polypodiidae 210
Polyporales 141
Polytrichaceae 149
Pomacentridae 257
Pomatiidae 216
Pomphorhynchidae 221
Pontederiaceae 209
Pontogammarus robustoides 86
Porcellidiidae 224
Porcellionides pruinosus 65
Porifera 210
Portulacaceae 168
Portunidae 224
Portunus pelagicus 292
Potamidae 224
Potamogetonaceae 209
Potamopyrgus antipodarum 83
Pottiaceae 149
Pratylenchidae 220
prickly-pear cactus 352

Primulaceae 171
Pristilomatidae 216
Procambarus clarkii 87, 316
Procyon lotor 124, 368
Procyonidae 263
Prodidomidae 252
Promesostomidae 212
propagule 36, 55, 83, 377
Prorocentraceae 145
Prosthogonimidae 212
Proteales 188
Proteidae 258
Protista 133
Protostrongylidae 220
Prunus serotina 355
Psammobiidae 216
Pseudococcidae 65, 233
Pseudocumatidae 224
Pseudodiaptomidae 225
Pseudorasbora parva 88, 317
Psittacidae 261
Psittakula krameri 112, 369
Psocoptera 65, 227
Psoquilidae 228
Psyllidae 234
Psyllipsocidae 228
Pteridaceae 148
Pteridophyta 147
Pteriidae 216
Pteroclididae 261
Pteromalidae 251
Pterophoridae 245
Pteropodidae 263
Ptiliidae 240
Ptilodactylidae 241
Pulex irritans 66
Pulicidae 248
Punctidae 216
Punicaceae 188
Pycnonotidae 261
Pyralidae 245
Pyramidellidae 216
Pyrenulales 141
Pyrocystaceae 145
Pyroglyphidae 253
Pythiales 134
Pyuridae 254

R
rabbitfish 296
raccoon dog 365
raccoon 368
Rachycentridae 257

Ralfsiaceae 144
Ranidae 114, 258
Raninidae 225
Ranuncales 188
Ranunculaceae 189
Ranunculales 47, 48
Rapana venosa 293
Rathkeidae 210
Rattus norvegicus 2, 124, 370
Rattus rattus 119
red king crab 289
red swamp crayfish 316
red-gilled mud worm 285
Reductoniscus costulatus 65
Reduviidae 234
reintroduction 4, 51f, 123, 376
release 52, 58, 68, 86f, 99f, 105, 109f, 113ff, 124, 378
Reptilia 105, 258
Resedaceae 162
residence time 44ff, 56, 377
Reticulitermes flavipes 72
Retusidae 216
Rhabdoweisiaceae 149
Rhamnaceae 190
Rhaphidophoridae 226
Rheidae 261
Rhinolophidae 263
Rhinotermitidae 226
Rhizogoniaceae 149
Rhizophydiales 141
Rhizosoleniaceae 144
Rhizostomatidae 210
Rhododendron ponticum 2, 53, 56, 356
rhododendron 356
Rhodomelaceae 147
Rhodophyta 146
Rhodymeniaceae 147
Rhopilema nomadica 100, 294
Rhynchodemidae 212
Rhynchophorus ferrugineus 76
Rhytismatales 141
Ricciaceae 149
Ricciocarpos natans 36
Riodinidae 246
Ripiphoridae 241
Rissoidae 216
Robinia pseudoacacia 49, 56f, 357
Roeslerstammiidae 246
Rorippa austrica 57
Rosa rugosa 50, 57, 358
Rosaceae 47ff, 190
Rosales 190
rose-ringed parakeet 369

Rotifera 221
round goby 315
Rousettus aegyptiacus 121, 127
Rubiaceae 176
ruddy duck 367
rugosa rose 358
Ruscaceae 209
Russulales 141
Rutaceae 194
Rutelidae 241

S
Sabellariidae 219
Sabellidae 219
Sabelliphidae 225
Sacculinidae 225
sacred ibis 373
Sagartiidae 210
Sagittidae 219
Salamandridae 114, 258
Saldidae 234
Salicaceae 48, 184
Salifidae 217
sally lightfoot crab 290
Salmon fluke 313
Salmonidae 257
Salticidae 71, 252
Salvelinus fontinalis 83, 318
Salvelinus namaycush 84
Salviniaceae 148
Samia cynthia 69
Santalaceae 194
Sapindaceae 50, 195
Sapindales 194
Sapotaceae 171
Saprolegniales 134
Sarcomeniaceae 147
Sargassaceae 144
Sargassum muticum 98
Sarracenia purpurea 51
Sarraceniaceae 171
Sarsiellidae 225
Saturniidae 246
Saurida undosquamis 295
Saururaceae 188
Saxifragaceae 196
Saxifragales 195
Scaphoides titanus 72
Scarabaeidae 241
Scaridae 257
Schizopetalidae 225
Schizoporellidae 220

Sciaenidae 257
Sciaridae 248
Scincidae 115
Sciuridae 263
Sciurus carolinensis 1, 126, 371
Sclerosporales 135
Scolytidae 241
Scombridae 257
Scorpaenidae 257
Scrophulariaceae 182
Scrupariidae 220
Scrupocellariidae 220
Scydmaenidae 241
Scyllaridae 225
Scyphidiidae 133
Scytodidae 252
Scytosiphonaceae 144
sea walnut 314
Seiridium cardinale 22, 322
Selaginellaceae 148
Sematophyllaceae 149
Semelidae 216
Sepiidae 216
serpentine leaf miner 338
Serpulidae 219
Serranidae 257
Sesiidae 246
Siberian chipmunk 372
Sicariidae 253
Sididae 225
Siganidae 258
Siganus rivulatus 296
sika deer 361
Sillaginidae 258
Silphidae 241
Siluridae 258
Silurus glanis 84, 86
Silvanidae 241
Simaroubaceae 50, 195
Simmondsiaceae 168
Siphonaptera 65, 248
Siphonariidae 216
Siphonocladaceae 145
Siricidae 251
Sitophilus granarius 66
Skeletonemaceae 144
slipper limpet 280
Smilacaceae 210
Sminthuridae 226
Smittinidae 220
Solanaceae 48ff, 197
Solanales 47f, 196
Solanum elaeagnifolium 2

Solenoceridae 225
Soricidae 263
Spanish slug 328
Sparassidae 253
Sparidae 258
Spartina x townsendii 56
Spartina alterniflora 51, 56
Spartina anglica 297
Sphaerocarpaceae 149
Sphaeroceridae 248
Sphaerococcaceae 147
Sphaeromatidae 225
Sphaeropsis sapinea 21
Sphecidae 251
Spheniscidae 261
Sphindidae 241
Sphyraenidae 258
Spionidae 219
Spiophanicolidae 225
Spirobolellidae 225
Spodoptera littoralis 339
Spondylidae 216
Spongillidae 211
Squamariaceae 147
Squillidae 225
Staphyleaceae 169
Staphylinidae 241
Stathmopodidae 246
Stegobium paniceum 66
Stenothoidae 225
Stephanopyxidaceae 144
Sternorrhyncha 65
Stone moroko 317
stowaway 36, 51f, 69, 86, 114, 378
Stratiomyidae 248
Streptaxidae 216
Strigidae 261
Stromateidae 258
Strombidae 217
Strongyloididae 220
Stropharia aurantiaca 21
Sturnidae 261
Styela clava 298
Styelidae 254
Stylochidae 212
Styloniscidae 225
Styracaceae 171
Suberitidae 211
Subulinidae 217
Sycettidae 211
Syllidae 219
Symmocidae 246

Synaptidae 254
Synchytrium endobioticum 23
Syngnathidae 258
Synodontidae 258
Syrphidae 248

T
Tachinidae 248
Taeniidae 212
Talitridae 225
Talpidae 263
Tamaricaceae 168
Tamias sibiricus 126, 372
Tanaidae 225
Taphrinales 141
Tarsonemidae 253
Taxaceae 150
Taxodiaceae 150
Tayassuidae 263
Techisporales 141
Tellinidae 217
Telotylenchidae 220
Temoridae 225
Tenebrio molitor 66
Tenebrionidae 72, 243
Tenthredinidae 251
Tenuipalpidae 253
Tephritidae 248
Teraponidae 258
Terebellidae 219
Teredinidae 217
Teredo navalis 299
Tergipedidae 217
Termitidae 226
Testacellidae 217
Testudinidae 259
Testudo graeca 112
Tethinidae 248
Tethyidae 217
Tetillidae 211
Tetraclitidae 225
Tetragnathidae 253
Tetragoniaceae 168
Tetranychidae 253
Tetranychus evansi 72
Tetraodontidae 258
Tetraonidae 261
Tettigoniidae 226
Thalassiosiraceae 144
Theaceae 171
Thecadiniaceae 145
Thelastomatidae 220

Thelypteridaceae 148
Theridiidae 65, 71, 253
Thiaridae 217
Thomisidae 253
Thorictidae 243
Threskiornis aethiopicus 111, 373
Threskiornithidae 261
Thripidae 228
Throscidae 243
Thuidiaceae 149
Thymelaeaceae 186
Thysanoptera 65, 228
Tiliaceae 186
Tilletiales 141
Timaliidae 261
Tinamidae 262
Tineidae 246
Tingidae 234
Torpedinidae 254
Tortricidae 246
Torymidae 251
Trachelipodidae 225
Trachemys scripta 374
Trapeziidae 217
tree of heaven 342
Tremellales 141
Tremoctopodidae 217
Tricellaria inopinata 300
Triceratiaceae 143, 145
Trichinellidae 220
Trichodectidae 227
Trichodoridae 221
Trichogrammatidae 252
Tricholomatales 141
Trichoniscidae 225
Trichopsocidae 228
Trichoptera 243
Trichorphina tomentosa 65
Trichosphaeriales 141
Trichuridae 221
Trimenoponidae 227
Trionychidae 259
Triophidae 217
Triozidae 234
Triphoridae 217
Trissexodontidae 217
Trochidae 217
Trogidae 243
Trogiidae 228
Trogossitidae 243
Tropaeolaceae 162
Tropiduchidae 234

Truncatellidae 217
tube worm 282
Tubificidae 218
Tubulariidae 210
Tullbergiidae 226
Tulostomatales 141
Tunicata 254
Turdidae 262
Turridae 217
Typhaceae 210

U
Ulidiidae 248
Ulmaceae 194
Ulmaridae 210
Uloboridae 253
Ulvaceae 145
Umbridae 258
unaided dispersal 378
Undaria pinnatifida 98, 301
Ungulinidae 217
unintentional introduction 378
Unionidae 217
Uredinales 141
Urocoptidae 217
Urocystidales 143
Urosalpinx cinerea 2
Urticaceae 194
Ustilaginales 143

V
Vacuolariaceae 145
Valerianaceae 170
Valloniidae 217
Varanidae 259
Varroidae 254
Varunidae 225
vascular plant 8, 15, 31ff, 43ff
veined rapa whelk 293
Veneridae 217
Verbenaceae 183
Veronica persica 49
Veronicellidae 217
Verrucidae 225
Vertiginidae 217
Vesicularia reticulata 30
Vespa velutina 73
Vespidae 252
Vibrionaceae 133
Victorellidae 220

Violaceae 185
Viperidae 259
Vitaceae 198
Vitales 198
Viverridae 263
Viviparidae 217

W
wakame 301
Walkeriidae 220
Watersiporidae 220
western corn rootworm 333
western flower thrips 334
wild cucumber 347
Winteraceae 162
Woodsiaceae 148

X
Xanthidae 225
Xanthium strumarium 49

Xiphinematidae 221
Xylariales 143

Y
Yponomeutidae 246

X
Zamiaceae 150
zebra mussel 310
Zingiberaceae 210
Zodariidae 253
Zonitidae 217
Zonitoides arboreus 65
Zoopsis liukiuensis 30
Zopheridae 243
Zoropsidae 253
Zygaenidae 247
Zygentoma 65
Zygophyllaceae 198
Zygophyllales 198